Landfilling of Waste: Biogas

Landfilling of Waste: Biogas

Edited by

T. H. CHRISTENSEN
Department of Environmental Engineering,
Technical University of Denmark,
Lyngby, Denmark

R. COSSU
Department of Land Engineering,
University of Cagliari,
Cagliari, Sardinia, Italy

R. STEGMANN
Institute of Waste Management,
Technical University Hamburg-Harburg,
Hamburg, Germany

 **CRC Press**
Taylor & Francis Group
Boca Raton London New York

CRC Press is an imprint of the
Taylor & Francis Group, an **informa** business

A TAYLOR & FRANCIS BOOK

CRC Press
Taylor & Francis Group
6000 Broken Sound Parkway NW, Suite 300
Boca Raton, FL 33487-2742

First issued in paperback 2019

© 1996 by Taylor & Francis Group, LLC
CRC Press is an imprint of Taylor & Francis Group, an Informa business
Typeset in 10/12 Plantin by WestKey Limited, Falmouth, Cornwall

No claim to original U.S. Government works

ISBN-13: 978-0-419-19400-2 (hbk)
ISBN-13: 978-0-367-44898-1 (pbk)

A catalogue record for this book is available from the British Library
Library of Congress Catalog Card Number: 95–71860

Visit the Taylor & Francis Web site at
http://www.taylorandfrancis.com

and the CRC Press Web site at
http://www.crcpress.com

Preface

During the last couple of decades landfilling of waste has developed dramatically, and today in some countries involves fully engineered facilities subject to extensive environmental regulations. Although much information and experience in landfill design and operation has been obtained in recent years, only a few landfills meet the current standards for an environmentally acceptable landfill. In view of the increasing public awareness and new scientific understanding of waste disposal problems, the standards for landfilling of waste may improve even further within the years to come.

In view of this great demand for information on landfills we decided to establish, as editors, a series of international reference books on landfilling of waste. No worldwide, long tradition of publishing information on landfilling exists, and this book series is seen as an attempt to establish a common platform representing the state of the art, useful for implementing improvements at actual landfill sites and for identification of new directions in landfill research.

The first book deals with leachate in terms of its generation, composition, treatment and potential environmental effects, and the second book deals with barriers in landfills in terms of principles and materials for bottom lining, top capping and draining. This book, being the third book in the series, is devoted to biogas generated in landfills. Aspects such as gas generation, composition and collection, gas disposal, gas utilization, gas treatment, gas migration, safety and several case studies are discussed in great detail. The introductory chapter presents the issues concerning landfill gas as well as the utilization of landfill gas. Some of the contributions in this book may express different opinions and represent different experiences, allowing the readers to develop their own balanced views.

This book on gas from landfills consists of edited, selected contributions to the International Landfill Symposia held in Sardinia (Italy) every second year and of chapters specially written for this book. The responsibility for the technical content of the book primarily rests with the individual authors and we cannot take any credit for their work. Therefore reference should be made directly to the authors of chapters. Our role as editors has been one of reviewing and homogenizing the chapters and of making constructive suggestions during the preparation of the final manuscripts.

We would like to thank all our contributors for allowing us to edit their manuscripts, and ask for everybody's forbearance with our unintended mistreatment of the English language. For this reason we would like to give credit to the publisher, E & FN Spon for correcting the English where necessary and for giving shape to the book.

Thomas H. Christensen
Raffaello Cossu
Rainer Stegmann

Contents

List of Contributors

G. Andreottola
Department of Civil and Environmental Engineering, University of Trento, Via Mesiano 77, 38050 Trento, Italy

S. Ashton
Thomas Graveson Ltd, Keer Bridge House, Warton, Carnforth, Lancashire LA5 9HA, UK

G. Baldwin
Site Remediation and Waste Management, National Environmental Technology Centre, Culham, Abingdon, Oxfordshire OX14 3DB, UK

S. Blackley
Environmental Resources Management, Eaton House, Wallbrook Court, North Hinksey Lane, Oxford OX2 0QS, UK

N. C. Blakey
WRc Environmental Management, Medmenham, Buckinghamshire SL7 2HD, UK

D. J. V. Campbell
AEA Technology, National Environmental Technology Centre, Culham, Abingdon, Oxfordshire OX14 3DB, UK

S. Cernuschi
DIIAR – Environmental Section, Technical University of Milan, Piazza L. da Vinci 32, 20133 Milan, Italy

T. H. Christensen
Institute of Environmental Science and Engineering, Technical University of Denmark, Building 115, DK-2800, Lyngby, Denmark

R. Cossu
DIGITA – Department of Geoengineering and Environmental Technologies, Faculty of Engineering, University of Cagliari, Piazza D'Armi, 09123 Cagliari, Italy

B. Couth
Aspinwall and Company, Walford Manor, Baschurch, Shrewsbury, Shropshire SY4 2HH, UK

A. Deipser
Institute of Waste Management, Technical University of Hamburg-Harburg, 21071 Hamburg, Germany

H.-J. Ehrig
Bergische Universität, Gesamthochschule Wuppertal, Pauluskirschstr. 7, 42285 Wuppertal, Germany

T. F. Eikmann
Institute of Hygiene and Environmental Medicine, Justus-Liebig-University, Friedrichstrasse 16, 35385 Giessen, Germany

U. Eymann
WAT Wasser- und Abfalltechnik Ingenieurgesellschaft mbH, Kleinoberfeld 5, 76135 Karlsruhe, Germany

R. A. Figueroa
Institute of Waste Management, Technical University of Hamburg-Harburg, Harburger Schlossstrasse 37, 21071 Hamburg, Germany

F.-B. Frechen
Aqua System Consult, Am Hackenbruck 47, 40231 Düsseldorf, Germany

K. Giessler
CarboTech Anlagenbau GmbH, Franz Fischer-Weg 61, 45307 Essen, Germany

M. Giugliano
DIIAR – Environmental Section, Technical University of Milan, Piazza L. da Vinci 32, 20133 Milan, Italy

R. K. Ham
Department of Civil and Environmental Engineering, University of Wisconsin-Madison, 3232 Engineering Building, 1415 Johnson Drive, Madison, WI 53706, USA

P. L. A. Henigin
WAT Wasser- und Abfalltechnik Ingenieurgesellschaft mbH, Kleinoberfeld 5, 76135 Karlsruhe, Germany

K.-D. Henning
CarboTech-Aktivkohlen GmbH, Franz Fischer-Weg 61, 45307 Essen, Germany

H. B. Kerfoot
Kerfoot and Associates, 2200 E. Patrick Lane, Suite 23, Las Vegas, NV 89119, USA

P. Kjeldsen
Institute of Environmental Science and Engineering, Technical University of Denmark, Building 115, DK-2800 Lyngby, Denmark

A. Leach
82 Bower Street, Bedford, Bedfordshire MK40 3QZ, UK

B. Lindhardt
Institute of Environmental Science and Engineering, Technical University of Denmark, Building 115, DK–2800, Lyngby, Denmark

R. J. Lofy
Lofy Engineering, PO Box 5335, Pasadena, CA 91117, USA

B. J. W. Manley
Environmental Resources Management, Eaton House, Wallbrook Court, North Hinksey Lane, Oxford OX2 0QS, UK

G. Massacci
DIGITA – Department of Geoengineering and Environmental Technologies, University of Cagliari, Piazza D'Armi, 09123 Cagliari, Italy

H. D. T. Moss
Shanks & McEwan (Energy Services) Ltd, The Cottage, Church Road, Woburn Sands, Milton Keynes, Bucks MK17 8TA, UK

G. M. Motzo
DIGITA – Department of Geoengineering and Environmental Technologies, Faculty of Engineering, University of Cagliari, Piazza D'Armi, 09123 Cagliari, Italy.

A. Muntoni
DIGITA – Department of Geoengineering and Environmental Technologies, Faculty of Engineering, University of Cagliari, Piazza D'Armi, 09123 Cagliari, Italy.

U. Neumann
Garden and Landscape Architect, Falterweg 17, 14055 Berlin-Eichkamp, Germany

R. Paparella
AMIU, Via Morandi 54, 41100 Modena, Italy

J. H. Penido Monteiro
Rua dos Parecis 15, Cosme Velho, Rio de Janeiro, Brazil

A. Peroni
AMIU, Via Morandi 54, 41100 Modena, Italy

S. Picchiolutto
Department of Environment, Municipality of Modena, Via Santi 40, 41100 Modena, Italy

T. Poller
Institute of Waste Management, Technical University of Hamburg-Harburg, 21071 Hamburg, Germany

B. Reinicke
Energie-Versorgung Schwaben, Kriegsbergstrasse 32, 70174 Stuttgart, Germany

M. Reiter
CISA – Environmental Sanitary Engineering Centre, Via Marengo 34, 09123 Cagliari, Italy

G. Rettenberger
Fachhochschule Trier, Schneidershof, 54293 Trier, Germany

M. Schäfer
RWE Energie AG, Kruppstrasse 5, 45128 Essen, Germany

M. J. J. Scheepers
GASTEC NV, Postbus 137, 7300 AC Apeldorn, The Netherlands

J. Schneider
DEPOGAS GmbH, Glienicker Str. 100, 14109 Berlin, Germany

W. Schreier
Ingenieurgruppe RUK, Schockenriedstrasse 4, 70565 Stuttgart, Germany

P. E. Scott
Site Remediation and Waste Management, National Environmental Technology Centre, Culham, Abingdon, Oxfordshire OX14 3DB, UK

R. Stegmann
Institute of Waste Management, Technical University of Hamburg-Harburg, 21071 Hamburg, Germany

S. A. Thorneloe
US Environmental Protection Agency, Air and Energy Engineering Laboratory, Research Triangle Park, NC 27711, USA

S. Tillotson
Environmental Resources Management, Eaton House, Wallbrook Court, North Hinksey Lane, Oxford OX2 0QS, UK

R. Tipping
Environmental Resources Management, Eaton House, Wallbrook Court, North Hinksey Lane, Oxford OX2 0QS, UK

G. Widder
Amt für Abfallwirtschaft und Stradtreinigung, Mainzer Str. 97, 65186 Wiesbaden, Germany

K. Wiemer
Gesamthochschule Kassel, Fachgebiet Abfallwirtschaft und Recycling, Nordbahnhofstr. 1a, 37213 Witzenhausen, Germany

V. Wilhelm
Tiefbau Berufsgenossenschaft, Vollmoellerstr. 11, 70563 Stuttgart, Germany

H. C. Willumsen
Hedeselskabet, Klostermarken 12, DK-8800 Viborg, Denmark

C. P. Young
*WRc Environmental Management, Medmenham, Buckinghamshire
SL7 2HD, UK*

1. INTRODUCTION

1.1 Gas Issues in Landfill Management

THOMAS H. CHRISTENSEN[a], RAFFAELLO COSSU[b] &
RAINER STEGMANN[c]

[a] Institute of Environmental Science and Engineering, Technical
University of Denmark, Building 115, DK-2800 Lyngby, Denmark
[b] DIGITA, Department of Geoengineering and Environmental
Technologies, University of Cagliari, Piazza d'Armi, I-09123
Cagliari, Italy
[c] Institute of Waste Management, Technical University of
Hamburg-Harburg, Harburger Schlossstrasse 37, D-21071
Hamburg, Germany

INTRODUCTION

Landfill plays a most important role in the framework of solid waste
disposal and will remain an integral part of the new strategies based
on integrated solid waste management. Landfill design concepts are
mainly devoted towards ensuring minimal environmental impact in
accordance with observations made concerning old landfills, but in
recent years landfill design usually also includes facilities for controlling gas migration and utilization of the energy associated with
the biogas.

Gas generated in the landfill (LFG) is a result of a mass transfer
process. Waste entering the landfill undergoes biological, chemical
and physical transformation governing the relationship between the
solid phase (the waste), the liquid phase (leachate) and the gas
phase. The main components of the landfill gas (CH_4, CO_2) are a
result of biological processes, while the trace components, which are
numerous, are generated by biological processes as well as by
volatilization. Gas in landfills has existed as long as landfills have

Landfilling of Waste: Biogas. Edited by T. H Christensen, R. Cossu and
R. Stegmann. Published in 1996 by E & FN Spon, London.
ISBN 0 419 19400 2.

existed, but the realization of landfill gas causing problems to the environment is relatively recent.

HISTORY

For a long time, people working on landfills knew that gas is inside a landfill. This could be proven by simply pushing a steel pipe into the landfill and lighting the open end of the pipe. The flame may burn over long time until blown out by the wind.

Migration of gas into the surrounding areas was observed in terms of dying trees and mal-growth of crops on agricultural fields. This was seen in particular at landfills located in gravel pits. Explosions occurred, where gas accumulated in buildings and manholes in or adjacent to the landfills and was ignited by the use of open fires or electrical sparks.

Landfill gas control started in the late 1960s and early 1970s in the USA, where huge landfills had been created. The first plant in Europe came into operation in Germany in the mid-1970s, incorporating a great deal of experience gained in the USA. Later, LFG technology spread all over Europe and into other countries.

In Germany, gas utilization focused from the beginning on direct thermal utilization (e.g. in industry) and electricity production using gas engines. These were, in general, Otto engines that were developed for natural gas. By contrast, in the USA LFG was at first upgraded to natural gas quality by removing CO_2. Substantial experience has been gained since then. LFG plant operators had to face severe corrosion problems e.g. in the gas pretreatment plants for CO_2 removal in the USA and in gas engines in Germany. In addition, it was found out that there are a great number of organic trace components in LFG that are responsible for these corrosion problems. At first, no explosion control standards existed, and the emissions from thermal treatment plants were unknown. With time, more and more regulations were made regarding the planning, construction and operation of LFG extraction and utilization plants. This also influenced the economy of gas utilization. While in the USA, in the early days, LFG utilization had been seen as a big business (there was a big run on the gas utilization rights at large landfills), this has changed with time. In Germany the main emphasis for LFG utilization was also commercial, until it was realized that no big profits (if any) could be made. The economics of LFG utilization are of course also closely linked to energy prices. The reasons for LFG abstraction and utilization changed. While in some states in the USA gas abstraction was done to control releases of the carcinogenic trace component vinyl chloride, explosion control and vegetation protection became more and more relevant in other countries. Today also the contribution of LFG emissions to the greenhouse effect is a reason in many countries that LFG extraction and utilization plants are mandatory at all new landfills.

Landfill gas technology initially was developed by pioneers: people who operated a landfill and who became enthusiastic about LFG. The scientific investigations followed with a certain time lag. Today, LFG technology is a proven technology, but many a lesson has been learned in the past. This book offers a comprehensive state of the art of landfill gas and its utilization.

There are still areas in this field where scientific and practical experiences are at a low level. This is the case, for example, for the prediction of long-term gas production in combination with active extraction, the effects of different kinds of surface liners on the LFG recovery rate, the extraction of LFG at small and old landfills, and the use of biofilters for the treatment of LFG at low rates. Also a new, but not yet proven, technology with artificial aeration of old landfills must be closely studied.

GAS CONTROL STRATEGIES

Gas control strategies may involve control of the waste input, control of the landfill reactor, and control of gas discharges to the environment.

Control of Waste Input

The first step in the waste input control strategy should be that of reducing to a minimum the amount of waste to be landfilled. This could be achieved by the introduction of cleaner technology, waste minimization programmes, separate collection, recycling, incineration, digestion or composting. In particular, exclusion of organic waste such as household waste (or the organic fraction of household waste), garden waste, wastewater sludge and organic industrial waste will reduce the gas generation dramatically and may reduce the gas control to natural venting. Separation of hazardous components of the waste, such as expired medicines, mercury lamps, pesticides, paints and solvents, would significantly reduce the trace content of the landfill gas.

Pretreatment in terms of incineration will reduce the organic content of the residue to be landfill to less than a few percent of carbon and thereby prevent significant gas generation. Pretreatment by digestion (anaerobic) or composting (aerobic) prior to landfilling will reduce the gas potential of the landfill, but will not make gas control measures redundant.

Control of Landfill Reactor

Controlling the processes in the landfill to avoid gas generation is not feasible, if organic waste has been accepted at the landfill. Gas generation could be kept at a minimum if the moisture content was kept low by avoiding water into the landfill. However, this would also prevent loss of organic carbon by leachate removal and as such conserve the organic carbon in the landfill, allowing a low gas-generation rate over an extended time period, which may involve several decades.

Most often, controlling the processes in the landfill means enhancing the degradation processes, in order to obtain a significant and usable methane production rate that can be exploited technically and economically. This approach also has positive effects on the leachate quality and will shorten the period needed to stabilize the landfill.

Control of Gas Discharge into the Environment

Control of the gas discharge into the environment may involve several elements in combination:

Lining. A lining system is usually installed in order to control the leachate, but may also act as a control for the gas. A lining system for control of gas must in particular pay attention to sidewards and upwards migration of gas.

Gas abstraction and collection. A rational gas abstraction and collection system is important to avoid emissions or accumulation of gas inside the landfill. This usually involves a dense system of abstraction shafts or wells and, where the gas generation is significant, a transportation system to a central treatment and utilization plant. Owing to the explosive nature of the landfill gas, these systems are equipped with safety controls.

Treatment and utilization. Direct discharge of landfill gas usually should not take place owing to odour, trace components and greenhouse effects. At small facilities flaring of the gas may be a sufficient treatment, but where the gas is intended for utilization some kind of treatment for removal of water vapours and corrosive elements, and maybe carbon dioxide, may be needed. Methods of utilizing the gas for steam, heat or electricity production are commercially available.

Environmental monitoring. To ensure the efficiency of the gas control measures some kind of environmental monitoring, e.g. above or in nearby soil strata, should be performed.

1.2 Landfill Gas Utilization: An Overview

RAINER STEGMANN

Institute of Waste Management, Technical University of Hamburg–Harburg, Harburger Schlossstrasse 37, D-21071 Hamburg, Germany

INTRODUCTION

Since landfill gas (LFG) is an emission that has to be controlled at each landfill, its utilization should be seriously considered. Experiences in many countries of the world show that LFG can be successfully used to replace other energy sources (Gendebien *et al.*, 1992). Since about 120–150 m^3 of LFG/t dry MSW with a calorific value of 5.9 kWh/m^3 (energy value about two-thirds that of natural gas) are produced there is a great energy potential available. LFG can be used for different purposes, such as burning to produce heat, supplying gas engines or turbines producing electricity, and upgrading to natural gas quality by removing CO_2 and trace gases. In order to make LFG utilization economically and ecologically feasible, high efficiencies of energy production and utilization should be aimed at. These goals may not be reached in many cases, owing to the location of the landfill and relatively low gas production rates at smaller landfills. But also, if no economical profits can be expected, LFG utilization should be practised for ecological reasons; it should therefore also be considered that possible deficits may be paid through the disposal fees.

The problem of potential dioxin production during and after combustion is now better understood, and emission measurements have been made (see also Chapters 8.5 and 8.6). In conclusion there is a potential for dioxin production, which has to be minimized or

Landfilling of Waste: Biogas. Edited by T. H Christensen, R. Cossu and R. Stegmann. Published in 1996 by E & FN Spon, London. ISBN 0 419 19400 2.

avoided by using appropriate flares and burners (see Chapters 6.1 and 6.2). In addition, emission control monitoring is obligatory on a routine basis. Respecting these measures, dioxin emissions are not a problem.

It is often mentioned that the varying methane concentration of LFG is a problem, e.g. by feeding gas engines with LFG. The concentration changes may cause severe operational problems, although certain variations may be tolerated. In general, LFG has a methane concentration of about 55 ($\pm$ 5)%. Lower concentrations are a result of overpumping the landfill, pulling air from outside into the landfill and diluting the LFG. If overpumping takes place at a high level and/or over longer periods, so that high amounts of oxygen enter the landfill, no or only little methane production takes place owing to toxic effects of the oxygen on the methane-forming bacteria. It may take months until the landfill has recovered. For this reason, overpumping should be avoided by installing sufficient wells and controlling the LFG extraction rate by measuring the methane concentration and keeping it at a constant level.

If LFG is sold to a company or other institutions, the contract should be based on a conservative LFG prediction (see Section 4). Gas delivery rates should not be guaranteed, unless there is a huge landfill with back-up facilities. LFG operation may fail during certain periods of time, so that either the landfill gas supplier or the user needs a reserve unit.

LFG AS A FUEL FOR DIRECT USE

As already mentioned, the utilization of LFG directly as fuel is most efficient and economic if it is practised throughout the year. The investment costs are relatively low, as in most cases a modified burner, which is available on the market, can be used. For emission control an afterburner may be necessary, e.g. if the LFG is used for steam production, in order to burn the exhaust gases at temperatures >1000 °C. Using LFG as direct fuel leads to high efficiency rates, up to 90%. As the LFG should be extracted 24 h per day the efficiency goes down to 45% if only a 12 h per day utilization (e.g. in a factory) is possible. The situation is similar if the LFG is used for heating in houses or greenhouses, as the LFG has to be flared during the summer months. In order to increase the efficiency in those cases the aim should be to find other users. Although this is difficult in many cases, there might be possibilities: for example, to dry sludges, or to produce steam for small units that may operate on or near the landfill. LFG can also be stored for shorter periods of time in pressure tanks. All the different possibilities have to be evaluated on the basis of reliability and economics.

It is always the best method of LFG utilization if the producer is also the user. This is advantageous in terms of the economics, and it guarantees that the LFG extraction system is supervised and – if necessary – repaired.

There are many different kinds of examples where LFG is successfully used as fuel: the gas is burnt in different kinds of factories in food production, in cement and brick kilns (Sperl, 1988), in bitumen production, ore processing, knackery, sludge drying facility, leachate treatment plant (reverse osmosis, condensate drying), etc. (see also Gendebien *et al.*, 1992). There is one example in Austria where LFG is used for the heating of private houses (Tscherner, 1985); in Sweden LFG is also used for district heating.

PRODUCTION OF ELECTRICITY AND HEAT IN INTERNAL COMBUSTION ENGINES

As direct use of LFG may only be possible at certain landfill sites, utilization for the production of electricity might be an alternative. Although the investment costs are in general significantly higher than for direct use and the efficiency is lower (electricity ≈ 30%, electricity and heat up to 80%) it should be kept in mind that electricity production can be practised all year round. Of course it is the most economic way if the producer of the electricity is also the user, as under these circumstances the value of the electricity gained from LFG is the same as that paid to the utility company. If the electricity is sold to the utility company, only about half the price can be obtained (in Germany). This situation may be different in other countries: thus it is very difficult to make overall statements on economics. From the economical point of view, the delivery of electricity in the peak hours of the day would be interesting.

Production of electricity in internal combustion engines is practised all over the world in many plants. In Germany about 50 are in operation, where the oldest plants have now been running over a period of 11 years. Utilization of the heat produced by the cooling water of the engines is not possible at all the plants. Experience shows that this kind of LFG utilization is a proven option and can be recommended; of course a number of specific points have to be respected. Electricity can be either used by the producer or sold to the public utility company. Small systems, which are installed in containers, can be bought off the shelf.

Landfill gas is very corrosive and water saturated. This has to be borne in mind when the materials for gas pipes, valves and seals are chosen. In addition, the water has to be knocked out. Monitoring programmes show that also in MSW landfills fluorinated and chlorinated hydrocarbons may reach concentrations of about 200 mg/Nm^3 (Laugwitz *et al.*, 1988). These concentrations decrease progressively, so that at MSW landfills older than 3 years the concentrations are in general $\leq$ 50 ppm. The halogenated hydrocarbons may be converted into a strong acid in the engine (Reinicke, 1988). The first problem occurred at the sanitary landfill at Braunschweig (co-disposal of industrial and municipal waste), where LFG is used in internal combustion engines, and one engine broke completely (Dernbach, 1985). At that time there was no information about the concentrations of halogenated hydrocarbons available and only the H$_2$S concentrations were considered with regard to corrosion (H$_2$S can be converted in the engine to H$_2$SO$_4$). After this event the problem of trace organics became obvious, as in the landfill gas at Braunschweig approximately 700–1000 mg/Nm^3 of halogenated hydrocarbons have been detected.

Corrosion problems can be avoided when specific oils developed for LFG engines are used, and the oil is measured for metal content and pH as well as acidity on a routine basis. Oil exchange rates should be higher at the beginning, and the total oil volume of the engines may be extended (Reinicke, 1988).

As emission control becomes more significant in order to meet the standards for NO$_x$ and CO (in Germany e.g. 500 mg NO$_x$/Nm^3 etc and 650 mg NO$_x$/Nm^3 if engines produce more than 1 MW electricity and heat respectively), in Germany in most cases lean gas engines are used, where the ratio of fuel to air is low. These engines are in most cases turbocharged and meet the above-mentioned emission standards without using a catalyst (Reinicke, 1988).

It has been proven that discontinuous gas extraction from the landfill (e.g. only during peak hours) may cause many problems, such as uncontrolled gas emissions combined with odour production and entry of air into the landfill. The best way to achieve discontinuous gas utilization is to install and operate a pressurized storage tank, e.g. with a capacity of 12–14 h.

As already mentioned, the production of electricity using LFG-powered gas engines is standard procedure, but there are a lot of points that have to be respected. Among others it is important to

- have a constant LFG quality;
- respect and react on the problems associated with H_2S, halogenated hydrocarbons and other trace constituents;
- observe the oil quality in the engine and exchange at relevant intervals;
- keep the engine at constant temperature also during breaks or maintenance and repair;
- observe the emissions in the exhaust gas;
- do general maintenance and inspection of the engine at shorter intervals (compared to natural gas powered engines).

Engines should only be purchased from those manufacturers that have sufficient experience with landfill gas utilization and can supply good service. A flat-rate maintenance contract on the basis of produced kilowatt hours should be considered, so that the manufacturer also has a commercial interest in good maintenance work.

PRETREATMENT OF LANDFILL GAS

For LFG with higher concentrations of halogenated hydrocarbons (due to a specific waste composition), test results as well as practical experience exist on the use of activated carbon columns or organic solvents to remove the trace organics from LFG before it is fed into the internal combustion engines (see Sections 6 and Chapter 9.3). As the activated carbon is loaded after 12–24 h, the desorption of the adsorbed halogenated hydrocarbons has to occur at this frequency. In general, the desorbed halogenated hydrocarbons are concentrated by means of cooling. Owing to the loss in effectiveness with each desorption cycle and the high price of the activated carbon, pretreatment costs are high. In addition, the pretreatment becomes even more ineffective if low concentrations of halogenated hydrocarbons (< 50 ppm) have to be treated, as the maximum loading rate of the activated carbon is in this case also lower. Very few full-scale facilities are or have been in operation e.g. in Germany, where LFG is pretreated (Kewitz, 1988; Schneider, 1988). Much more experience has been gained in the USA (Snyder, 1984; Gendebien *et al.*, 1992).

Pretreatment may also be necessary if high H_2S concentrations are present. Treatment has been practised successfully in test facilities by using activated carbon and iron compounds (Mollweide and Rettenberger, 1988) (see Chapter 6.5).

Pretreatment should only be considered in specific situations, since as already mentioned the halogenated hydrocarbon concentrations decrease with the time of active gas extraction, and in some cases the concentrations have decreased dramatically after the pretreatment plant was built. A way of dealing with the high concentrations in the early phase of gas extraction may be the installation of two parallel gas transportation pipes, one connected to those wells with high, and the other to those with low trace component concentrations. As long as high concentrations (> 100 mg/m^3) are extracted, this gas portion should be flared, whereas the other part – the 'clean gas' – can be used.

OTHER METHODS OF LANDFILL GAS UTILIZATION

In the USA, at huge landfills, LFG is treated to pipeline quality using CO_2-absorption processes. There are quite a few plants in operation where 'Selexol' is mainly used for the absorption of CO_2. These processes are expensive and only economical if landfill gas from comparatively young landfills of more than 10 million tonnes is used (see also Snyder, 1984, and Chapter 6.4).

In the USA and Switzerland, membranes are or have been used for the separation of CO_2 and CH_4 (Gandolla & Simonet, 1985). As a result of these experiences, this process is already operative, but very few plants have been constructed. It is not clear why this is the case. It might be the price of the membranes and the presence of chlorinated hydrocarbons, which may cause operational problems.

The separation of CO_2 and CH_4 has also been tested in activated carbon columns using the differential pressure method. Because of the relatively high investment and operation costs, this process may only be useful at very large landfills.

Gas turbines are in operation at bigger landfills in the USA. The advantage may be that this method creates less problems with chlorinated hydrocarbons and is characterized by a simpler technique. On the other hand, if normal gas turbines are used, the electrical efficiency is lower than that of internal combustion engines. Experience and economics will show in the future if this is a real alternative to internal combustion engines under European conditions. In Germany tests at the landfill in Wiesbaden were not successful.

Another possibility is the production of steam using LFG, where the steam runs a turbine, which drives an electrical generator. This system operates at a relatively low electrical efficiency, but it could be economically feasible if the excess heat could be utilized. This development has to be seen in the light of the problem of the existence of chlorinated and fluorinated hydrocarbons in the LFG. When the steam is used to run a turbine, the problem of corrosion in internal combustion engines does not exist. At one landfill in Germany (Karlsruhe) a steam motor has been installed (see Chapter 7.3). There is only little information available regarding the experience (Henigin, 1991).

SAFETY ASPECTS

If the methane concentrations in the atmosphere rise to between 5 and 12%, explosive mixtures are present. This is also true if the oxygen concentration in LFG is greater than 11% vol. Precautionary measures therefore have to be taken so that explosions can be excluded. The best way to do this is to avoid the build-up of explosive mixtures, which can be done by measuring the gas quality in different sections of the plant; if set concentrations are exceeded an alarm starts and/or the plant is shut down automatically. Further installations may have to be made depending on the specific situation (see also Chapters 8.3 and 8.4).

CONCLUSIONS

Landfill biogas utilization is a proven technology and is practised at many landfills all over the world. The technologies for electricity production and heat production have proven controllable and the existing techniques used for natural gas have been adapted to LFG. Further development is under way, especially in the area of gas engines, in order to meet lower emission standards for exhaust gases. Lean gas engines are mainly used in Germany for these reasons; experimental work is also being done on the use of catalysts.

The discussion about organic trace components and their effect on gas utilization and flare emissions has slowed down, owing to the results of monitoring programs. Flares are constructed to have more controlled burning as well as higher temperatures and detention times of the gas in the combustion area.

Gas purification before utilization is practised in several cases, but only limited results are available. Investigations show that the elimination of organic trace components using organic solvents and activated carbon is possible but expensive. However, in most cases purification as a pretreatment before utilization may not be needed.

The separation of CO_2 from LFG to produce gas of pipeline quality is only economical at very large landfills; this technique is used in the USA at a couple of landfills.

Landfill gas extraction and utilization should be practised at each landfill, including those where no revenue from energy utilization can be expected.

Although LFG utilization is a standard procedure, there are many aspects that have to be respected. For this reason only experienced consultants using reliable LFG equipment should be involved.

REFERENCES

Dernbach, H. (1985). *Korrosionsprobleme beim Betrieb der Blockheizkraftwerke auf der Deponie Braunschweig*, Dokumentation einer Fachtagung, Veröffentlichung des Bundesministers für Forschung und Technologie (ed.).

Gandolla, M. and Simonet, R. A. (1985). *Einsatz der Membrantechnologie zur Gastrennung*. BMFT-Forschungsreihe, *Deponiegasnutzung*, Dokumentation einer Fachtagung 1984, Umweltbundesamt (ed.).

Gendebien, A., Pauwels, M., Constant, M. *et al.* (1992). *Landfill Gas, From Environment to Energy*, Commission of the European Communities.

Henigin, P. (1991). *Deponiegasbehandlung und -verwertung am Beispiel der Deponien Wiesbaden und Karlsruhe*, Trierer Berichte zur Abfallwirtschaft 2, Economica-Verlag.

Kewitz, J. J. (1988). *Gasvorreinigung und -nutzung in Gasottomotoren auf der Deponie Kapiteltal/Kaiserslautern*, Hamburger Berichte 1, Abfallwirtschaft, Economica-Verlag, Bonn 2.

Laugwitz, R., Poller, T. and Stegmann, R. (1988). *Entstehen und Verhalten von Spurenstoffen im Deponiegas sowie umweltrelevante Auswirkungen von Deponiegasemissionen,* Hamburger Berichte 1, Abfallwirtschaft, Economica-Verlag, Bonn 2.

Mollweide, S. and Rettenberger, G. (1988). *Versuche zur Entfernung von Kohlenwasserstoffen aus Deponiegasen,* Hamburger Berichte 1, Abfallwirtschaft, Economica-Verlag, Bonn 2.

Müller, K. G. and Rettenberger, G. (1986). *Anleitung zur Entwicklung sicherheitstechnischer Konzepte für Gasabsuag- und -verwertungsanlagen aus Mülldeopnien.* Forschungsbericht 1430293, BMFT and Umweltbundesamt (eds).

Reinicke, B. (1988). *Testprogramm mit deponiegasbetriebenen Gasmotoren – Erfahrungsbericht,* Hamburger Berichte 1, Abfallwirtschaft, Economica-Verlag, Bonn 2.

Schneider, J. (1988). *Gaserfassung und -verwertung auf den Deponien Hailer-Gelnhausen und Lampertheimer Wald,* Hamburger Berichte 1, Abfallwirtschaft, Economica-Verlag, Bonn 2.

Snyder, N. W. (1984). *Biogas treatment to high-BTU gas. Technical and financial – analysis,* in *Proceedings of the 7th GRCDA International Landfill Gas Symposium,* April 1984, Piscataway, USA.

Sperl, J. G. (1988). *Gasverwertung in einer Ziegelei, Deponie Buckenhof/Erlangen und Neunkirchen. Deponiegasnutzung,* Hamburger Berichte 1, Abfallwirtschaft, Economica-Verlag, Bonn 2.

Tscherner, C. (1985). *Anlage zur Gasnutzung auf der Deponie Halbenrain/ Steiermark,* Dokumentation einer Fachtagung, Veröffentlichung des Bundesministers für Forschung und Technologie (ed.).

1.3 Landfill Gas Utilization: Statistics of Existing Plants

HANS C. WILLUMSEN

Hedeselskabet, Klostermarken 12, DK-8800 Viborg, Denmark

INTRODUCTION

During the last 15 years more than 300 landfill gas utilization plants have been established all over the world. In addition, numerous landfills have been supplied with simple gas collection and flaring systems in order to improve security on the site and reduce gas emissions. However, these systems do not utilize the energy and are not discussed any further in this chapter.

Accurate statistics on existing landfill gas utilization plants are difficult to compile from the literature for several reasons: plant descriptions are not always complete in the literature; sometimes it is difficult to determine whether reported data are design parameters or actual performance data; and plants may have been closed or enlarged since being described in the literature. In addition, literature data may not be in accordance with information revealed by an on-site investigation. Therefore this overview does not claim to be complete and devoid of false or erroneous data. Some dubious information has been sorted out, however, and plants in the planning stage or under construction are not included.

This overview covers numbers of plants, their energy potential, extraction systems, utilization system and, for some of the plants, economical information. Details on actual gas production rates are found in Chapter 4.2.

Landfilling of Waste: Biogas. Edited by T. H Christensen, R. Cossu and R. Stegmann. Published in 1996 by E & FN Spon, London. ISBN 0 419 19400 2.

EXISTING PLANTS AND ENERGY PRODUCTION

The total number of landfill utilization plants is supposedly around 300 plants. Table 1 lists 246 plants from 18 countries, where actual information was available as per 1990. Most of the literature information has been collected from Berenyi and Gould (1989), Richards (1989), Richards and Alston (1990) and Gendebien *et al.* (1992). In addition, about 50 of the plants have been visited in connection with tasks for the European Community and the World Bank.

The first plant was established in 1978, but the majority of the existing plants were established in 1985 or later. The gas and energy production of the existing plants is shown by country in Table 1. The total gas production amounts to approximately 1.2×10^9 m^3 CH$_4$/year corresponding to approximately 1.1×10^6 TOE/ year (ton oil equivalent per year) or 12000 GWh/year (gigawatt-hours per year).

The energy produced from landfill gas constitutes in Europe about 0.03% and in the USA about 0.044% of the total energy consumption.

TABLE 1. Gas and Energy Production from Existing Landfill Gas Utilization Plants (1990). TOE = Tonne oil equivalent

Country	*Number of* plants	*Production* 1000 m^3 CH$_4$/year	*Production* TOE/year	*Production* GWh/year
USA	79	843 570	746 512	8 419
Germany	80	149 298	132 000	1 490
UK	27	126 653	112 000	1 264
Netherlands	10	19 539	17 250	195
Canada	9	6 520	5 763	65
Sweden	8	12 040	10 640	120
Italy	7	11 322	10 000	113
Denmark	5	4 108	3 640	41
France	5	4 709	4 150	47
Others[a]	16	51 191	45 310	511
Total	246	1 228 950	1 087 265	12 265

[a]Switzerland 4, Brazil 3, Australia 2, Norway 2, Spain 1, Austria 1, South Africa 1, India 1, Japan 1.

EXTRACTION SYSTEMS

The most common method of extraction of gas from landfills is vertical wells (about 80% of all specified plants). The predominance of vertical wells may partly be a consequence of the fact that most landfill gas utilization plants were not included in the original landfill design, but have been established after landfilling had taken place. Only a few landfills have horizontal wells for gas extraction, while about 10% of the landfills have combination of vertical wells and horizontal pipes. Table 2 presents data on the extraction systems.

Top covers in terms of impermeable membranes may be part of the extraction system in order to improve the extraction efficiency. However, very few sites currently have top covers as a part of the gas extraction system.

TABLE 2. Extraction Systems at Landfill Gas Utilization Plants

	Wells	*Pipes*	*Combined Wells/Pipes*	*No Information available*
USA	69	0	9	1
Germany	35	4	9	32
UK	5	0	2	20
Others[a]	23	1	3	33
Total	132	5	23	86

[a] See Table 1.

ENERGY UTILIZATION SYSTEMS

The energy associated with the methane in the landfill gas can be utilized as fuel for engines, as gas in boiler plants or, after purification, can be fed into the natural gas distribution network. Table 3 summarizes the current methods for landfill gas utilization.

The most common utilization method is to use the gas in gas engines with electricity production (53%). About a quarter of these plants are combined heat and power plants, while three-quarters produce electricity only. Only 6% of the plants upgrade the landfill gas to natural gas quality, while a quarter of all utilization plants (23%) use the landfill gas for heat production. Specific utilizations

TABLE 3. Landfill Gas Utilization Systems

	Heat	*Power*	*Combined Heat/Power*	*Upgraded*	*Other or unknown*
USA	5	55	0	10	9
Germany	26	31	22	0	1
UK	10	8	3	0	6
Others*	16	2	9	4	29
Total	57	96	34	14	45

* See Table 1.

as fuels for vehicles, direct burning in kilns etc. have also been reported (Genebien *et al.*, 1992).

ECONOMY OF LANDFILL GAS UTILIZATION

Economic information, also including operation and maintenance costs, has been available for only 14 of the plants and is presented in the following to give a first appreciation of the landfill gas utilization economy. The economic data for the 14 plants, of which 9 are USA plants, are presented in Table 4.

The investment varies greatly, and so do the size and standard of the plants. No simple relationship exists between investments and landfill size, although the specific costs (e.g. per m^3 CH_4 recovered) tend to be smaller for large facilities. Based on investment data for 42 USA plants, it was calculated that the investment per MW extracted energy (megawatt) on average was of the order of US $270 000, while the electricity-producing plants on average had investments of US $1 000 000 per MW produced electricity.

The income from sale of gas and/or energy varies dramatically depending on local conditions, and no general rates can be referred to.

Operation and maintenance cost (O&M) is to some extent related to the size and the standard of the plant, but the current data do not allow for estimating any specific operation and maintenance cost.

Table 4 also presents a coarsely calculated payback period for the 14 plants in order to give a first appreciation of the profitability of landfill gas utilization plants. The payback period is here calculated

TABLE 4. Summary of Economical Investigation of 14 Landfill Gas Utilization Plants (1990)

Countries	Investment (US$ × 10⁶)	Sales per year (US$ × 10⁶)	O&M per year (US$ × 10⁶)	Payback (Years)
USA				
Duarte	2.5	0.91	0.15	3.3
Marina	1.3	0.29	0.12	7.6
Palos Verdes	18	9.4	1.8	2.4
Puente Hills	27.8	34	3.6	0.9
Brown St Road	5.9	1.31	0.28	5.7
Gude Southlawn	5.0	1.53	0.51	4.9
Roland	0.15	0.089	0.002	1.7
Greater Libanon	1.4	0.66	0.075	2.4
Brattleboro	0.37	0.13	0.096	10.9
Germany				
Hailer	3.0	0.8	0.25	5.5
Hohberg	0.9	0.18	0.03	6
Canada				
Richmond	0.4	0.2	0.06	2.9
Denmark				
Grinsted	0.83	0.09	0.02	11.9
Viborg	1.5	0.15	0.05	15

as the investment divided by the difference between income from energy sale and operation and maintenance cost without paying attention to financial aspects such as interest and inflation rates. The calculated payback period varies from 1 to 15 years, with no simple relationship to the size of the landfill or the amount of energy recovered.

CONCLUSION

Plants for landfill gas utilization, amounting to approximately 300 on a world basis, are an established technology, although only 10–15 years old. This, however, does not necessarily mean that all problems have been solved. The economic information available indicated that the technology is a viable method for recovering the energy in landfill gas and simultaneously solving an environmental problem.

More landfill gas utilization plants are expected in the future as the technology spreads from the leading countries (USA and Germany), and as further improvement of emission standards for landfill gas substitute simple flaring techniques with more advanced technology including energy recovery.

The technology, at least with respect to gas extraction, may also develop in the future as more landfill gas utilization systems will be an integral part of new landfills from the early planning and design.

REFERENCES

Berenyi, E. & Gould, R. (1989). *1988–89 Methane Recovery from Landfill Yearbook. Directory & Guide*, Governmental Advisory Associates Inc., New York.

Gendebien, A., Pauwels, M., Constant, M., Ledaut-Damanet, M.J., Willumsen, H.C., Butson, J., Fabey, R., Ferraro, G.-L. and Nyns, E.-J. (1992). *Landfill Gas. From Environment to Energy*, Office for Official Publications of the European Communities, Luxembourg.

Richards, G.-E. & Alston, Y.-R. (1990). *Proceedings of International Conference Landfill Gas: Energy and Environment '90*, Bournemouth, England, 16–19 October 1990.

Richards, K.-M. (1989). Landfill gas: working with Gaia. *Biodeterioration Abstracts*, 3(4), 317–331; and additions (1990) presented to the Methane Emissions Workshop, Washington, March.

2. GAS GENERATION AND COMPOSITION

2.1 Gas-Generating Processes in Landfills

THOMAS H. CHRISTENSEN, PETER KJELDSEN &
BO LINDHARDT

*Institute of Environmental Science and Engineering, Technical
University of Denmark, Building 115, DK-2800 Lyngby, Denmark*

INTRODUCTION

The gaseous emissions from landfills are a result of processes taking
place in the landfill and as such are related to the waste landfilled
and the landfill technology used. In landfills receiving organic waste,
the microbial conversion of organic carbon to methane and carbon
dioxide is the dominating gas-generating process responsible for the
main gas components and the gas flux also transporting the trace
components out of the landfill. The trace components in the landfill
gas are a result of microbiological as well as physico-chemical
processes.

This chapter describes the gas-generating processes in landfills
with respect to microbial conversion of organic matter into methane
and carbon dioxide, physico-chemical volatilization of organic trace
components, and microbially mitigated conversion of trace compounds (reductive dehalogenation and methylation). With respect
to the methane-generating processes the governing factors and
landfill operation procedures are also briefly summarized.

MICROBIAL CARBON CONVERSION

The predominant part of the landfilled waste will soon after disposal
become anaerobic, and microbial processes will degrade the organic

Landfilling of Waste: Biogas. Edited by T. H Christensen, R. Cossu and
R. Stegmann. Published in 1996 by E & FN Spon, London.
ISBN 0 419 19400 2.

waste, eventually converting the solid organic carbon to methane (CH_4) and carbon dioxide (CO_2). However, this is no simple process, and to understand the overall process attention must be paid to the microbial consortium performing the carbon conversion, the basic parameters influencing the process, and how actual landfilling technology affects the methane formation. Methane formation is usually seen as beneficial in terms of improving both the leachate quality and the efficiency of the landfill gas utilization.

Microbial Methane Formation

Figure 1 illustrates the most important interactions in an anaerobic landfill between the bacterial groups involved, the substrates involved, and intermediate products. The anaerobic degradation can be viewed as consisting of three stages. In the first stage, solid and complex dissolved organic compounds are hydrolysed and fermented by the fermenters to primarily volatile fatty acids, alcohols, hydrogen and carbon dioxide. In the second stage, an acetogenic group of bacteria converts the products from the first stage to acetic acid, hydrogen and carbon dioxide. In the final stage, methane is produced by the methanogenic bacteria. This may be done by acetophilic bacteria converting acetic acid to methane and carbon dioxide or by hydrogenophilic bacteria converting hydrogen and carbon dioxide to methane. The overall process of converting organic compounds to methane and carbon dioxide may stoichio-metrically be expressed by (Buswell and Mueller, 1952):

$$C_nH_aO_b + \left(n - \frac{a}{4} - \frac{b}{2}\right)H_2O \rightarrow \left(n - \frac{n}{2} - \frac{a}{8} + \frac{b}{4}\right)CO_2 +$$
$$\left(\frac{n}{2} + \frac{a}{8} - \frac{b}{4}\right)CH_4$$

But, as indicated by Figure 1, the actual process is not at all a simple reaction.

The hydrolysis process is a very important process in the landfill environment, as the solid organic waste must be solubilized before the micro-organisms can convert it. After the smaller, easily soluble part of the organic matter has been converted, the hydrolysis may prove to be the overall rate-limiting process in the landfill environment (Leuschner, 1983; McInerney and Bryant, 1983; Barlaz et al., 1989; El-Fadel et al., 1989). The hydrolysis is caused by extra-

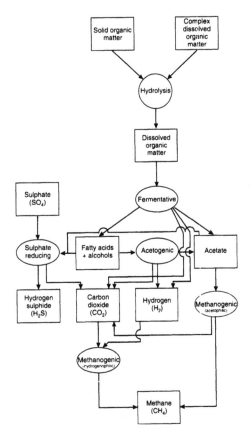

Figure 1. Substrates and major bacterial groups in the methane-generating ecosystems (from Christensen and Kjeldsen, 1989).

cellular enzymes produced by the fermenting bacteria (Jones *et al.*, 1983).

The fermenters are a large, heterogeneous group of anaerobic and facultatively anaerobic bacteria. Some of the important reactions are shown in Table 1. The acetogenic bacteria are also a large heterogenic group. The acetogenic bacteria produce acetic acid, hydrogen and also carbon dioxide (McInerney and Bryant, 1983), if the volatile fatty acid being converted contains an odd number of carbon atoms. The acetogenic bacteria may also convert aromatic compounds containing oxygen (e.g. benzoic acid and phenols), while aromatic hydrocarbons (e.g. benzene and toluene) are

Christensen, Kjeldsen, Lindhardt

TABLE 1. Examples of Important Reactions for Four Groups of Bacteria Involved in Anaerobic Waste Degradation, Based on Hansson and Molin (1981), McInerney and Bryant (1983), Zehnder (1978) and Postgate (1979), as Presented by Christensen and Kjeldsen (1989).

Reactants converted to products

Fermentative processes

$C_6H_{12}O_6 + 2H_2O$	$\rightarrow$	$2CH_3COOH + 4H_2 + 2CO_2$
$C_6H_{12}O_6$	$\rightarrow$	$CH_3C_2H_4COOH + 2H_2 + 2CO_2$
$C_6H_{12}O_6$	$\rightarrow$	$2CH_3CH_2OH + 2CO_2$

Acetogenic processes

$CH_3CH_2COOH + 2H_2O$	$\rightarrow$	$CH_3COOH + CO_2 + 3H_2$
$CH_3C_2H_4COOH + 2H_2O$	$\rightarrow$	$2CH_3COOH + 2H_2$
$CH_3CH_2OH + H_2O$	$\rightarrow$	$CH_3COOH + 2H_2$
$C_6H_5COOH + 6H_2O$	$\rightarrow$	$3CH_3COOH + CO_2 + 3H_2$

Methanogenic processes

$4H_2 + CO_2$	$\rightarrow$	$CH_4 + 2H_2O$
CH_3COOH	$\rightarrow$	$CH_4 + CO_2$
$HCOOH + 3H_2$	$\rightarrow$	$CH_4 + 2H_2O$
$CH_3OH + H_2$	$\rightarrow$	$CH_4 + H_2O$

Sulphate-reducing processes

$4H_2 + SO_4^{2-} + H^+$	$\rightarrow$	$HS^- + 4H_2O$
$CH_3COOH + SO_4^{2-}$	$\rightarrow$	$CO_2 + HS^- + HCO_3^- + H_2O$
$2CH_3C_2H_4COOH + SO_4^{2-} + H^+$	$\rightarrow$	$4CH_3COOH + HS^-$

$HCOOH$, formic acid; CH_3COOH, acetic acid; CH_3CH_2COOH, propionic acid; $CH_3C_2H_4COOH$, butyric acid; $C_6H_{12}O_6$, glucose; CH_3OH, methanol; CH_3CH_2OH, ethanol; C_6H_5COOH, benzoic acid; CH_4, methane; CO_2 carbon dioxide; H_2, hydrogen; SO_4^{2-}, sulphate; HS^-, hydrogen sulphide; HCO_3^-, hydrogen carbonate; H^+, proton; H_2O, water.

apparently not degraded. Some of the important acetogenic processes are also shown in Table 1.

The methanogenic bacteria are obligate anaerobic and require very low redox potentials. One group, the hydrogenophilic, converts hydrogen and carbon dioxide to methane, while another group, the acetophilic, converts primarily acetic acid to methane and carbon dioxide. The methanogenic bacteria may also convert formic acid and methanol. Some of the important reactions are shown in Table 1. The conversion of acetic acid to methane is believed to be the most important part of the methane-forming process.

Finally, the sulphate-reducing bacteria, dominated by *Desulfovibrio* and *Desulfotomaculum* (Postgate, 1979), play an important role, as this group of bacteria in many ways resembles the methanogenic group and as sulphate is a major compound of many waste types (demolition waste, incinerator slag, fly ashes). The sulphate-reducing bacteria are obligate anaerobic and may convert hydrogen, acetic acid and higher volatile fatty acids during sulphate reduction. However, the organic carbon is always oxidized to carbon dioxide as opposed to the conversion by the methanogenic group of bacteria. A high activity of sulphate reducers may hence decrease the amount of organics available for methane production. Some of the sulphate-reducing reactions are shown in Table 1.

A summary of the chemistry of organic compounds in landfilled waste and their degradation pathways is presented by Senior & Balba (1990), to whom the reader is referred.

The overall influence of these microbial degradation processes on the main composition of the landfill gas is illustrated for a homogeneous landfill cell in Figure 2. The gas composition can be related to eight distinct phases according to the progress of the waste degradation:

- *Phase I.* A short aerobic phase depleting O_2 by composting of easily degradable organic matter to CO_2.
- *Phase II.* Fermentative and acidogenic bacteria produce under anaerobic conditions volatile fatty acids, CO_2 and H_2. The presence of these gases reduces the content of N_2.
- *Phase III.* In a second anaerobic phase, methanogenic bacteria start to grow producing CH_4, while CO_2 and H_2 decrease.
- *Phase IV.* The stable methanogenic phase is characterized by 50–60% CH_4 and low concentrations of H_2. The latter being oxidized by CO_2 to CH_4.

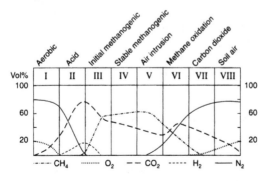

Figure 2. Illustration of developments in gas composition in a landfill cell.

- *Phase V.* Air starts to intrude into the outer part of landfill body reducing the formation of CH_4. The lower rates lead to a relatively more significant washout of CO_2 and a relative increase in CH_4 content of the gas.
- *Phase VI.* Methane produced in the centre of the waste is oxidized to CO_2 as it migrates through the outer part of the landfill body. N_2 is now present in significant concentrations in the gas.
- *Phase VII.* Methane formation is now negligible and intruding air now oxidizes solid organic carbon (and reduced inorganic species) yielding CO_2.
- *Phase VIII.* The rates of the processes now approach the rates found in an active soil and the landfill gas starts to resemble soil air.

The idealized gas composition sequence in Figure 2 purposely presents no estimates of the length of the phases involved, owing to their dependence on abiotic factors and local conditions such as waste composition and landfilling procedure. After the initial aerobic phase – which lasts only for days or a few weeks – months, years and decades are the time units applying to the other phases.

Influence of Abiotic Factors

The many different kinds of waste received at the landfill, the variable degradability of the organic fraction, and the heterogeneity of the waste cells, may make the landfill a highly diverse, but rather inefficient ecosystem. For a given organic waste, the gas-generating processes will be influenced by the local environment as characterized by the abiotic factors: oxygen, hydrogen, pH and alkalinity, nutrients, inhibitors, temperature and water content. The influence of these factors is briefly described in the following paragraphs.

Oxygen. The absence of free oxygen is essential for the anaerobic bacteria to grow and perform the conversion of the solid carbon to methane and carbon dioxide. In particular, the methanogenic bacteria are sensitive, requiring very low redox potentials (below -330 mV). Oxygen may enter the landfill by diffusion from the atmosphere, but this is supposedly limited to only the uppermost part of the landfill. If a substantial vacuum is created in the landfill by extensive gas extraction, air may enter the landfill by advection, inhibiting the methane formation in the influenced areas. Although no spore-forming methanogenic bacteria are known (Zehnder,1978), the methanogenic community is not completely wiped out by the introduction of oxygen and will recover upon the depletion of the introduced oxygen.

Hydrogen. Hydrogen is produced by both the fermentative and the acetogenic bacteria. The fermentative bacteria yield hydrogen, carbon dioxide and acetic acid at low hydrogen pressures, but hydrogen, carbon dioxide and ethanol, butyric acid and propionic acid at high hydrogen pressures (McInerney & Bryant, 1983). The last three organic compounds may be further converted by the acetogenic bacteria provided the hydrogen pressure is low. The conversion of propionic acid requires hydrogen pressures below 9×10^{-5} atmospheres. This means, in the case that the hydrogen pressure increases, propionic acid and to some extent also butyric acids will be generated, but not further converted, potentially leading to an accumulation of volatile organic acids, a decrease in pH and possibly inhibition of the methane formation.

pH and alkalinity. The methanogenic bacteria operate efficiently only within a narrow pH-range of 6–8 (Zehnder *et al.*, 1982). The pH range for the fermentative and acetogenic bacteria is much wider than for the methanogenic bacteria. If the methanogens are stressed by other factors, their conversion of hydrogen and acetic acids decreases, leading to an accumulation of volatile organic acids and a decrease in pH, which furthermore may inhibit the methane formation and lead to a further decrease in pH. Eventually, the methane generation may stop. The methanogenic ecosystem in the landfill is rather delicate, and a balanced relation between bacterial groups is crucial for good methane production. The sulphate- reducing bacteria have a somewhat wider pH range than the methanogenic bacteria, less than pH 5 to pH 9 according to Postgate (1979).

Sulphate. Both the sulphate-reducing bacteria and the methanogenic bacteria convert acetic acid and hydrogen. Several experiments, both batch experiments and laboratory landfill simulators, have shown that when sulphate is present in substantial concentrations the methane production is dramatically reduced (e.g.

Stegmann and Spendlin, 1985). The suppression of methane production by sulphate is not related to any toxic effects of sulphate on
the methanogenic bacteria, but due to simple substrate competition.
The higher energy yielded by sulphate reduction favours the growth
of sulphate reducers (Zehnder *et al.*, 1982).

Nutrients. The anaerobic ecosystem must, besides organic matter,
have access to all required nutrients, in particular nitrogen and
phosphorus. All the necessary micronutrients, e.g. sulphur, calcium,
magnesium, potassium, iron, zinc, copper, cobalt, molybdenite and
selenium, are considered to be fully available in most landfills. The
anaerobic ecosystem assimilates only a very small part of the substrate
into new cells and therefore requires much less nitrogen and phosphorus than the aerobic system. Optimal ratios between organic
matter (expressed as chemical oxygen demand), nitrogen and phosphorus are listed by McCarty (1964) as 100 : 0.44 : 0.08. On average, the mixed waste landfill will not be limited by nitrogen and
phosphorus, but insufficient homogenization of the waste may result
in nutrient-limited environments. Phosphorus is, if any, the nutrient
most likely to limit the anaerobic degradation processes.

Inhibitors. The methane-forming ecosystem is considered to be
rather sensitive to inhibitors. The inhibitory effects of oxygen,
hydrogen, proton activity and sulphate have already been discussed
separately because of their special significance, but other inhibitors
may also appear in the landfill environment. The inhibitory effects
of volatile fatty acids have been studied by McCarty and McKinney
(1961a) and Kugelman and Chin (1971). Heuvel (1985).
Kugelman and Chin did not find any inhibitory effects of total
concentrations of acetic acid, propionic acid and butyric acid up to
6000 mg/l. Inhibitory effects of volatile organic acids are expected
to be rare in landfills. Carbon dioxide, being a product of many of
the degradation processes taking place in the landfill, may exhibit
inhibitory effects on the acetic acid conversion to methane in
particular. Hansson and Molin (1981) found inhibition of acetic
acid conversion at carbon dioxide partial pressure as low as 0.2.
However, acetic acids were also converted in a pure carbon dioxide
atmosphere although the rate was significantly reduced. In the acid
phase of the landfill stabilization, carbon dioxide partial pressures
may reach 0.9 and will typically be around 0.5 in the methanogenic
phase. Macro-ions such as sodium, potassium, calcium, magnesium

and ammonium may have inhibitory effects on methane formation when exceeding a couple of thousands mg/l each (McCarty and McKinney, 1961b), but such high levels are rarely found in landfill leachate. Data on inhibition by specific organic compounds, as summarized by Johnson (1981), indicate that most of the investigated compounds should be present in concentrations above 10 mg/l and many above 100 mg/l to cause substantial inhibition of methane formation. Such high concentrations are only expected rarely and only in landfills receiving substantial amounts of industrial waste. This short summary on compounds potentially inhibiting methane formation indicates that in most landfill environments the inhibitors will be present in concentrations too low to be detrimental to the methane formation, although they may cause a reduction in the methane formation rate.

Temperature. Like all other microbial processes, methane formation is highly affected by temperature. In laboratory landfill simulators, the methane formation rate has been shown to increase significantly (up to 100 times), when the temperature is raised from 20 to 30 and 40 °C (e.g. Buivid, 1980; Scharf, 1982; Ehrig, 1984). Although the anaerobic degradation of organic compounds yields much less heat than the aerobic process (about 7% according to Christensen and Kjeldsen, 1989), elevated temperatures (30–45 °C) in landfills have been reported (e.g. Rees, 1980b). Elevated temperatures primarily develop in landfills with a good methane production, a moderate water flux and substantial waste thickness providing good insulation. At elevated temperatures, the methane production is more vigorous and produces more heat, and as such is self-enhancing.

Water content. Several laboratory investigations have shown that the methane production rate increases with increasing moisture content of the waste (e.g. Buivid, 1980; Rees, 1980a). Rees (1980a) summarized findings from the literature suggesting an exponential increase in gas production rates between 25 and 60% water content. The main effect of the increased water content, besides limiting the oxygen transport from the atmosphere, is probably the facilitated exchange of substrate, nutrients, buffer, and possibly dilution of inhibitors and spreading of micro-organisms between the micro-environments.

Influence of Landfill Operation Procedures

Landfill operation as practised at actual full-scale landfills cannot control the above-mentioned abiotic factors individually. Although a specific operation procedure may aim at improving one of the abiotic factors, the procedure will usually affect several abiotic factors simultaneously, and the overall effect may be difficult to predict from knowledge of the individual factors. In addition, some operation procedures may have been introduced due to other aspects than improvement of the gas generation. Recently Christensen *et al.* (1992) reviewed the literature with respect to the effects of landfill operation on landfill stabilization and leachate and gas quality. The most significant results as to effects on gas generation are summarized in the following paragraphs.

Waste composition. The composition of the waste landfilled is usually governed by the disposal needs of the community and has only in a few cases been studied as a controllable factor. German investigations by Stegmann and Spendlin (1986) and Wolffson (1985) showed that an increased content of newspaper did not increase the methane production in laboratory landfill simulators over a 250 day period, while magazine printing did have a positive effect. Wet organic matter from kitchen and garden delayed the methane production, supposedly owing to a more intensive acid generation, but in the long run the total gas production was increased by the presence of the putrescibles.

Sewage sludge addition. The effects of sewage sludge addition on waste degradation and methane formation have been studied in several cases. Potential positive effects of sewage sludge may be attributed to increasing water content, supply of readily available nutrients and supply of an active anaerobic biomass: factors that all would increase methane formation. The beneficial effects of this potential, however, seem to be conditioned by two factors. First, if methanogenic conditions are already present, addition of sewage sludge may have only a limited effect. Second, the influence of sewage sludge on the pH of the landfill seems crucial. Sludge low in pH, e.g. septic sludges, may have a negative effect on methane production (Leckie *et al.*, 1979; Leuschner and Melden, 1983) while neutral, well-buffered sewage sludge may have positive effects (Buivid, 1980; Leuschner and Melden, 1983; Kinman *et al.*, 1987; Leuschner, 1989).

Buffer addition. The detrimental effects of low pH values on methane formation have frequently led to the concept of adding buffer to the landfill, either by incorporating a solid buffer material (e.g. calcium carbonate) or by adding a base or buffer solution (e.g. NaOH or $NaHCO_3$) to the leachate prior to recirculation. Most of the reported studies, which have been of laboratory scale, found positive effects on methane formation by buffer addition (Pohland and Kang, 1974; Buivid, 1980; Tittlebaum, 1982; Leuschner and Melden, 1983; Barlaz *et al.*, 1987; Kinman *et al.*, 1987). The methane formation is by itself producing buffer capacity and increasing pH, and only where this process cannot outweigh the acidity produced by the acid phase has buffer addition an effect. This means that buffer addition can be used as a prophylactic to avoid the development of pH values that are too low, if the acid phase should become very vigorous, or as cure in landfills where methane formation is inhibited by low pH values. However, addition of buffer may be redundant at many well-managed landfills.

Shredding. Shredding of waste prior to landfilling potentially may have advantageous effects on waste stabilization in terms of increasing the homogeneity of the waste by size reduction and mixing, increasing the specific surface area of the waste, removing water barriers caused by plastic bags and foil, and improving the water content and distribution in the landfilled waste. However, shredding may primarily intensify the acid phase of the landfill stabilization, preventing or postponing the start of the methane formation (EMCON Associates, 1975; Eifert, 1976; Chian *et al.*, 1977; Buivid, 1980). If the negative effects of shredding in the initial phase of the landfill stabilization could be circumvented, shredding could prove beneficial in the later phases by improving the hydrolysis and the acid formation. Both Barlaz *et al.* (1989) and El-Fadel *et al.* (1989) stated that hydrolysis tends to limit the methane formation in the later phases of the landfill stabilization. Thus shredding alone cannot be recommended, but in combination with measures to control the acidic phase shredding may be beneficial.

Compaction. Compaction as performed by a compactor at the tipping front is a very common operation, argued by the need for optimum use of the landfill capacity and for obtaining geotechnical stability of the landfilled waste. Thus it is expected that all waste sooner or later will be compacted at a modern landfill. The question is merely how an initial or delayed compaction may affect the methane formation. A thorough compaction of the waste leads to some homogenization and mixing of the waste landfilled, and in practice these factors cannot be evaluated separately. Experimental data on the effects of compaction on methane formation are scarce (Buivid, 1980; Buivid *et al.*, 1981; Rees and Grainger, 1982) and distinct conclusions cannot be justified. However, the few results available indicate that a loose compaction or lack of compaction may initially make the acid phase less vigorous and hence improve the start of the methane formation, at least in relatively wet waste. In dry waste, compaction may on the other hand improve methane formation by improving the water content and by reducing the intrusion of oxygen.

Soil cover.   Very little information is available on the effect of daily soil covers on waste stabilization (Ham and Bookter, 1982; Lee *et al.*, 1986). Negative effects of soil covers may be expected if the upper layer is supposed to undergo aerobic degradation. A soil cover will here decrease the diffusion of oxygen into the waste layer and thus decrease the composting rate: the heavier the soil, the more adverse the effect expected. Use of heavy clayish soils as daily covers may at a later stage cause heterogeneous water distribution in the landfill, and eventually perched water-tables in the waste or very dry zones below soils of low permeability. Positive effects of daily soil covers may be expected if the soil provides important buffer capacity to the landfill, avoiding low pH values inhibitory to methane formation.

Recirculation of leachate.   Recirculation of leachate is the most investigated landfill management procedure. Reports are found on positive aspects of recirculation on gas production and composition (EMCON Associates, 1975; Augenstein *et al.*, 1976; Leckie *et al.*, 1979; Pohland, 1980; Buivid, 1980; Klink and Ham, 1982; Leuschner and Melden, 1983; Stegmann, 1983; Doedens and Cord-Landwehr, 1984; Christiansen *et al.*, 1985; Stegmann and Spendlin, 1986; Barlaz *et al.*, 1987; Beker, 1987; Kinman *et al.*, 1987), on no effects (Barlaz *et al.*, 1987) and on negative effects (Leuschner and Melden, 1983; Beker, 1987; Kinman *et al.*, 1987). The reported arguments for introducing leachate recirculation are, besides providing internal leachate treatment: enhancing methane formation by increase in the water content and circulation in the waste; by supply and distribution of nutrients and biomass; and by dilution of locally high concentrations of inhibitors. Although the experiences with recirculation of landfill leachate are not straightforward, a properly balanced view of its possibilities for improving gas generation must pay attention to leachate composition and waste dryness. In a relatively wet climate, recirculation is supposedly only beneficial in the first years of the landfill life, by improving the water content and distribution in the waste cells by returning stabilized leachate to biological active cells. However, if the leachate exhibits low pH values and the methanogenic conditions are not well established, recirculation may be detrimental to the methane formation unless the leachate is pH-adjusted and buffered. When the waste cells have proper water content and the leachate is low in BOD_5, recirculation supposedly has very little effect on methane formation. In dry climates, recirculation may be beneficial for longer periods in order to improve the water content and distribution in the landfill. But here also the recirculation should be handled carefully, so that unfavourable conditions are not created for methane formation.

Precomposting. In recent years, the concept of *in situ* pre-composting of the bottom layer of the landfill in order to prevent too vigorous an acid phase degradation has developed in Germany (Stegmann, 1983). By allowing partial stabilization of the waste through aerobic processes, the acid phase becomes less vigorous and the methanogenic consortia establish faster, when the waste is made anaerobic by compaction and additional waste layers. This results in overall enhancement of the methane formation. The engineering approaches of this concept are described by Spillmann and Collins (1981), Ehrig (1982), Stegmann (1983) and Stegmann and Spendlin (1989). When the anaerobic methane-forming filter is established in the bottom of the landfill, the technology for landfilling the layers on top is less critical, because the bottom filter will have a very high capacity for methane production (Doedens and Cord-Landwehr, 1984).

PHYSICO-CHEMICAL VOLATILIZATION

Landfill gas usually contains a variety of trace organics (see Chapter 2.2) in the range 0.05–1000 mg/m^3. These organic trace components primarily originate from the waste deposited, although some of them also may be formed in the landfill as intermediary or final products of microbial or abiotic degradation processes. The organic trace components important in the context of landfill gas are all rather volatile and enter easily into the gas void in the landfill. However, these components may also dissolve in the leachate and sorb (partition) into the organic waste in the landfill cell. If the organic compounds have entered the landfill as a chemical waste (a free phase or a mixture of free phases), a substantial reservoir of these compounds may be present in the landfill unless the compounds are susceptible to degradation. The following paragraphs describe the physico-chemical behaviour of these trace components in the landfilled waste and provide some estimates of their distribution and significance over time.

Phase Distribution in Landfilled Waste

A compound j present in the landfill in a total contration of $C_{t,j}$ (for example expressed in g of compound per m^3 of landfill) may be present in the leachate (w), gas (a), solid phase (s) and in a free separate phase (f) as expressed by:

$$C_{t,j} = v_w C_{w,j} + v_a C_{a,j} + b C_{s,j} + v_f C_{f,j} \qquad (1)$$

where $C_{t,j}$ is the total concentration of compound j in the landfill (g compound/ m^3 landfill); v_w is the volumetric content of water in the landfill (m^3 of water/ m^3 of landfill); $C_{w,j}$ is the concentration of compound j in the water (leachate) (g compound/ m^3 water); v_a is the volumetric content of air in the landfill (m^3 of air/ m^3 of landfill); $C_{a,j}$ is the concentration of compound j in the air (gas) (g compound/ m^3 air); b is the bulk density of waste in the landfill (metric tonnes of dry waste/ m^3 of landfill); $C_{s,j}$ is the concentration of compound j in the dry waste (g compound/ metric tonne of dry waste); v_f is the volumetric fraction of free separate organic phase in the landfill (m^3 free separate phase/ m^3 of landfill); $C_{f,j}$ is the concentration of compound j in the free separate phase (g compound/ m^3 of free separate phase).

Considering 1 m^3 of landfilled waste compacted to a density of 1 tonne per m^3 at 40% moisture content, the volume and mass of each phase defined as water, air and dry waste can be estimated assuming a particle density of the dry waste of approximately 1.2 tonne per m^3:

1 tonne of wet waste = 0.4 tonne of water + 0.6 tonne of dry waste

equal to (1 tonne of wet waste equals 1 m³ of landfill):

1 m³ of wet waste = 0.4 m³ of water + 0.1 m³ of gas + 0.5 m³ of dry solids

The above simplifications tacitly assume that the mass of the gas and eventual free phases of organic chemicals per m³ of wet waste are negligible.

The total concentration of an organic chemical j per volume of landfill, according to the general phase distribution equation presented above, may therefore be expressed by:

$$C_{t,j} = 0.4\ C_{w,j} + 0.1\ C_{a,j} + 0.6\ C_{s,j} + v_f\ C_{f,j} \tag{2}$$

The relationships between the different terms in Equation (2) and methods for their estimation are discussed in the next section.

Phase Equilibrium Distribution: Free Phase Present

If chemical waste has been disposed of in the landfill and is in contact with the waste (containers broken or corroded), the presence of a free phase will control the concentration of the compound in the gas phase. If the amount of free phase is small, the influence on the gas composition may only be local and may not be reflected in the content of the compound in the extracted gas.

In the presence of a free phase, all the other phases will be saturated: i.e. the concentration of the compound in the dry waste, in the water and in the gas phases will be constant no matter how much free phase is present. The gas phase concentration will be governed (under ideal conditions) by Raoult's law (3) and the concentration in the solid phase and in the water will be related to the gas concentration according to the equations described in the section on phase distribution without a free phase present.

The concentration in the gas phase can under ideal conditions be estimated from Equation (3):

$$C_{a,j} = \frac{x_j\ p_j\ MW_j}{RT} \tag{3}$$

where x_j is the mole fraction of compound j in the free phase (mol/mol); p_j is the vapour pressure of compound j (Pa); MW_j is the

TABLE 2. Vapour Pressures (Mackay et al., 1992), Saturated Gas Phase Concentrations at 25 °C for Selected Volatile Organic Compounds and Typical Trace Gas Concentrations Measured in Landfill Gas (partly based on Chapter 2.2)

Volatile components	Vapour pressure (Pa)	Saturated gas concentration (g/m³)	Typical LFG concentration (g/m³)
n-Hexane	20 200	700	0.05
Benzene	12 700	400	0.005
Toluene	3 800	140	0.2
m-Xylene	1 100	50	0.1
1,3,5-Trimethylbenzene	325	165	0.1
Chloroethane	16 000	415	0.01
1,1-Dichloroethane	30 260	1 210	0.01
1,1,1-Trichloroethane	16 500	890	0.001
Chloroethene (vc)	354 600	8 940	0.01
cis-1,2-Dichloroethene	27 000	1 050	0.02
Trichloroethene	9 900	525	0.01
Tetrachloroethene	2 415	160	0.02
Trichlorofluoromethane	102 200	5 650	0.02
Freon 113	483 200	36 500	0.005

mole weight of compound j (g/mol); R is the gas constant (8.31 Pa m³ mol^{-1} K^{-1}); T is the temperature (K).

Table 2 shows some calculated gas concentrations for selected landfill gas trace components assuming that x_j is 1.0 for each component. This assumption is of course not valid in a real landfill, but allows for estimating the maximum concentrations of the components. Table 2 also presents some typical concentrations measured in landfill gas (actual concentrations may vary by two orders of magnitude). The typical concentrations are much lower than the maximum concentrations estimated from a pure free phase ($x_j = 1.0$). This indicates that either no free phase is present or the free phase present only contains the individual components in very small fractions (x_j is very low).

Assuming that 150 m³ LFG is generated per tonne of dry waste (Chapter 1.2), 1 m³ of waste will generate 90 m³ of LFG or approximately 900 porevolumes. If 20 m³ is generated within the first 5 years, about 0.5–10 kg of most of the compounds listed in Table 2 could be volatilized, assuming that x_j is equal to 0.1. The volatilization of the most volatile compounds (vinyl chloride, Freon 113 and Freon 21) would be larger and the volatilization of the less volatile components (toluene, xylene, trimethylbenzene) would be smaller. These estimates show that free separate phases, which eventually are present at the disposal of the waste, in most cases will be depleted very rapidly by volatilization, except possibly for toluene, xylene and trimethylbenzene. For example, the proposed gas generation within the first 5 years would remove only 1 kg of xylene from a pure separate phase of xylene ($x_j = 1.0$) in 1 m^3 of waste.

Phase Equilibrium Distribution: No Free Phase Present

When no free phase is present in the waste, the equation describing the distribution is simplified to

$$C_{t,j} = 0.4 \ C_{w,j} + 0.1 \ C_{a,j} + 0.6 \ C_{s,j} \tag{4}$$

At equilibrium, two equations describe the relationship between the phases. The relationship between the gas phase and the water phase is described by Henry's constant ($K_{H,j}$, m^3 water/ m^3 air, Equation (5)) and the relationship between the solid phase and the water phase is described by the distribution coefficient ($K_{d,j}$, m^3 water/ tonne dry waste, Equation (6) and Equation (7).

$$C_{a,j} = K_{H,j} \ C_{w,j} \tag{5}$$

$$C_{s,j} = K_{d,j} \ C_{w,j} \tag{6}$$

K_d may be estimated according to the following relation, which is often used for soils and sediments:

$$K_{d,j} = f_{OC} \ K_{OC,j} \tag{6}$$

where f_{OC} is the fraction of organic carbon in the dry waste; $K_{OC,j}$ is the distribution coefficient for compound j onto solid organic carbon (m^3 water/ tonne carbon). These values are often reported in the literature or can be estimated (Abdul et al., 1987) from the octanol–water partitioning coefficients of the compounds, available in chemical handbooks.

Substituting these relationships into Equation (4), the following relationship is found:

$$C_{t,j} = 0.4 \ (K_{H,j})^{-1} \ C_{a,j} + 0.1 \ C_{a,j} + 0.6 \ f_{OC} \ K_{OC,j} \ (K_{H,j})^{-1} \ C_{a,j}$$

Based on this equation, the relative distributions of selected volatile components among the phases are estimated in Table 3, assuming a fraction of organic carbon of 0.2 of the dry waste.

Emptying of the landfill volume of volatile trace gases by volatilization (corresponding to 90% removal) is estimated to require between 45 (vinyl chloride) and 65 000 (1,3,5-trimethylbenzene) porevolumes, corresponding to between 4.5 and 6500 m^3 LFG produced per m^3 of waste. This shows, assuming as earlier that 20 m^3 of LFG is produced within the first 5 years, that some of the trace compounds (vinyl chloride (unless continuously generated by degradation of higher chlorinated compounds), Freon 113 and trichlorofluoromethane) will be present in the LFG for only a short period of time (less than 5 years), while others (benzene, toluene, xylene, trimethylbenzene, trichloroethene and tetrachloroethene) will be sorbed strongly to the solid organic carbon and most likely will not be stripped off completely with the gas produced by the degradation processes in the waste.

The above calculations assume a linear distribution between the three phases and do not depend on the actual concentrations present, as long as no free phases are present. This implies that if the initial concentration of a volatile component is low in one phase, then the concentrations in the other phases are low as well, and although it may take many years to strip off the component, the concentrations in the gas may be negligible.

TABLE 3. Estimated Distribution between Solid Phase, Water Phase (Leachate) and Gas Phase (Gas) of Selected Volatile Components Assuming the Absence of Any Free Phases (f_{OC}=0.2; K_H calculated from Mackay et al. (1992) and K_{OC} from Abdul et al. (1987)).

Volatile component	K_{OC}	K_H	Percentage distribution		
			Solid	Water	Air
n-Hexane	13 400	73.92	99.52	0.02	0.46
Benzene	139	0.22	97.47	2.39	0.14
Toluene	508	0.27	99.30	0.65	0.05
m-Xylene	1 647	0.29	98.74	1.17	0.09
1,3,5-Trimethylbenzene	3 953	0.32	99.91	0.08	0.01
Chloroethane	27	0.07	87.42	12.35	0.32
1,1-Dichloroethane	63	0.25	94.40	5.27	0.33
1,1,1-Trichloroethane	321	0.59	98.81	1.04	0.15
Chloroethene (vc)	24	3.24	77.31	12.54	10.15
cis-1,2-Dichloroethene	74	0.30	95.19	4.47	0.34
Trichloroethene	352	0.48	98.94	0.95	0.11
Tetrachloroethene	788	1.08	99.46	0.42	0.12
Trichlorofluoromethane	352	5.20	97.84	0.94	1.22
Freon 113	1 502	45.93	97.30	0.22	2.48

MICROBIALLY MITIGATED CONVERSION OF TRACE COMPONENTS

The importance of the microbial processes for the emission of trace organics with the LFG is not fully understood, but it is known that microbial processes as well as abiotic processes may influence the fate of chlorinated aliphatic compounds and lead to the generation of less chlorinated compounds, eventually vinyl chloride. Also, mercury and arsenic may be affected by microbial processes, resulting in methylated and volatile species, which may be found in the gas phase.

Dehalogenation

The chlorinated aliphatic compounds (-methanes, -ethanes, -ethenes), and in particular the most-chlorinated compounds (tetra- (or per-), tri- and eventually di-) may under strongly reducing conditions, as found in a methanogenic landfill, potentially undergo reductive dehalogenation. By this process a chlorine atom is substituted by a hydrogen atom. As a 'rule of thumb', the dechlorination rate is higher, the higher the number of chlorine atoms in the compound. Several studies are reported for microbial cultures in the laboratory (e.g. Freedman and Gossett, 1989), but only a few reports exist related to landfills. Christensen *et al.* (1993) reported on some preliminary experiments involving leachate from a landfill. TCA (1,1,1-trichloroethane) degraded 70–90% within a 2 month period, while TeCM (tetrachloromethane) and PCE (per-chloroethene) showed less degradation.

As the dechlorination rate decreases with the number of chlorine atoms in the compound, accumulation of the less chlorinated compounds may occur. For example, vinyl chloride may accumulate, and high concentrations may be found in the LFG. Several factors, such as microbial activity, hydrogen pressure and the presence of catalytic metal surfaces, may affect these processes, which supposedly involve both biological as well as abiotic processes. With respect to the trace composition of the LFG, these processes may decrease the concentrations of the higher chlorinated aliphatic compounds, but lead to substantial concentrations of, for example, vinyl chloride for longer periods than would have been predicted from the high volatility of vinyl chloride.

Mercury

Mercury (Hg) has been detected in elevated concentrations above four landfills in Sweden ($10-24$ ng/m^3) corresponding to 2–4 times the background values (Wallin, 1989). Gendebien *et al.* (1992) states that elevated concentrations of mercury have been found in landfill gas as well. However, the reports are few, and the general importance of these single observations is difficult to evaluate.

Mercury may evaporate slowly (as elementary mercury) at elevated temperature inside the landfill, and may be transported with the gas. However, the volatilization may be increased by microbial formation of methylated mercury. This process has been documented in anaerobic aquatic environments (Jensen and Jernelöv, 1969) and could supposedly also take place in the landfill environment.

Arsenic

Arsenic (As) may also form volatile methylated species under anaerobic conditions, in particular trimethylarsine (As $(CH_3)_3$) according to Cope *et al.* (1983). Blakey (1984), in a study on co-disposal of arsenical waste, found that a substantial fraction of the arsenic could volatilize at neutral pH. No attempts were made to determine the volatile species.

Although no reports have be found on arsenic in LFG, these observations may indicate that As could be found in LFG in some cases.

REFERENCES

Abdul, A. S., Gibson, T. L. and Rai, D. N. (1987). Statistical correlations for predicting the partition coefficient for nonpolar contaminants between aquifer organic carbon and water. *Hazardous Waste and Hazardous Materials*, **4**, 211–22.

Augenstein, D. C., Wise, D. L. and Wentworth, R. L. (1976). Fuelgas recovery from controlled landfilling of municipal wastes. *Resource Recovery and Conservation*, **2**, 103–17.

Barlaz, M. A., Milke, M. W. and Ham, R. K. (1987). Gas production parameters in sanitary landfill simulators. *Waste Management & Research*, **5**, 27–40.

Barlaz, M. A., Schaeffer, D. M. and Ham, R. K. (1989). Bacterial population development and chemical characteristics of refuse decomposition in a simulated sanitary landfill. *Applied and Environmental Microbiology*, 55, 55–65.

Beker, D. (1987). Control of acid phase degradation, in *Process Technology and Environmental Impact of Sanitary Landfill*, Proceedings of an International Symposium, Cagliari, Sardinia 19–23 October, vol. I, ISWA, pp. V-1–V-10.

Blakey, N. C. (1984). Behaviour of arsenical wastes co-disposed with domestic solid wastes. *Journal of Water Pollution Control Federation*, 56, 69–75.

Buivid, M. G. (1980). *Laboratory simulation of fuel gas production enhancement from municipal solid waste landfills*, Dynatech R & D Co., Cambridge, MA.

Buivid, M. G., Wise, D. L., Blanchet, M. J., *et al.* (1981). Fuel gas enhancement by controlled landfilling of municipal solid waste. *Resources and Conservation*, 6, 3–20.

Buswell, A. M. and Mueller, H. F. (1952). Mechanisms of methane fermentations. *Industrial and Engineering Chemistry*, 44, 550–52.

Chian, E. S. K., DeWalle, F. B. and Hammerberg, E. (1977). Effect of moisture regime and other factors on municipal solid waste stabilization, in *Management of Gas and Leachate in Landfills*, Proceedings of the Third Annual Municipal Solid Waste Research Symposium, EPA-600/9-77-026, US Environmental Protection Agency, Cincinnati, OH, pp. 73–86.

Christensen, T. H. and Kjeldsen, P. (1989). Basic biochemical processes in landfills, in *Sanitary Landfilling: Process, Technology and Environmental Impact* (eds T. H. Christensen, R. Cossu & R. Stegmann), Academic Press, London, UK, pp. 29–49.

Christensen, T. H., Albrechtsen, H.-J., Kromann, A., Ludvigsen, L. and Skov, B. (1993). The degradation of chlorinated aliphatic compounds in a sanitary landfill, in *Sardinia 93*, Fourth International Landfill Symposium, Proceedings, vol. II. CISA – Environmental Sanitary Engineering Centre, Cagliari, Italy, pp. 1087–92.

Christensen, T. H., Kjeldsen, P. and Stegmann, R. (1992). Effects of landfill management procedures on landfill stabilization and leachate and gas quality, in *Landfilling of Waste: Leachate*, (eds T. H. Christensen, R. Cossu & R. Stegmann), Elsevier, London, UK, pp. 119–137.

Christiansen, K., Prisum, M. and Skov, C. (1985). *Undersøgelse af lossepladsers selvrensende effekt ved recirkulering af perkolat*, Enviroplan A/S, Lynge, Denmark.

Cope, C. B., Fuller, W. H. and Willetts, S. L. (1983). *The Scientific Management of Hazardous Waste*, Cambridge University Press, Cambridge, UK.

Doedens, H. and Cord-Landwehr, K. (1984). Sickerwasserkreislaufführung auf Deponien – neue Erkenntnisse und betriebliche Varianten. *Müll und Abfall*, 16, 68–77.

Ehrig, H.-J. (1982). Auswirkungen der Deponietechnik auf die Umsetzungsprozesse im Deponiekörper – Einführung in die Thematik sowie Untersuchungen der Deponie Venneberg/Lingen, in *Gas- und Wasserhaushalt von Mülldeponien*, Internationale Fachtagung 29.9–1.10.1982, Braunschweig. Technische Universität Braunschweig, Deutschland (*Veröffentlichungen des Instituts für Stadtbauwesen*, Heft 33), pp. 124–44.

Ehrig, H.-J. (1984). Laboratory scale tests for anaerobic degradation of municipal solid waste, in *Proceedings from the International Solid Wastes and Public Cleansing Association Congress*, Philadelphia, 15–20 September, ISWA, 16 pp.

Eifert, M. C. (1976). Variations in gas and leachate production from baled and non-baled refuse, in *Gas and Leachate from Landfills: Formation, Collection and*

Treatment, Proceedings of a research symposium held at Rutgers University, New Brunswick, NJ, March 1975, EPA-600/9-76-004, US Environmental Protection Agency, Cincinnati, OH, pp. 55–72.

El-Fadel, M., Findikakis, N. and Leckie, J. O. (1989). A numerical model for methane production in managed sanitary landfills. *Waste Management & Research,* **7,** 31–42.

EMCON Associates (1975). *Sonoma County Solid Waste Stabilization Study,* EPA SW-65d.1, PB 239 778, US Environmental Protection Agency, Cincinnati, OH.

Farquhar, C. J. and Rovers, F. A. (1973). Gas production during refuse decomposition. *Water, Air, and Soil Pollution,* **2,** 483–95.

Freedman, D. L. and Gossett, J. M. (1989). Biological reductive dechlorination of tetrachloroethylene and trichloroethylene to ethylene under methanogenic conditions. *Applied and Environmental Microbiology,* **55,** 2144–51.

Gendebien, A., Pauwels, M., Constant, M. *et al.* (1992). *Landfill Gas. From Environment to Energy,* Commission of the European Communities, Brussels, Belgium.

Ham, R. K. and Bookter, T. J. (1982). Decomposition of solid waste in test lysimeters. *Journal of Environmental Engineering,* **108,** 1147–70.

Hansson, G. and Molin, N. (1981). End product inhibition in methane fermentations: effects of carbon dioxide and methane on methanogenic bacteria utilizing acetate. *European Journal of Applied Microbiology and Biotechnology,* **13,** 236–41.

Heuvel, J. C van der (1985). *The acidogenic dissimilation of glucose: a kinetic study of substrate and product inhibition,* Laboratory of Chemical Engineering, University of Amsterdam, The Netherlands.

Jensen, S. and Jernelöv, A. (1969). Biological methylation of mercury. *Nature,* **223,** 753–4.

Johnson, L. D. (1981). Inhibition of anaerobic digestion by organic priority pollutants. PhD Thesis, Iowa State University, Ames, IA.

Jones, K. L., Rees, J. F. and Grainger, J. M. (1983). Methane generation and microbial activity in a domestic refuse landfill site. *European Journal of Applied Microbiology and Biotechnology,* **18,** 242–5.

Kinman, R. N., Nutini, D. L., Walsh, J. J. *et al.* (1987). Gas enhancement techniques in landfill simulators. *Waste Management & Research,* **5,** 13–26.

Klink, R. E. and Ham, R. K. (1982). Effects of moisture movement on methane production in solid waste landfill samples. *Resources and Conservation,* **8,** 29–41.

Kugelman, I. J. and Chin, K. K. (1971). Toxicity, synergism and antagonism in anaerobic waste treatment processes. Anaerobic biological treatment processes. *Advances in Chemistry Series,* **105,** 55–90.

Leckie, J. O., Pacey, J. G. and Halvadakis, C. (1979). Landfill management with moisture control. *ASCE, Journal of Environmental Engineering Division,* **105,** 337–55.

Lee, G. F., Jones, R. A. and Ray, C. (1986). Sanitary landfill leachate recycle. *BioCycle,* **27,** 36–8.

Leuschner, A. P. (1983). *Feasibility study for recovering methane gas from the Greenwood Street sanitary landfill, Worcester, Mass., Vol. I. Task 1 – Laboratory feasibility,* Dynatech R & D Co., Cambridge, MA.

Leuschner, A. P. (1989). Enhancement of degradation: laboratory scale experiments, in *Sanitary Landfilling: Process, Technology and Environmental Impact*, (eds T. H. Christensen, R. Cossu & R. Stegmann), Academic Press, London, UK, pp. 83–102.

Leuschner, A. P. and Melden, Jr., H. A. (1983). Landfill enhancement for improving methane production and leachate quality. Presented at the 56th Annual Conference of the Water Pollution Control Federation, 2–7 October, Atlanta, GA.

Mackay, D., Shiu, W. Y. and Ma, K. C. (1992). *Illustrated Handbook of Physical-Chemical Properties and Environmental Fate for Organic Chemicals*, Vols I–III, Lewis Publishers, Chelsea, MI.

McCarty, P. L. (1964). Anaerobic waste treatment fundamentals. Parts 1, 2, 3 and 4. *Public Works*, **95**(9), 107–12, (10), 123–6, (11), 91–4, (12), 95–9.

McCarty, P. L. and McKinney, R. E. (1961a). Volatile acid toxicity in anaerobic digestion. *Journal, Water Pollution Control Federation*, **33**, 223–32.

McCarty, P. L. and McKinney, R. E. (1961b). Salt toxicity in anaerobic digestion. *Journal, Water Pollution Control Federation*, **33**, 399–415.

McInerney, M. J. and Bryant, M. P. (1983). Review of methane fermentation fundamentals, in *Fuel Gas Production from Biomass*, (ed. D. L. Wise), CRC Press, Boca Raton, FL, Chapter 2.

Pohland, F. G. (1980). Leachate recycle as landfill management option. *ASCE, Journal of Environmental Engineering Division*, **106**, 1067–9.

Pohland, F. G. and Kang, S. J. (1974). Sanitary landfill stabilization with leachate recycle and residual treatment, *AIChE Symposium Series*, **71**(145), 308–18.

Postgate, J. R. (1979). *The Sulphate-Reducing Bacteria*, Cambridge University Press, Cambridge, UK.

Rees, J. F. (1980a). The fate of carbon compounds in the landfill disposal of organic matter. *Journal of Chemical Technology and Biotechnology*, **30**, 161–75.

Rees, J. F. (1980b). Optimisation of methane production and refuse decomposition in landfills by temperature control. *Journal of Chemical Technology and Biotechnology*, **30**, 458–65.

Rees, J. F. and Grainger, J. M. (1982). Rubbish dump or fermenter? Prospects for the control of refuse fermentation to methane in landfills. *Process Biochemistry*, **17**(6), 41–4.

Rettenberger, G. and Stegmann, R. (1995). Landfill gas components. This book (Chapter 2.2).

Scharf, W. (1982). Untersuchungen zur gemeinsamen Ablagerung von Müll und Klärschlamm im Labormasstab, in *Gas- und Wasserhaushalt von Mülldeponien*, Internationale Fachtagung 29.9–1.10.1982, Braunschweig, Technische Universität Braunschweig, Deutschland (*Veröffentlichungen des Instituts für Stadtbauwesen*, Heft 33), pp. 83–98.

Senior, E. and Balba, M. T. M. (1990). Refuse decomposition, in *Microbiology of Landfill Sites* (ed. E. Senior), CRC Press Inc., Boca Raton, FL, pp. 17–57.

Spillmann, P. and Collins, H. J. (1981). Das Kaminzug-Verfahren. *Forum Städtehygiene*, **32**, 15–24.

Stegmann, R. (1983). New aspects on enhancing biological processes in sanitary landfills. *Waste Management & Research*, **1**, 201–11.

Stegmann, R. and Spendlin, H.-H. (1985). Research activities on enhancement of biochemical processes in sanitary landfills. Paper presented at the Conference New Directions and Research on Enhancement of Biochemical Processes in Sanitary Landfills, 23–28 June, University of British Columbia, Vancouver, Canada.

Stegmann, R. and Spendlin, H.-H. (1986). Research activities on enhancement of biochemical processes in sanitary landfills. *Water Pollution Research Journal Canada*, **21**(4), 572–91.

Stegmann, R. and Spendlin, H.-H. (1989). Enhancement of degradation: German experiences, in *Sanitary Landfilling: Process, Technology and Environmental Impact*, (eds T. H. Christensen, R. Cossu & R. Stegmann), Academic Press, London, UK, pp. 61–82.

Tittlebaum, M. E. (1982). Organic carbon content stabilization through landfill leachate recirculation. *Journal, Water Pollution Control Federation*, **54**, 428–33.

Wallin, S. (1989). *Avgång av kvicksilverånga från avfallsupplag*, Depå 90, Statens Naturvårdsverk, Solna, Sweden.

Wolffson, C. (1985). Untersuchungen über den Einfluss der Hausmüllzusammensetzung auf die Sickerwasser- und Gasemissionen – Untersuchungen im Labormasstab, in *Sickerwasser aus Mülldeponien – Einflüsse und Behandlung*, Fachtagung März 1985, Braunschweig. Technische Universität Braunschweig, Deutschland (*Veröffentlichungen des Instituts für Stadtbauwesen*, Heft 39), pp. 119–46.

Zehnder, A. J. B. (1978). Ecology of methane formation, in *Water Pollution Microbiology*, vol. 2 (ed. R. Mitchell), John Wiley, New York, pp. 349–76.

Zehnder, A. J. B., Ingvorsen, K. and Marti, T. (1982). Microbiology of methane bacteria, in *Anaerobic Digestion*, Proceedings of the 2nd International Symposium of Anaerobic Digestion, Travemünde, 6–11 September 1981 (ed. D. E. Hughes), Elsevier Biomedical Press, BV, Amsterdam, The Netherlands, pp. 45–68.

2.2 Landfill Gas Components

GERHARD RETTENBERGER[a] & RAINER STEGMANN[b]

[a] *Fachhochschule Trier, Schneidershof, 54293 Trier, Germany*
[b] *Technical University of Hamburg-Harburg, 21071 Hamburg, Germany*

INTRODUCTION

The status of the anaerobic degradation processes reflects biogas composition (Christensen *et al.*, 1989). The different phases of waste degradation can be observed in lysimeters although not to the same degree in actual landfills (Stegmann & Spendlin, 1989). The reason for this is partly non-homogeneity of waste composition and water content. Therefore, one part of the landfill may be at the acid phase while methane production takes place in others; the remaining areas may be too dry for any biological activity to take place.

MAIN COMPONENTS

Although leachate quality indicates that a landfill is in the acid phase, the landfill may already exhibit active gas production with methane concentrations >50%. As has been measured in test lysimeters, H_2 can also be found in full-scale landfills, but only over short periods of time.

The main components of landfill gas after relatively short times after disposal (weeks to months) are 55 (±5)% of methane and 45 (±5)% of carbon dioxide. These concentrations remain relatively constant whereas higher methane concentrations can be observed

Landfilling of Waste: Biogas. Edited by T. H Christensen, R. Cossu and R. Stegmann. Published in 1996 by E & FN Spon, London. ISBN 0 419 19400 2.

to a larger degree in older landfills. A change of LFG composition inside the landfill will take place when oxygen enters the landfill; this has been observed only during gas extraction when air has been sucked into the fill. Whether at some time in the far future a landfill or part of it may turn aerobic is difficult to predict (see also Christensen and Kjeldsen, 1989).

TRACE COMPONENTS

Besides the main components (CH_4 and CO_2) the landfill gas also contains a certain amount of trace components. The type and concentration of these trace components depends on the composition of the landfilled wastes (see also Chapter 2.3). They may change owing to biological and chemical processes that take place inside the landfill. These trace components may cause damage to the technical equipment used for gas extraction and utilization (gas motors, for example) and have an adverse effect on the environment, especially the atmosphere, and on the health of human beings and animals.

Owing to their generation, a differentiation into two types of trace components can be made:

- trace components generated during anaerobic degradation in the landfill;
- trace components generated by man (anthropogenic trace components) and deposited together with the wastes.

Trace Elements Generated During Biological Degradation Processes in the Landfill

Three main groups of biologically generated trace components can be differentiated:

- oxygen compounds;
- sulphur compounds;
- hydrocarbons.

Oxygen compounds.   Compounds containing oxygen are generated mainly during the degradation of the organic waste components.

TABLE 1. Range of Concentrations of Detected Oxygen-Containing Components in Landfill Gases according to Brookes and Young (1983), Young and Parker (1983), Young and Heasman (1985), and Höfler *et al.* (1987)

Substance	Concentration (mg/m^3)
Ethanol	16–1 450
Methanol	2.2–210
1-Propanol	4.1–630
2-Propanol	1.2–73
1-Butanol	2.3–73
2-Butanol	18–626
Acetone	0.27–4.1
Butanone	0.078–38
Pentanal	0.8
Hexanal	4.04
Acetic ester	2.4–263
Butyric ester	<0.9–350
Acetic butyl ester	60
Butyric propyl ester	<0.1–100
Acetic propyl ester	<0.5–50
Acetic acid	<0.06–3.4
Butyric acid	<0.02–6.8
Furan	0.1–2.4
Methylfurans	0.06–170
Tetrahydrofuran	<0.5–8.8

They appear in the gas only in the early phase of gas generation.

Table 1 gives an overview of the oxygen-containing compounds detected in landfill gases (Laugwitz *et al.*, 1988).

Sulphur components. The sulphur components in landfill gas cause two main effects:

- they are mainly responsible for the odour of the gas;
- some of these components, like hydrogen sulphide and the mercaptans, belong to the more toxic landfill gas components.

Owing to the origin from easily degradable material, some sulphur components, such as mercaptans, occur mainly during the operation phase of the landfill. Others, especially hydrogen sulphide, are generated in all phases of landfill gas production. In general the

TABLE 2. Concentration Ranges of Some Sulphur Components Detected in Landfill Gases according to Brookes and Young (1983), Young and Parker (1983), Young and Heasman (1985), and Rettenberger (1986)

Substance	Concentration (mg/m^3)
Methyl cercaptan	0.1–430
Ethyl mercaptan	0–120
Dimethyl sulphide	1.6–24
Dimethyl disulphide	0.02–40
Carbon disulphide	<0.5–22
Carbon oxysulphide	<0.1–1.9

hydrogen sulphide concentrations range between 0 and 70 mg/m^3.

In Table 2 some of the sulphur components that have been detected in landfill gas are listed.

Hydrocarbons.   Table 3 shows the terpene hydrocarbons detected in landfill gas.

These hydrocarbons may as well be naturally generated in the landfill as deposited with the landfilled waste.

Anthropogenic Trace Components

The anthropogenic trace components in landfill gas can be differentiated into two groups:

TABLE 3. Concentration Ranges of Terpene Hydrocarbons in Landfill Gas according to Brookes and Young (1983), Young and Parker (1983), Young and Heasman (1985), and Rettenberger (1986)

Substance	Concentration (mg/m^3)
Limonene	3.3–269
Menthene	14
Camphor/fenchene	3–13
Others	5.5–503

- aromatic hydrocarbons;
- chlorinated hydrocarbons.

Aromatic hydrocarbons. Aromatic hydrocarbons, owing to their widespread utilization, are widely detected in landfill gas. Compared with the generation of aromatic hydrocarbons from other sources, such as traffic and chemical industries, the emissions of aromatic hydrocarbons from landfills are negligible with regard to their environmental effects. Nevertheless, these components are of importance because of the possible effects they may have on those who work on the landfill. In particular, the contents of benzene in landfill gas should be carefully monitored owing to its proven carcinogenic effect.

In Table 4 concentration ranges for aromatic hydrocarbons detected in landfill gas are listed.

Chlorinated hydrocarbons. Chlorinated hydrocarbons are used for cleaning purposes (metal industry, laundries, etc.). Freons (chlorinated and fluorinated hydrocarbons) may be used in refrigerators, sprays and in the chemical industry. The toxicity of the majority of these substances is rather low. With regard to environmental impact due to their persistence, these substances are the most significant trace components found in landfill gas.

When landfill gas is used in gas motors, corrosion problems may occur, as hydrogen chloride (HCl) may be generated from the chlorinated hydrocarbons contained. Furthermore, chlorinated hydrocarbons represent a considerable hazardous potential to the

TABLE 4. Concentration Ranges of Aromatic Hydrocarbons in Landfil Gas according to Laugwitz *et al.* (1988), Brookes and Young (1983), Young and Parker (1983), Young and Heasman (1985), and Rettenberger (1986)

Substances	Concentration (mg/m^3)
Benzene	0.03–15
Toluene	0.2–615
Ethylbenzene	0.5–236
Xylene	0.2–383
Styrol	0.1–10
Propylbenzene	1.5–173

TABLE 5. Concentration of Chlorinated Hydrocarbons in Landfill Gas (mg/m^3)

	Rettenberger (1986)	Brookes and Young (1983)	Young and Heasman (1985)	Arendt (1985)	Pruggmayer et al. (1982)	Schilling and Hinz (1987)	Laugwitz et al. (1988)
Dichlorodifluoromethane	4–119	10	6–602	5–700	6.4–107	99–149	4–145
Trichlorofluoromethane	1–84	20	0.4–185	1–500		11–56	0–220
Trichlorotrifluoromethane				1–30			0–6
Chlorotrifluoromethane	0–10						
Chlorodifluoromethane		4	2–276				3–28
Dichlorofluoromethane	5						0.4–14
Chlorofluoromethane							
Dichloromethane	0–6	1	0.1–110	1–100		36–684	0–51
Trichloroethylene	0–182	140	7.7–490	3–150	0.52–8.8	36–684	0–14
Tetrachloroethylene	0.1–142	10	1.2–116	10–100	0.67–31	26–312	0–10
Vinyl chloride	0–264	250	0.3–110	1–30	1.3–23	6.9–104	0–22
Dichloroethylene	0–294	16	0.03–3			0–43	
1,2-dichloroethylene	2–100	68	0.07–28				
1,1-dichloroethylene				≈1	19–138	0.4	
1,1,1-trichloroethane	0.5–4	29	<0.1–3.7	1–5		0–5.4	0–9
Dichlorotetrafluoroethane		1	<2–20				0.4–14
1,1-dichloroethane			<0.5–21				
Trichloromethane	0–2	<1	<0.2–1	1–50			0–3
Dichlorobenzene			<0.1–5.3				

environment. For these reasons, the concentration of chlorinated hydrocarbons in landfill gas has been extensively investigated. In Table 5 values of the concentration of chlorinated hydrocarbons in landfill gas are listed.

Biochemical degradation of anthropogenic trace components. Biochemical degradation of aromatic hydrocarbons under methanogenic conditions is very limited, and most of the aromatic hydrocarbons will persist in the landfill.

Biochemical degradation of chlorinated hydrocarbons under methanogenic conditions generates non-polar readily volatile substances which appear in the landfill gas.

Out of the chlorinated hydrocarbons listed in Table 5 only tetrachloroethylene and trichloroethylene imply a certain toxic risk. Far more important are the toxic effects caused by substances generated by the biochemical degradation of these and other chlorinated hydrocarbons.

The most important means of biochemical degradation is anaerobic halogen–hydrogen substitution (reductive dehalogenation). By this mechanism, starting from tetrachloroethylene, the far more toxic trichloroethylene is generated. In a second step a mixture of dichloroethylene (*cis*-1,2 and *trans*-1,2-dichloroethylene) is generated, from which the carcinogenic vinyl chloride ultimately develops. Besides benzene and methylmercaptan, vinyl chloride is the most hazardous substance emitted from landfills.

Chlorinated and fluorinated hydrocarbons can also be transformed, according to Laugwitz *et al.* (1988) and Deipser & Stegmann (1993). Thus from dichlorodifluoromethane (R12) the far more toxic chlorodifluoromethane (R22) is generated. From trichlorofluoromethane it is presumed that, as a result of degradation, the carcinogenic dichlorofluoromethane (R21) is generated.

REFERENCES

Arendt, G. (1985). Untersuchung von Spurenstoffen im Deponiegas hinsichtlich technischer Nutzungskonzeptionen, in Umweltbundesamt (Ed.) Deponiegasnutzung – Dokumentation einer Fachtagung 1984, Berlin, pp. 5–11.

Brookes, B. I. and Young, P. J. (1983). The development of sampling and gas chromatography – mass spectrometry analytical procedures to identify and determine the minor organic components of landfill gas. *Talanta*, **30**(9), 665–76.

Christensen, T. H. and Kjeldsen, P. (1989). Basic biochemical processes in landfill, in *Sanitary Landfilling, Process, Technology and Environmental Impact.* (eds T. H. Christensen, R. Cossu and R. Stegmann), Academic Press, pp. 29–49.

Deipser, A. and Stegmann, R. (1993). The origin and fate of volatile trace components in MSW-landfills, *Waste Management and Research*, accepted for publication.

Höfler, F., Schöneich, Ch. and Möckel, H. J. (1987). Analytical investigations on the composition of gaseous effluents from a garbage dump. II. Analysis of carbonyl compounds by derivatization and reversed phase liquid chromatography. *Fresenius Zeitschrift für Analytische Chemie*, **328**, 244–6.

Laugwitz, G., Poller, T. and Stegmann, R. (1988). Entstehen und Verhalten von Spurenstoffen im Deponiegas sowie umweltrelevante Auswirkungen von Deponiegasemissionen, in *Deponiegasnutzung – Dokumentation einer Fachtagung 1988*, Hamburg, pp. 153–63.

Pruggmayer, D. *et al.* (1982). Identifizierung chemischer Stoffe in Deponien, Bericht für den Regierungspräsidenten Köln, BF-R-64. 965–3.

Rettenberger, G. (1986). Spurenstoffe im Deponiegas. Auswirkungen auf die Gasverwertung. GIT Supplement 1/86, 53–57.

Schilling, H. and Hinz, W. (1987). Konzeption der Gasreinigungsanlage für die Deponie Kapiteltal, Firmenschrift Leybold.

Stegmann, R. and Spendlin, H. H. (1989). Enhancement of degradation: German experiences, in *Sanitary Landfilling: Process, Technology and Environmental Impact* (eds T. H. Christensen, R. Cossu and R. Stegmann), Academic Press, pp. 61–82.

Young, P. J. and Heasman, L. A. (1985). An assessment of the odour and toxicity of the trace compounds of landfill gas, in *Proceedings of the GRCDA 8th International Landfill Gas Symposium*, San Antonio, April.

Young, P. J. and Parker, A. (1983). The identification and possible environmental impact of trace gases and vapours in landfill gas. *Waste Management and Research*, **1**, 213–26.

2.3 Emissions of Volatile Halogenated Hydrocarbons from Landfills

ANNA DEIPSER, THOMAS POLLER & RAINER STEGMANN

Institute of Waste Management, Technical University of Hamburg-Harburg, D-21071 Hamburg, Germany

INTRODUCTION

Various landfill gas (LFG) analyses have shown that considerable concentrations of trace organics are present in LFG (see Chapter 2.2; Laugwitz *et al.*, 1988; Hagendorf *et al.*, 1989). Owing to their characteristics, the readily volatile chlorinated hydrocarbons (VCCs) and the fluorinated hydrocarbons (CFCs) pose a special problem. Most VCCs are toxic, and some are suspected of being carcinogenic. Owing to their chemical stability the CFCs reach the stratosphere, where the chlorine atom is separated and the radical causes the ozone to break down (Bräutigam, 1988).

Emissions of halogenated hydrocarbons from landfills as a source to (of) atmospheric pollution have just recently become of interest. Very little is known about the origin of halogenated hydrocarbons in municipal solid waste (MSW) and the fate of these compounds in the landfill. Data on halogenated hydrocarbons in LFG have only been available for the last 10 years.

This chapter presents information about the origin of halogenated hydrocarbons in waste components as well as measured contents in waste, gas and leachate. All values refer to products containing CFCs and VCCs. The Montreal Protocol that regulates the production utilization and disposal of substances that deplete the ozone layer is not considered in this chapter. Implications for waste management and landfilling strategies are also discussed.

Landfilling of Waste: Biogas. Edited by T. H Christensen, R. Cossu and R. Stegmann. Published in 1996 by E & FN Spon, London. ISBN 0 419 19400 2.

HALOGENATED HYDROCARBONS IN MUNICIPAL SOLID WASTE

CFCs are in most cases non-toxic and non-inflammable. Owing to their low boiling point and their low thermal conductivity coefficient they are frequently utilized as propellants in spray cans and in foam plastics for insulation purposes.

Technically the CFCs are described by a special nomenclature. The term is composed of a registered trade mark and an identification number. Trade marks are normally company-specific names, e.g. Frigen, Kaltron, Freon. The neutral mark is the character R for the word 'refrigerant'. The identification number is a three-figure number, which is read from left to right and has the following significance: first figure, number of C atoms minus 1; second figure, number of H atoms plus 1; third figure, number of fluorine atoms. For only one C atom the first digit is omitted (instead of zero).

The CFCs most commonly occurring in waste are:

- trichlorofluoromethane, R11;
- dichlorodifluoromethane, R12;
- dichlorofluoromethane, R21;
- chlorodifluoromethane, R22;
- trifluorotrichloroethane, R113;
- tetrafluorodichloroethane, R114.

VCCs are characterized by their lipophilic properties and their relatively high volatility in steam. Owing to these properties these compounds are often used as solvents and extracting agents.

The VCCs most commonly found in waste are the following substances:

- chloromethanes:
  tetrachloromethane
  trichloromethane
  dichloromethane
- chloroethenes:
  tetrachloroethene
  trichloroethene
  cis-dichloroethene
- chloroethanes:
  1,1,1-trichloroethane

Use and Application of Halogenated Hydrocarbons

CFCs are, owing to their high volatility and inactivity, mainly used as (Ullmann, 1976a, b and c; Aurand *et al.*, 1978; Bauer, 1985; LAWA, 1988; Deipser, 1989):

- operating media in refrigerating aggregates;
- propellants in aerosol production for spray cans;
- aerating agents to foam up plastics in heat-insulation engineering;
- detergents in the technical area of the electronics industry;
- solvents in the production of plastics.

The consumption and use of CFCs are changing these days because of the intensive focus on the effects of CFCs on the ozone layer. Until 1990, trichlorofluoromethane (R11) and dichlorodifluoromethane (R12) were the CFCs most often used. In the 1980s approximately 55 000 t of R11 and R12 were used annually in Germany (FRG) for the following purposes: production of spray cans (50%), foaming-up of plastics (32%, of which about 85% were used for the production of polyurethane foam plastics), utilization as refrigerant (12%), utilization as solvent, (6%).

VCCs are utilized for a multitude of products:

- dichloromethane:
    paint-stripping pastes and baths (for polyester and epoxy lacquers);
    solvents for the fabrication of plastics (e.g. for PVC);
    solvent component in plastic adhesives;
    industrial detergent and degreasing agent;
    refrigerant in low-pressure refrigerating aggregates;
- trichloromethane:
    base product for the production of chlorodifluoromethane (R22);
    extracting agent for grease, oil and wax;
    paint stripper (in combination with dichloromethane);
- tetrachloromethane:
    in the production of tetraethyl lead;
    refrigerating agent (with dichloromethane);
    solvent for oil, grease and resin;
    extracting agent for heat-sensitive substances;

- 1,2-dichloroethane:
  production of vinyl chloride;
  raw material for tri- and tetrachloroethene, trichloroethane;
  solvent and extracting agent (combined with other solvents);
  as an insecticide and fungicide;
- 1,1,1-trichloroethane:
  cold detergent for electric machines, fittings and electronic devices;
  stain remover for textiles;
  as domestic detergent in spray cans;
- chloroethene/vinyl chloride:
  base product in PVC production;
- cis-1,2-dichloroethene:
  extracting agent for heat-sensitive substances;
  production of rubber solutions;
  additive in paints and lacquers;
- trichloroethene:
  degreasing of metals in closed steam, dip or spray degreasing apparatus;
  degreasing of textiles;
  extracting agent for oil fruit, pressed pulps, bones, glue, meat, fish meal, residues from animal bodies, wax;
  as stain remover (combined with other solvents);
  cold and hot dipping lacquer coating;
  solvent for paints and lacquers;
  solvent for crude rubber and sulphur;
  cleaning of glass and optical lenses;
- tetrachloroethene:
  solvent for dry cleaning;
  degreasing agent for light metals and their alloys;
  extracting agent in animal body utilization plants;
  for the production of trichloroacetic acid, rubber solutions;
  paint strippers, printing colours;
  production of fluorinated hydrocarbons.

In Germany (FRG) at the beginning of the 1980s the production of VCCs (chlorinated aliphatic compounds) amounted to about 680 000 t annually, of which dichloromethane, tetrachloroethene and tetrachloromethane were the most important compounds. As vinyl chloride is mainly used as the starting product in PVC production it occurs in products only as polymerized VC. In 1985, the production of PVC amounted to approximately 1.2×10^6 t in Germany (FRG).

Contents of Halogenated Hydrocarbons in Various Products

The halogenated hydrocarbons mentioned above are present in a multitude of products used in the household (Ullmann, 1976a, b and c; Kehrmann, 1983; Brun, 1984; Öko-Institut, 1984; Reimann, 1984; Wingert, 1987):

Spray cans contain mainly R11 and R12. The concentrations, however, can vary considerably. Spray cans with insecticides, odorous substances, disinfectants, and deodorants contain up to 80% by weight of CFCs. Spray cans with lacquers and hairsprays contain 30–60% by weight of CFCs; stain removers, polishing materials and detergents about 10–30% by weight. Cold detergents contain at least 15% by weight of trichloroethene, tetrachloroethene, dichloromethane, and 1,1,1-trichloroethane. Refrigerators and freezers often contain R12 in the refrigerant circuit. R11 is present in the insulation (polyurethane hard foam). It may be assumed that approximately 120–160 g of CFCs are present in the refrigerant and 400–600 g in the insulation. Foam plastics that are foamed up with CFCs during production often contain R11 or R12. The content is approximately 10–35% by weight in soft foam plastics and up to 70% by weight in hard foam plastics. Paint strippers may contain up to 80% by weight of dichloromethane. Stain removers, impregnating agents, and universal detergents often contain 1,1,1-trichloroethane, tetra- or trichloroethene. Adhesives and lacquers contain both tetra- or trichloromethane and 1,1,1-trichloroethane.

PVC plastics in municipal solid waste mostly consist of the following items: tablecloths and floor coverings (PVC, soft); carrier bags (PVC, soft); parts for building and coverings (PVC, hard); packaging bodies as canisters, bottles and containers (PVC, soft or hard).

Estimates on Contributions of Halogenated Hydrocarbons to Municipal Solid Waste and Bulky Waste

Based on the above presented data and on data on waste quantity and composition in Germany (FRG), theoretical maximum estimates are made of the contents of halogenated hydrocarbons in municipal solid waste (MSW) and in bulky waste (furnitures, refrigerators, etc.).

Spray cans are for the most part disposed of with the MSW. Based on a figure of 20 million t/a of MSW (according to an MSW analysis in the FRG from 1986) and a residue in the spray cans of about 10% (figure from the recycling centre at Hamburg-Harburg, Okeke, 1988) the resulting concentration of R11/R12 in the MSW is estimated to be 150 mg/kg from spray cans.

Annually about 1.8–2.5 million refrigerators and freezers are disposed of in the bulky waste in Germany (FRG). As regards the CFC content of refrigeration equipment used in households, Jacobi (1988) distinguishes between: CFCs refrigerants (mainly R12), 400 t/a; insulation material of polyurethane foam, 500 t/a; and cooling oil with dissolved refrigerant (no data). If these figures are compared with the annual quantity of bulky waste (approx. 2.5 million t) this results in approximately 350 mg/kg R11 and R12 in bulky waste.

Soft foam plastics (e.g. used in furnitures and mattresses) contain about 10–35% by volume of R11, corresponding to 500–2000 g/m^3, while hard plastics (e.g. polyurethane foam for insulation and packaging foam plastics) contain 70% by volume corresponding to 400 g/m^3. About 90% of the foaming agents used are estimated to stay in the product, while 10% escape to the atmosphere during production (Jacobi, 1988). Assuming that PU foam plastics in furniture have a useful life of about 10 years and that there is a residual content of CFCs of 10%, then 400 t/a of CFCs get into the bulky waste, resulting in 160 mg/kg. An increasingly large quantity of foam plastics containing R11 are being used in the car industry for moulded parts (Jacobi, 1988). The quantity of foam plastics from car utilization (shredded waste from seats, instrument panelling), which in Germany are partly used as cover material in landfills, is difficult to assess. Assuming a content of 100 l of foam plastics per car and approximately 500 g R11 per m^3 of foam plastics, results in approximately 100 t/a.

The VCCs cannot be evaluated in detail, but assuming that 10% remains in the products disposed of, the following estimates are made for MSW: dichloromethane approximately 100 mg/kg; trichloroethene, approximately 28 mg/kg; tetrachloroethene, approximately 55 mg/kg; and 1,1,1-trichloroethane, approximately 52 mg/kg.

Maximum Theoretical Contents of Halogenated Hydrocarbons in Municipal Solid Waste and Bulky Waste

With the values mentioned above it is possible to calculate a theoretical maximum content of halogenated hydrocarbons in MSW and bulky waste, as presented in Fig. 1. The chlorine and fluorine contents of MSW and bulky waste are on average 288 mg org. Cl/kg and 30 mg org. F/kg for MSW and 322 mg org. Cl/kg and 92 mg org. F/kg for bulky waste respectively.

Using an annual quantity of MSW and bulky waste of 22.5 million t as a calculation base, values of 292 mg org. Cl/kg and 40 mg org. F/kg are obtained. The chlorine content in R11/R12 amounts to 53%; the proportional chlorine content of dichloromethane is 29%. Consequently about 74% of the organic chlorine in MSW is caused by R11/R12 and R30.

Measured Concentrations of Halogenated Hydrocarbons in Municipal Solid Waste

The content of halogenated hydrocarbons in six German MSW samples was determined by Deipser and Stegmann (1993). The analysis was performed on 15 kg (dry weight) samples, but the replicates showed high variations. Waste components specifically

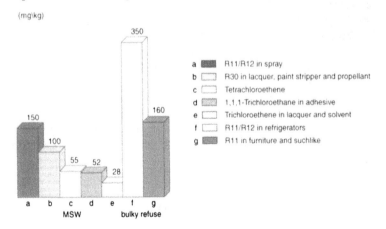

Figure 1. Estimated theoretical maximum contents of halogenated hydrocarbon in MSW and bulky waste.

TABLE 1. Concentration Range of Halogenated Hydrocarbons and Aromatic Hydrocarbons in Six German MSW Samples (Deipser and Stegmann, 1993)

Components in MSW	Concentration range (mg/kg MSW)
Dichlorodifluoromethane (R12)	0.01–10.79
Trichlorofluoromethane (R11)	0.006–0.219
Dichlorofluoromethane (R21)	0.002–0.082
Chlorodifluoromethane (R22)	ND–0.89
1,1,2-Trichlorotrifluoroethane (R113)	0.005–0.24
Dichlorotetrafluoroethane (R114)	ND–0.007
Dichloromethane	0.006–2.67
Trichloromethane	0.001–0.07
Vinyl chloride	ND–0.4
cis-1,2-Dichloroethene	ND–4.98
1,1,1-Trichlorethane	0.008–3.64
Trichloroethene	0.002–0.58
Hexane	0.015–0.224
Benzene	0.024–0.68
Toluene	ND–0.388
o-Xylene	ND–1.207
m.p-Xylene	0.004–2.932

ND = not detectable.

containing halogenated hydrocarbons were removed manually prior to the analysis. The chlorine content ranged from 0.153 to 7.16 mg org. Cl/kg MSW and the fluorine content from 0.004 to 3.87 mg org. F/kg MSW. These concentration intervals are well below the maximum theoretical content estimated in the previous sections: 288 mg org. Cl/kg MSW and 30 mg org. F/kg MSW.

As the analysed samples were without products specifically containing halogenated hydrocarbons it is likely that MSW will have a content of halogenated hydrocarbons between the measured values and the maximum theoretical values.

The measured concentrations of the individual halogenated hydrocarbons are shown in Table 1 (including some data on aromatic hydrocarbons as well).

VOLATILIZATION FROM LANDFILLED WASTE

MSW samples were studied in laboratory landfill lysimeters in terms of emissions of halogenated hydrocarbons as well as aromatic

TABLE 2. Trace Gas Concentrations Measured in Six Laboratory Landfill Lysimeter Studies and in Eight Different Actual Landfill Gas Samples. Based on Deipser *et al.* (1991)

Gas component	Concentration (mg/m³)		
	Min	*Max*	*Average*
Laboratory landfill lysimeters gas (6 samples)			
Dichlorodifluoromethane (R12)	0.25	3207.12	535.17
Chlorodifluoromethane (R22)	ND	389.50	64.92
Dichlorofluoromethane (R21)	0.02	13.37	2.65
Trichlorofluoromethane (R11)	0.54	103.57	19.3
Dichlorotetrafluoroethane (R114)	ND	0.03	0.01
1,1,2-Trichlorotrifluoroethane (R113)	0.17	33.73	6.82
Trichloromethane	0.04	2.35	0.87
Dichloromethane	0.06	29.67	4.13
Tetrachloroethene	0.2	62.91	11.38
Trichloroethene	0.07	30.13	5.45
cis-1,2-Dichloroethene	ND	76.65	16.35
Monochloroethene	ND	30.49	8.82
1,1,1-Trichloroethane	0.02	174.35	29.25
Actual landfill gas (8 samples)			
Dichlorodifluoromethane (R12)	10.3	111.0	51.5
Chlorodifluoromethane (R22)	1.9	30.7	13.4
Dichlorofluoromethane (R21)	0.7	28.0	11.8
Trichlorofluoromethane (R11)	0.3	35.0	9.7
Dichlorotetrafluoroethane (R114)	2.3	8.9	4.1
Trichlorotrifluoroethane (R113)	0.07	1.7	1.0
Trichloromethane (R20)	ND	0.2	0.06
Dichloromethane (R30)	0.01	57.3	27.0
Tetrachloroethene	0.5	10.5	3.3
Trichloroethene	0.04	8.6	2.3
cis-1,2-Dichloroethene	2.4	14.7	6.7
Monochloroethene	0.3	5.7	3.4
1,1,1-Trichloroethane	ND	2.4	0.7

ND = not detected.

hydrocarbons in the gas and in the leachate (Deipser *et al.* 1991). The gas concentrations measured in the laboratory experiments are shown in Table 2, together with gas concentrations measured at eight actual landfills. The same gas components were identified in the laboratory experiments and in the actual landfill gas samples, but very large variations were observed in the laboratory experiments as compared with the field data.

During the acid phase of the landfill lysimeters, where the gas production is low, substantial fractions of VCCs and CFCs were emitted with the leachate: about 28% of the organic chlorine emission, 3% of the organic fluorine emission and about 29% of the BTEX emission. The distribution of emission between gas and leachate is shown in Fig. 2. In the methanogenic phase, where the gas production is substantial, close to 100% of the remaining VCCs and CFCs were emitted with the gas. The concentrations of VCCs and CFCs in the leachate were below the actual detection limits in the methanogenic phase of the laboratory lysimeters.

Considering that about 13.5 million t of waste annually are landfilled in the former West Germany, the presented data indicate that the annual emissions from the landfills are of the order of 37 t of organic chlorine, 13 t of organic fluorine and 10 t of BTEX. Assuming that all landfills have a substantial acid phase (several years), the distribution between gas and leachate emissions is estimated in Fig. 3.

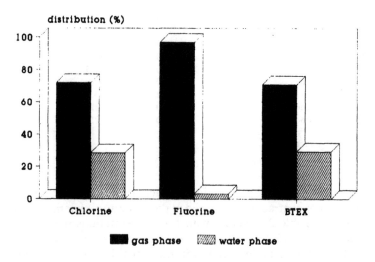

Figure 2. Distribution of the emissions of organic chlorine, fluorine, and BTEX by gas and leachate in the acid phase as measured in laboratory landfill lysimeters.

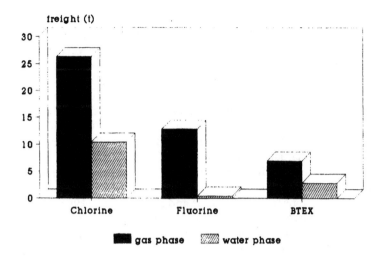

Figure 3. Estimated organic chlorine, fluorine, and BTEX emissions from 13.5 million t of MSW annually landfilled in the former West Germany.

DISCUSSION AND CONCLUSION

CFCs reach and will continue to reach landfills, mainly through spray cans and old refrigerating equipment. The problematic fraction of old refrigerating equipment, besides the refrigerant in the compressor, is the insulation. This is mainly polyurethane soft foam, which may contain up to 70% by volume of the propellant trichlorofluoromethane. In addition, polyurethane foam plastics are used as packaging material, which also reaches landfills in large quantities. Investigations of MSW showed that R12 is the dominating halogenated hydrocarbon, primarily from spray cans, as they are disposed of directly with MSW. The polyurethane foams are presumed to reach landfills with MSW-like trade waste. VCCs occur as solvents in spray cans: in most cases dichloromethane. Traces of this substance may also be contained in polyurethane plastics. The amount of VCCs is negligibly small compared with the CFCs dichlorodifluoromethane and trichlorofluoromethane.

The VCCs/CFCs are released from both polyurethane foam plastics and spray cans by mechanical destruction. In the products mentioned above, the CFCs are always present in gaseous form, meaning that high volatility exists when the porespace of foam plastic material is damaged. The actual potential presence of CFCs in spray cans can only be roughly estimated. Owing to the high volatility of the CFCs, accurate measurements are very difficult. Laboratory experiments showed that the highest concentrations of VCCs/CFCs in landfill gas and leachate occur in the acid phase, i.e. in the period of time immediately after the deposition of the waste. The highest VCC/CFC emissions take place when the methane gas production becomes substantial. If no further VCC/CFC entry occurs, the concentrations decrease relatively rapidly, depending on the total potential. However, the VCC/CFC emissions via the leachate decreased dramatically during the methanogenic phase of the waste stabilization.

Investigations by Poller (1990) and Deipser *et al.* (1991) showed that the stable methane phase did not start until only low VCC/CFC concentrations were detectable in the gas. This behaviour could be observed in waste with both added polyurethane foam plastics and added VCCs. Consequently, even at relatively low concentrations, the VCCs/CFCs may have the property to inhibit or even interrupt methane production (Poller, 1990). This means that if there is a high VCC/CFC charge in the MSW, high concentrations of these trace components will be present in the landfill gas. High VCC/CFC concentrations can be measured in the landfill gas, especially in the starting phase after the deposition of the waste. In the acid phase the concentration of organic chlorine in the gas can range from 100 to 1000 mg org. Cl/m^3. Once the landfilling is completed the content steadily decreases. This state will be reached after about 5–10 years when the total VCC/CFC potential has then been released. The organic chlorine concentration in landfill gas amounts to 50 mg/m^3 or less in most cases.

The content and release of halogenated hydrocarbons in landfilled waste are relatively high, and the presence of these compounds in landfilled waste should be minimized. This could be done either by sorting out products containing these compounds or by reducing or terminating the production of these products.

REFERENCES

Aurand, K., Hässelbarth, U., Müller, G., Lahmann, E. and Niemitz, W. (1978). Organische Verunreinigungen in der Umwelt, *Erkennen – Bewerten – Vermindern,* E. Schmidt Verlag, Berlin.

Bauer, U. (1985). Bewertung der Emissionspfade am Beispiel der halogenierten Kohlenwasserstoffe, *Materialien* 1/85 Umweltbundesamt, pp 281–297

Bräutigam, R. (1988). FCKWs: production, emission, trends, in (eds D. Behrens, J. Wiener and R. Zellner), *Anthropogene Beeinflussung der Ozonschicht*, 6. Dechema-Fachgespräch Umweltschutz, Frankfurt/Main, 16–17 December 1987.

Brun, R. (1984). *Der Fischer Öko-Almanach 84/85*, Fischer Taschenbuch Verlag, Frankfurt/Main, pp 243–263.

Deipser, A. (1989). Quellen leichtflüchtiger halogenierter Spurenstoffe im Hausmüll und hausmüllähnlichem Gewerbemüll. Diplomarbeit, FH Bergedorf.

Deipser, A., Poller, T. and Stegmann, R. (1991). Untersuchungen zum Verhalten von ausgewählten organischen Schadstoffen unter kontrollierten Deponiemilieubedingungen in Laborlysimeters. DFG-Abschlußbericht, unpublished.

Deipser, A. and Stegmann, R. (1993). Untersuchungen von Hausmüll auf leichtflüchtige Spurenstoffe. *J. Müll und Abfall*, **25**, pp 69–81.

Hagendorf, U. and Krafft, H. (1989). Belastungen von Abfallplätzen mit leichtflüchtigen Halogenkohlenwasserstoffen und ihre Auswirkungen auf Grundwasser. *Zeitschrift Wasser und Boden*, **5**, 291–297.

Jacobi, H. (1988). *Fluorchlorkohlenwasserstoffe (FCKW) – Verwendung und Vermeidungsalternativen*, E. Schmidt Verlag, Berlin.

Kehrmann, K. (1983). Gift im Mülleimer. *Umweltmagazin*, **2**, 40–46.

LAWA (Länderarbeitsgemeinschaft Wasser) (1988). *Organische Halogenverbindungen*, E. Schmidt Verlag, Berlin.

Laugwitz, R., Poller, T. and Stegmann, R. (1988). *Entstehen und Verhalten von Spurenstoffen im Deponiegas sowie umweltrelevante Auswirkungen von Deponiegasemissionen, Deponiegasnutzung, Hamburger Berichte*, Heft 1, Economica Verlag, Bonn.

Öko-Institut Freiburg (ed.) (1984). *Chemie im Haushalt*, Wohwohlt Verlag, Reinbek.

Poller, T. (1990). *Hausmüllbürtige LCKW/FCKW und deren Wirkung auf die Methangasbildung, Hamburger Berichte 2*, Economica Verlag, Bonn.

Reimann, D. O. (1984). Chlorverbindungen im Müll und in der Müllverbrennung – Einfluß des Kunststoffs PVC, *J. Müll und Abfall*, **2**, 169–176.

Ullmann, Encyklopädie der technischen Chemie (1976a). *Chlorkohlenwasserstoffe, aliphatische*, Band 9, Verlag Chemie, Weinheim.

Ullmann, Encyklopädie der technischen Chemie (1976b). *Aerosole*, Band 7, Verlag Chemie, Weinheim.

Ullmann, Encyklopädie der technischen Chemie (1976c). *Fluorverbindungen, organische*, Band 11, Verlag Chemie, Weinheim.

Wingert, H. (1987). *Der Haushalts-Knigge*, Rowohlt Taschenbuch Verlag, Reinbek.

2.4 Physico-Chemical Characteristics and Toxicology of Landfill Gas Components

GIORGIO MASSACCI

DIGITA – Department of Geoengineering and Environmental Technologies, University of Cagliari, Piazza D'Armi, 09123 Cagliari, Italy

INTRODUCTION

In landfill gas (LFG) more than 100 different compounds have been detected, most of which are organic (see Chapter 2.2).

Many components of LFG may be considered dangerous to human health, as they originate toxic, irritative or narcotic effects. Some of them have carcinogenic action. The possibility of formation of dioxines, furans and other substances in combustion products from gas flaring or in the exhaust gas of motors has to be considered as well. Moreover, fire or explosion hazards are attributed to LFG (see Chapter 3.2).

In the present chapter the physico-chemical characteristics of the most important substances detected in LFG are summarized; the main characteristic parameters, as far as safety and health protection are concerned, are identified.

Landfilling of Waste: Biogas. Edited by T. H Christensen, R. Cossu and R. Stegmann. Published in 1996 by E & FN Spon, London. ISBN 0 419 19400 2.

TABLE 1. Physico-Chemical Properties of LFG Components

Chemical name	Formula	Molecular weight (g/mol)	State	Volumetric weight (kg/Nm³)	Volumetric weight (kg/l)	Relative vapour density	Boiling point (°C)	Melting point (°C)	Vapour pressure (kPa)	Water solubility (g/l)
Ammonia	NH₃	17.00	Gas	0.77	–	0.60	–33.00	–78.00	8500	520.00
Carbon monoxide	CO	28.10	Gas	1.25	–	0.97	–191.00	–205.00	–	0.03
Carbon dioxide	CO₂	44.01	Gas	1.98	–	1.53	–78.50	–	5720	1.69
Methane	CH₄	16.00	Gas	0.72	–	0.55	–161.00	–182.00	–	0.0251/0.064
Ethane	C₂H₆	30.10	Gas	1.36	–	1.05	–88.60	–183.00	–	0.06
Ethylene	CH₂=CH₂	28.10	Gas	1.26	–	0.97	104.00	–169.00	2600	insoluble
Butane	CH₃–CH₂–CH₂–CH₃	58.10	Gas	2.70	–	2.10	–0.50	–135.00	2400	>500
Butylene	CH₃–CH₂–CH=CH₂	56.10	Gas	2.50	–	1.94	–6.30	–186.00	2500	0.43
Propane	CH₃–CH₂–CH₃	44.10	Gas	2.00	–	1.60	–42.00	–190.00	1	50<x<500
Propylene	CH₂=CH–CH₃	42.10	Gas	1.91	–	1.45	–48.00	–185.00	1020	>500
Pentane	C₅H₁₂	72.20	Liquid	–	0.63	2.50	36.00	–131.00	57	>500
Hexane	C₆H₁₄	86.20	Liquid	–	0.66	2.97	58 ÷ 63	(–98 to –154)	20	Insoluble
Cyclohexane	C₆H	84.16	Liquid	–	0.78	2.91	80.80	6.50	10	Insoluble
Heptane	C₇H₁₆	100.20	Liquid	–	0.68	3.60	98.00	–90.00	5	Insoluble
Octane	C₈H₁₈	114.23	Liquid	–	0.7	3.95	125.80	–56.80	1	Insoluble
Nonane	C₉H₂₀	128.26	Liquid	–	0.72	4.43	150.80	–53.52	0	Insoluble
Xylene	C₆H₄(CH₃)₂	106.20	Liquid	–	0.86	3.70	135.00	–50.00	700	0.14
Toluene	C₆H₅(CH₃)	92.14	Liquid	–	0.87	3.20	110.80	–95.00	29	0.47
Benzene	C₆H₆	78.11	Liquid	–	0.88	2.70	80.20	5.50	10	1.80
Ethyl benzene	C₆H₅–C₂H₅	106.17	Liquid	–	0.87	3.67	136.20	–94.40	1	0.20
Tetrahydrobenzene	C₆H₁₀	82.15	Liquid	–	0.81	2.84	83.30	–103.50	9	Insoluble
Isopropylbenzene	C₆H₅CH(CH₃)₂	120.20	Liquid	–	0.86	4.20	153.00	–97.00	1	Insoluble
Chloromethane (methyl chloride)	CH₃Cl	50.50	Gas	2.30	–	1.79	–23.70	–98.00	490	<50

TABLE 1. Continued

Chemical name	Formula	Molecular weight (g/mol)	State	Volumetric weight (kg/Nm³)	Volumetric weight (kg/l)	Relative vapour density	Boiling point (°C)	Melting point (°C)	Vapour pressure (kPa)	Water solubility (g/l)
Dichloromethane (methylene chloride)	CH_2Cl_2	85.00	Liquid	–	1.3	2.90	40.00	–95.00	44	20.00
Trichloromethane (chloroform)	$CHCl_3$	119.38	Liquid	–	1.48	4.12	61.70	–63.50	21	8.20
Tetrachloromethane (carbon tetrachloride)	CCl_4	153.82	Liquid	–	1.59	5.30	76.70	–22.90	12	0.80
1,1-Dichloroethane	CH_3CHCl_2	99.00	Liquid	–	1.18	3.40	57.25	–96.60	24	6.00
Trichloroethane	CH_3-CCl_3	133.40	Liquid	–	1.32	4.60	74.00	–30.40	13	0.10
Dichlorotetrafluoroethane	$CClF_2-CClF_2$	170.92	Gas	7.67	–	5.89	3.50	–94.00	183	1.30
Trichlorotrifluoroethane	$CCl_2F-CClF_2$	187.40	Liquid	–	1.58	6.47	47.60	–35.00	36	0.17
Chloroethylene (vinyl chloride)	$CH_2=CHCl$	62.50	Gas	0.92	–	2.20	–14.00	–160.00	340	<50
1,1-Dichloroethylene	CH_2CCl_2	97.00	Liquid	–	1.25	3.50	32.00	–122.00	67	<50
1,2 cis-Dichloroethylene	$CHCl=CHCl$	97.00	Liquid	–	1.28	3.40	60.00	–80.00	22	<50
1,2 trans-Dichloroethylene	$CHCl=CHCl$	97.00	Liquid	–	1.26	3.40	48.00	–50.00	22	<50
Trichloroethylene	$CHCl=CCl_2$	131.39	Liquid	–	1.46	4.54	87.00	–73.00	8	1.20
Tetrachloroethylene	$CCl_2=CCl_2$	165.80	Liquid	–	1.62	5.73	121.10	23.50	2	0.40
Chlorotrifluoroethylene	$CClF=CF_2$	116.47	Gas	5.34	–	4.02	–28.40	–158.00	560	0.10
Chlorodifluoromethane	$CHClF_2$	86.47	Gas	3.88	–	2.98	–40.80	–160.00	910	1.50
Chlorotrifluoromethane	$CClF_3$	104.46	Gas	4.67	–	3.63	–81.90	–181.00	3180	0.09
Dichlorofluoromethane	$CHCl_2F$	102.90	Gas	4.62	–	3.55	8.90	–135.00	150	6.90
Dichlorodifluoromethane	CCl_2F_2	120.90	Gas	5.51	–	4.17	–29.80	–155.00	570	0.28
Sulphidic acid (hydrogen sulphide)	H_2S	34.10	Liquid	–	0.8	1.20	–60.00	–86.00	1810.00	4.19
Thioglicolic acid	$CH_2SHCHOOH$	92.12	Liquid	–	1.33	3.3	104	–16.5	0.02 (30°C)	Insoluble
1-Propyl alcohol	C_3H_8O	60.09	Liquid	–	0.78		97.10	–90	3.00	>500
2-Butyl alcohol	$C_4H_{10}O$	74.12	Liquid	–	0.81		117.50	–	0.90	50<x<500
Carbon disulphide	CS_2	76.14	Liquid	–	1.21	2.67	46.5	–111.6	100	Insoluble
Methyl mercaptan (methanethiol)	CH_4S	48.11	Liquid	–	0.87		6.00	–123	0.87	<50
Ethyl mercaptan (ethanethiol)	C_2H_6S	62.13	Liquid	–	0.84		37.00	–121	0.84	<50

PHYSICO-CHEMICAL CHARACTERISTICS OF LFG COMPONENTS

LFG primarily contains methane (usually between 50% and 60% by volume) and carbon dioxide (40–50%). The most common values of volumetric ratio CH_4/CO_2 range from 1.2 to 1.5 (Wilhelm, 1993). The volatile organic compounds (VOCs) represent less than 1% by volume, but are qualitatively important as the majority of them are potentially toxic and, in some cases, even carcinogenic.

Table 1 summarizes the main physico-chemical properties of the most important compounds.

Carbon dioxide is heavier than air, having a density (relative to air) of 1.53; methane is lighter, as its relative density is 0.55. LFG is therefore lighter than air when the methane content is in excess of 54%, and heavier if the methane content is less than 54%. As LFG is actually a mixture saturated by water vapour, the resulting volumetric weight is commonly greater than that of air.

At the temperatures normally reached inside a landfill (30–40 °C) some LFG compounds are present as vapours. Gas cooling inside the gas network determines condensate formation, which will be contaminated by some of these compounds (see Chapter 5.5).

FLAMMABILITY AND THERMAL PROPERTIES

Table 2 refers to the thermal and flammability properties of the main LFG components. Ignition temperature is the minimum temperature at which a combustible, in the presence of air or oxygen, begins to burn without the necessity of ignition by flame or spark. The lower (upper) limit of flammability in air of a substance is the minimum (maximum) concentration in air, by volume, of that substance at the state of vapour or gas under (over) which a flame does not propagate from an ignition source.

Moreover, Table 2 reports, for each substance, the values of two hazard indexes: the fire hazard index and the reactivity index.

The fire hazard index identifies the susceptibility of a substance to burn, and is defined as follows (CRC, 1986):

4   materials that will rapidly or completely vaporize at atmospheric pressure and normal ambient temperature, or which are readily dispersed in air and which will burn readily;

3 liquids and solids that can be ignited under almost all ambient temperature conditions;

2 materials that must be moderately heated or exposed to relatively high ambient temperature before ignition can occur;

1 materials that must be preheated before ignition can occur;

0 materials that will not burn.

The reactivity index identifies the susceptibility of a substance to release of energy, and is defined as follows:

4 materials that are readily capable of detonation or of explosive decomposition or reaction at normal temperature and pressures;

3 materials that are capable of detonation or of explosive reaction but require a strong initiating source, or which must be heated under confinement before initiation, or which react explosively with water;

2 materials that are normally unstable and readily undergo violent chemical change but do not detonate; also materials that may react violently with water or which may form potentially explosive mixtures with water;

1 materials that are normally stable, but which can become unstable at elevated temperatures and pressures, or which may react with water with some release of energy but not violently;

0 materials that are normally stable, even under fire exposure conditions, and which are not reactive with water.

HEALTH HAZARD

Table 3 summarizes the most relevant data regarding the hazard potential of the main LFG components as far as human health is concerned. In addition to the maximum detected concentrations found in the literature, the threshold limit values (TLVs) adopted for 1993–94 by the American Conference of Governmental Industrial Hygienists (ACGIH, 1993) are listed. TLV-TWA and TLV-STEL are reported; they are respectively defined as follows.

TLV-TWA (threshold limit value, time-weighted average) is the time-weighted average concentration for a normal 8 h workday and a 40 h workweek, to which nearly all workers may be

TABLE 2. Flammability Risk of LFG Components

Chemical name	Formula	Ignition point (°C)	Flammable limits in air (% by vol)	Flammability hazard index	Reactivity index
Ammonia	NH_3	630	15.0–30.0	1	0
Carbon monoxide	CO	605	12.5–74.0	4	0
Carbon dioxide	CO_2	–	–	–	–
Methane	CH_4	537	5.0–15.0	4	2
Ethane	C_2H_6	515	3.0–15.5	4	2
Ethylene	$CH_2=CH_2$	425	2.7–3.4	4	0
Butane	$CH_3-CH_2-CH_2-CH_3$	365	1.5–8.5	4	0
Butylene	$CH_3-CH_2-CH=CH_2$	384	1.6–9.3	4	0
Propane	$CH_3-CH_2-CH_3$	470	2.1–9.3	4	0
Propylene	$CH_2=CH-CH_3$	460	2.0–11.1	4	1
Pentane	C_5H_{12}	285	1.4–8.0	4	0
Hexane	C_6H_{14}	260	1.0–7.5	3	0
Cyclohexane	C_6H	260	1.2–8.3	3	0
Heptane	C_7H_{16}	215	1.0–6.7	3	0
Octane	C_8H_{18}	210	0.8–6.5	3	0
Nonane	C_9H_{20}	205	0.83–5.6	–	–
Xylene	$C_6H_4(CH_3)_2$	470	1.0–8.0	3	0
Toluene	$C_6H_5(CH_3)$	535	1.2–7.0	3	0
Benzene	C_6H_6	555	1.2–8.0	3	0
Ethyl benzene	$C_6H_5-C_2H_5$	430	1.0–7.8	3	0
Tetrahydrobenzene	C_6H_{10}	310	–	3	0
Isopropylbenzene	$C_6H_5CH(CH_3)_2$	420	0.8–6.0	3	0
Chloromethane (methyl chloride)	CH_3Cl	625	7.1–18.5	4	0
Dichloromethane (methylene chloride)	CH_2Cl_2	605	13.0–22.0	0	0

TABLE 2. Continued

Chemical name	Formula	Ignition point (°C)	Flammable limits in air (% by vol)	Flammability hazard index	Reactivity index
Trichloromethane (chloroform)	$CHCl_3$	—	—	—	—
Tetrachloromethane (carbon tetrachloride)	CCl_4	—	—	—	0
1,1-Dichloroethane	CH_3CHCl_2	660	5.6–16	3	—
Trichloroethane	CH_3-CCl_3	537	8.0–15.5	—	—
Dichlorotetrafluoroethane	$CClF_2-CClF_2$	—	—	—	—
Trichlorotrifluoroethane	$CCl_2F-CClF_2$	—	—	—	—
Chloroethylene (vinyl chloride)	$CH_2=CHCl$	415	3.8–31.0	4	2
1,1-Dichloroethylene	CH_2CCl_2	440	5.6–13	4	2
1,2 cis-Dichloroethylene	$CHCl=CHCl$	175	6.2–16	4	2
1,2 trans-Dichloroethylene	$CHCl=CHCl$	175	9.7–12.8	3	2
Trichloroethylene	$CHCl=CCl_2$	410	7.9	3	—
Tetrachloroethylene	$CCl_2=CCl_2$	—	—	—	—
Chlorotrifluoroethylene	$CClF=CF_2$	—	8.4–39.0	—	2
Chlorodifluoromethane	$CHClF_2$	—	—	4	—
Chlorotrifluoromethane	$CClF_3$	—	Weakly flammable	—	—
Dichlorofluoromethane	$CHCl_2F$	552	Weakly flammable	—	—
Dichlorodifluoromethane	CCl_2F_2	—	Weakly flammable	—	—
Sulphidic acid (hydrogen sulphide)	H_2S	270	4.3–45.5	—	0
Thioglicolic acid	$CH_2SHCOOH$	112	—	4	—
1-Propyl alcohol	C_3H_8O	404	2.5–13.5	3	0
1-Butyl alcohol	$C_4H_{10}O$	365	1.7–18	3	0
Carbon disulphide	CS_2	100	1.3–44	3	0
Methyl mercaptan	CH_4S	—	3.9–21.8	4	0
Ethyl mercaptan	C_2H_6S	299	2.8–18	4	0

TABLE 3. Health Risk (S = skin allergic reaction) of LFG Components

Chemical name	Formula	Maximum detected concentration (mg/m^3)	ACGIH TLV (1993-94) TLV-TWA (mg/m^3)	TLV-STEL (mg/m^3)	Odour threshold (ppm)
Ammonia	NH_3	–	17	24	1–50
Carbon monoxide	CO	–	29	–	–
Carbon dioxide	CO_2	–	9000	54000	–
Methane	CH_4	–	–	–	–
Ethane	C_2H_6	–	–	–	–
Ethylene	$CH_2=CH_2$	–	–	–	500
Butane	$CH_3-CH_2-CH_2-CH_3$	560	1900	–	–
Butylene	$CH_3-CH_2-CH=CH_2$	10	n.d.	n.d.	–
Propane	$CH_3-CH_2-CH_3$	100	–	–	–
Propylene	$CH_2=CH-CH_3$	28	–	–	–
Pentane	C_5H_{12}	613	1770	2210	3000
Hexane	C_6H_{14}	76	176	–	–
Cyclohexane	C_6H_{12}	40	1030	–	$1.4-1000$ mg/m^3
Heptane	C_7H_{16}	47	1640	2050	930
Octane	C_8H_{18}	99	1400	1750	$0.09-713$ mg/m^3
Nonane	C_9H_{20}	270	1050	–	–
Xylene	$C_6H_4(CH_3)_2$	383	434	651	1 mg/m^3
Toluene (S)	$C_6H_5(CH_3)$	615	188	–	$0.6-153$ mg/m^3
Benzene	C_6H_6	15	32 A2	–	–
Ethyl benzene	$C_6H_5-C_2H_5$	236	434	543	$8.7-870$ mg/m^3
Tetrahydrobenzene	C_6H_{10}	–	–	–	$0.6-1023$ mg/m^3
Isopropyl benzene	$C_6H_5CH(CH_3)_2$	173	–	–	–
Chloromethane (methyl chloride) (S)	CH_3Cl	10	103	207	–
Dichloromethane (methylene chloride)	CH_2Cl_2	684	174 A2	–	$550-880$ mg/m^3

TABLE 3. Continued

Chemical name	Formula	Maximum detected concentration (mg/m^3)	ACGIH TLV (1993-94) TLV-TWA (mg/m^3)	ACGIH TLV (1993-94) TLV-STEL (mg/m^3)	Odour threshold (ppm)
Trichloromethane (chloroform)	$CHCl_3$	2	49 A2	–	50–200 ml/m^3
Tetrachloromethane (carbon tetrachloride) (S)	CCl_4	n.d.	31 A3	63 A3	–
1,1-Dichloroethane	CH_3CHCl_2	21	405	2460	–
Trichloroethane	CH_3-CCl_3	29	1910	–	–
Dichlorotetrafluoroethane	$CClF_2-CClF_2$	20	6990	9590	–
Trichlorotrifluoroethane	$CCl_2F-CClF_2$	2	7670	–	–
Chloroethylene (vinyl chloride)	$CH_2=CHCl$	264	13 A1	–	50
1,1-Dichloroethylene	CH_2CCl_2	138	20	79	500–1000
1,2-cis-Dichloroethylene	$CHCl=CHCl$	294	793	–	500–1000
1,2 trans-Dichloroethylene	$CHCl=CHCl$	–	–	–	–
Trichloroethylene	$CHCl=CCl_2$	312	269 A5	537 A5	8.3–345 mg/m^3
Tetrachloroethylene	$CCl_2=CCl_2$	250	170 A3	685 A3	–
Chlorotrifluoroethylene	$CClF=CF_2$	2.3	–	–	–
Chlorodifluoromethane	$CHClF_2$	276	3540	–	–
Chlorotrifluoromethane	$CClF_3$	10	–	–	–
Dichlorofluoromethane	$CHCl_2F$	5	42	–	–
Dichlorodifluoromethane	CCl_2F_2	602	4950	–	–
Hydrogen sulphide	H_2S	19.8	14	21	0.025–0.1
Thioglicolic acid	$CH_2SHCOOH$	n.d.	3.8	–	–
1-Propyl alcohol	C_3H_8O	630	492	614	–
2-Butyl alcohol	$C_4H_{10}O$	626	303	455	–
Carbon disulphide	CS_2	22	31	–	–
Methyl mercaptan	CH_4S	430	0.98	–	–
Ethyl mercaptan	C_2H_6S	120	1.3	–	–

repeatedly exposed, day after day, without adverse health effects.

TLV-STEL (threshold limit value, short-time exposure limit) is the time-weighted average concentration to which workers can be exposed continuously for a short period of time (no longer than 15 min and no more than four times per day) without suffering from (1) irritation, (2) chronic or irreversible tissue damage, or (3) narcosis of sufficient degree to increase the likelihood of accidental injury, impair self-rescue or materially reduce work efficiency, and provided that the daily TLV-TWA is not exceeded.

The letter S stands for skin, and applies where there is a potential significant contribution to the overall exposure by the cutaneous route, including mucous membranes and the eyes, either by contact with vapours or by direct skin contact with the substance.

The potential carcinogenicity of the substances is classified in the following five categories:

A1   confirmed human carcinogen;
A2   suspected human carcinogen;
A3   animal carcinogen;
A4   not classifiable as a human carcinogen;
A5   not suspected as a human carcinogen.

Potentially dangerous substances to health are generally present in the LFG in relatively reduced concentrations. For a significant number of components, however, concentrations far over the respective TLV have been detected. In particular, the possibility of concentrations over their TLV-TWA are pointed out for mercaptans, chloroethylene (carcinogenic of A1 class), hydrogen sulphide, 1,1-dichloroethylene, dichloromethane (carcinogenic of A2 class), toluene, 2-butyl alcohol, tetrachloroethylene, trichloroethylene, and 1-propyl alcohol. The literature reports concentrations over their TLV-STEL for 1,1-dichloroethylene, 2-butyl alcohol and 1-propyl alcohol.

In the case of inhalation, the above-mentioned substances may generate toxic (ethyl- and methylmercaptan, chloroethylene, hydrogen sulphide, 1,1-dichloroethylene, trichloroethylene), narcotic (methylmercaptan, chloroethylene, dichloromethane, tetrachloroethylene, trichloroethylene, 1-propyl alcohol, 2-butyl alcohol) or irritant effects (toluene).

Even if the individual concentrations of compounds were lower than respective TLV, adverse consequences to health cannot be excluded a priori owing to the lack of toxicology knowledge on the co-presence of chemical hazard agents.

In the case of exposure to two or more substances that are potentially dangerous to health, the combined effects instead of the independent effects of each one of them have to be considered. In the absence of information to the contrary, the effects of the different hazards should be considered as additive. If C_i indicates the observed gas concentration of the i_{th} substance and T_i the corresponding threshold limit, if the sum

$$\frac{C_1}{T_1} + \frac{C_2}{T_2} + \dots + \frac{C_n}{T_n}$$

exceeds unity, then the threshold limit of the mixture should be considered as being exceeded.

Finally, the presence even in traces of some compounds can generate nuisance (for instance unpleasant odours) in the landfill area and in the neighborhood. To this end, Table 3 indicates the known values of odour threshold.

CONCLUSIONS

Most compounds detectable in LFG are classifiable as volatiles, as they are characterized at atmospheric pressure by high values of vapour pressure and by low values of boiling temperature. About ten of the volatile compounds have been found in concentrations higher than their TLV-TWA and for some of them even higher than their TLV-STEL. Even in relation to the possibility of combined dangerous effects, a situation of potential hazard for workers cannot be excluded especially in the case of activity in places of possible accumulation. In such situations, moreover, explosion and fire hazard also exist.

As far as the risks are concerned for the population in the neighborhood of the landfill area, it is usually believed that air dilution of LFG emissions is widely sufficient to guarantee their protection. Substances that could cause the greatest worry are represented by some mercaptans, hydrogen sulphide, benzene, chloroethylene (vinyl chloride), tetrachloroethylene (perchloro-ethylene) and trichloroethylene.

REFERENCES

ACGIH (1993). *1993–1994 Threshold Limit Values for Chemical Substances and Physical Agents in the Work Environment*, American Conference of Governmental Industrial Hygienists, Cincinnati, OH.

CRC (1986). *Handbook of Laboratory Safety*, 2nd edn, (ed. N. V. Steere), CRC Press, Boca Raton, FL.

Wilhelm, V. (1993). Occupational safety at landfill sites – hazards and pollution due to landfill gas, in *Contaminated Soil '93* (eds F. Arendt, G. J. Annokkée, R. Bosman and W. J. van den Brink), Kluwer Academic Publishers, Dordrecht, pp. 449–58.

3. ENVIRONMENTAL ASPECTS

3.1 Landfill Gas Migration in Soil

PETER KJELDSEN

Institute of Environmental Science and Engineering, Technical University of Denmark, Building 115, DK-2800 Lyngby, Denmark

INTRODUCTION

General Perspectives

Organic wastes disposed of in landfills undergo anaerobic degradation (Chapter 2.1), which leads to production of biogas consisting of methane, carbon dioxide and trace gases (hydrogen, hydrogen sulphide, volatile organic compounds etc.; Chapter 2.2). The produced biogas is utilized at some larger landfills (Chapter 1.2), but the extraction is never 100% effective. Besides, at most landfills the biogas is not utilized, owing to low economical profitability, and only on a minor number of landfills is the biogas flared. Therefore, on most landfills, landfill gas is emitted to the atmosphere either directly from uncovered waste, through top covers, or by migrating through the surrounding soil layers.

Potential Effects of Migrating Landfill Gas

The migration and emission of landfill gas potentially lead to different effects in the surroundings. The most important of these effects are:

- fire and explosion hazards;
- health risks;
- damages to vegetation;
- groundwater pollution;

Landfilling of Waste: Biogas. Edited by T. H Christensen, R. Cossu and R. Stegmann. Published in 1996 by E & FN Spon, London.
ISBN 0 419 19400 2.

- global climate effects;
- odour nuisances.

The various effects are connected to different scales (Fig. 1). A short description of the effects is given in the following; they are covered in more detail in the following chapters on environmental aspects.

Fire and explosion hazards.   Landfill gas is explosive mainly owing to the methane content. The explosive limit range for methane lies between 5 and 15% in air at atmospheric pressure and ambient temperature. The limits are only slightly affected by the presence of other constituents (Gendebien *et al.*, 1992). If landfill gas is vented directly to the atmosphere, no explosion hazard exists (Hoather and Wright, 1989), but surface fires have been observed.

The main environmental hazard related to landfill gas is believed to be the explosion hazard caused by landfill gas entering houses through cracks in foundations, penetrating services etc. After mixing of the gas with air, an energizer (spark in electrical components, striking of a match etc.) can initiate the explosion.

Many cases of elevated methane concentrations in houses due to

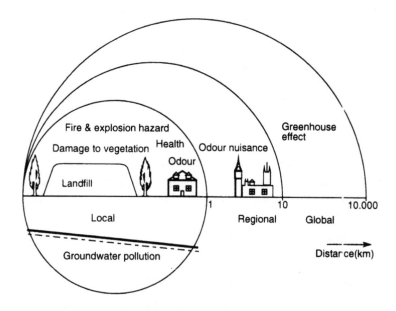

Figure 1. The different scales of landfill gas effects (after Luning and Tent, 1993).

landfill gas are reported in the literature. Gendebien *et al.* (1992) described 60 cases from the UK, USA, Germany and Canada of elevated methane concentrations, and 55 cases from the USA, UK and Canada involving explosion, fire and human injuries. Details concerning explosion and fire hazard associated with landfill gas are given in Chapter 3.2.

Health risks.   Carbon dioxide displaces oxygen in the respiratory system and causes indisposition even at low concentrations. Landfill gas has high concentrations of carbon dioxide as compared to the threshold limit value (TVL) of 0.5% or 5000 ppm (Hoather and Wright, 1989). This means that in general a ten times higher dilution factor is needed not to exceed the TVL value for CO_2 than is needed not to exceed the lower explosion limit (LEL) of 5% methane.

   The health effects from the organic trace components were discussed by Young and Parker (1983). They concluded that no severe health effects were related to the organic trace components in landfill gas. This conclusion was, however, reached on the basis of relative high air quality standards for the organic compounds. Little *et al.* (1992) state that vinyl chloride is a health risk, especially for houses closer than 100 m or built on top of the landfill. Also, Stephen *et al.* (1986) found the concentration of vinyl chloride in ambient air exceeding air quality standards in the vicinity of a landfill that had received large amounts of liquid hazardous waste. Also, Ward *et al.* (1992) found that vinyl chloride, among 77 identified trace organics, had moved the furthest from the landfill (50 m). Petersen (1988) stated that, in addition to vinyl chloride, benzene can also exhibit a health risk at landfills. Assessments of health risk related to landfill gas are given in Chapter 3.3.

Damages to vegetation.   Many cases of damage to vegetation in the vicinity of landfills are reported in the literature. Gendebein *et al.* (1992) described 31 different cases from the UK, USA, Germany, Canada and Japan. The main reason for damage to vegetation from landfill gas is asphyxia by removal of oxygen in the root zone. This removal can be either due to a displacement of the oxygen by landfill gas or to oxidation of methane. High concentrations of carbon dioxide (>20%) are also toxic to plants (Rettenberger, 1985), and some trace compounds (hydrogen sulphide, halo-organic compounds, etc.) are toxic to plants as well. In Chapter 3.4 further details of the effects of landfill gas on vegetation are given.

Groundwater pollution. Landfill gas migrating in the surrounding unsaturated zone is exposed to infiltrating water. Some of the components in the gas are highly water-soluble (the solubility of carbon dioxide is 2320 mg l^{-1} at 10 °C and only 30 mg l^{-1} for methane). Many of the trace organics in the landfill gas are also highly water-soluble, and can be leached out by infiltrating water, thereby contaminating the underlying groundwater. Chapter 3.5 describes further the effect of landfill gas on groundwater quality.

Global climate effects. The global importance of landfill gas emission to the atmosphere has been overlooked until recently. Methane and possibly some of the trace gases (e.g. CFCs) from landfills are considered to be 'greenhouse gases' and may also influence the ozone layer. Chapter 3.6 describes further the greenhouse effect related to landfill gas.

Odour nuisances. Complaints from local residents due to unpleasant odours emitted from landfills are often put forward (Gendebein *et al.*, 1992). The unpleasant odour does originate from the principal components of landfill gas, as both methane and carbon dioxide are odourless. The main problems of odour are connected to the initial acid fermentation stage of a landfill, where mercaptans and volatile acids are formed. A 1 000 000 dilution is needed for some of these compounds to bring them below the odour threshold (Young and Parker, 1983). Details of odour emission and control are given in Chapter 3.7.

Objectives

To make a proper evaluation of the potential effects of landfill gas, a detailed understanding of the migration of the gas in soil is very important. The objective of this chapter is to describe and, where possible, to quantify the factors affecting the migration of landfill gas through soil surrounding or covering landfills. Atmospheric dispersion behaviour of landfill gas is not dealt with (see Chapter 3.8), nor are methods to monitor or prevent gas migration.

Only the open literature has been consulted. Commercial reports dealing with the subject of landfill gas migration are numerous, but their accessibility is limited.

CONCEPTUAL MODEL OF LANDFILL GAS MIGRATION

The conceptual framework of landfill gas migration is given in Fig. 2. The figure shows the landfill gas transport directions by arrows, the basic processes governing the migration and the factors affecting the basic processes.

The basic processes are divided into transport processes and sink (or source) processes:

- transport processes:
  diffusion,
  advection,
  dilution;
- sink processes;
  dissolution of landfill gas constituents in water,

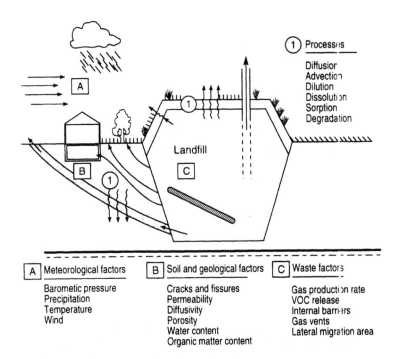

A	Meteorological factors	B	Soil and geological factors	C	Waste factors
	Barometic pressure		Cracks and fissures		Gas production rate
	Precipitation		Permeability		VOC release
	Temperature		Diffusivity		Internal barriers
	Wind		Porosity		Gas vents
			Water content		Lateral migration area
			Organic matter content		

Figure 2. The conceptual framework of landfill gas migration. The processes governing the migration are shown together with the factors affecting the governing processes.

sorption to soil particles,
oxidation (and other degradation processes).

The factors affecting the processes can be divided into three classes:

- meteorological conditions;
- soil conditions;
- waste and landfill conditions.

BASIC THEORY FOR GAS MIGRATION THROUGH SOIL

Many investigations of landfill gas migration have demonstrated that both diffusive and advective transport can be important processes (Ghabaee and Rodwell, 1989).

Diffusion

Diffusional fluxes are caused by variation in gas concentrations within the porous medium due to the Brownian movement of the gas molecules. The diffusional flux can be described by Fick's law:

$$\mathcal{J}_D = - \varepsilon_a\, D\, \frac{\partial C}{\partial x} \tag{1}$$

where $\mathcal{J}_D$ is the diffusional flux (g m^{-2}s^{-1}), ε_a is the gas-filled porosity (fraction), D is the diffusion coefficient in the soil (m^{2}s^{-1}) and $\partial C/\partial x$ is the concentration gradient (g m^{-4}).

Millington and Quirk (1961) give further details for calculation of the diffusion coefficient:

$$D = D_0\, \tau \tag{2}$$

where D_0 is the diffusion cofficient in air at given temperature (m^{2}s^{-1}) and τ is the tortuosity:

$$\tau = \frac{\varepsilon_a^{\frac{7}{3}}}{\varepsilon^2} \tag{3}$$

where ε is the total porosity of the soil (dimensionless). Sallam *et al.* (1984) found in laboratory experiments that this expression gives a reasonable description of the changes in diffusion as a result of changing air-filled porosity (obtained by changing the water content).

For diffusion through unsaturated soils, Fick's law only holds if the mean free path of a gas molecule is smaller than the pore size (i.e. the gas molecules collide with each other within the pore space) (Williams and Aitkenhead, 1991). If this is not the case, the diffusional flux is also governed by molecule collision with the pore walls: the Knudsen diffusion (Ghabaee and Rodwell, 1989). For methane at atmospheric pressure the mean free path is about 500 Å (Williams and Aitkenhead, 1991). Most pores even in clay are much bigger than 500 Å (small pores will be water-filled), which means that the Knudsen diffusion in most cases has no importance.

Advection

Advective flux of a gas constituent through a porous medium is caused by a pressure gradient and is described by Darcy's law:

$$\mathcal{J}_A = -C \frac{k_a}{\mu} \frac{\partial p}{\partial x} \qquad (4)$$

where $\mathcal{J}_A$ is the advective flux (g m^{-2}s^{-1}), C is the gas concentration of the constituent (g m^{-3}), k_a is the gas permeability of the soil (m^2), μ is the viscosity of the gas (N s m^{-2}), and $\partial p/\partial x$ is the pressure gradient (Pa m^{-1} = N m^{-3}).

Lagerkvist (1986) gave the following expression for calculation of the gas permeability from the hydraulic permeability, k (m^{-2}):

$$k_a = k \frac{\varepsilon_a}{\varepsilon} \qquad (5)$$

Physical Constants

Diffusion coefficients in air, D_0. The diffusion coefficient in air is not constant for a specific gas, but is a function of pressure and temperature. At normal atmospheric pressures the pressure dependence is insignificant. The temperature dependence is given by the following equation (Lagerkvist, 1986):

$$D_0 = (D_0)_{273K}(T/273K)^{1.75} \qquad (6)$$

where $(D_0)_{273K}$ is the diffusion coefficient in air at 273 K (0 °C) and T is the gas temperature in kelvin.

Table 1 gives diffusion coefficients in air at 20 °C for different

TABLE 1. Data for Diffusion Coefficients in Air D_0, Viscosities μ and Densities ρ for different Gas Constituents at 20 °C (Lagerkvist, 1986)

Gas	D_0 $(m^2 s^{-1})$	μ $(N s^{-1} m^{-2})$	ρ $(g m^{-3})$
CH_4	2.0×10^{-5}	11.0×10^{-6}	670
CO_2	1.5×10^{-5}	14.7×10^{-6}	1800
O_2	2.2×10^{-5}	20.4×10^{-6}	1330
N_2	2.2×10^{-5}	17.6×10^{-6}	1160
H_2	8.2×10^{-2}	–	–

gas constituents. In comparison the diffusion coefficient for methane in water is 1.5×10^{-9} m^{2}s^{-1}, which is a factor of 10^4 lower than the diffusion coefficient in air.

Viscosity and densities of gas mixtures μ and ρ. Viscosity is a function of temperature and gas composition. Ghabaee and Rodwell (1989) provide expressions for the temperature dependence and the dependency on gas composition. Also, density is a function of temperature, pressure and gas composition and can be calculated from basic thermodynamics (Dalton's law and the ideal gas law).

Table 1 presents also viscosities and densities for different gas constituents. For a landfill gas with a composition of 50% CH_4 and 50% CO_2 at 10 °C and 1 atm, the following values can be used (Farquhar and Metcalfe, 1982):

$\mu = 13.2 \times 10^{-6}$ N s^{-1} m^{-2}
$\rho = 1300$ g m^{-3}

Permeabilities, k. Table 2 gives typical permeabilities for a range of different soils (Lagerkvist, 1986). The variability of permeability for real soils is highly significant and depends on the grain size distribution, organic matter content, etc.

Evaluation of the Significance of Diffusive and Advective Fluxes

The use of the above-mentioned equations is demonstrated by calculating the diffusive and advective methane fluxes for a 1 m deep top cover of a landfill. Three different soil types all with a water content at field capacity, are considered. The methane concentration in the landfill gas is set to 50% (v/v) and the pressure above

TABLE 2. Typical Ranges for the Permeability, *k*, for Different Soils (Lagerkvist, 1986)

Soil type	Permeability (m^2)
Moraine clay	10^{-18}–10^{-16}
Clay	10^{-17}–10^{-15}
Clay loam	10^{-17}–10^{-15}
Silt loam	10^{-16}–10^{-14}
Silt	10^{-16}–10^{-14}
Sandy loam	10^{-15}–10^{-13}
Loamy sand	10^{-14}–10^{-12}
Fine sand	10^{-13}–10^{-11}
Medium sand	10^{-12}–10^{-10}
Coarse sand	10^{-11}–10^{-9}
Fine gravel	10^{-10}–10^{-8}

barometric pressure in the landfill is set to 20 cm H_2O. For comparison, the diffusive flux through a 1 mm thick top-liner of polyethylene (PEL) is calculated as well. The data used and the results are shown in Box 1. The box shows that the diffusive fluxes are of the same order of magnitude for all four cases, even for diffusion through the PEL liner. The sample calculation also shows that the advective fluxes are negligible for the PEL and the clay, but orders of magnitude larger than the diffusive flux for the sandy top layer. The diffusive and advective flux for the loam are of the same magnitude. Calculations like these are of course to be used with care, as the assumptions are simplified, and no other processes and factors are considered (such as cracks and fissures, oxidation processes, etc.).

The use of the basic theory is discussed further in the section on landfill gas migration models.

FACTORS AFFECTING GAS MIGRATION THROUGH SOIL

Many studies have dealt with the different factors that can affect the migration of landfill gas either vertically through the top cover of the landfill or laterally through the geological strata surrounding the landfill. All these factors together control the emissions from the landfill and the maximal distances that landfill gas can migrate laterally.

BOX 1. Calculation of Diffusional and Advective Flux of Methane Through Top Covers

Methane		
Diffusion coefficient in air	D_o (m^2 s^{-1})	2.00×10^{-5}
Viscosity of landfill gas	μ (N s^{-1} m^{-2})	1.32×10^{-5}
Concentration in atm.	C_o (g m^{-3})	0

Landfill			*PEL top-liner*	
Pressure above barometric pressure	P_L (N m^{-2})	1960	Thickness L(m)	1.00E-03
Methane concentration	C_L (gm)$^{-3}$	340	Dif.coef. D(m^2 s^{-1})	7.63E-11[a]

Top cover		
Depth	L (m)	1

Soils			
Soil type	Sand	Loam	Clay
Total porosity, ε	0.26	0.45	0.60
Water-filled porosity at field cap., ε_w	0.10	0.34	0.52
Gas-filled porosity, ε_a	0.16	0.11	0.08
Total permeability, k (m^2)	1.00×10^{-11}	1.00×10^{-15}	1.00×10^{-17}
Gas permeability,[b] k_a (m^2)	6.15×10^{-12}	2.44×10^{-16}	1.33×10^{-18}
Tortuosity,[c] τ	2.06×10^{-1}	2.86×10^{-2}	7.66×10^{-3}
Soil diffusion coefficient,[d] D (m^2 s^{-1})	4.11×10^{-6}	5.7×10^{-7}	1.53×10^{-7}

Fluxes through top soil			
Diffusive flux,[e] J_D (g m^{-2} s^{-1})	2.24×10^{-4}	2.14×10^{-5}	4.17×10^{-6}
Diffusive flux – yearly basis, J_D (g m^{-2} yr^{-1})	7.05×10^{3}	6.75×10^{2}	1.31×10^{2}
Advective flux[f] J_A (g m^{-2} s^{-1})	3.11×10^{-1}	1.23×10^{-5}	6.73×10^{-8}
Advective flux – yearly basis J_A (g m^{-2} yr^{-1})	9.80×10^{6}	3.89×10^{2}	2.12×10^{0}

Flux through PEL top-liner	
Diffusive flux,[g] J_D (g m^{-2} s^{-1})	2.59×10^{-5}
Diffusive flux – yearly basis, J_D (g m^{-2} yr^{-1})	8.18×10^{2}

[a] Obtained from Kjeldsen(1993)
[b] Calculated by use of equation 5
[c] Calculated by use of equation 3
[d] Calculated by use of equation 2
[e] Calculated by use of equation 1
[f] Calculated by use of equation 4
[g] $J_D = D \cdot (\partial C / \partial x)$, (Kjeldsen. 1993)

Very few studies have tried to investigate the maximal distances to which the gas can migrate. This is also a difficult task, as both high spatial and temporal variability in the migration distances are expected. In Table 3 the observed or stated maximal migration distances are shown. The criteria for measurable landfill gas used is different, but it seems that migration distances over 300 m are

TABLE 3. Summary of Observed Maximal Migration Distances of Methane at Landfills

Reference	*Docu-mented case*	*Undocu-mented state-ment*	*Maximal migration distance*	*Remarks*
Ghabaee and Rodwell (1989)		x	300–400 m	
Hodgson *et al.* (1992)	x		70 m	
Kjeldsen and Fischer (1995)	x		90 m	5% CH4
O'Leary and Tansel (1986)		x	>300 m	
Raybould and Anderson (1987)	x		300 m	Unpleasant odour
Williams and Aitkenhead (1991)	x		100 m	Measurable CH4
Wittman (1985)		x	approx. 100 m	Vegetation stress
Wood and Porter (1987)	x		180 m	1 % CH4 and measurable VOCs

seldom observed. Kjeldsen and Fisher (1995) intensively monitored the lateral migration from a 20 m deep landfill by measuring a time series of gas composition in 30 wells over a 35-day period. The LEL value of methane, 5% (v/v), was never measured more than 90 m away from the landfill border.

Gas Composition

The composition of the landfill gas is, of course, important for the significance of landfill gas migration, for both the main components (methane and carbon dioxide) and the trace components. High concentrations need more attenuation in the surroundings to reach a 'no-effect-level'.

The concentration of the main components depends on the degradation stage of the waste, and on several environmental factors (see Chapters 2.1 and 2.2). The composition of landfill gas with respect to trace components, which can either be products from the degradation of the organic waste fraction or can be volatilized from co-disposed chemical waste, is very diverse, with large concentration differences between different landfills (see Chapter 2.2).

Gas Pressure in the Landfill

Gas pressures above barometric pressure inside landfills have been observed several times in the literature. Table 4 shows the observed gas pressures measured in wells screened in the waste. Large variations are observed, but pressures up to 20 cm H_2O above barometric pressures in landfills with low permeability top-covers seem normal.

The gas pressure observed is not directly a measure for the gas production activity in the landfill. The gas pressure in a certain location in the landfill builds up until the gas transport (mainly by advective forces) outbalances the gas produced. This means that insignificant pressure build-up is observed if the landfill is uncovered, or covered with high-permeability soils, and the resistance towards migration is low. The gas production activity is therefore of primary importance for the advective migration in relation to the diffusive migration. The gas production activity depends on several factors (waste composition, age, water content, landfill size, operational procedures, etc.) and is thoroughly covered in Chapter 2.1.

A high degree of variability in gas pressure within a landfill can often be expected. The use of low-permeability soil as daily cover

TABLE 4. Observed Pressures above Barometric Pressure Observed in Landfills

Reference	Docu- mented case	Undocu- mented state- ment	Observed pressure above barometric (cm H_2O)	Remarks
Bogner et al. (1988)	x		5–20	Silty clay top cover
Campbell (1989)		x	>80	
Jones et al. (1988)	x		0.1–0.3	Top cover: inert fill
Kerfoot (1993)	x		20–25	Clay top cover
Kjeldsen and Fischer (1995)	x		2–20	Clay top cover
McOmber et al. (1982)		x	<10	
Williams and Aitkenhead (1991)	x		20	Clay top cover
Wittman (1985)		x	3–30	Max. value: 250 cm

of the waste will impede uniform gas permeabilities in the landfill and create a wide diversity of gas pressures. Conversely, the use of very permeable materials, temporary road constructions and venting systems might create localized zones of high permeability where gas pressures are low (Campbell, 1989).

In general the probability of lateral migration within landfill sites will be higher than that of vertical migration. Waste is commonly placed in a layered formation, which will tend to produce horizontal zones within the site, with higher compaction and hence lower porosity achieved primarily at the surface of each waste layer. The use of daily covers will also tend to create horizontal layers of low permeability, either due to the use of clay or silty soils or due to the formation of perched water-tables in permeable cover soils placed on top of highly compacted waste. The different aspects of gas pressure and transport behaviour are summarized in Fig. 3.

Clay soil has often been used for construction of final covers on landfills in order to minimize the leachate formation. This might, however, have a very significant effect on the lateral migration of landfill gas, owing to pressure build up within the landfill. Several

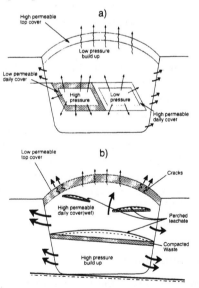

Figure 3. The gas pressure and gas transport behaviour within landfills: (a) landfill with high permeable top cover, general low-pressure build up and local high gas pressures due to the presence of low permeable layers; (b) landfill with low permeability top cover with cracks, and compacted waste layers within the waste.

authors report that a clay soil top cover has been placed immediately before significant landfill gas effects were observed. Both Kjeldsen and Fischer (1995) and Williams and Aitkenhead (1991) reported that landfills were covered with a clay soil cover a few months before serious explosion accidents occurred. Wittman (1985) observed that the trees surrounding a landfill died off one month after a clay cover was constructed. Also, off-site odour problems have been observed as a result of covering a landfill with clay soil (Raybould and Anderson, 1987).

The presence of gas-venting installations or gas-recovery wells in the landfill has a great effect on the lateral migration and the vertical migration through the top cover, as the installations effectively lower the pressure in the landfill. Bogner *et al.* (1993) found gas emission rates through the top cover of a landfill that were 10 000 times lower far from gas recovery wells (>30 m) than rates measured in the vicinity of the wells (<5 m). Also, cracks formed by differential settlement in the underlying waste, 'rodent diggings' etc. could lead to higher emission rates locally.

Landfills are usually either placed above ground or contained within quarries, pits etc. Landfills contained in deep gravel pits without any liners at the side walls have large areas open for lateral gas migration. In these sites, the gas pressure might be only slightly above the barometric pressure, as the gas effectively migrates to the sides. For 'above-ground' landfills the lateral migration is probably of minor importance as the gas has to cross the bottom layer of the landfill to escape laterally through adjacent soil layers. The side slopes on 'above-ground' landfills are often cut through by small erosion gullies formed by surface runoff or by cracks formed by earth slips. This means that the gas emission rates at the side slopes might be higher than at the top of the landfill. Bogner *et al.* (1988) found higher methane pore gas concentration at the side slopes (40% v/v on average) as compared with the concentration measured at the top part of the landfill (10% v/v on average).

Properties of the Surrounding Strata

The lateral migration of the landfill gas is highly dependent on the geology of the surrounding strata. In Fig. 4(a) a simple hypothetical situation is shown where a completely homogeneous stratum exists and is subjected to a given gas pressure. In this case the potential for gas to migrate will be directly related to the gas permeability of the stratum. The distance that gas migrates from the site boundary

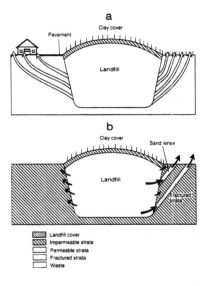

Figure 4. Lateral migration of landfill gas depending on surrounding strata. (a) A homogeneous sand surrounding the landfill covered with a low permeable clay. The vertical migration is large close to the landfill and decreases with distance. If the ground surface is paved, longer lateral migration distances are obtained. (b) Sloping layers of sand lenses or fractured strata lead to high spatial variability in the vertical migration (after Campbell, 1989).

of the landfill is in particular related to the possibility for the gas to escape through the surface soils. Thus a clay soil overlying a sand stratum (or pavement as shown in Fig. 4(a) might result in increased migration distances.

The very simple situation shown in Fig. 4(a) is very unlikely to exist at many sites. In most cases, the formation of the strata through the geological ages will lead to different strata with very different permeabilities. The variation in permeabilities of different soil types is very large (see Table 2). This means that most of the landfill gas tends to migrate through, sometimes thin, distinct layers: for instance, sand lenses in clay deposits or fractured sandstone strata as shown in Figure 4(b).

In many cases the strata of the geological formations may not be horizontal (as indicated in Figure 4(b)), leading to more permeable horizons intersecting the land surface at discrete distances from the site. Williams and Aitkenhead (1991) observed fractured sandstone strata and coal seams with a dip of 5–11 ° at the Loscoe Landfill, as shown in Fig. 5. The major part of the gas was migrating through Bed 16 and 17 shown on the figure, but also migration through lower

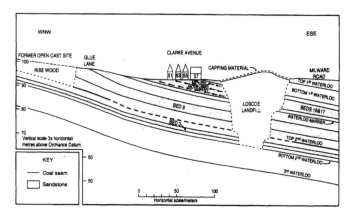

Figure 5. Geological cross-section through the Loscoe Landfill showing sloping layers of fractured sandstone and coal seams, which outcrop close to the landfill (Williams and Aitkenhead, 1991).

layers of sandstone might have occurred. Also, Hill (1986) found that an outcrop of fractured sandstone in the vicinity of a landfill site was the main exit point for the gas.

Variations in the groundwater level over time could also have an effect on the potential for gas migration. Lowering of a groundwater-table due to drought or increased extraction of groundwater could expose highly fractured strata to migrating landfill gas. Local perched water-tables may also dry out, which might give higher vertical permeabilities of the strata.

Potential Pathways Made by Man

The presence of mining shafts in the surroundings, or man-made fissures as a consequence of previous rock-blasting operations, could significantly increase the permeability of the strata. Raybould and Anderson (1987) found that some old worked coal seams were the main migration pathways at a landfill in St Helens, UK.

Kerfoot (1993) mentioned that groundwater and gas-monitoring wells, which are screened in the unsaturated zone, can be an ideal vent for landfill gas, increasing the direct atmospheric emission of landfill gas.

Leaking sewers, drain pipes etc. could also serve as migration pathways and lead to elevated lateral migration distances.

The potential for indoor air effects from landfill gas exists not only if landfill gas has migrated to the adjacent soil layers underneath a specific building, but also requires that the gas can enter the house. Entrance of landfill gas into a building demands a leakage in the subsurface structure and sufficient permeability of the soil. The entrance of the gas is often governed by advective flow and can be driven by the presence of a gas pressure gradient within the soil strata controlled by the gas pressure in the landfill. This will only be the case if the building is placed on top of or close to the landfill. The advective flow can also be controlled by indoor–outdoor thermal differences, windloads on the building or an unbalanced ventilation system (Hodgson *et al.*, 1992).

Meteorological Conditions

Several meteorological parameters can have a decisive importance for the migration of landfill gas. These parameters will be discussed in the following.

Barometric pressure changes. Perhaps the most important meteorological parameter is changes in the barometric pressure over time. Williams and Aitkenhead (1991) found at the Loscoe Landfill, where a bungalow was completely destroyed by a landfill gas explosion, that at the time of explosion an intense atmospheric depression was passing over the region. The event occurs on average only once every 6 years. The depression resulted in a drop of 40 mbar in 10 h, but the explosion occurred after a fall of only 20 mbar. Similar observations have been reported in relation to other explosion accidents (Franklin and Vetter, 1987; Kjeldsen and Fischer, 1995).

Pirkle *et al.* (1993) investigated the emission of methane through the top cover of a landfill in South Carolina by measuring the methane content under 2.5 m × 2.5 m ground sheets installed on the top cover. Figure 6 shows the methane concentrations under a ground sheet. The barometric pressure is shown as well. The figure shows the 'barometric pumping' of methane from the landfill: methane is low or undetectable in stable pressure situations or at increasing pressure. When the pressure decreases, significant releases of methane through the top cover are observed. Even very small pressure decreases instantaneously lead to higher methane concentrations. Close study of the data reveals that even short periods of stable pressure in an overall falling pressure regime cause cessation of the methane flux.

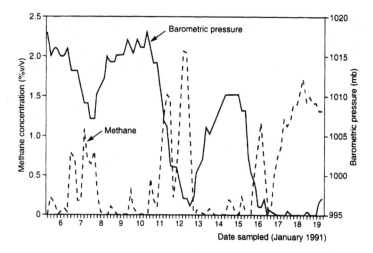

Figure 6. Methane concentration measured under a groundsheet placed on a landfill top cover and barometric pressure vs time (Pickle *et al.*, 1993).

Bogner *et al.* (1988) studied the landfill gas emission through the top cover of a landfill in Illinois. The barometric pressure as well as the pressure in probes installed within the top cover were monitored. Figure 7 shows the barometric pressure as a function of time over an 8-day period, together with the pressure difference (soil gas pressure minus barometric pressure) measured in one of the probes.

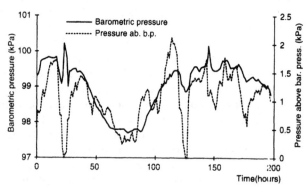

Figure 7. Barometric pressure as a function of time together with the pressure difference measured in a probe installed in the top cover of a landfill site in Illinois (after Bogner *et al.*, 1988).

The figure shows that the pressure difference gradually increases under stable or increasing barometric pressure regimes. Very rapid decreases in pressure difference are observed at every fall in barometric pressure, indicating that substantial venting of landfill gas is occurring under these events.

Kjeldsen and Fischer (1995) studied, at a landfill in Denmark, the variations in soil gas composition as a function of time and distance from the landfill boundary, and related the variations to the changes in barometric pressure (see Fig. 8). For well 0601, placed only 10 m from the boundary, no effect from changing pressures is seen. At this short distance, the gas pressure gradient between the landfill and the soil is headed from the landfill towards the atmosphere, and is not changed in direction by any barometric pressure changes. This means that no atmospheric air is observed in the well at any moment. For well 0201, placed 60 m from the boundary, a distinct relationship between methane concentration and the barometric pressure is observed: the highest methane concentrations are observed for decreasing barometric pressure and vice versa. For oxygen the opposite is the case. In well 0101, placed 120 m from the landfill, no elevated methane concentrations are observed at any time.

Kjeldsen and Fischer (1995) also studied the changes in soil gas composition under a decreasing barometric pressure regime. Figure 9 shows the methane concentration as a function of time and depth. The measurement was carried out over a 33 h period. The barometric pressure gradually dropped from 1022 mbar to 1010 mbar over the period. The figure shows that the methane concentration increases at 80 cm depth from below 1% at $t = 0$ h to near 40% at $t = 33$ h. At 2.5 m depth, however. the concentration changes are smaller. Latham and Young (1993) found also, by model calculations, that the effects of barometric pressure changes were smoothed out by increasing depth. They found that the effects were insignificant below 3 m. Similar results were found by Ward *et al.* (1992) at a landfill in the UK: very substantial variations in concentration over time 0–2 m below ground surface and much more stable concentrations below 2 m depth.

Precipitation. The water content of the soil, as previously discussed, influences both the gas permeability and the diffusivity of the soil. The amount of precipitation therefore has an effect on the gas migration, both in the long and the short term. In general, the

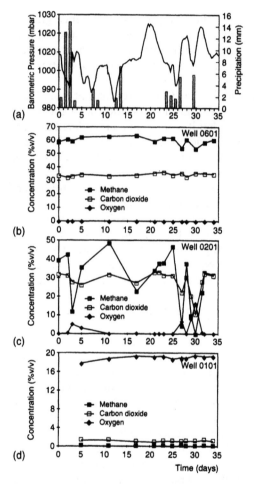

Figure 8. (a) Daily precipitation (columns) and barometric pressure (line) vs time at Skellingsted Landfill, Denmark, together with (b)–(d) measured concentrations of methane, carbon dioxide and oxygen in Wells 0601, 0201 and 0101, respectively 10 m, 60 m and 120 m from landfill border (Kjeldsen and Fischer, 1995).

water content of the upper soil layer is higher in the winter period, owing to higher precipitation and lower evapo-transpiration. This means that the vertical gas permeability and the diffusivity of the upper soil layers are significantly lowered through the winter period.

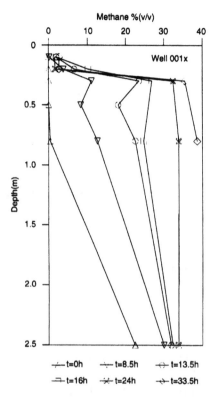

Figure 9. Methane concentration as a function of time and depth in a well at Skellingsted Landfill, Denmark (Kjeldsen and Fischer, 1995).

In the short term, large amounts of precipitation can lead to water saturation of the top soil, which totally blocks the vertical gas migration. Large amounts of precipitation often follow the passage of atmospheric depressions. This was also the case at the Skellingsted Landfill, around 21 March 1991, where a gas explosion accident occurred in a house close to the landfill. Figure 10 shows the barometric pressure and the precipitation for March 1991 (Kjeldsen and Fischer, 1995). Heavy rain and a substantial drop in barometric pressure are seen the days before the accident. The heavy rain filled the pores of the upper soil layer and decreased the vertical gas transport. The soil below the house was not sealed by water. The barometric pumping enhanced the horizontal migration of

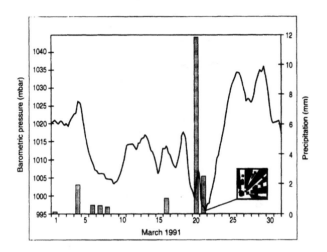

Figure 10. Daily precipitation (columns) and barometric pressure (line) for March 1991 as measured at a meteorological station close to Skellingsted Landfill, Denmark. An explosive accident happened on 21 March as indicated (Kjeldsen and Fischer, 1995).

landfill gas, and the only place where the gas could escape was through the dry soil layers below the house, and through sewer and water ducts into the house.

The unfavourable combination of high precipitation and barometric pressure drop has also been reported as the main factor at a landfill gas explosion close to a landfill in Wisconsin (Franklin and Vetter, 1987).

Bogner (1992) studied the effect of soil moisture changes on the gas composition in a clay soil top cover of a landfill site in Illinois. After a very dry summer, approximately 25 cm of rain fell in a 2 month period (from 15 September to 15 November). During this period (see Fig. 11) the moisture content in the top cover increased, the oxygen concentration decreased, and the methane concentration increased slowly as the gas-filled porosity was reduced owing to increased soil moisture, effectively sealing the cover soil. Bogner (1992) also carried out a rapid sprinkling experiment on a sandy silt top cover at a landfill in California. Five cm of water were evenly infiltrated during a 10 h period. Methane concentrations in soil gas at a depth of 7 cm increased from <200 ppm to 35% (vol) over the 10 h period, indicating rapid sealing of shallow gas-filled pore spaces.

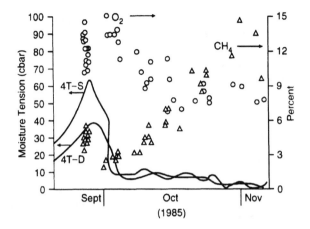

Figure 11. Effect of increased soil moisture on soil gas composition in a top cover on a landfill site in Illinois. The development in moisture tension over time in two tensiometers is shown together with oxygen and methane concentrations in the cover (Bogner, 1992).

Thibodeaux *et al.* (1982) reported on seasonal changes of top cover permeabilities. They found for 12 waste trenches that the average permeabilities of the top covers in April were a factor of 20 lower than the values measured in July.

The dynamics of water infiltration into the soil has, as shown in the above-mentioned cases, a major influence on the vertical transport of methane in cover soils and in the surrounding soil layers.

Other meteorological parameters. In some areas through the winter period a significant amount of the precipitation falls as snow. Permeabilities of snow covers can be drastically decreased by changing periods of frost and thaw, owing to formation of a tight layer of ice. In contrast to soil sealing by heavy rain, sealing by a layer of ice can last for a long time. Thomas and Salmon (1993) reported a landfill gas explosion accident at a landfill in eastern France where the soil was frozen to a depth of 50 cm, which led to an enhanced lateral migration of gas.

Bogner *et al.* (1988) and Figueroa (1993) stated that changes in wind speed and temperature gradients have an important effect on the exchange of soil gas with the atmosphere for the top few centimetres of the soil layer.

In general, meteorological conditions seem to have a very important role in controlling the migration of landfill gas, both laterally and through top covers. As most meteorological parameters show very stochastic behaviour, the prediction of gas compositions at specific locations is difficult. Gas explosion accidents seem to be connected with worst-case scenarios with unfavourable combinations of different factors (especially meteorological factors).

Sinks and Sources

The following paragraphs describe the different sinks (and sources) that can affect the composition of the migrating landfill gas.

Dilution. Studies of the spreading of pollutants in groundwater have shown that the dispersion is important for the dilution of the pollutants along the flow direction. The dispersion is introduced by variations in permeabilities at all scales. For an introductory description of the dispersion term see, for instance, Fetter (1993).

Also, for the transport of landfill gas in unsaturated soil, the dispersion process is expected to be of major importance. An even higher dispersion can be expected as compared with dispersion in the groundwater zone, as a larger variability in gas permeabilities exists owing to the highly variable water content (directly affecting the gas permeability). However, no studies directly headed toward the understanding of the dispersion process in landfill gas migration have been reported in the literature.

Gas-to-water transfer. When landfill gas migrates it flows in direct contact with infiltrating water. Under these circumstances a transfer of the gas constituents into the water phase takes place. At equilibrium it is controlled by Henry's law:

$$C_a = K_H C_w \qquad\qquad (7)$$

where C_a is the gas-phase concentration (mg l^{-1}), C_w is the water-phase concentration (mg l^{-1}), and K_H is the Henry's law constant (dimensionless). Under non-equilibrium, kinetic factors control the transfer process. In such a case, the departure of conditions from equilibrium determines the magnitude of the gas/water phase transfer (see Chapter 3.5). The transfer process can run in each direction: a sudden decrease in gas concentration can 'retransfer' compounds back into the gas phase from the water phase.

The significance of the transfer process for a certain gas constituent is determined by the magnitude of the Henry's law constant. For methane and carbon dioxide the constant is 29 and 1.1 respectively. This means in general that a higher proportion of the carbon dioxide in migrating landfill gas tends to be 'absorbed' by the water content in the soil, than is the case for methane. The Henry's law coefficients for many of the trace organics in landfill gas are low as well, which means that these compounds also have the tendency to be transferred to infiltrating water: see also Chapter 3.5.

Latham and Young (1993) made model calculations of the surface flux of landfill gas through a top cover of a hypothetical landfill site taking into account barometric pressure changes and gas–water transfer, but no degradation processes. The results of the model calculations are shown in Fig. 12. The simplified barometric pressure profile used is shown in Fig. 12a and the resulting surface flux of methane, carbon dioxide and air in Fig. 12b. The figure shows, as has been discussed earlier, that the surface flux of all three constituents is highly dependent on the barometric pressure changes with high landfill gas emissions at pressure drop and vice versa. However, the figure also shows that the amplitude of the carbon dioxide oscillation is lower than for the methane oscillation. This is due to a higher transfer of carbon dioxide from gas to water when the surface flux increases and from water to gas when the flux decreases. This is even more obvious in Fig. 12c, where the predicted soil gas concentrations at a depth of 1 m below the surface are shown: the carbon dioxide concentration is nearly constant as compared with that of methane because the relatively high carbon dioxide content of the porewater serves as a buffer. Exactly the same observations were made by Kjeldsen and Fischer (1995), as shown in Fig. 8c. The attenuation of the carbon dioxide oscillations can, however, also be explained by methane oxidation, as the oxygen pressed into the soil at increasing barometric pressures oxidizes the methane to carbon dioxide (see also later).

Sorption. Landfill gas components can be sorbed onto the soil particles, mainly by the organic matter content of the soil. A simple model, as presented by Little *et al.* (1992), is given in Box 2. The model calculates the retardation of the gas transport, and also takes the gas-to-water transfer into account, as this process serves as an extra retarding process in addition to the sorption process. The retardation factor is equal to 1 if the component does not participate in any reaction. A retardation factor of, for instance, 2 means that

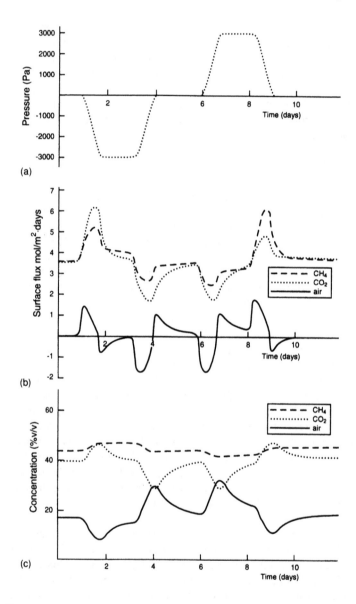

Figure 12. The effect of barometric pressure changes on the surface flux of landfill gas through a hypothetical landfill top cover as determined by model calculations: (a) the used simplified barometric pressure profile; (b) resulting surface flux of methane, carbon dioxide and air; (c) predicted soil gas concentrations at a depth of one metre below surface (Latham and Young, 1993).

BOX 2. Simple Model for the Description of Sorption between Gas and Soil (after Little *et al.*, 1992)

No kinetic effects in the sorption process and linear sorption are assumed:

$$C_s = K_d \cdot C_a$$

C_s = mass fraction of contaminant sorbed to soil (mg kg^{-1})
C_a = gas-phase concentration of contaminant (mg l^{-1})
K_d = air-to-unsaturated-soil partition coefficient (cm^3 g^{-1})

$$K_d = \frac{K_d^{sat}}{K_H} + \frac{\varepsilon_w}{\rho_b \cdot K_H}$$

K_d^{sat} = water-to-soil partition coefficient (cm^3 g^{-1})
K_H = Henry's law constant (dimensionless, concentration basis)
ε_w = water-filled soil porosity (dimensionless)(must be >0.05)
ρ_b = bulk density of soil (g cm^{-3})

$$K_d^{sat} = f_{oc} \cdot K_{oc}; \quad K_{oc} = \text{function } (K_{ow})$$

f_{oc} = fraction of organic carbon in soil (g g^{-1}) (must be >0.001)
K_{oc} = water-to-soil-organic-matter partition coefficient (cm^3 g^{-1})
K_{ow} = octanol–water partition coefficient (cm^3 g^{-1})
(for relationships between K_{oc} and K_{ow} see Fetter (1993)

$$R = 1 + \rho_b \cdot K_d / \varepsilon_a$$

R = Retardation factor (dimensionless)
ε_a = gas-filled soil porosity (dimensionless)

---

the velocity of the moving front (transported by advection) is a factor of 2 lower than calculated by the simple Darcy's equation.

In Table 5 typical retardation factors for methane, carbon dioxide, trichloroethylene (TCE), and o-xylene are calculated by use of the equations in Box 2. The table shows that methane is only slightly retarded; carbon dioxide is slightly more retarded, solely due to its high water solubility. TCE and o-xylene exhibit significant retardation.

Ward *et al.* (1992) studied the behaviour of trace organics in migrating landfill gas at a landfill site in the UK and found that the less lipophilic (and thereby less sorbable compounds) vinyl chloride and dichlorofluoromethane peak furthest away from the landfill (>50 m) compared with chloroethane and dichloromethane (30 m) and 1,2-dichloroethane (20 m).

Brusseau and Rao (1989) described kinetic effects on the sorption of organics in aquifer materials. They generally found that, for small scales and high flow velocities, kinetics could have a significant effect on the sorption. Similar conclusions might relate to the

TABLE 5. Calculation of Retardation Factors, _R_, for Different Landfill Gas Constituents Using the Equations given in Box 2

Component	Henry's law constant, K_H	$logK_{ow}$	K_{oc} (cm^3g^{-1})	K_d^{sat} (cm^3g^{-1})	K_d (cm^3g^{-1})	R
CH_4	29	1.1	19	0.19	0.009	1.06
CO_2	1.1	–	0	0	0.057	1.36
TCE	0.37	2.3	140	1.4	4.0	26.3
o-Xylene	0.21	3.1	527	5.3	25	161

$f_{oc} = 0.01$, $\varepsilon_w = 0.1$, $\varepsilon_a = 0.25$, $\rho_b = 1.6$ g cm^{-3}.
The estimation formula: $logK_{oc} = 0.72logK_{ow} + 0.49$ is used (Schwarzenbach and Westall, 1981)

sorption of organics in sandy top covers, where the transport is controlled by advection, owing to a relatively small scale (cover thickness of 1–2 m) and high flow velocities. No such research has, however, been reported in the literature.

Methane oxidation. The diffusion and dilution process that landfill gas undergoes while migrating through soil expose the methane to oxygen from the atmosphere. The methane may then be oxidized by methane-consuming (methanotrophic) bacteria by the reaction:

$$CH_4 + 2O_2 \rightleftharpoons CO_2 + 2H_2O + heat \qquad (8)$$

The reaction shows that a volume or pressure reduction takes place: 3 mol of gas are transformed to only 1 mol (plus water). The reaction is exothermic, and yields approx. 890 kJ/mol CH_4.

Many different strains of bacteria are able to oxidize methane,

TABLE 6. The average CH_4/CO_2 Ratio in Wells Grouped according to Distances from the Skellingsted Landfill (Kjeldsen and Fischer, 1995)

Distance from landfill border (m)	Average CH_4/CO_2 ratio	No. of single measurements
<0	1.86	35
0–60	1.73	270
60–100	0.06–1.17[a]	90

[a] The range of averages for each well is given, as large variations between wells were observed.

and many of these have been isolated from soils covering landfills (Jones and Nedwell, 1990; Nozhevnikova *et al.*, 1993a, b).

A decreasing CH_4/CO_2 ratio over time in a specific location can be used as an indicator of methane oxidation. The gas–water transfer processes are also able to make changes in the CH_4/CO_2 ratio, but on average the ratio tends to be increased by the transfer processes, owing to wash-out of CO_2, by infiltration.

Many workers have studied the influence of methane oxidation on the methane emission through top covers of landfills (Mancinelli and McKay, 1985; Jones and Nedwell, 1990, 1993; Whalen *et al.*, 1990; Nozhevnikova, 1993a, b; Figueroa, 1993; Bergman *et al.*, 1993). In contrast, only a few have studied the influence of methane oxidation on the lateral migration of landfill gas.

Kjeldsen and Fischer (1995) studied the lateral migration of landfill gas from a landfill site in Denmark. They monitored the soil gas composition over a 35-day period in more than 30 wells within a distance of 0–120 m from the landfill boundary. The wells were screened in depths ranging from 100 to 250 cm. They found decreasing average CH_4/CO_2 ratios with distance, as shown in Table 6, showing that methane oxidation effectively attenuates the methane. Similar results were found by Ward *et al.* (1992), who showed that the average CH_4/CO_2 ratio in samples taken below 2 m depth gradually decreased from 1.8 at the border of the landfill to 1.0 50 m away (Fig. 13). Kjeldsen and Fischer (1995) calculated the average CH_4/CO_2 ratios from wells screened less than 25 cm below ground surface and found, even close to the landfill, ratios less than 1. The results from the UK and the Danish landfill show that methane oxidation also plays an important role for the lateral spreading of methane from landfills.

The methane oxidation is controlled by different environmental factors (temperature, water content, nutrients, substrate concentration, etc.). Both Whalen *et al.* (1990) and Figueroa (1993) found that the optimal temperature for the oxidation rate was in the range 30–36 °C. Whalen *et al.* (1990) found that the rate was reduced to only 25% of the maximal rate at 5 °C; Figueroa (1993) found that the rate at 10 °C was only 5% of the rate at optimal temperature. Nozhevnikova (1993b) found that the methane oxidation rate at 6 °C was 2.5 times lower than that at 25 °C.

Whalen *et al.* (1990) and Figueroa (1993) also determined the optimal water content. For a sand till, Whalen *et al.* (1990) found that the optimal water content was 11% (w/w), probably due to a

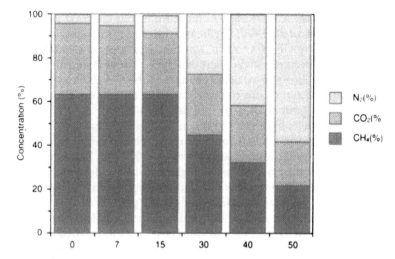

Figure 13. Average soil gas composition below 2 m depth as a function of distance, from a UK landfill (Ward *et al.*, 1992).

high diffusion rate and the absence of desiccation at this water content. Figueroa (1993) found, for a sand till and a humic soil, an optimal water content of 13 and 20% (w/w) respectively.

Bergman *et al.* (1993) found by investigating top covers at ten Swedish landfills that most of the top covers showing signs of methane oxidation, also had the highest nutrient contents.

Oxidation of methane produces heat. This has been observed at several landfill sites. Williams and Aitkenhead (1991) found temperatures higher than 18 °C down to 2 m below the surface. Temperatures higher than 50 °C have also been reported (Hill, 1986; Bergman *et al.*, 1993). At most places a desiccation of the soil was observed, owing to evaporation of water by the heat produced. The heating effects of methane oxidation could in some cases follow a cyclic pattern (Williams and Aitkenhead, 1991). The net loss of moisture leads to higher permeability of the soil, especially if desiccation cracks develop in the soil, enhancing the vertical transport of landfill gas, and counteracting the oxygen diffusion into the soil profile. The methane oxidation then stops and the temperature drops. After new infiltration of rainwater the cycle is repeated.

The gas volume reduction followed by methane oxidation could have an enhancing effect on the transport of air and landfill gas to the location where the oxidation takes place. The volume reduction (2 mol of oxygen plus 1 mol of methane gives 1 mol of carbon dioxide) leads to a pressure drop, which is compensated by advective flow of air from one side and landfill gas from the other. The pressure drop could further be enhanced by the higher water solubility of the produced carbon dioxide in comparison to oxygen and methane (see Box 3). Owing to the volume reduction. the soil gas concentration of nitrogen can exceed its normal air concentration (see Box 3). Observation of nitrogen enrichment is perhaps the most apparent indication of methane oxidation, a fact that is largely overlooked (Bergman *et al.*, 1993).

BOX 3. A Simple Example of the Effect of Methane Oxidation on Pressure Conditions and Gas Composition

---

1 l of soil in a closed vessel is considered. Atmospheric air and landfill gas is added to the vessel in such a ratio that both the methane and the oxygen are used up by methane oxidation. The total pressure in the vessel at the start is 1 atm.

Soil conditions
ε_w = water-filled soil porosity = 0.1
ε_a = gas-filled soil porosity = 0.3
ρ_b = bulk density of soil = 1.6 g cm^{-3}

Air	*Landfill gas*
80% N_2	60% CH_4
20% O_2	40% CO_2

The oxygen and methane content must be in a 2 : 1 ratio at the start. The total gas volume is 300 ml and is then composed of:

26 ml CH_4, 51 ml O_2, 17 ml CO_2, 206 ml N_2

After oxidation:

0 ml CH_4, 0 ml O_2, 43 ml CO_2, 206 ml N_2

A part of the CO_2 is transferred to the water phase to equilibrium (Henry's law constant = 1.1). The gas volume composition after equilibrium is reached:

0 ml CH_4, 0 ml O_2, 33 ml CO_2, 206 ml N_2

The vessel is closed. The final total pressure in the vessel is then:

(33 ml + 206 ml)/300 ml = 0.8 atm

The N_2 concentration in the gas phase is:

206 ml/239 ml = 86%

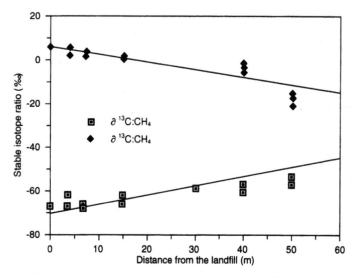

Figure 14. Changes in the stable isotope ratio of carbon (^{13}C/^{12}C) as a function of distance, from a UK landfill site (Ward *et al.*, 1992).

Another indication of methane oxidation is obtained by analysis of stable carbon isotopes (^{13}C/^{12}C) in the methane and associated carbon dioxide. Ward *et al.* (1992) showed at the Foxhall Landfill changes of the stable isotope ratios in both methane and carbon dioxide with distance. The changes for methane and carbon dioxide are approximately equal but opposite (see Fig. 14): methane becomes heavier, i.e. enriched in ^{13}C, while carbon dioxide becomes lighter, i.e. enriched in ^{12}C. This is consistent with carbon isotope fractionation, in which ^{12}C is preferentially metabolized by methane-oxidizing bacteria in the soil, thereby making residual methane heavier (Barker and Fritz, 1981). This is consistent with findings by Nozhevnikova *et al.* (1993b) in samples from a top cover on a Russian landfill.

Table 7 presents a summary of observed methane oxidation rates in soils sampled from landfill top covers. The results are highly variable, probably partly owing to very different experimental conditions in the four investigations.

Whalen *et al.* (1990) found a maximal methane oxidation rate for the top cover of 45 g CH$_4$ m^2 d^{-1}. By comparing the rate with normal methane production rates in landfills, Whalen found that up to 50%

TABLE 7. Summary of Observed Methane Oxidation Rates for Single Soil Samples or Integrated for a Soil Profile Representing the Total Landfill Top Cover

Reference	Soil	Temperature ($°C$)	Methane oxidation ($\mu gCH_4(gDS)^{-1}d^{-1}$)	Integrated methane oxidation ($gCH_4m^{-2}d^{-1}$)
Whalen *et al.* (1990)	Sandy till	5–36	20–70	13–45[a]
Jones and Nedwell (1993)	–	15–25	12–50[c]	0.2–7.2[b]
Nozhevnikova *et al.* (1993b)	Refuse	25	1900–9600	–
Figueroa (1993)	Till (disturbed)	20	115–770	–
	Humic soil (disturbed)	20	270–2100	
	Undisturbed samples	20	7–550	–

[a] For a 12 cm long soil profile.
[b] For a 36 cm long soil profile.
[c] A soil bulk density of 1.6 g cm^{-3} is used for recalculation.

of the methane produced can potentially be oxidized in the top cover. Understanding of the methane oxidation in soils surrounding or covering landfills is very important for proper estimations of the total emission of methane from landfills to the atmosphere.

Methane oxidation could be of major importance for the migration of halogenated aliphatics contained in landfill gas. Broholm *et al.* (1991) showed that certain chlorinated aliphatics were being co-metabolized in unsaturated soil under methane-oxidizing conditions. Similar observations could be expected in landfill gas-affected soil, but no such observations has been reported in the literature.

Other degradation processes. Nozhevnikova *et al.* (1993b) found, besides high rates of methane oxidation, that hydrogen and carbon monoxide were also oxidized with high rates in the landfill top cover. Hydrogen is found in trace levels in landfill gas, especially in the acidogenic phase of the anaerobic waste degradation process.

The variety of VOCs found in landfill gas could probably nearly all be degraded by different degradation processes. Degradation of chlorinated aliphatics by methanotrophic bacteria has already been mentioned. Some of these are also degradable under anaerobic conditions. In general it is known from groundwater pollution research that many organic compounds are degradable under different redox environments (Christensen *et al.*, 1994). Similar environments are very likely to exist in landfill gas-affected soil. However, the retention times for migrating landfill gas might be very much lower than the retention times relevant for groundwater. No specific research on the degradation of VOCs in landfill gas affected soil has been reported in the literature.

OBSERVED LANDFILL GAS EMISSIONS THROUGH TOP COVERS

Landfill gas emissions through top cover soil have been determined in several ways according to the literature. The emission has been measured directly by flux chambers, which are half-open chambers placed with the open end into the cover soil. Both static, closed-flux chambers and dynamic flux chambers are used. In the dynamic chamber, the gas volume in the chamber is continuously replaced by atmospheric air to simulate the real situation, where emitted gas is instantaneously diluted into the atmosphere (Reinhart and Cooper, 1992). In the static flux chamber a sample from the chamber volume is withdrawn after a relatively short period (15 min–3 h) to avoid influence on the emission rate from gas pressure build-up within the chamber (Rolston, 1986).

The emission rate has also been calculated from soil gas profiles measured in the top cover, assuming only diffusive flux (Bogner *et al.*, 1989; Bogner and Spokas, 1993; Bogner *et al.*, 1993). The Equations 1–3 were used for the calculations.

A summary of measured landfill gas emission rates is given in Table 8. Only maximal values found by Nozhevnikova (1993a, 1993b) are shown, but they found considerable spatial variation in emission rates. Jones and Nedwell (1990, 1993) found significantly higher emission rates in the summer than in the winter, owing to the lower water content of the top cover in the summer. In contrast, Bogner *et al.* (1993) found higher emission rates in the winter period, for some unexplained reasons.

Bogner and co-workers compared, in investigations carried out at two landfills, measured and calculated emission rates. Reasonable agreement between the two methods is seen. However, the calculations only take into account diffusional flow, and overlook any influence from advective flow or methane oxidation.

The investigations of methane emission from landfills only deal

TABLE 8. Summary of Observed Gas Emission Rates from Top Covers. All rates have been recalculated to gCH$_4$ m^{-2} d^{-1}

Reference	Landfill name	Landfill thickness (m)	Soil cover	Method	Time of observation	CH$_4$ emission rate ($gm^{-2}d^{-1}$)	CO$_2$ emission rate ($gm^{-2}d^{-1}$)
Nozhevnikova et al. (1993a,b)	Kuchino, Moscow	8–20	?	DFC[a]	?	33[d]	78[d]
	Ramenki, Moscow	2–5	?	DFC[a]	?	~0	10[d]
Jones and Nedwell (1990, 1993)	Martin Farm, UK	10–15	Miscell.	SFC[b]	May–October	15–77	n.a
					November–April	230[d]	
					Annually	4–11	n.a.
						22, 40[e]	n.a.
Bogner et al. (1989) Bogner (1990) Bogner and Spokas (1993)	Olinda, California	?	Sandy silt, unvegetated	SFC[b]	March	350–1900	9000–40 700
Bogner et al. (1993)	Mallard Lake, Illinois	6	Silty clay	CSP[c]	March	1660–1900	2700–2900
				SFC[b]	June–July	0.003[f]	n.a.
						11[g]	
				SFC[b]	August–January	21[g]	n.a.
				CSP[c]	June–July	0.023[f]	n.a.
						23[g]	
				CSP[c]	August–January	0.21[f]	n.a.
						220[g]	
Bogner et al. (1988)	Mallard North, Illinois	10–20	Clay	CSP[c]	April–July	app. 9–860	n.a.

[a] DFC = dynamic flux chamber.
[b] SFC = static flux chamber
[c] CSP = calculation from soil profiles.
[d] Maximum value.
[e] Annual average values for two locations.
[f] <5 m from gas recovery wells.
[g] >30 m from gas recovery wells.

with transport through the top covers. For landfills placed in gravel pits and covered with a clay top cover, significant lateral migration is expected. Observations of the methane emission through the soil surface of the adjacent areas of such landfills could be very valuable.

The importance of barometric pressure changes has been stressed previously, for instance by Latham and Young (1993). No field investigations on the influence of this effect on emission rates have, however, been found in the literature.

MODELLING OF GAS MIGRATION

Efforts have been made to describe the gas migration in soil by mathematical models ever since the early start of landfill gas migration research in the late 1970s. Table 9 gives a summary of the different models presented in the literature with respect to dimension (one- or two-dimensional models), method of solution (analytic or numeric) and time dependence (steady-state or dynamic model). The different processes included in each model, and the different factors that can be varied in the models, are also described.

Table 9 shows that the early models were quite complicated dynamic two-dimensional models. One type of model was developed at the Ohio State University, USA, and is described in Alzaydi *et al.* (1978), Moore (1979), Moore *et al.* (1979), Moore *et al.* (1982), and McOmber *et al.* (1982a, b). The model simulated methane migration using two-dimensional finite difference techniques. The porous medium was assumed to consist of a bundle of parallel uniform tubes of various radii. The equation of flow in a single capillary tube was applied to the individual tubes. Diffusion fluxes included the effect of Knudsen diffusion (see previous discussion) and intermolecular diffusion, and the advective flow (which was omitted in one of the models) was assumed to be Poiseuille flow. The fluxes were coupled using the so-called 'dusty gas' model.

The model was used to simulate the gas migration from a landfill in Ontario (McOmber *et al.*, 1982a) and from a landfill in Kentucky (McOmber *et al.*, 1982b). At the landfill in Ontario, the computer predictions of methane gas concentration in the region up to 25 m from the landfill agreed with field observations. However, beyond 25 m the computer predictions showed large discrepancies. At the landfill in Kentucky good agreement was found for steady-state

TABLE 9. Summary of Mathematical Models of Landfill Gas Migration

	Reference[a]											
	1	2	3	4	5	6	7	8	9	10	11	12
Dimension	2D	2D	2D	2D	2D	1D	1D	1D	1D	1D	1D	1D
Analytic/numeric	N	N	N	N	N	A	A	A	N	A	A	A
Steady/unsteady	U	U	U	U	U	S	U	S	U	S	U	S
Processes												
Diffusion	x	x	x	x	x	x	x	x	x	x	x	x
Advection	x	x	x	x	x			x	x	x	x	x
Dilution			x									
Dissolution					x		x					
Sorption							x				x	
Degradation											x	
Factors												
Gas production									x			
Pressure changes					x				x			
Water content			(x)			x						
Variable permeability				x								
Variable gas density	x	x	x									
Variable gas viscosity	x	x	x									

[a] (1) McOmber et al. (1982a); (2) Mohsen et al. (1978 & 1980); (3) Metcalfe and Farquhar (1987); (4) Moore et al. (1979, 1982); Moore (1979); (5) Latham and Young (1993); (6) Bogner and Spokas (1993), Bogner et al. (1993); (7) Little et al (1992); (8) Bergman et al. (1993); (9) Thibodeaux et al. (1982); (10) Herrera et al. (1989); (11) Ward et al. (1992); (12) Koo et al. (1991).

methane distributions. The model has, however, many adjustable parameters, which make the model validation rather questionable.

Other types of model were developed by the University of Waterloo, Canada. A finite element model, as described by Mohsen *et al.* (1978, 1980), was developed to simulate the migration of gases in soil from a buried landfill. The model is capable of incorporating soil anisotropy and inhomogeneity in an axisymmetric configuration. The model has facilities for including time-varying fluid properties and boundary conditions. The model is used to display the influence of seasons on the migration of gas. Metcalfe and Farquhar (1987) developed another finite element model by redefinition of a groundwater solute transport model including variable density conditions. The model accounts for gas migration due to gas pressure, concentration and velocity gradients. The model includes dilution effects introduced by dispersion, and is the only gas migration model that takes this process into account. The model successfully reproduced observed historical gas pressure and concentration data at two landfill sites.

The model by Latham and Young (1993) incorporates effects from barometric pressure changes and gas–water transfer. The model is very inadequately described by Latham and Young (1993), but seems promising in comparison with the older two-dimensional models, which are not able to consider the effect of barometric pressure changes and gas–water transfer.

Recently, a major effort has been made to develop simpler analytical one-dimensional models, which mainly simulate the migration of gas through landfill top covers. One model, which is described by several authors (Thibodeaux *et al.*, 1982; Herrera *et al.*, 1989; Koo *et al.*, 1991; Bergman *et al.*, 1993), is the steady-state one-dimensional model with both diffusive and advective transport. This model will be described in the following.

One-Dimensional Steady-State Diffusive Advective Transport Model

In this model no gas–water transfer, sorption or degradation in the soil are considered. The definitions for the model are shown in Fig. 15. The flux of a landfill gas constituent is described by

$$J_2 = \varepsilon_a D_2 \frac{dC_2}{dz} - v\, C_2 \tag{9}$$

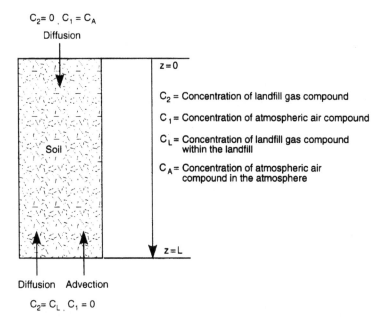

$C_2 = 0$, $C_1 = C_A$

Diffusion

z = 0

C_2 = Concentration of landfill gas compound

C_1 = Concentration of atmospheric air compound

C_L = Concentration of landfill gas compound within the landfill

C_A = Concentration of atmospheric air compound in the atmosphere

Soil

z = L

Diffusion   Advection

$C_2 = C_L$, $C_1 = 0$

Figure 15. Gas flow directions and boundary conditions for the one-dimensional steady state diffusive-advective migration model (after Bergman *et al.*, 1993).

where D_2 is the soil diffusion coefficient of the landfill gas constituent ($m^2 \, s^{-1}$), and v is the Darcy velocity in the soil ($m \, s^{-1}$). The flux of an atmospheric air constituent is

$$\mathcal{J}_1 = \varepsilon_a \, D_1 \, \frac{dC_2}{dz} - v \, C_1 \tag{10}$$

where D_1 is the soil diffusion coefficient of the atmospheric air constituent ($m^2 \, s^{-1}$). With the boundary conditions given in Fig. 15 the solution is

$$\mathcal{J}_2 = v \, C_L \, \frac{\exp\left(\dfrac{Lv}{\varepsilon_a D_2}\right)}{\exp\left(\dfrac{Lv}{\varepsilon_a D_2}\right) - 1} \tag{11}$$

The concentration profile for the landfill gas constituent is

$$C_2\,(z) = C_L \frac{\exp\left(\dfrac{zv}{\varepsilon_a D_2}\right) - 1}{\exp\left(\dfrac{Lv}{\varepsilon_a D_2}\right) - 1} \tag{12}$$

The concentration profile for the atmospheric air constituent is

$$C_1\,(z) = C_A \frac{\exp\left(\dfrac{Lv}{\varepsilon_a D_1}\right) - \exp\left(\dfrac{zv}{\varepsilon_a D_1}\right)}{\exp\left(\dfrac{Lv}{\varepsilon_a D_1}\right) - 1} \tag{13}$$

The soil diffusion coefficients can be calculated if the water-filled porosity, ε_w, is known by use of Equations 2 and 3 as presented earlier. The Darcy velocity can be estimated by use of the previously presented Equations 4 and 5:

$$v = \frac{k\,\varepsilon_a}{\mu\,\varepsilon} \cdot \frac{p_L - p_A}{L} \tag{14}$$

where p_L and p_A are the pressure in the landfill and in the atmosphere (Pa) respectively.

Concentration profiles for a landfill gas constituent are shown in Fig. 16 for three different soil types. The landfill concentration is set to 100%. See Bergman *et al.* (1993) for details of the parameters chosen.

Problems in Modelling Landfill Gas Migration

Comparing landfill gas migration with, for instance, contaminant transport in groundwater, it appears that the gas migration is much more complicated: it involves flow of two phases (gas and water), and is highly dynamic, with barometric pressure changes and varying methane oxidation as important factors. In addition, lateral migration will often be a three-dimensional problem. Considering the variety of migration models described in the literature, a comprehensive model to be used for prognostic simulations has still to be developed. It is important to stress that the data requirement for the use of a comprehensive model is very high and costly, and that computational speed could also be a limiting factor.

For description of migration through top covers the models available seem much more promising, owing to the smaller scale and

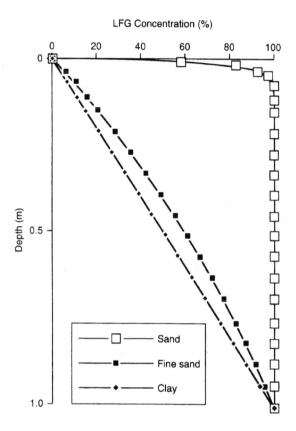

Figure 16. Soil profiles for a landfill gas constituent calculated by the one-dimensional steady state diffusive-advective migration model. Concentration below the top cover is set to 100% (after Bergman *et al.*, 1993).

relatively homogeneous soil properties. Models accounting for methane oxidation, however, have not been developed.

CONCLUSIONS

The migration of landfill gas potentially leads to many different environmental effects. The effects are related to different scales: from local explosion hazards to contributions to global climate changes.

The gas migration is governed by several processes and factors in an often complicated network of connections. The transport of landfill gas is controlled by both diffusive and advective forces. Which of the two forces dominates in a specific case can be very difficult to evaluate, as only very small pressure gradients are needed to give significant advective transport in some cases.

Some of the most important factors for the migation of methane, besides the gas production intensity, are barometric pressure changes and methane oxidation. In many cases, the barometric pressure changes control, in combination with other factors (high permeable layers, ground surface sealing by rainwater or snow), the maximal migration distance. This effect makes the migration very dynamic and difficult to predict. However, effects of landfill gas in terms of vegetation damage and explosions have rarely been observed at distances from the landfill greater than 300 m.

Methane oxidation significantly decreases the overall methane emission to the atmosphere. The oxidation process is in some cases very effective in landfill top covers. The importance of methane oxidation to the emission through soil surfaces of adjacent areas affected by lateral migration is only partly understood.

The migration of the volatile organic compounds present in the landfill gas is, besides the factors controlling the main constituents, controlled by sorption and degradation processes. Hardly any studies on this subject are reported in the literature, but by transferring findings from similar groundwater environments, a significant effect from sorption and degradation on the migration of organic compounds is expected. The retention times in the gas-affected soil systems are, however, in many cases much lower than for polluted aquifers.

The lateral migration of landfill gas currently is very difficult to simulate or predict by mathematical models, owing to the complicated dynamics of the problem, high data requirements and possibly demands for computational power.

REFERENCES

Alzaydi, A. A., Moore, C. A. and Rai, I. S. (1978) Combined pressure and diffusional transition region flow of gases in porous media. *American Institute of Chemical Engineers Journal,* **24,** 35–42.

Barker, J. F. and Fritz, P. (1981) Carbon isotope fractionation during microbial methane oxidation. *Nature,* **293,** 289–291.

Bergman, H., Jacobsson, A. and Lagerkvist, A. (1993) Investigations regarding biofilters for methane oxidation in landfill top covers, in *Proceedings Sardinia 93*, Fourth International Landfill Symposium, S. Margherita di Pula, Cagliari, Italy, 11–15 October 1993. CISA, Cagliari, Italy, 717–728.

Bogner, J., Rose, C., Vogt, M. and Gartman, D. (1988) Understanding landfill gas generation and migration, in *Proceedings of the 11th Annual International Landfill Gas Symposium*, 21–24 March, Houston, Texas. GRCDA, Silver Spring, MD, USA, 225–242.

Bogner, J., Vogt, M. and Piorkowski, R. (1989) Landfill gas generation and migration: review of current research II, in *Proceedings Anaerobic Digestion Review Meeting*, 25–26 January, Golden, Colorado, USA. SERI/SP-231-3520, DE89 009453, Solar Energy Research Institute, Golden, Colorado, USA, 139–157.

Bogner, J. (1992) Anaerobic burial of refuse in landfills: increased atmospheric methane and implications for increased carbon storage. *Ecological Bulletins*, **42**, 98–108.

Bogner, J. and Spokas. K. (1993) Landfill CH_4: rates, fates, and role in global carbon cycle. *Chemosphere*, **26**, 369–386.

Bogner, J., Spokas. K. and Jolas. J. (1993) Comparison of measured and calculated methane emissions, in *Proceedings Sardinia 93*, Fourth International Landfill Symposium, S. Margherita di Pula, Cagliari, Italy, 11–15 October 1993. CISA, Cagliari, Italy, 605–616.

Broholm, K., Christensen, T. H. and Jensen, B. J. (1991) Laboratory feasibility studies on biological in-situ treatment of a sandy soil contaminated with chlorinated aliphatics. *Environmental Technology*, **12**, 279–289.

Brusseau, M. L. and Rao, P. S. C. (1989) Sorption nonideality during organic contaminant transport in porous media. *CRC Critical Reviews in Environmental Control*, **19**, 33–99.

Campbell, D. (1989) Landfill gas migration, effect and control, in *Sanitary Landfilling: Process, Technology and Environmental Impact* (eds T. H. Christensen, R. Cossu and R. Stegmann), Academic Press, London, xxxv-1–xxxv-28.

Christensen, T. H., Kjeldsen, P., Albrechtsen, H.-J., *et al.* (1994) Attenuation of pollutants in landfill leachate polluted aquifers. *Critical Reviews in Environmental Science & Technology*, **24**, 119–202.

Farquhar, G. J. and Metcalfe, D. E. (1982) Modelle zur Gasmigration, in *Gas und Wasserhaushalt von Mülldeponien*, Heft 33, Veröffentlichungen des Instituts für Stadtbauwesen, Technische Universität Braunschweig, Braunschweig, Germany (in German), 393–418.

Fetter, C. W. (1993) *Contaminant Hydrology*, Macmillan Publishing Company, New York, p. 458.

Figueroa, R. A. (1993) Methane oxidation in landfill top soils, in *Proceedings Sardinia 93*, Fourth International Landfill Symposium, S. Margherita di Pula, Cagliari, Italy, 11–15 October 1993. CISA, Cagliari, Italy, 701–716.

Franklin, J. D. and Vetter, R. J. (1987) Emergency response to landfill gas migration. *Public Works*, (May), 84–85.

Gendebien, A., Pauwels, M., Constant. M. *et al.* (1992) *Landfill gas – From Environment to Energy*, Commission of the European Communities, Directorate-General Telecommunications, Information Industries and Innovation, Luxembourg.

Ghabaee, K. and Rodwell, W. R. (1989) *Landfill gas modelling – A literature survey of landfill gas generation and migration*, Petroleum Reservoir Technology Division, Winfrith Technology Centre, Winfrith, Dorchester, Dorset, UK.

Herrera, T. A., Lang, R. and Tchobanoglous, G. (1989) A study of the emissions of volatile organic compounds found in leachate gas, in *43rd Purdue Industrial Waste Conference Proceedings*, Lewis Publishers, Chelsea, MI, USA, 229–238.

Hill, C. P. (1986) Landfill gas migration from operational landfill sites – monitoring and prevention. *Wastes Management* (April), 169–178.

Hoather, H. A. and Wright, P. A. (1989) Landfill gas: site licensing and risk assessment, in *Landfill Gas and Anaerobic Digestion of Solid Wastes*, (eds Y. R. Alston and G. E. Richards), Proceedings Conference Chester, UK. Harwell Laboratories, UK.

Hodgson, A. T., Garbesi, K., Sextro, R. G. and Daisey, J. M. (1992) Soil-gas contamination and entry of volatile organic compounds into a house near a landfill. *Journal of Air & Waste Management Association*, **42**, 277–283.

Jones, D. L., Crowcroft, P. and Pritchard, B. N. (1988) Design of a motorway service station on a landfill site, in *Proceedings of the 11th Annual International Landfill Gas Symposium*, 21–24 March, Houston, Texas. GRCDA, Silver Spring, MD, USA, 32–63.

Jones, H. A. and Nedwell, D. B. (1990) Soil atmosphere concentration profiles and methane emission rates in the restoration cover above landfill sites: equipment and preliminary results. *Waste Management & Research*, **8**, 21–31.

Jones H. A. and Nedwell, D. B. (1993) Methane emission and methane oxidation in land-fill cover soil. *FEMS Microbiology Ecology*, **102**, 185–195.

Kerfoot, H. B. (1993) Landfill gas effect on groundwater samples at a municipal solid waste facility, in *Proceedings Sardinia 93*, Fourth International Landfill Symposium, S. Margherita di Pula, Cagliari, Italy, 11–15 October 1993. CISA, Cagliari, Italy, 1141–1154.

Kjeldsen, P. (1993) Evaluation of gas diffusion through plastic materials used in experimental and sampling equipment, *Water Research*, **27**, 121–131.

Kjeldsen, P. and Fischer, E. (1995) Landfill gas migration – field investigations at Skellingsted Landfill, Denmark. *Waste Management & Research*, **13**, 467–484.

Koo, J. K., Yoon, S. P., Shin, H. S., *et al.* (1991) Emission and control of VOC under influence of landfill gas flow, in *Proceedings Sardinia 91*, Third International Landfill Symposium, S. Margherita di Pula, Cagliari, Italy, 14–18 October 1991. CISA, Cagliari, Italy, 179–188.

Lagerkvist, A. (1986) Om nedbrytnings- och transportprocesser i avfallsupplag. PhD Thesis, University of Luleå, Luleå, Sweden, p.129 (in Swedish).

Latham, B. and Young, A. (1993) Modelization of the effects of barometric pressure on migrating landfill gas, in *Proceedings Sardinia 93*, Fourth International Landfill Symposium, S. Margherita di Pula, Cagliari. Italy, 11–15 October 1993. CISA, Cagliari, Italy, 681–690.

Little, J. C., Daisey, J. M. and Nazaroff, W. W. (1992) Transport of subsurface contaminants into buildings. *Environmental Science & Technology*, **26**, 2058–2066.

Luning, L. and Tent, J. (1993) Gaseous emission of landfill sites, in *Proceedings Sardinia 93*, Fourth International Landfill Symposium, S. Margherita di Pula, Cagliari, Italy, 11–15 October 1993. CISA, Cagliari, Italy, 657–668.

Mancinelli, R. L. and McKay, C. P. (1985) Methane-oxidizing bacteria in sanitary landfills, in *Proceedings First Symposium on Biotechnological Advances in Processing Municipal Wastes for Fuels and Chemicals*, August, Minneapolis, MN, USA (ed. A. Antonopoulos), Argonne National Laboratory Report, pp. 437–450.

Metcalfe, D. E. and Farquhar, G. J. (1987) Modelling gas migration through unsaturated soils from waste disposal sites. *Water, Air and Soil Pollution*, **32**, 247–259.

McOmber, R. M., Moore C. A. and Beatty, B. W. (1982a) Field evaluation of methane migration predictions. *Canadian Geotechnical Journal*, **19**, 239–249.

McOmber, R. M., Moore, C. A., Massmann, J. W. and Walsh, J. J. (1982b) Verification of gas migration at Lees Lane Landfill, in *Land Disposal of Hazardous Waste*. Proceedings 8th Annual Research Symposium, Ft Mitchell, Kentucky, USA, 150–160.

Millington, R. J. and Quirk, J. P. (1961) Permeability of porous solids. *Transactions of the Faraday Society*, **57**, 563–568.

Mohsen, M. F. N., Farquhar, G. J. and Kouwen. N. (1978) Modelling methane migration in soil. *Applied Mathematical Modelling*, **2**, 294–301.

Mohsen, M. F. N., Farquhar, G. J. and Kouwen. N. (1980) Gas migration and vent design at landfill sites. *Water, Air and Soil Pollution*, **13**, 79–97.

Moore, C. A. (1979) Landfill gas generation, migration, and control. *CRC Critical Reviews in Environmental Control*, **9**, 157–183.

Moore, C. A., Rai, S. and Alzaydi, A. A. (1979) Methane migration around sanitary landfills. *ASCE, Journal of the Geotechnical Engineering Division*, **105**, 131–144.

Moore, C. A., Rai, S. and Lynch, A. M. (1982) Computer design of landfill methane migration control. *ASCE, Journal of the Geotechnical Engineering Division*, **108**, 89–106.

Nozhevnikova, A. N., Nekrasova, V. K., Lebedev, V. S. and Lifshits, A. B. (1993a) Microbiological processes in landfills. *Water Science & Technology*, **27**, 243–252.

Nozhevnikova, A. N., Lifshits, A. B., Lebedev, V. S. and Zavarzin, G. A. (1993b) Emission of methane into the atmosphere from landfills in the former USSR. *Chemosphere*, **26**, 401–417.

O'Leary, P. and Tansel, B. (1986) Landfill gas movement, control and uses. *Waste Age*, (April), 104–116.

Petersen, T. N. (1988) Human health risk assessment of landfill gas emissions, in *Proceedings 11th Annual Madison Waste Conference – Municipal & Industrial Waste*, 13–14 September. Department of Engineering Professional Development, University of Wisconsin-Madison, Madison, WI, USA, 164–177.

Pirkle, R. J., Wyatt, D. E. and Looney, B. B. (1993) The effect of barometric pumping on the migration of volatile organic compounds from the vadose zone into the atmosphere, in *Proceedings National Symposium on Measuring and Interpreting VOCs in Soils: State of the Art and Research Needs*, 12–14 January, Las Vegas, Nevada. USEPA, Las Vegas, Nevada.

Raybould, J. G. and Anderson, D. J. (1987) Migration of landfill gas and its control by grouting – a case story. *Quarterly Journal of Engineering Geology*, **20**, 75–83.

Reinhart, D. R., Cooper. D. C. and Walker, B. L. (1992) Flux chamber design and operation for the measurement of municipal solid waste landfill gas emission rates. *Journal of the Air Waste Management Association*, **42**, 1067–1070.

Rettenberger, G. (1985) Gasförmige Emissionen aus Abfalldeponien im Hinblick auf Umweltrelevante Schadstoffe, in *Umwelteinflüsse von Abfalldeponien und Sondermüllbeseitigung*, Proc. Mülltechnisches Seminar und Fachgespräch Sondermüll, November 1984, München, Germany, 40–56 (in German).

Rolston, D. E. (1986) Gas diffusivity, in *Methods of Soil Analysis, Part I, Physical and Mineralogical Methods*, Agronomy Monographs No. 9, 2nd edn, American Society of Agronomy, Soil Science Society of America, Madison, Wisconsin, USA, 1089–1102.

Sallam, A., Jury, W. A. and Letey, J. (1984) Measurement of gas diffusion coefficient under relatively low air-filled porosity. *Soil Science Society of America, Journal*, **48**, 3–6.

Schwarzenbach, R. P. and Westall, J. (1981) Transport of nonpolar organic compounds from surface water to groundwater. Laboratory sorption studies. *Environmental Science & Technology*, **15**, 1360–1367.

Stephens, R. D., Ball, N. B. and Mar, D. M. (1986) A multimedia study of hazardous waste landfill gas migration, in *Pollutants in a Multimedia Environment*, (ed. Y. Cohen), Plenum Press, New York, USA, pp. 265–287.

Thibodeaux, L. J., Springer, C. and Riley, L. M. (1982) Models of mechanisms for the vapor phase emission of hazardous chemicals from landfills. *Journal of Hazardous Materials*, **7**, 63–74.

Thomas, S. and Salmon, P. (1993) Lateral migration of biogas in an old landfill site, in *Proceedings Sardinia 93*, Fourth International Landfill Symposium, S. Margherita di Pula, Cagliari, Italy, 11–15 October 1993. CISA, Cagliari, Italy, 691–700.

Ward, R. S., Williams, G. M. and Hills, C. C. (1992) *Changes in gas composition during migration from the Foxhall Landfill, Suffolk*, Technical Report WE/92/22, Fluid Processes Group, British Geological Survey, Nottingham, UK.

Whalen, S. C., Reeburgh, W. S. and Sandbeck. K. A. (1990) Rapid methane oxidation in a landfill cover soil. *Applied and Environmental Microbiology*, **56**, 3405–3411.

Williams, G. M. and Aitkenhead, N. (1991) Lessons from Loscoe: the uncontrolled migration of landfill gas. *Quarterly Journal of Engineering Geology*, **24**, 191–207.

Wittmann, S. G. (1985) Landfill gas migration: early warning signs, monitoring techniques and migration control systems, in *Proceedings 23rd Annual International Seminar, Equipment, Services, and Systems Show*, 27–29 August, Denver, Co. GRCDA, Silver Spring, MD, USA, 317–328.

Wood, J. A. and Porter, M. L. (1987) Hazardous pollutants in class II landfills. *Journal of the Air Pollution Control Association*, **37**, 609–615.

Young, P. J. and Parker, A. (1983) The identification and possible impact of trace gases and vapours in landfill gas. *Waste Management & Research*, **1**, 213–226.

3.2 Explosion and Fire Hazards Associated with Landfill Gas

DAVID J. V. CAMPBELL

AEA Technology, National Environmental Technology Centre, E6 Culham, Abingdon, Oxfordshire OX14 3DB, UK

INTRODUCTION

'Landfill gas' is a collective term used to describe the complex mixture of gaseous components evolved during the degradation and stabilization of waste materials deposited in landfill sites. In Chapter 2 detailed discussion identified the various mechanisms and factors that control the production and composition of landfill gas. Emissions of methane-rich landfill gas, whether observed directly from surface environments or indirectly via subsurface lateral migration beyond site boundaries, can be a source for both fire and explosion hazards, thus adversely impacting on persons and property. While most of the following discussion will concentrate on the fire or explosive potential and hazards associated with methane-rich gaseous emissions and atmospheres, the production of hydrogen gas, as generated in the early stages of organic waste fermentation, should not be ignored. Concentrations of up to 20% hydrogen have been recorded in otherwise largely carbon dioxide gas mixtures.

As wastes degrade microbially, and/or when chemical interactions take place, gases are generated or released, which result in a build-up of pressure within the deposits. Pressure gradients, both within sites and across site boundaries, result in gas migration away from the source, both to the surface of the site and laterally into strata beyond site boundaries. In many modern landfill sites low-permeability barriers and installed gas abstraction/control systems

Landfilling of Waste: Biogas. Edited by T. H Christensen, R. Cossu and R. Stegmann. Published in 1996 by E & FN Spon, London.
ISBN 0 419 19400 2.

restrict migration within the confines of the site, but in many former sites no such barriers or controls existed. Substantial temperature gradients are also often established between sites and adjacent rock strata, and where rapid degradation is occurring the high temperatures present may also be a significant cause of migration even where pressure gradients are very small.

In recognizing that a very wide spectrum of gaseous component concentrations may exist at the source of production, the component concentrations as measured at the monitoring or emission point more remote from the source will equally be very variable. More importantly, the measured concentrations and ratios of some components may bear little or no relationship to those previously measured at the source. Many different factors can be responsible for the changing ratios and concentrations, including dilution, methane oxidation, absorption/adsorption, and the presence and/or mixing of landfill gas with other non-landfill gas sources, e.g. marsh gas, mines gas or natural gas.

Measured concentrations may also vary widely over time, and sometimes over very short time intervals, owing to the influences of such factors as climate change (pressure, rainfall, temperature, wind speed and direction), subsurface groundwater/leachate level changes, and adjacent man-made activities (e.g. mines gas pumping).

As a consequence of these, and no doubt many other, site-specific factors, the identification of specific gas component concentrations and subsequent interpretation of the potential risks due to them must be treated with considerable caution. If little or no gas is observed on one single monitoring occasion, this does not mean that the same situation will apply during the next sampling visit.

FLAMMABILITY AND EXPLOSION LIMITS

As previously intimated, landfill gas mixtures may contain an almost infinitely variable range of component concentrations. Thus while concentrated landfill gas mixtures, generated from 'maturing' waste, will generally contain mostly methane (50–60%) and carbon dioxide (50–40%) along with many other trace gas components, any dilution of that gas or mixing with gas generated from less mature waste can substantially alter the concentrations or ratios of the principal components present. It is important to recognize that at monitoring locations increasingly removed from the original source of the gas, high methane concentrations *will* become depleted and fall within the lower flammability/explosive range concentrations at some point following air dilution.

The most commonly accepted flammability ranges (Fig. 1) for methane in air mixtures are 5.3–14% (by volume in air) where normal pressures are maintained during the passage of the flame (Coward and Jones, 1952). The flammability range becomes slightly extended to 5.0–15% when mixtures of methane in air are retained within a small vessel or void, such as might be applicable to gas collected within an enclosed void within buildings. Flammability implies that the homogeneous and combustible gas–air mixture can propagate a flame freely within a limited range of composition. Such a mixture will be explosive when contained in a small volume from which the products of combustion cannot readily escape.

Methane gas potential for flammability or explosivity becomes

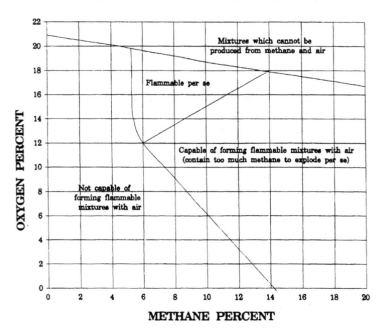

Figure 1. Relationship between quantitative composition and flammability of mixtures of methane, air and nitrogen. (Reproduced from Coward and Jones, 1952).

much more complex in mixtures containing other gases than normal air components, including carbon dioxide and/or nitrogen of variable concentration when present with or without oxygen (from air dilution). Thus no flammable mixtures can exist where nitrogen or carbon dioxide is the principle diluent of methane present in the mixture *and* where oxygen levels fall below about 12.8%. This is particularly relevant where landfill gas migration may have occurred and where, through methane oxidation for example, the oxygen concentrations may be negligible in a gas mixture that nevertheless contains methane in a potentially flammable concentration. Of course, as the gas subsequently mixes with air following surface emission, a flammable hazard may be reintroduced. Hydrogen flammability ranges (4–74% by volume in air) similarly require minimum concentrations of oxygen to be present (5% with nitrogen as diluent, and 5.9% with carbon dioxide as diluent).

While situations are unlikely to involve methane gas (alone) from landfills being present in mixtures containing only pure oxygen, it should be noted that the flammability range alters markedly to 5.1–61% (hydrogen 4.0–94%).

Ignition

For a gas mixture to be flammable or explosive there is a need to establish a suitable ignition source. With methane–air mixtures within the flammability range, the necessary energy provided by the ignition source will be variable depending on the methane concentration. As indicated in Fig. 2 a spark with an energy of about 0.3 mJ can ignite a flammable mixture only when containing methane concentrations close to 8.5%, while an energy of greater than 1 mJ would be required for concentrations of methane below 6% *or* above 11.5%. As the figure indicates, a flame may be propagated beyond the concentration limits indicated for ignitability, i.e. up to the flammability limits.

TYPES OF HAZARD

The types of gas-related hazard that may arise at or in the vicinity of landfill sites are many and varied. They are, however, very site

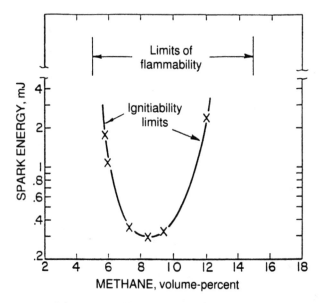

Figure 2. Ignitability curve and limits of flammability for methane–air mixtures at atmospheric pressure and 26 °C. (Reproduced from Zabetakis, 1965).

specific, given the wide diversity of site types, locations and waste deposits and ages that exist. In general, problems and hazards will be most likely at former landfills where few if any specific controls were introduced and designed to prevent migration and emissions. Figure 3 provides an indication of the potential locations and hazards that may exist at landfill sites. While an hypothetical site is depicted as being located within a former mineral excavation void, the issues identified might equally apply to a mounded landfill located solely above ground levels. Some 19 separate locations in, on and/or around the site are shown where a hazard could exist and/or where monitoring may be necessary to ascertain the extent and concentrations of migrating gas. This single illustrative example demonstrates that the following may be of concern at any individual site:

- gas migration from wastes via porous or fissures strata towards property, services, or valued vegetation, crops etc.;
- gas entering property (basements, cavity walls etc.) to collect in confined spaces where an ignition source may exist;
- gas collecting in building services, e.g. telephone, electricity,

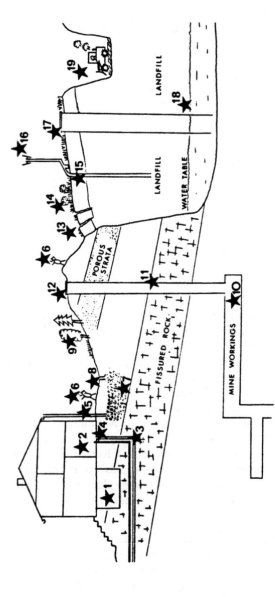

Figure 3. Landfill gas monitoring requirements (Campbell and Young, 1985): 1, gas in basements; 2, gas in buildings; 3, interception of gas by underground services; 4, entry of gas in services; 5, gas in gutter drainage pipes; 6, odour nuisance from gas; 7, collection of gas by soakaways; 8, gas in land drains; 9, vegetation damage (off site); 10, irrigation through mine workings; 11, interception of gas by mine shafts; 12, gas in mine shafts; 13, gas release through cracked capping; 14, vegetation damage (restored site); 15, release of gas around or via boreholes; 16, efficiency of gas flaring; 17, gas in manhole wells; 18, gas pressure affected by water-table; 19, health and flammability risks from gas in confined spaces or poorly ventilated areas on site. Monitoring point locations: site layout (monitoring points outside the area of investigation are omitted for clarity).

water, sewers, which have not been sealed from the strata, or at the entry point into property;
- gas collecting in rainwater collection pipes linked to soakaways (in commercial premises, non-sealed rainwater piping has been known to be conduited *inside* building structures!);
- gas migrating into adjacent man-made subsurface workings or cavities, e.g. mine workings;
- gas emissions at site boundaries, especially via fissures/faults in site capping (due to settlement);
- gas emissions from or around monitoring boreholes or freely venting on-site wells;
- gas emissions from faulty pipework associated with gas migration/recovery/flare plants;
- gas emissions through site surfaces causing damage to vegetation and giving rise to risks to on-site buildings or equipment.
- Each of the above sources of gas emissions, especially in the case of concentrated gas plumes, may equally give rise to health risks associated with toxicity thresholds of some landfill gas components.

The above does not attempt to provide an exhaustive list of the risks or hazards of landfill gas and potential sources for fire or explosion. The installation, at most modern landfills, of comprehensive gas-monitoring plans and facilities, and management systems to recover and utilize or flare generated gas, will undoubtedly minimize if not totally eliminate such risks. Nevertheless all due care must be taken, as would be required with the use of natural gas for domestic heating or cooking, to adhere to all necessary health and safety requirements, and thus to treat landfill gas with caution.

CASE STUDIES

In identifying below a few typical examples of the problems and hazards that have been encountered in the UK, it is perhaps pertinent to emphasize that almost all recorded incidents have been associated with former sites where, prior to the incident, little or no monitoring or recognition of a concern had been noted, even less where appropriate control measures had been introduced. It is also important to recognise that the term 'incident', as often used in connection with many recorded gas problems (about 100 or so recorded cases exist in the UK alone over the last two decades), usually refers only to the *potential* for observed migrating gas to give rise to a risk of fire or explosion, or possibly a lesser problem associated with vegetation/crop damage. In other words, most examples of incidents are simply records of observed gas migration identified through monitoring programmes, rather than examples of actual harm or damage caused. Despite this there are several salutatory lessons that can be learned from some genuine incidents, perhaps more correctly termed accidents, which have occurred and have been reported from several countries in Europe and North America. On a worldwide basis about 55 reports exist on explosion damage caused by landfills (Gendebien *et al.*, 1992). Below, four typical UK examples are presented.

Case 1

This incident, generally referred to as the 'Loscoe Explosion', took place in 1986 at a site in Derbyshire, England. A bungalow, one of many properties surrounding a significant and recently completed landfill site, blew up as a consequence of the ignition of landfill gas, which had migrated through fissured strata and had collected in the building close to a central heating boiler. The bungalow was completely destroyed, and the two occupants, though hurt, were fortunate to walk from the building.

Following an inquiry into the incident the cause was related to several issues. The site had relatively recently been capped with low-permeability clay materials, which had reduced the opportunity for continued surface emissions of gas and had thus further encouraged lateral migration beyond site boundaries. An extreme climatic event – a very low pressure zone – had occurred just prior to the incident, and this was believed to have contributed significantly by aiding surface release of a gas reservoir, which had collected in subsurface strata during prior migration. A further important issue identified was that while gas emissions in the vicinity of the property had previously been observed, they were not at the time attributed to the nearby landfill but to probable coal seam gas. More particularly, once this source had been discounted, little attention was paid to identification of alternative sources and hence to early prevention of further emissions.

Case 2

In April 1979 an explosion occurred in the ladies' changing room of an indoor bowling club and squash courts built over and adjacent to former landfilled areas. The buildings were constructed on reinforced concrete floors supported on piles and ring beams. No specific measures had been taken to vent the underside of the floor slab. The explosion was caused by a lighted cigarette, and resulted in severe burns to a lady's face as well as causing the collapse of an internal concrete block wall. Significant concentrations of methane, in excess of 1% gas, were subsequently measured beneath and in the vicinity of the building, with methane concentrations up to 0.06% recorded inside shower facilities and in the squash court areas.

Case 3

A large landfill site with a capacity of some 1.7×10^6 m^3 was constructed in a disused opencast colliery which was filled, between 1978 and 1986, with both high-density baled domestic waste and non-baled commercial and non-toxic industrial waste. The southern and eastern boundaries of the site were surrounded by houses some 50 m from the waste limits. During the mid-1980s significant concentrations of gas were detected both in the gardens and in property services. It was thought initially that this gas was related to a leaking natural gas main. However, gas was confirmed to originate from the nearby landfill, and necessitated the installation, following extensive monitoring and on-site pumping trials, of a comprehensive gas migration control system via a series of gas wells located in waste close to site boundaries.

Case 4

In November 1980, two electricians were injured following an explosion in a dry, shallow manhole chamber. This chamber was close to a further manhole used as a sump from which leachate was extracted via a submersible pump. The electricians were working on the pump's electrical supply switchgear when a spark was generated between electrical contacts and ignited accumulated methane gas. This incident emphasizes the need for proper codes of practice when working in confined spaces as well as the need to isolate electrical switchgear from potential sources of gas.

CONCLUSION

Explosion and fire hazards associated with landfill gas production and emissions from landfill sites are relatively rare occurrences. While in the past this has perhaps been due more to good luck than to good management, the issues and problems of landfill gas are now well recognized. There should be little or no cause to add to the small but nevertheless serious catalogue of former incidents.

REFERENCES

Campbell, D. J. V. and Young, P. J. (1985). Landfill monitoring – landfill gas, in *Landfill Monitoring Symposium Proceedings*, Harwell, May 1985. Harwell Laboratory, UK.

Coward, H. F. and Jones, G. W. (1952). *Limits of Flammability of Gases and Vapours*, Bulletin 503, Bureau of Mines, US Department of the Interior, US Government Printing Office.

Gendebien, A., Pauwels, M., Constant, M. *et al.* (1992). *Landfill Gas. From Environment to Energy*, Office for Official Publications of the European Communities, Luxembourg.

Zabetakis, M. G. (1965). *Flammability Characteristics of Combustible Gases and Vapours*, Bulletin 627, Bureau of Mines, US Department of the Interior, TN 23.0.4, No 627 622.06173.

3.3 Health Aspects of Gaseous Emissions from Landfills

THOMAS F. EIKMANN

*Institute of Hygiene and Environmental Medicine,
Justus-Liebig-University, Friedrichstrasse 16, D-35385 Giessen,
Germany*

INTRODUCTION

People living in the neighbourhood of landfill sites often complain of being irritated by odours and gases emitted from these sites. The local residents are frequently afraid that the exposure to landfill gas represents not only a potential environmental nuisance but also a serious toxic hazard. It is the responsibility of the authorities and of experts to assess the possible intake of toxic substances by the exposed persons and to calculate the real health risk of the population in the neighbourhood of landfill sites.

The main components of landfill gas, such as carbon dioxide and hydrogen sulphide, are potentially toxic to humans (see Chapter 2.4, and Gendebien *et al.*, 1992), but most concern is generally assumed to be related to organic trace components in the gas.

Experience has shown that exposure concentrations of toxic substances in air relevant to health arise especially when the landfill gas emissions can reach confined spaces where the available renewable air is limited (Gendebien *et al.*, 1992). The results of several epidemiological and other medical studies provide a survey of the possible qualitative and quantitative health effects on man of volatile contaminants from waste disposal sites. On the basis of the results of these studies it is possible to show which of the different intake paths are relevant to human health and which kinds of effects are to be expected.

Landfilling of Waste: Biogas. Edited by T. H Christensen, R. Cossu and R. Stegmann. Published in 1996 by E & FN Spon, London. ISBN 0 419 19400 2.

The composition of the trace substances in landfill gas is dependent on the composition of landfilled waste. Volatile organic compounds (VOC) are the most harmful constituents (see Chapter 2.2); several of the gaseous substances are even considered to be mutagenic or carcinogenic (see Chapter 2.4). From a medical toxicological point of view, in assessing volatile contaminants present in landfill gas, different criteria have to be applied from those for semi- and non-volatile contaminants. The latter contaminants have recently gained much focus in the context of soil quality criteria. Gases and vapours escaping from landfilled waste can easily migrate into the surrounding environment and can be inhaled by anybody.

PRINCIPLES OF ENVIRONMENTAL TOXICOLOGY

Environmental toxicology is defined as a subsection of toxicology (differentiating it from ecotoxicology) with the aim of evaluating risks to health from chemicals in the environment. Human health can only be assumed to be completely out of danger if no exposure occurs. A health hazard can no longer be ruled out the moment a substance or a physical factor exists in the human environment. The tasks and aims of environmental toxicology are to recognize, estimate and evaluate and when necessary to limit or avoid the risks.

Harmful substances in the environment are those substances that have the potential of adversely affecting man, other life forms, the ecosystem and material goods, and they can originate from natural or anthropogenic sources. An effect of a harmful substance is assumed when the normal physiological processes of an organism or part of an organism are reversibly or irreversibly changed and the alterations of normal physiological functions can be related to the substance in question. Substances with reversible effects are those with a dose or exposure concentration below which there is no effect observable. The intake of a substance in concentrations below the effect threshold can therefore be classified as completely harmless to health. However, with carcinogenic and mutagenic substances an effect threshold does not exist; here, because of the presumed risks, even with low concentrations particular standards for the reduction or limitation of such harmful substances have to be imposed.

The results of animal experiments, *in vitro* experiments on biological material, human case reports and epidemiological investigations have been the main sources for the identification and evaluation of the effects of environmental pollutants.

The characteristic effects of a substance result essentially from the combined effect of dose, exposure time and the underlying effective mechanisms. An effect, in regard to an environmental pollutant, is every change caused by a substance that occurs following acute or chronic uptake. Damage to health is understood as irreversible or reversible undesirable changes. Toxicity is the capacity of a substance or a factor to cause such damage, depending on the applied dose and duration of exposure. Risk is the probability that a given damage occurs in an exposed section of a population.

NOAEL (No Observed Adverse Effect Level)

The NOAEL (no observed adverse effect level), is the dose of a substance at which no statistically or biologically significant difference can be observed between an exposed group and a control group in the frequency or strength of adverse effects. The NOAEL is a basis for the estimation of a dose range that is classified for humans as completely harmless.

ADI (Acceptable Daily Intake)

The ADI (acceptable daily intake), was introduced in 1961 by the Joint FAO/WHO Expert Committee on Food Additives (JECFA); it is an important toxicological basis for the limitation of potentially harmful substances, particularly in the area of food and drinking water. The ADI value indicates the quantity of a substance (in mg/kg body weight and day) at which no recognizable health risk is observed when a consumer consumes the given substance lifelong. The ADI value is derived from the NOAEL divided by an established safety factor (SF).

Carcinogenicity

Carcinogenicity is the capacity of a substance or a physical factor to cause cancer in humans or animals. As a result of the influence of cancer risk factors, uncontrollable and irregular proliferation of cells occurs (neoplasia); neoplasia refers to both benign and malignant tumours. Physical factors such as ionizing and UV radiation, a number of chemical substances and also oncogenic viruses can have a carcinogenic effect; further, immune defects, hormonal influences and hereditary factors can influence the formation of cancer. The development of tumours – corresponding to the multiple-stage model – occurs in three phases:

1. *Initiation.* Essentially a sudden appearance of persistent cell changes that are transmitted to the progeny of this cell. This enables the affected cell, with the interaction of a promotor, to react with tumour development.
2. *Promotion.* A slow continual process (weeks to years) that with the interaction of the promotors (cancer risk factors) causes the initiated cells to multiply. Promotion does not directly or inevitably lead to the formation of malignant cells but increases the risk of a progression to a malignant tumour.
3. *Progression.* An increase in growth autonomy and malignancy (tumour manifestation).

Agents that increase the initiated effect of a substance are co-carcinogens; syncarcinogenicity is the cooperation of carcinogenic substances whereby the effect of the individual substances is increased in an additional or superadditional fashion; anti-carcinogens are agents that inhibit the formation of cancer following the influence of chemical factors.

As a measure of the carcinogenity of a substance the unit factor is applied. This factor indicates which cancer risk occurs with a lifelong exposure to 1 $\mu g/m^3$ air of the compound inhaled by a 70 kg individual. Here the upper-bound risk value is generally given: i.e. the actual risk is, with a probability of 95%, below the given unit risk.

Mutagenicity

Mutagenicity is the capacity of a substance or physical factor to cause permanent structural changes in the genetic information of cells *in vivo* and *in vitro*. The changes can affect one or more genes, one or more chromosomes or chromosome segments. Mutations of the gametes are passed on to the progeny; mutations of the somatic cells remain in the organism in which the mutation occurs.

Teratogenicity

Teratogenicity is the ability of a particular chemical substance, micro-organism or ionizing radiation to cause non-inherited malformation of embryos, especially mammalian. Teratogenic agents affect the prenatal life of humans so that there is a permanent (irreversible) disruption of embryonic development in the uterus or a permanent (irreversible) impairment of postnatal development of the child. Theoretically, substances can injuriously intervene in all reproductive phases. Reproduction toxicology deals with the effects of compounds on sperm and ova, conception and implantation, organogenesis and fetal maturity. It is not yet exactly known how embryonic malformations occur. It is assumed that the teratogenic effects are the result of chromosome aberrations, impairment of mitosis and inhibition or interference with nucleic acid synthesis and repair.

EPIDEMIOLOGICAL INVESTIGATIONS IN CONNECTION WITH GAS EMISSIONS FROM LANDFILL SITES AND TOXIC WASTE SITES

The assessment of possible effects on health from contaminants in soil show large discrepancies between the assumed danger and the actual effects on the persons examined. An overestimation of the real dangers from semi- and non-volatile substances is observed quite often. The generally wide safety limits set for soil contaminant concentration, which also take into consideration the most sensitive group (e.g. children playing), adequately cover these dangers. The use of less specific parameters can lead to an incorrect estimation of the existing risks (Eikmann, 1988a).

A number of epidemiological investigations of the exposure of persons to volatile contaminants from waste disposal sites exist, whose results allow for the development of criteria and guidelines for these substances (Eikmann, 1988b).

Chemical Waste Site: Love Canal (Niagara Falls, USA)

The chemical waste site in Love Canal near Niagara Falls (New York) is one of the particularly well-known examples made public where the normal population was exposed to toxic volatile pollutants. Between 1942 and 1953, 22 000 tons of chemicals (benzene, chlorobenzene and others) were dumped in the Canal, mainly residual substances from the production of the pesticide lindane. In 1953 the Canal was covered with clay, and later the surrounding area residentially developed.

In the mid-1970s the release of volatile contaminants from the hazard waste was evident (including intense odours), and the residents of the housing estate (approximately 300 inhabitants) were examined in an epidemiological study to establish possible health impairment. According to the authors, owing to the small group studied, only a few effects could be observed, such as pregnancy outcomes (abortions) and deficiencies in newborns (low birthweight). Later examinations of the same risk group could not confirm these observations, and gave no indication of an increased health risk for the residents (Heath, 1987).

The absence of specific biological monitoring investigations related to the different contaminants in the chemical waste site impairs the strength of the conclusions of the various studies. There were also no adequate data on the exposure concentrations.

Chemical Waste Site: Hardeman County (Tennessee, USA)

In the years between 1964 and 1972 liquid and solid waste from a nearby pesticide manufacturer were landfilled. In the year 1977 the residents complained of the unpleasant taste and foul odour of the water from their wells and reported unusually frequent illnesses. These included mainly skin and eye irritation, weakness in arms and legs, shortness of breath, and gastro-intestinal symptoms such as abdominal pains, diarrhoea and vomiting. It was discovered that a number of toxic substances (tetrachloromethane, trichloromethane, tetrachloroethene etc.) had reached residential water wells.

Two independent medical investigations in 1978/79 and 1982 arrived at the same conclusion: that the contaminated well water was the cause of the illnesses. An increase in the alkaline phosphatase and various enzymes (SGOT, SGPT), indicating liver function impairment, was shown in the exposed individuals. A few months after cessation of exposure, enzyme concentrations in the serum of the affected persons returned to normal (Meyer, 1983).

Hazardous Waste Site: Bielefeld-Brake (North Rhine-Westphalia, Germany)

A clay quarry of a former brickworks was used until 1974 as a special hazardous waste landfill for sludge from metal manufacturing industries, carbide and marble sludge, and to a great extent for municipal solid waste. From 1975 to 1978 the quarry was filled with soil and demolition waste in order to develop the land. No consideration was given to the landfill in the planning or the actual building on the site, which was completed in 1983.

Soil samples were taken and an epidemiological investigation was instigated following the massive complaints from the new residents of health problems, in particular when excavation was carried out, and when methane was detected in the houses.

Soil analyses showed increased to greatly increased concentrations of lead, cadmium, mercury, chromium and nickel. Dichloromethane, trichloroethene and tetrachloroethene were found in relatively high concentrations, and hydrocyanic acid, 2,4-dichlorophenol, trichloromethane, 1,1,1-trichloroethane and tetrachloroethene were measured in lower concentrations in the soil–air and to some extent in the emission measurements.

The results of the epidemiological studies of the persons exposed indicated no health impairment of the children. However, 24% of the adults had liver enzymes (SGOT, SGGT, SGPT) in the serum exceeding the reference levels, and 7% were even in the diseased range. Formic acid in urine, one of the biological monitoring parameters used to indicate an exposure to chlorinated methane, was the only parameter with a clearly higher excretion in comparison with control groups.

To avoid a possible permanent danger to health the affected persons were resettled. A control investigation approximately a year after resettlement showed a tendency towards normal enzyme values (Heil *et al.*, 1989).

Chemical Waste Site: Grevenbroich Neuenhausen (North Rhine-Westphalia, Germany

A number of residents living 500 m west of a chemical waste site complained of foul odours and, in their opinion, resulting health complaints (such as headaches and vertigo). The technical deficiencies of the waste site were remedied. Following the epidemiological investigation of the affected persons and measurements of gaseous emission from this site, no lasting effects were observed.

The results of the general medical examinations (blood counts, serum values and lung function) and the biological monitoring parameters (benzene, toluene, xylene and chlorinated hydrocarbons) of all those examined were within the normal reference limits and showed no noticeable deviations.

The ambient air measurements made at the periphery of the waste site, which in all probability included the emission from the working section of the waste site, revealed values similar to those expected from city background pollution. Towards the residential area a further decrease in concentrations would be expected so that the actual exposure here could be assumed to be considerably lower.

On the basis of the results of the epidemiological study and the air pollutant measurements a connection between the health complaints described by the residents and the gaseous emission from the waste site area could not be shown (Einbrodt *et al.*, 1990).

Assessment of the Results of the Epidemiological Studies

In the assessment of the results of the different studies mentioned, some exposure situations involving the intake of volatile contaminants can be classified as particularly relevant to human toxicology.

In general, inhalation of volatile contaminants by the affected persons is seen as considerably more important than inhalation of dust carried slightly or non-volatile contaminants. Increased exposure, or the appearance of symptoms or illnesses, can generally only be observed if relatively high concentrations of contaminants are directly ingested from drinking water (Hardeman County) or inhaled owing to the accumulation of gases in indoor air (Bielefeld-

TABLE 1. Concentration (μg/m^3) of Highly Volatile Chlorinated Hydrocarbons in Soil–Air and Air at and around the Hazardous Waste Site in Bielefeld-Brake, Germany.

		Landfill area		*Non-contaminated area*
	Soil–air	*Cellars (approx. 150 m away)*	*1.50 m above surface*	
Trichloromethane	26	15	3	2
Tetrachloromethane	30	160	10	7
Trichloroethene	152000	5100	22	3
Tetrachloroethene	83500	5000	140	5
Dichloromethane	1111	200	12	3

Brake). Owing to the volatility of the contaminants in the drinking water from Hardeman County, it is likely that inhalation was also a relevant intake path here.

In all studies, foul odours were the warning signs of possible contaminant exposure. Although the odour threshold of many substances is in a concentration range that is assumed not to affect health, an increase in complaints about foul odours should always warrant an investigation of the exposure situation.

The Bielefeld-Brake study also shows a possible exposure of residents who were not in direct contact with the landfill area. In this case exposure occurred at relatively great distances away from the waste site (several hundred metres) (Table 1). The migration of gases in soil can be increased by the presence of utility installations in the ground.

The measurements at the chemical waste site in Grevenbroich-Neuenhausen further showed that the concentrations of volatile substances, even within close range of the waste site, were generally comparable with background levels in local and inner city areas. Generally, compared with the soil–air values, distinct concentration gradients are postulated in the vicinity of toxic waste sites.

RISK ASSESSMENT

Assessing the risk of landfill gas to public health has been performed in several cases in the recent years, but this has certainly not yet become a routine at all landfills. Some specifications have been issued at state level in the USA, but no standard procedures are generally accepted, although the points described below must usually be addressed.

Identification of Health Risks

Exposure to landfill gas is generally dominated by the inhalation exposure route. Exposure by ingestion and by dermal absorption is usually much lower, and will not be addressed further in this context. Explosion hazards, potentially detrimental to those actually exposed, are dealt with in Chapter 3.2.

The volatile compounds of relevance for the inhalation pathway involve both carcinogenic compounds and non-carcinogenic compounds. The carcinogenic compounds are: benzene, carbon tetrachloride, chloroform, 1,2-dichloroethane, 1,1-dichloroethene, ethylene dibromide, methylene chloride, 1,1,2,2-tetrachloroethane, tetrachloroethene, 1,1,2-trichloroethane, trichloroethene and vinyl chloride (also called chloroethene and chloroethylene). The non-carcinogenic compounds that may be included in the health risk assessment are: chlorobenzene, 1,1-dichloroethane, ethylbenzene, methyl ethyl ketone, tetrachloroethylene, toluene, 1,1,1-trichloroethane, 1,1,2-trichloroethane) and xylenes. Others may also be of relevance, depending on the type of landfill. In particular, if emissions from gas treatment facilities are also included in the health risk assessment, the list of compounds to consider may be expanded, e.g. with dioxins and furans.

In general, the most critical compounds in landfill gas are believed to be vinyl chloride and benzene (Petersen, 1988).

Estimation of Dosage

At existing landfills, the exposure concentrations should be determined by measuring the air concentrations, indoors as well as outdoors, at critical receptor points. This could be at and in nearby houses and at houses on nearby hills facing the landfill. It is not necessarily at the closest houses that the highest exposure is found. The indoor measurements are particularly important since migration of landfill gas into confined rooms or cellars with little ventilation is of concern. Volatile substances can enter through fissures in foundations or cracks around utility installations. However, hobby glues, paints and other household chemicals may also be significant sources of some of the compounds that could have entered with the landfill gas.

At prospective landfills, measurements cannot be made, and modelling is the only approach to estimate potential exposure concentrations. Emission rates at point sources (gas flares, exhaust gas outlets at gas utilization installations) as well as area sources (the gas not collected) must be predicted. In particular, the area sources are difficult to quantify. Based on the emission rates, dispersion modelling accounting for climatic as well as topographic conditions can be made, although not by simple means (see Chapter 3.8). Both maximum concentrations and average concentrations are needed to address the health risks. At the not yet existing landfill, special care must be taken in defining the critical receptor (Wolff, 1990).

By using a standardized weight and breathing rate of an exposed individual, the exposure dose for the non-carcinogenic compounds can be estimated. By assuming also a life expectancy, the exposure dose of carcinogenic compounds can be estimated as well.

Estimation of Risks

For the non-carcinogenic compounds, the estimated dose can be compared directly with established health criteria (usually given as reference doses with no effects as mg/kg/day). Such values may not be available for all compounds in the landfill gas, and specific expert reviews of the toxicological literature must be performed.

For the carcinogenic compounds, the exposure concentrations are transformed into expected lifetime dosage per kilogram of body weight and multiplied by the potency factor of the compound to yield a risk of cancer to an individual exposed over a lifetime. Unit factors are available in some countries for selected compounds (Wolff, 1990) and quantify the risk of cancer per $\mu g/m^3$ of the substance inhaled. While the criteria for the non-carcinogenic compounds indicate doses that are below doses causing any known effects, a risk of cancer is always estimated for the carcinogenic compounds, and hence must be compared with an acceptable risk. This risk is often set at 1 additional cancer case per lifetime in 10 000 or in 100 000 exposed persons.

REFERENCES

Eikmann, Th. (1988a). Allgemeine Gesundheitsrisiken durch Altlasten, in *Handbuch Altlastensanierung* (eds V. Franzius, R. Stegemann, and K. Wolf) 1.3.2.2. R. v. Decker's Verlag, Heidelberg, pp. 1–8.

Eikmann, Th. and Kloke, A. (1991) Nutzungs- und schutzgutbezogene Orientierungswerte für (Schad-) Stoffe in Böden, in *Handbuch Bodenschutz* (eds D. Rosenkranz, G. Einsele and H. M. Harress), Erich Schmidt Verlag, Berlin, pp. 1–19.

Einbrodt, H. J., Eikmann, Th., Krieger, Th. and Michels, S. (1990). *Beurteilung der Gefährdungssituation der Bewohner von Grevenbroich-Neuenhausen durch Gasemissionen der Sondermülldeponie Neuenhausen.* Institut für Hygiene und Arbeitsmedizin, Aachen.

Gendebien, A., Pauwels, M., Coustent, M. *et al.* (1992). *Landfill Gas. From Environment to Energy*, Office for Official Publications of the European Communities, Luxembourg.

Heath, C. W. (1987). Assessment of health risks at Love Canal, in *Health Effects from Hazardous Waste Sites*, (eds J. B. Andelman, and D. W. Underhill), Lewis Publishers, Michigan, pp. 211–220.

Heil, H., Eikmann, Th., Einbrodt, H. J. *et al.* (1989). Konsequenzen aus dem Altlastenfall Bielefeld-Brake. *Vom Wasser*, **72**, 321–348.

Meyer, Ch. R. (1983). Liver dysfunction in residents exposed to leachate from a toxin waste dump. *Environmental Health Perspectives*, **48**, 9–13.

Petersen, T. N. (1988). Human health risk assessment of landfill gas emissions, in *Proceedings 11th Annual Madison Waste Conference*, 13–14 September, University of Wisconsin, Madison, USA.

Schlipköter, H.-W. (1986a). *Gutachten zur Frage des Gesundheitsrisikos durch Bodenverunreinigungen in Dortmund-Dorstfeld*, Medizinisches Institut für Umwelthygiene, Düsseldorf.

Schlipköter, H.-W. (1986b). *Gutachten zur Frage des Gesundheitsrisikos durch Bodenverunreinigungen in Dortmund-Dorstfeld – Untersuchung der Bevölkerung*, Medizinisches Institut für Umwelthygiene, Düsseldorf.

Van den Berg, R. and Roels, J. M. (1991) *Evaluation of the risk for man and environment by exposure to soil pollutants – interpretation of various aspects.* Reichsinstitut für Volksgesundheit und Umwelthygiene, Bilthoven, Bericht Nr. 725201007.

Wolff, S. K. (1990). Health risk assessment for landfill gas emissions from solid waste landfills. *GRCDA Journal of Municipal Waste Management*, **1** (August), 67–74.

3.4 Effects of Landfill Gas on Vegetation

UWE NEUMANN[a] & THOMAS H. CHRISTENSEN[b]

[a]Garden and Landscape Architect, Falterweg 17, 14055
Berlin-Eichkamp, Germany
[b]Institute of Environmental Science and Engineering, Technical
University of Denmark, Building 115, DK-2800, Lyngby, Denmark

INTRODUCTION

Damage to vegetation on landfills and adjacent to landfills has been
reported in a number of cases. Gendebien *et al.* (1992) found 31
cases described or mentioned in the world literature, but supposedly
many more cases exist. In fact, vegetation damage caused by landfill
gas (LFG) is usually very limited in space, and not all cases are
published in the open literature. Still, many landfills exist with
properly developed vegetation covers, and extensive damage is
supposedly more an exception than a rule.

Vegetation damaged by LFG exhibits symptoms such as chlorosis, defoliation and branch die-back. These symptoms are not
specific to LFG damages, but may also be ascribed to drought, salt
damage and, in the case of chlorosis, to nutrient deficiency. The
effects of LFG may also appear as dwarf growth, surfacial root
developments and, in severe cases, as plant death.

The damage caused by LFG has been observed on vegetation on
completed landfill sites as well as on adjacent fields. However, in
most cases not much had been done to provide the vegetation with
proper growth conditions and to divert the gas away from the
vegetation. Proper design of gas control systems, completion of the
landfill with a proper final soil cover and selection of tolerant plant
species should prevent significant damage to vegetation in the
future.

Landfilling of Waste: Biogas. Edited by T. H Christensen, R. Cossu and
R. Stegmann. Published in 1996 by E & FN Spon, London.
ISBN 0 419 19400 2.

EFFECTS OF LANDFILL GAS COMPONENTS ON VEGETATION

The effects of LFG on plants are associated with the migration of gas into the root zone, and displacement of the natural soil air. This will usually lead to anoxic conditions (at least in part of the root zone) and may affect the plant by asphyxiation, by the presence of toxic gases or by changes in pH and composition of the soil solution. In actual situations, the observed effect may be the result of several of these factors and of external stresses as drought and strong wind.

Asphyxiation

Asphyxiation is caused by the lack of sufficient oxygen in the root zone as a result of physical displacement by LFG or by microbial oxidation of methane in the LFG, depleting the soil air oxygen. According to the literature, as reviewed by Gendebien *et al.* (1992), most plants grow normally at 5–10% oxygen in the soil air, but more demanding species, in particular woody species, may require more than 12–14% oxygen. A great variability exists among plant species as to tolerance for low oxygen concentrations, and advantage of this could be taken in selection of species for landfill completion, as discussed later.

The root zone may not be completely depleted of oxygen, as re-aeration from the atmosphere may provide some oxygen to the upper part of the soil layer. This may allow some species to grow, although optimal conditions are not present. A thin aerated soil layer may lead to the development of a shallow root system limited to this zone. Figure 1 shows a poplar, which naturally develops a tap root, that on a landfill site has developed a very shallow, horizontal root system. This situation leads to limited access to water and nutrients, resulting in dwarf growth and possibly death in extended dry periods. Figure 2 shows dwarf growth of 20-year-old alder trees on a landfill. The shallow root system may also make trees vulnerable to wind-fall because of the limited strength of this root system compared to deep root systems.

The heterogeneity of landfill covers and variations in LFG escape routes may lead to highly variable vegetation covers on landfills: some areas may be completely bare, while others may host tall woody species (Fig. 3). This observation could also be utilized in the design of landfill covers by providing good, deep soils for

Figure 1. Poplar tree with shallow root system developed in the upper part of a landfill soil cover.

vegetation in some areas while thin layers of coarser soils in a neighbouring area allow the landfill gas to escape.

Effects of Methane

Methane (CH_4) is usually not considered to be toxic to plants. Low concentrations of methane ($< 5\%$) have been reported to be beneficial to plants (Ernst, 1976), while very high concentrations (45%) may partly damage some plants.

The major effect of methane present in the landfill soil cover is believed to be related to microbial oxidation of methane. This leads to depletion of oxygen, to increased carbon dioxide concentrations, and possibly to increased temperatures in the soil, thereby enhancing asphyxiation.

Effects of Carbon Dioxide

The carbon dioxide (CO_2) content of normal soil air may range up to a few percent, owing to organic matter degradation in the soil and

Figure 2.   Dwarf growth of 20-year-old alder trees on a completed landfill with limited soil cover.

respiration processes. Growth inhibition may take place for some species above this level, although many species will function normally up to 5% carbon dioxide. At higher levels phytotoxic effects may appear. However, substantial variability exists among plant species as to carbon dioxide tolerance (see Gendebien *et al.* 1992, for details).

Indirect effects of carbon dioxide may be caused by a decrease in soil pH and the consequent changes in soil solution composition.

Other Effects

The trace gases in LFG may potentially be toxic to some plant species, but the significance of this in actual cases still remains to be demonstrated. Usually, asphyxiation is considered to be a much more dominating aspect.

If LFG creates anoxic conditions in a soil, this may significantly affect the soil structure. If the gas migration is remedied in order to revegetate the soil, this may demand a special effort to improve the structure of the soil, e.g. by mechanical means.

Figure 3. Section of completed landfill showing bare as well as vegetation covered areas.

DAMAGE TO VEGETATION: CASES

As indicated above, LFG has the potential to damage vegetation if LFG is present in the root zone in sufficient concentrations, and, as mentioned, numerous cases exist. It should be emphasized that most of the reported cases are associated with landfills that do not meet modern standards and where very little consideration has been given to avoid damaging the vegetation. The four selected examples below therefore serve primarily as illustrations of how damage can happen, if landfill gas is not properly controlled and managed. More examples can be found in Gendebien *et al.* (1992).

Case 1 (Germany)

At the Berlin-Wannsee Landfill an approximately 100 000 m^2 completed landfill section was supplied with grass and trees to constitute a recreational area. Severe damage to the vegetation was soon observed, and a few years later trees as far as 50 m away from the landfill were damaged as well. These trees on the adjacent land had not previously shown signs of damage.

Case 2 (USA)

In Gloucester County, New Jersey, a peach plantation neighbouring a 250 000 m^2 large landfill experienced tree death about 3 years after the landfilling had started. Sealing of the landfill with a soil cover apparently increased the lateral migration of LFG, and a few years later a total of 70 trees had died.

Case 3 (USA)

In Burlington County, New Jersey, a corn field located approximately 150 m away from a landfill established in an abandoned gravel pit was damaged (chlorosis and decreased growth) 5 years after the landfill operation had begun. LFG was traced as far as 250 m away from the landfill.

Case 4 (UK)

In Kent, UK, damage to commercial crops up to 100 m away from an abandoned landfill was consistently reported. The LFG had migrated away from the landfill through ancient mine workings and had escaped through access holes and fractures. Owing to the special conditions no remedial actions were taken and the damage persisted for several years.

REVEGETATION OF LANDFILL SITE

Revegetation of completed landfills is essential in order to adapt the site to the surrounding environment, to improve public acceptance, to minimize erosion on slopes and to minimize leachate production by increasing the evapotranspiration.

Several procedures have been proposed to improve the revegetation of completed landfill sites (e.g. Gilman *et al.*, 1985). The main aspects to observe are gas migration control, soil cover, plant species selection and planting strategies for woody species.

Gas migration control in terms of abstraction of gas for utilization or controlled venting is mandatory in order to prevent substantial amounts of LFG from migrating through the rooting zone. At new landfills this will usually be compulsory just to reduce the emissions of greenhouse gases and to avoid uncontrolled migration of gas to adjacent property.

Soil covers providing sufficient depth for proper root development (minimum 0.5 m for grass covers and 1.5 m for trees), sufficient water storage capacity to survive drought periods, and sufficient nutrients to sustain growth, are mandatory for developing adequate vegetation.

Species selection should always pay attention to the local conditions, such as climate, soil types, depth of soil layer and wind-exposed areas. The species should also be tolerant to low oxygen concentrations and to dryness, because such conditions may develop at the landfill because of landfill gas migration.

Planting strategy. The stability to strong wind can be improved if the landfill vegetation is staggered: i.e. bushes are planted at the wind front, small trees like *Carpinus betulus* behind, and further behind large trees like *Betula pendula, Alnus incana, Robinia pseudacacia* and *Acer pseudoplatanus.* In this way the wind forces on the individual trees are minimized. It may be worth considering not planting tree species on the top of hilly landfills, because here it may be difficult to ensure stability towards the wind. However, increasing the thickness of the soil layer may improve the stability. In addition, areas with bushes and dry meadows, such as may develop on top of a completed landfill, are today rare elements in the cultivated landscape and may from a ecological point of view provide an important habitat for wild animals and flora. In the Berlin area, (Germany) a completed landfill proved to be the most important habitat for butterflies (Neumann, 1983).

REFERENCES

Ernst, W. (1976). Erfahrungen mit Pflanzenbegasung. *Gwf-gas/eragas*, **117**, 433–439.

Gendebien, A., Pauwels, M., Constant, M. *et al.* (1992). *Landfill Gas. From Environment to Energy*, Office for Official Publications of the European Communities, L-2985 Luxembourg.

Gilman, E. F., Flower, F. B. and Leone, I. A. (1985). Standardized procedures for planting vegetation on completed sanitary landfills. *Waste Management & Research*, **3**, 65–80.

Neumann, U. (1983): Gesichtpunkte des Arten schutzes bei der Rehultivierung von Mülldeponien. *Müll und Abfall*, **15**, 64–68.

3.5 Effects of Landfill Gas on Groundwater

HENRY B. KERFOOT

Kerfoot and Associates, 2200 E. Patrick Lane, Suite 23, Las Vegas, NV 89119, USA

INTRODUCTION

Because methane production from organic material in landfills results in elevated gas pressure within the landfill, pressure-driven subsurface gas migration can result. This is particularly true when a landfill is capped, eliminating the possibility of venting to the atmosphere. Subsurface gas migration ultimately leads to emission to the atmosphere, but the gas can contact groundwater as it travels, and such contact can affect groundwater quality. Such effects can occur when gas migration results in gas–water contact; because reports of them are rare in the literature and few studies of the process have been reported, the important factors have not been identified quantitatively.

The use of volatile organic compounds (VOCs) as indicators of leachate effects on groundwater has been proposed by numerous workers based upon their relatively high mobility in groundwater (due to minimal sorption and no ion-exchange interactions) and data on the frequency of detections of VOCs at landfills (Plumb and Pitchford 1985; Plumb, 1987). In the USA, monitoring for these compounds as indicator parameters for early warning of landfill releases is typically required. However, as analytical–chemical technology succeeds in obtaining lower and lower detection limits for these compounds in groundwater samples, interpretation of detections of them must incorporate consideration of more and more potential transport mechanisms.

Landfilling of Waste: Biogas. Edited by T. H Christensen, R. Cossu and R. Stegmann. Published in 1996 by E & FN Spon, London.
ISBN 0 419 19400 2.

The potential effects on groundwater quality of gas-to-water phase transfer of VOCs should also be considered in the light of the magnitude of such effects relative to those of leachate. Gas effects on groundwater differ from those of leachate owing to the lack of any density increase from gas effects. In the case of leachate, the higher density can result in significant vertical transport (downward) of contaminants (Deutsch, 1965; Bjerg and Christensen 1993). However, without any density-driven vertical transport, downward migration of contaminants from the gas/water interface in the case of gas effects is typically limited to the mechanisms of diffusion, dispersion, and (to an unknown extent) mixing due to fluctuations in water-table elevation.

Owing to the slow vertical transport of contaminants by these mechanisms relative to density-driven transport, the extent of vertical penetration of VOC contamination of groundwater can be limited by the rate of vertical mass transport in the saturated zone. This can result in a thin 'skin' of contamination in the capillary fringe and near the surface of the water-table, without significant VOC concentrations at greater depths. Because only limited data on the extent of vertical penetration of gas effects into the saturated zone are available and no detailed description of the process apparently has been presented, it is not possible to describe accurately exactly how much or little vertical penetration occurs from gas effects. Golwer and Matthes (1972) have considered gas exchange between groundwater and unsaturated-zone pore gases for carbon dioxide and oxygen. Mendoza and McAlary (1990) described a computer program to calculate the effects of unsaturated-zone gas contamination on groundwater, and mass transport into groundwater was dominated by dissolution in infiltrating water. Kerfoot (1994) has presented calculations using field data for unsaturated-zone/saturated-zone gas exchange based on a single-stage equilibrium batch contactor to simulate water-table fluctuation effects. Those models have not been validated.

This chapter describes the potential effects of landfill gas on groundwater. Potential effects of methane, carbon dioxide, and volatile organic compounds on groundwater quality are briefly discussed. Then methods to identify gas-to-water phase transfer as the source of groundwater contamination are discussed, case studies are presented, remedial measures are briefly described and the implications of gas effects on groundwater are discussed.

Methane (CH_4)

Methane, usually 50–60% (by volume) of landfill gas during stable gas production, has a relatively low solubility in water. At 25 °C, 24.1 mg l^{-1} methane is soluble, so that 12 mg l^{-1} would be in equilibrium with a 50:50 methane:carbon dioxide mixture. Once groundwater containing such a methane concentration was exposed to air or soil air containing no methane or a concentration below 50%, methane would partition out of the water. Methane can also be produced in strongly reduced or deep aquifers (including leachate plumes), and use of groundwater from such sources typically requires aeration to strip out the methane. Organic deposits such as peat can also result in localized methane production in shallow aquifers. In addition to potentially being produced in aquifers, methane can readily be oxidized by oxygen from soil gases or potentially by nitrate or sulphate, although the latter two oxidizing agents (or 'electron acceptors' in Lewis acid–base terminology) have not been unequivocally demonstrated. Because of the potential for such degradation of methane and its low solubility, methane concentrations in groundwater that are due to gas effects will typically limit any such effects to very close to the landfill. Methane production due to leachate plumes may extend further.

Carbon Dioxide (CO_2)

Carbon dioxide usually represents approximately 40–50% (by volume) of landfill gas during stable methane production. Owing to the greatly higher solubility of carbon dioxide over that of methane, landfill gas can result in larger effects on groundwater due to carbon dioxide than from methane. Bishop (1967) noted that carbon dioxide from decomposition of refuse can increase the corrosivity of groundwater. The gas/dissolved equilibrium is described by Henry's law:

$$CO_2(g) = K_H CO_2(aq)$$

with

$$K_H = \frac{P_{CO_2}}{[CO_2]}$$

where P_{CO_2} is the partial pressure of carbon dioxide (atm) and $[CO_2]$ is the dissolved carbon dioxide concentration (mol dm^3). K_H is 4.3×10^{-2} atm dm^3 mol^{-1} or 9.7×10^{-5} %vol dm^3 mg^{-1} at 15 °C and an ionic strength of 1.0 molal.

However, because carbon dioxide forms carbonic acid when dissolved in water, the situation is a bit more complicated. Acid-base equilibria involving carbonate and bicarbonate ions must be considered in evaluating partitioning of carbon dioxide from gas into water. For the sake of simplicity, we can use the sum of all the carbon-dioxide-related species, C_T, expressed as mol dm^{-3}, and write

$$CO_2(g) = H_{CO_2}C_T$$

where $CO_2(g)$ is the gas-phase carbon dioxide concentration in % by volume at sea level and H_{CO_2} is in %vol dm^3 mol^{-1}. The dissolved inorganic carbon concentration, or DIC, can be used for C_T, resulting in

$$CO_2(g) = 7.9 \times 10^{-2} \, DIC$$

for a pH of 7.0 and an ionic strength of 1.0 at 15 °C. Although temperature, pH, and ionic strength can all affect this relationship, it is most likely applicable for approximate calculations under most groundwater applications at landfills. More precise equations can be developed using data from the literature (Butler, 1982).

Gas effects on groundwater pH due to formation of carbonic acid can occur. For 50% carbon dioxide in the gas phase, the equilibrium pH of water in contact with that gas would be approximately 4 in the absence of buffering. However, owing to buffering by aquifer materials, pH changes from gas effects are very small or negligible. This buffering action can result in dissolution of aquifer materials, with concomitant changes in dissolved cations and anions, and that could potentially be interpreted as evidence of leachate effects without careful consideration and geochemical modelling. No evaluations of this aspect of landfill gas effects on groundwater have been performed yet.

Volatile Organic Compounds

Numerous minor components of landfill gas can exist, including sulphides and amines, but the ones that have received the most attention from the perspective of groundwater quality are volatile

TABLE 1. Landfill Gas Concentrations of Selected VOCs in Landfills in California, USA

Compound	Gas-recovery systems (n = 10)			Single-point samples (n = 44)	
	Detection limit (ppmv)	*Conc. range (ppmv)*	*Detection frequency (%)*	*Conc. range (ppmv)*	*Detection frequency (%)*
Benzene	0.30	0.52–3.6	100	ND[a]–39	80
Chlorobenzene	0.25	ND–1.6	60	ND–1.3	2
1,1-Dichloroethane	0.33	ND–3.7	60	ND–22	41
1,1-Dichloroethene	0.26	ND–0.26	10	ND–4	20
cis-1,2-Dichloroethene	0.55	ND–12	80	ND–20	52
trans-1,2-Dichloroethene	0.34	ND–0.85	20	ND–0.85	5
Ethylbenzene	0.26	ND–25	100	ND–8.8	89
Methylene chloride	0.35	ND–11	90	ND–230	84
Tetrachloroethene	0.18	ND–16	80	ND–180	59
1,1,1-Trichloroethane	0.27	ND–1.3	30	ND–15	23
Toluene	0.28	6.9–91	100	6.9–280	89
Trichloroethene	0.44	ND–12	80	ND–32	41
Vinyl chloride	0.43	1.2–8.5	100	1.2–32	59

[a] ND = not detected.

organic compounds (VOCs). Table 1 lists concentrations of selected volatile organic compounds that were determined in landfill gas samples in a study in the state of California in the USA. The table lists data from gas-recovery systems ($n = 10$), which are a more well-mixed representation of the overall average landfill gas concentrations of these specific VOCs, along with single-point data ($n = 44$). It should be noted that the landfills used in the California study were known to accept only municipal solid waste at the time of sampling but may have accepted industrial wastes earlier on in their operation.

From the data in Table 1 it can be noted that significant variation in concentrations of the various VOCs was observed between different landfills (gas recovery system data) and even more was observed among localized samples (single-point data). However, in 100% of the gas recovery system samples and most of the single-point samples, benzene, toluene, ethylbenzene, and vinyl chloride were observed. Based upon toxicity data, benzene and vinyl chloride

TABLE 2. Equilibrium Water Concentrations of VOCs Corresponding to Landfill-Gas Concentrations

| Compound | H (ppmvdm³µg) | Equivalent water concentration (µg/dm³)[a] | | |
		Detection limit	Gas-recovery systems	Single-point samples
Benzene	0.070	4.0	7.4–51	ND–560
Chlorobenzene	0.031	8.0	ND–53	ND–43
1,1-Dichloroethane	0.058	5.6	ND–64	ND–380
1,1-Dichloroethene	0.12	2.2	ND–2.2	ND–33
cis-1,2-Dichloroethene	0.077	7.1	ND–156	ND–260
trans-1,2-Dichloroethene	0.068	5.0	ND–12.5	ND–12.5
Ethylbenzene	0.074	3.5	54–340	ND–1180
Methylene chloride	0.030	12.0	ND–365	ND–7700
Tetrachloroethene	0.13	1.4	ND–120	ND–1400
Toluene	0.072	3.9	95–1270	95–3900
1,1,1-Trichloroethane	0.21	1.3	ND–6.4	ND–152
Trichloroethene	0.068	6.5	ND–180	ND–470
Vinyl chloride	11.1	0.039	0.11–0.77	0.11–2.9

[a] Equilibrium concentration, for temperatures of 10–25 °C, of C_g/H.

represent the greatest threat of the VOCs present in landfill gas as listed in Table 1, because both are known carcinogens.

The data in Table 1 can be used directly for assessment of the potential health risk due to emission of landfill gas into ambient air. However, for evaluation of the potential threat to groundwater quality posed by landfill gas the fact that the VOCs must partition from the gas into groundwater must also be taken into consideration. In order to consider the potential impact of VOCs from landfill gas on groundwater quality, the equilibrium groundwater concentrations corresponding to the gas concentrations in Table 1 can be calculated using Henry's law:

$$C_w = \frac{C_g}{H}$$

where C_w is the equilibrium water concentration (µg dm⁻³) and C_g is the landfill gas concentration (ppbv). Table 2 gives the Henry's law constant (ppbv dm³ µg⁻¹) and the equilibrium C_w values corresponding to the entries in Table 1. Values of H for numerous organic compounds are available (Mackay and Shiou, 1981).

The data in Table 2 can be used to evaluate the relative threat to groundwater of each of the landfill gas VOC concentrations in

Table 1. It must be noted that the equilibrium value of C_w, or C_g/H, represents the maximum water concentration that can result from a gas concentration equal to the C_g value in Table 1, and mass-transport limitations can result in a significantly lower value. However, such calculations can be used for a worst-case estimate. From Table 2 it can be seen that, owing to the very high equilibrium C_g/C_w value for vinyl chloride, a very high landfill gas concentration is required to result in a significant water concentration, while the lower equilibrium C_g/C_w value for benzene means that a much lower gas concentration can result in significant benzene concentrations.

IDENTIFICATION OF GAS EFFECTS ON GROUNDWATER

In order to develop appropriate remedial responses for groundwater contamination, the source of the detected contamination should be known. Although leachate releases can require so-called pump-and-treat remediation, source termination through installation or modification of a gas-collection system can be an effective remedial response to groundwater contamination by landfill gas. Because of this, and the potential for misinterpretation of VOC contamination of groundwater that is due to landfill gas emissions as being due to leachate, it is important to be able to differentiate between the two mass-transport mechanisms. There are three different techniques for identification of landfill gas effects. Each has applicability under different circumstances, so that all three are described below.

Non-Volatile Inorganic Parameters

The most common method of differentiation of landfill gas effects from leachate effects on groundwater is based upon the use of moieties present in leachate but not present in landfill gas. Specifically, the presence of volatile organic compounds in groundwater samples without elevated levels of parameters associated with leachate is an indication of landfill gas effects on groundwater. These parameters include specific electrical conductance, chloride, and often iron and manganese. Detection of volatile organic compounds in groundwater samples without increases in these leachate-related parameters indicates that the presence of the VOCs is due to landfill

TABLE 3. Groundwater Data from a Landfill in Illinois, USA

Well	Location[a]	Perchloro-ethene ($\mu g/dm^3$)	SO_4^{2-} (mg/dm^3)	Specific electrical conductance ($\mu S/cm$)	CO_2 (mg/dm^3)	Total dissolved solids (mg/dm^3)
1	Up	2.8	11	242	32	160
6	Up	ND	12	240	10	130
7	Up	1.1	12	240	10	140
3	Down	34	6	747	160	450

[a] Up = Hydraulically upgradient of landfill; Down = Downgradient of landfill.

gas and not leachate. Table 3 lists data from wells at a landfill in the state of Illinois in the USA. From these data it can be seen that the non-volatile parameters do not increase along with the detection of tetrachloroethene in Wells 1 and 7, which are hydraulically upgradient of the landfill, indicating landfill gas effects. Well 6 represents unaffected groundwater. However, significant increases in specific electrical conductance (SEC) and total dissolved solids (TDS) can be seen along with the downgradient well (3), which is affected by leachate.

The use of the absence of increases in non-volatile parameters associated with leachate to indicate gas effects as the source of VOCs in groundwater samples has limitations. High variability in background levels of these parameters, either spatially or temporally, can make it difficult to identify significant increases. Similarly, the presence upgradient of potential sources of reduced organic carbon, such as refineries, or of other non-volatile parameters associated with leachate, can complicate groundwater data sufficiently to make such an interpretation unclear or even impossible. Finally, as noted above in discussion of the potential effects of landfill gas CO_2 on groundwater, landfill gas can increase dissolved inorganic carbon or CO_2 concentrations, total dissolved solids (TDS), and specific electrical conductance (SEC). Such effects can be seen in the CO_2 and TDS data for Well 1 in Table 3.

Evaluation of Gas–water Disequilibrium

A second approach to identification of gas effects as the reason for detection of VOCs in groundwater samples at a landfill is through the evaluation of VOC concentrations in groundwater and head-space gas samples from monitoring wells thought to be affected by landfill gas. The rate of mass transport from gas into water is proportional to the difference between the gas concentration C_g, and the gas concentration that would be in equilibrium with the water concentration in contact with the gas, or HC_w:

$$\frac{dm}{dt} = K(C_g - HC_w)$$

This means that if the gas concentration C_g, is greater than the water concentration in contact with it multiplied by the Henry's law constant, HC_w, the direction of phase transfer will be from gas to water, while a negative value of $C_g - HC_w$ indicates water-to-gas phase transfer.

Figure 1 shows a plot of C_g and HC_w, with a solid line indicating equilibrium where $C_g = HC_w$. Departures from equilibrium where C_g is greater than HC_w, or data points above and to the left of the line, indicate gas-to-water mass transport, while data below and to

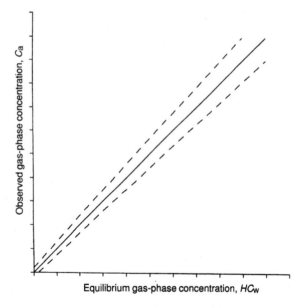

Figure 1. Gas concentrations (C_g) and equilibrium concentrations (HC_w) calculated from water concentratons (C_w).

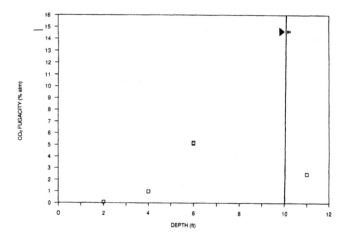

Figure 2. Unsaturated-zone and saturated-zone carbon dioxide data indicating gas-to-water transport.

the right of the line indicate water-to-gas mass transport. Through such comparisons, the direction of phase transfer can be established. The dashed lines in Fig. 1 are included to point out that uncertainty in analytical results can make data that lie too close to the equilibrium line inconclusive.

Figure 2 shows values of C_g from the unsaturated zone (as fugacity) and values of HC_w from groundwater samples (as fugacity) for carbon dioxide that were obtained in an evaluation of carbon dioxide production from hydrocarbon fuel (Kerfoot, 1994). The unsaturated-zone values of C_g greater than HC_w are consistent with gas-to-water transport of carbon dioxide. Figure 3, on the other hand, shows data from the same study where HC_w is greater than C_g, indicating water-to-gas transport of carbon dioxide.

Figures 2 and 3 demonstrate the use of actual field data from evaluation of C_g and HC_w at several depths to identify the direction of mass transport of carbon dioxide. An approach similar to that demonstrated in Fig. 1 can be used for VOC concentrations in groundwater and headspace gas samples from monitoring wells, where multidepth data is not available.

The use of evaluation of the direction of departure from gas–water equilibrium is limited to locations where gas-to-water transport is taking place; if the process is occurring upgradient but not in the well being evaluated, this technique will not indicate gas-to-

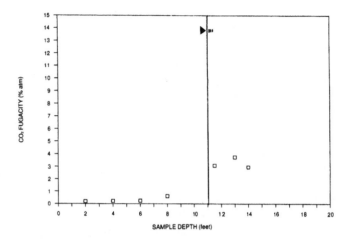

Figure 3. Unsaturated-zone and saturated-zone carbon dioxide data indicating water-to-gas transport.

water transport. In addition, the use of the data is limited to evaluation of the situation at the time of sampling. Because significant temporal and spatial variability can exist in landfill gas effects, these limitations should be kept in mind.

Radioisotopes

A third approach to identification of landfill gas as the source of VOCs detected in groundwater samples uses radioisotopes (Liu *et al.*, 1992). When groundwater is affected by either landfill gas or leachate, additional carbon dioxide will be added to that already present in the aquifer. The added carbon dioxide will have a ^{14}C content indicative of the refuse in the landfill, which will be carbon from the past several decades, while the carbon in the dissolved inorganic carbon in unaffected groundwater will typically be much older and thus have a lower ^{14}C content. Thus either leachate or landfill gas effects will result in increased ^{14}C in the dissolved inorganic carbon in groundwater samples.

In contrast to leachate, which is aqueous, landfill gas will often have a negligible water content. Because the water in landfill leachate is often high in ^{3}H relative to groundwater, the ^{3}H content of groundwater samples can be used to differentiate between landfill gas and leachate effects on groundwater. Therefore, landfill gas effects will be indicated by an increased ^{14}C content of the dissolved inorganic carbon in groundwater samples without an increase in the ^{3}H content of the water, while leachate effects will be indicated by an increase in both. Figure 4 shows how the ^{14}C content of dissolved inorganic carbon and the ^{3}H content of groundwater can be used to differentiate between leachate effects and landfill gas effects. ^{14}C data are reported as percent modern carbon (pMC), while tritium (^{3}H) data are reported in tritium units (TU).

The above approach to identification of landfill gas as the source of VOCs in groundwater can be used downgradient of the location where the actual phase transfer is taking place. However, spatial or temporal variations in these isotopes in groundwater can occur and confound interpretation of results. For example, artificial recharge using snow melt is common in arid areas such as California in the USA, and can alter both ^{3}H and dissolved inorganic carbon ^{14}C data on a seasonal basis, and localized recharge effects can result in spatial variability. In addition, CO_2 from landfill gas could dissolve

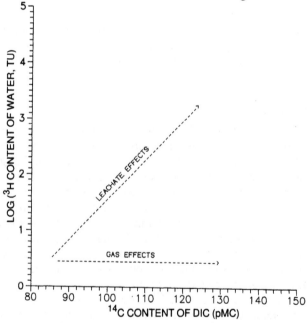

Figure 4. Use of ^{14}C and ^{3}H data to differentiate between leachate effects and landfill-gas effects on ground water.

carbonate minerals, increasing the apparent age of the dissolved inorganic carbon. It is not known how important this factor is in limiting the use of this approach.

CASE STUDIES

Two case studies demonstrating identification of landfill gas as the source of detected VOCs are summarized below. In one, monitoring well headspace gas and groundwater VOC concentrations was used to establish the direction of departure from phase equilibrium, as noted above. In the other, radioisotope data were used to establish landfill gas as the source of the VOCs detected in groundwater samples. Only a very limited number of other case studies are available in the literature, such as Kerfoot and Prattke (1992) and Liu *et al.* (1992). A recent review by Gendebien *et al.* (1992) reports two cases from California (USA) where carbon dioxide from landfill gases was claimed to have damaged groundwater quality; however, one merely establishes the release of gas from the landfill while the other does not prove that the observed groundwater contamination is not due to leachate.

Case study

In Wisconsin, USA, a programme to evaluate all the landfills in the state by low-detection-limit monitoring of groundwater for VOCs was implemented (Kerfoot and Prattke, 1992). At one landfill, where an older, unlined portion had recently had a cap installed but the gas collection system was not yet operating, concentrations of *cis*-1,2-dichloroethene (up to 10 µg dm^{-3}), 1,1-dichloroethane (up to 10 µg dm^{-3}), dichloromethane, trichloroethene, and tetrachloroethene were discovered in groundwater samples near the unlined portion. Although unsaturated zone gas-monitoring wells near those wells showed explosive conditions, indicating landfill gas effects, regulators wanted further evidence that the problem was due to gas migration and not leachate. For that reason, an evaluation of whether landfill gas was the cause of those detections was performed. It should be noted that evaluation of non-volatile parameters and the use of ^{14}C were not practical at the site, owing to the

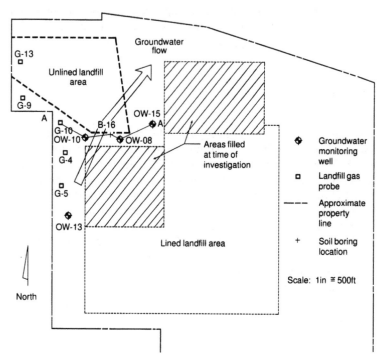

Figure 5. Landfill in Wisconsin, USA.

presence upgradient of an infiltration basin of an animal-rendering plant.

The site is shown in plan view in Fig. 5. The locations marked G-*xx* represent soil-gas sampling locations of one or two depths used to monitor for methane as an indicator of subsurface landfill gas migration. The stratigraphy of the site is described as a sandstone aquifer overlain by sand and silt layers. Table 4 lists groundwater data from the monitoring wells near the south side of the unlined portion of the facility. From that data it can be seen that OW-15 and OW-10 were the only wells with detectable VOCs in groundwater samples. OW-13 was a background well, remote from detectable soil-gas methane concentrations. Although OW-08 was in an area indicated to be affected by landfill gas, no VOCs were detected in OW-08 groundwater samples.

TABLE 4. Ground water VOC Concentrations ($\mu g/dm^3$) from a Landfill in Wisconsin, USA

Compound	H (dimensionless)	OW-08	OW-10	OW-13	OW-15
cis-1,2-dichloroethene	0.31	N(10)[a]	10	N(10)	10
1,1-Dichloroethane	0.23	N(7)	7	N(5)	10
Dichloromethane	0.11	N(5)	N(5)	N(5)	8
Trichloroethene	0.37	N(1.4)	2	N(1.4)	3
Tetrachloroethene	0.93	N(1.1)	3	N(1.1)	5

[a]N = not detected; value in parentheses is the detection limit.

Figure 6 shows the stratigraphy along the cross-section A–A indicated in Fig. 5 and the construction of the monitoring wells along that cross-section. OW-15 has the perforated interval extending above the water-table; such well construction is ideal to provide a vent to the atmosphere for migrating landfill gas, thus potentially exposing groundwater within the well to the gas. Indeed, if the migrating gas is in a higher-permeability layer that is not in contact with the capillary fringe, such well construction could result in the only significant landfill gas–groundwater contact that exists at a site.

It can also be seen that wells OW-08 and OW-10 are both in sandstone and differ mainly in the depth below the static water-table of the slotted interval through which water flows into the well. As OW-08 did not show VOCs in groundwater samples while OW-10 did, the depth of sampling may be very important for detection of VOCs due to landfill gas, owing to limited vertical penetration of gas effects.

Methane and oxygen concentrations in headspace gases were measured, and headspace gases and groundwater were sampled and analysed for VOCs. Well OW-15 showed approximately 50% methane and no detectable oxygen in the headspace gas, indicating totally landfill gas-affected headspace gases, while OW-08 and OW-10 showed 20% to 30% methane and approximately 10% to 12% oxygen, indicating partially gas-affected wells. OW-13, the background well, showed no methane and atmospheric oxygen concentrations in the headspace gas.

Figure 7 shows a plot of C_g and HC_w for OW-10 and OW-15, the two wells with VOCs detectable in groundwater samples. It can be seen that only one data point, tetrachloroethene for OW-15, indicates water-to-gas transport. Because the gas samples were taken through Teflon tubing, and tetrachloroethene shows a very high

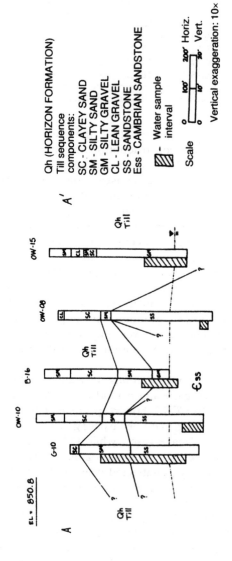

Figure 6. Cross-section A–A' (See Fig. 5) at a landfill in Wisconsin, USA.

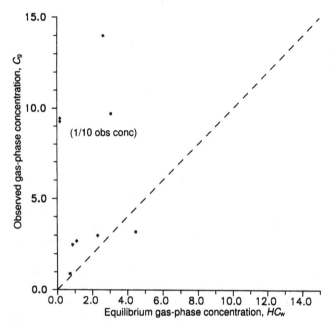

Figure 7. Monitoring-well headspace gas concentrations (C_g) and equilibrium gas concentrations for ground-water samples (HC_w).

tendency to partition from gas into Teflon (Kerfoot, 1992, unpublished data), the observed gas concentration of tetrachloroethene is suspected to be erroneously low owing to insufficient purging (approximately 40 volumes required).

The contrast between the *cis*-1,2-dichloroethene C_g/C_w ratios for OW-15 and OW-10 in Fig. 7 is of interest. Although C_w is equal for both wells, C_g is approximately ten times as high for OW-10 as C_g for OW-15. This means that gas-to-water phase transfer is approximately ten times as efficient in OW-15 as in OW-10. The greater efficiency of OW-15 in gas-to-water mass transport is understandable on the basis of the positions of the slotted intervals in the two wells relative to the water-table. The perforated portion of OW-15 extends into the unsaturated zone, and such well construction can enhance gas–water contact.

The differences in the groundwater VOC concentrations between OW-08 and OW-10 may also be significant. Although both wells had very similar gas concentrations of VOCs, one had detectable groundwater concentrations of all five VOCs in Table 3 while the other showed none. As both wells draw water from the same sandstone unit, the lower pump (relative to the water-table) in OW-08 and the resulting difference between the two wells in the depth below the water-table from which samples are taken may be the reason for the observed difference in groundwater concentrations. This is consistent with limited vertical penetration by landfill gas effects.

Case Study 2

At a landfill near Los Angeles, California, in the USA, detections of volatile organic compounds and increased alkalinity in groundwater samples during detection monitoring without concurrent increases in non-volatile parameters associated with leachate were observed. Because of the lack of significant increases in parameters associated with leachate, landfill gas was suspected as being responsible; however, regulators required more evidence. This was partly because of the high variability in background groundwater concentrations of these parameters due to seasonal changes in pumping and 'banking' of water by controlled infiltration. These processes could also greatly affect groundwater flow velocity and direction. Gas effects (as indicated by methane in monitoring wells) were quite temporally variable, as were VOC concentrations in groundwater samples. Owing either to temporal variability or to gas-to-water phase transfer occurring upgradient of the wells, an evaluation of the direction of departure from gas–water equilibrium that was performed did not give conclusive results. Because of that, and the high variability of background levels of non-volatile parameters, the above technique using radioisotopes was applied. Samples of landfill gas, leachate from the leachate-collection system, condensate from the gas-collection system, and groundwater were analysed for the ^{14}C content of CO_2, or dissolved inorganic carbon, and the ^3H content of water.

Groundwater is at a depth of approximately 50 m below land surface. The unsaturated zone and aquifer are described in drilling logs as cobbles and (occasionally) boulders, so that dual-wall percussion hammer methods are required for well installations. Because of the high cost of access to groundwater for sampling or treatment, unequivocal demonstration that a gas-control system would be sufficient for remediation was an important objective of work to evaluate the problem.

As noted above, both landfill leachate effects and landfill gas effects on groundwater will increase the ^{14}C content of the dissolved inorganic carbon in groundwater samples, while leachate effects will increase the ^{13}H content of the water and gas effects will not. Figure 8 shows the data for the liquid samples. It can be seen that all of the landfill-liquid samples (leachate, condensate) showed elevated ^3H while groundwater ^3H levels stayed essentially constant regardless of the ^{14}C level of dissolved inorganic carbon. This demonstrates that although landfill effects on groundwater samples were observed (as indicated by the ^{14}C data), no detectable leachate effects were detected, and thus the landfill effects must be due to landfill gas. Because of the low water content of landfill gas, no ^3H data are available; however, the landfill gas CO_2 ^{14}C content was approximately 120 pMC.

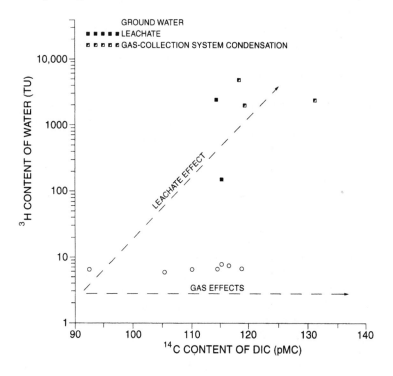

Figure 8. ^{14}C content of dissolved inorganic carbon and ^3H content of water from samples near a landfill in California, USA.

REMEDIAL MEASURES

If it is established that decreases in groundwater quality are due to landfill gas, it can be possible to control the source as a remedial measure. This is different from cases where leachate is reaching the groundwater, because the leachate stream must be intercepted to eliminate it, while gas control can be accomplished by decreasing the pressure within the landfill. Thus installation of a properly balanced gas-collection system can be sufficient in an unlined landfill, while gas effects on groundwater in a lined facility can often be ameliorated by decreasing the internal pressure near the location of the leak. However, it should be kept in mind that, even if the pressure difference within the landfill is eliminated through remedial measures, mass transport to groundwater by diffusion can still occur.

Upon implementation of steps to eliminate gas releases, decrease in the existing groundwater contamination will occur. The mechanism of this 'passive' remediation will include water-to-gas transport in a process that is the reverse of the gas-to-water process that contaminated the water. However, because of the nature of the dispersion and diffusion and other mixing processes that take place in the groundwater, such remediation will require more time than the gas-to-water transport did. Contaminant concentrations can also be significantly affected by groundwater flow and degradation of the contaminants. The contribution of the groundwater flow velocity to remediation will be site specific while the contribution of degradation to such self remediation will be compound specific and site specific.

Table 5 shows calculated groundwater concentrations of toluene at several depths below the gas–water interface based upon dilution due to volatilization and dispersion, using a calculation method similar to that of Mendoza and MacAlary (1990) without effects of contaminant degradation or fluctuations in water-table elevation. The initial concentrations are those calculated for 30 days' exposure to landfill gas with a VOC concentration equivalent to a groundwater concentration of 15 $\mu g\ dm^{-3}$ using the same approach. At time = 0, the unsaturated zone VOC concentration was set to zero, to simulate the end of landfill gas effects. From the day 0 data, the limited vertical penetration of landfill gas effects can be seen. Although the data in Table 5 carry through to the same number of days as that over which the gas effects occurred, it can be seen that the contamination has not been totally remediated.

The size of the numbers in Table 5 is noteworthy: the values demonstrate the sharp decrease with depth that can result from landfill gas effects. These results are only conceptual, and are presented to show that passive remediation will require significantly more time than the exposure period. The values that can result for specific VOCs in a particular situation will depend upon the concentration of the VOC in the gas, the Henry's law constant of the VOC, the diffusivity of the VOC in the unsaturated zone, the dispersivity of the VOC in groundwater, the exposure time, and other factors not known at present. Thus predictive models for gas effects or passive remediation must be qualitative until a better understanding of these processes is attained.

IMPLICATION

Because landfill gas can affect groundwater quality near landfills, it is important to consider this potential source of groundwater contamination in planning remedial responses. Detections of landfill gas effects on groundwater, including volatile organic compounds and changes in alkalinity, can take place during routine groundwater monitoring at landfills for detection of releases. As analytical–chemical techniques become more sensitive, effects of this transport mechanism will be observed more often. Because the remedial measures that are appropriate for gas effects on groundwater are different from those for leachate effects, identification of gas as the source of these detections and demonstration of that to regulatory authorities is important to avoid wasting resources.

TABLE 5. Groundwater VOC Concentrations ($\mu g/dm^3$) at Various Depths Below the Water Table Following Termination of Exposure to Landfill Gas

Depth (cm)				Day				
	0	4.3	8.6	12.9	17.2	21.5	25.8	30
0	15	8.9	5.7	4.0	3.0	2.3	1.9	1.6
30	0.10	0.11	0.12	0.13	0.13	0.13	0.13	0.13
60	3.6×10^{-4}	4.7×10^{-4}	5.9×10^{-4}	7.1×10^{-4}	8.4×10^{-4}	9.7×10^{-4}	1.1×10^{-3}	1.2×10^{-3}
90	8.0×10^{-7}	1.2×10^{-6}	1.8×10^{-6}	2.4×10^{-6}	3.1×10^{-6}	4.1×10^{-6}	5.1×10^{-6}	6.2×10^{-6}

From field data including the case studies above it can be seen that monitoring well design and construction can play an important role in determining whether gas can affect groundwater, and the magnitude of gas effects. Although gas effects on groundwater quality can occur at the gas–water interface in native material, a monitoring well design with a perforated interval that extends into the unsaturated zone can significantly increase gas effects on groundwater. However, numerous site-specific and compound-specific factors may also play a role in the process, so that generalizations have only limited applicability.

It should be kept in mind that computer models and field data indicate that gas effects on groundwater quality will have limited vertical penetration. This limits the severity of contamination through this mechanism. Because the contamination will typically be limited to near the surface of the water-table, passive remediation through water-to-gas phase transfer can remediate it, and for contaminants prone to oxidation by molecular oxygen, such as benzene, degradation supported by oxygen diffusion from the atmosphere may play an important role in passive remediation.

REFERENCES

Bishop, W. D. (1967). Water pollution hazards from refuse produced carbon dioxide, in *Advances in Water Pollution Research*, (eds O. Jaag and H. Liebman), Water Pollution Control Federation, Washington, DC, USA, 161–176.

Bjerg, P. L. and Christensen, T. H. (1993). A field experiment on cation exchange-affected multicomponent solute transport in a sandy aquifer. *Journal of Contaminant Hydrology*, **12**, 269–290.

Butler, J. N. (1982). *Carbon Dioxide Equilibria and Their Applications*, Addison-Wesley, London.

Deutsch, M. (1965). Natural controls in shallow aquifer contamination. *Groundwater*, **3**(3), 37–40; as cited in Matthes, G. (1978) *Properties of Groundwater* (translated by J. C. Harvey), John Wiley & Sons, New York, p. 159.

Gendebien A., Pauwels, M., Constant, M. *et al.* (1992). *Landfill Gas. From Environment to Energy*, Office for Official Publications of the European Communities, L-2985 Luxembourg.

Golwer, A. G. and Matthes, G. (1972). Die Bedeutung des Gasaustausches in der Grundluft fur die Selbstreingigungvorgange in verunreiningten Grundwassern. *Zeitschrift der Deutschen Geologischen Gesellschaft*, **123**, 29–38; cited in Matthes, G. (1982), *The Properties of Groundwater*, (translated by J. C. Harvey), Wiley-Interscience, New York, p. 115.

Kerfoot, H. B. (1994). In situ determination of the rate of unassisted degradation rate of subsurface hydrocarbon contamination. *J Air & Waste Manage Assoc* **44**, 877–880.

Kerfoot, H. B. and Prattke, M. J. (1992). False positive indications of leachate effects on groundwater due to the use of volatile organic compounds as indicator parameters: a field study of gas migration, in *Proceedings of the Sixth National Outdoor Action Conference on Aquifer Restoration, Groundwater Monitoring, and Geophysical Methods*, 11–13 May, Las Vegas, Nevada, USA, National Water Well Association, Dublin, OH, pp. 11–13.

Liu, C. L., Hackley, K. C. and Baker, J. (1992). Application of environmental isotopes to characterize landfill gases and leachate, Geological Society of America, Abstracts with Programs, 1992 Annual Meeting, Cincinnati, OH, USA, p. A35.

Mackay, D. M. and Shiou, W. Y. (1981). A critical review of Henry's law constants for chemicals of environmental interest. *Journal of Physical and Chemical Reference Data*, **10**(4), 1175–99.

Mendoza, C. A. and McAlary, T. A. (1990). Modeling of ground-water contamination caused by organic solvent vapors. *Ground Water*, **28**(2), 199–206.

Plumb, R. H. (1987). A comparison of monitoring data from CERCLA and RCRA sites. *Ground Water Monitoring Review*, **VII**(4), 94–100.

Plumb, R. H. and Pitchford, A. M. (1985). Volatile organic scans: implications for groundwater monitoring, in *Proceedings of the NWWA/API Conference on Chemicals in Ground Water*, National Groundwater Association, Dublin, OH, pp. 207–22.

3.6 Influence of Landfill Gas on Global Climate

SUSAN A. THORNELOE

US Environmental Protection Agency, Air and Energy Engineering Laboratory, Research Triangle Park, NC 27711, USA

INTRODUCTION

Methane (CH_4) produced by the decomposition of organic waste in landfills and open dumps is a significant contributor to global methane emissions. Previous estimates suggest that landfills may account for 8–20% of global anthropogenic methane emissions of 360 Tg/year or 360 million t per year (IPCC, 1992). As methane is considered a significant greenhouse gas, control of methane emissions from landfills has been targeted as part of the greenhouse gas reduction programmes.

Trace gases emitted from landfills, in particular the CFC components (chlorofluorocarbons), may also influence global climate in terms of infrared radiation absorption (greenhouse effect) and destruction of ozone. Although the importance of the trace components on global climate is well accepted (e.g. Ramanathan *et al.*, 1985), the relative contributions by trace components in landfill gas have not been quantified yet.

This chapter provides an estimate of the significance of landfill methane to the global emission of greenhouse gases.

METHANE AND ITS IMPORTANCE TO CLIMATE CHANGE

The Intergovernmental Panel on Climate Change (IPCC) has concluded that the average global temperature has increased

Landfilling of Waste: Biogas. Edited by T. H Christensen, R. Cossu and R. Stegmann. Published in 1996 by E & FN Spon, London. ISBN 0 419 19400 2.

between 0.3 and 0.6 °C (0.5–1.1 °F) over the last 100 years (IPCC, 1992). This could be attributed to climate change or to natural climate variability. With our limited understanding of the underlying phenomena, neither can be ruled out. The IPCC also concluded that emissions resulting from human activities are substantially increasing the atmospheric concentrations of the greenhouse gases (CO_2, CH_4, chlorofluorocarbons, nitrous oxide) (IPCC, 1990, 1992). General circulation models project that an increase in the concentrations of greenhouse gases, equivalent to a doubling of the pre-industrial level of atmospheric CO_2, would produce global average temperature increases between 1.9 and 5.2 °C (3.4–9.4 °F). If the higher projection proves to be accurate, substantial responses would be needed, and the stresses on this planet and its inhabitants would be serious (NAS, 1991). Currently, there are many uncertainties in the predictions, particularly with regard to timing, magnitude and regional patterns of climate change. Despite the great uncertainties, the general consensus at the United Nations Conference on Environment and Development (UNCED) held in 1992 was that greenhouse warming is a potential threat sufficient to justify action now.

Methane is a potent greenhouse gas, owing to its radiative forcing ability. Of the anthropogenic emissions, methane is estimated to contribute with 18% of the global integrated radiative forcing by gas (Table 1). The table is provided to present a general understanding of the contribution by methane to future warming based on the global warming potentials for a 100-year time horizon as presented in IPCC (1990). The global warming potential reflects the effect that releasing 1 kg of the gas would have over a specified time horizon, relative to releasing 1 kg of CO_2. However, these global warming potentials are continually being revised, owing to a variety of scientific and methodological issues. It is likely that the contribution of CFC will decrease and that the contribution of other gases will be about the same or greater upon further investigation.

Methane is about 20 times more effective at trapping heat in the atmosphere than carbon dioxide (USEPA, 1993). Methane is reported with a global warming potential of 11 over a 100-year period, but the indirect effects are comparable in magnitude to the direct effects (IPCC, 1992).

The atmospheric concentration of methane was 1.72 ppmv (parts per million on a volume basis) in 1990 or slightly more than twice that of 1750. It is rising at a rate of 0.9% per year. The

TABLE 1. Global Contribution to Integrated Radiative Forcing by Gas (USEPA, 1993)

Carbon dioxide	66%
Methane	18%
CFC	11%
Nitrous oxide	5%

doubling of the methane concentration over the last 200 years is attributed to increasing emissions from anthropogenic sources. Anthropogenic emissions currently constitute about 70% of total emissions. The contribution of major anthropogenic methane sources to global emissions is provided in Table 2.

GLOBAL LANDFILL METHANE EMISSIONS

Previous estimates for landfill methane are believed to overstate the emissions, primarily owing to limitations in available data for waste quantities being landfilled and the use of optimistic assumptions regarding anaerobic decomposition within a landfill. Using data from sites that are collecting and controlling the gas resulting from landfilled waste, the USEPA's Air and Energy Engineering Research Laboratory (AEERL) has developed a methodology for estimating global landfill methane emissions (Peer *et al.*, 1992, 1993). The country-specific estimates using this methodology are presented in Table 3. The information used to develop these estimates is also identified in Table 3. These estimates will be revised as additional data and information are being collected. For example, the estimates in Table 3 do not adjust for the type of waste being landfilled.

TABLE 2. Contribution of Major Methane Sources to Global Anthropogenic Emissions (estimates are from IPCC (1992), except for Waste Disposal Estimates from Thorneloe *et al.* (1993))

	(million tonnes/yr)	*(%)*
Coal mining, natural gas and petroleum industry	100	28
Enteric fermentation	80	23
Waste disposal (landfills, sewage, animal waste)	72	21
Rice paddies	60	17
Biomass burning	40	11

TABLE 3. Country-Specific Methane Emission Estimates from Landfills and Open Dumps

Country	Waste generated (Tg/yr)	EPA/AEERL's regression model		
		Lower bound (Tg/yr)	Midpoint (Tg/yr)	Upper bound (Tg/yr)
Africa (Ref 1–15)[a]				
Congo	0.24	0.00	0.01	0.01
Egypt	6.99	0.08	0.13	0.17
Gambia	0.08	0.00	0.00	0.00
Ghana	2.35	0.03	0.05	0.06
Kenya	2.28	0.04	0.06	0.08
Liberia	0.32	0.01	0.01	0.01
Morocco	3.12	0.05	0.08	0.11
Nigeria	10.61	0.18	0.28	0.38
South Africa	11.17	0.11	0.18	0.24
Sudan	2.79	0.03	0.05	0.07
Tanzania	2.29	0.03	0.04	0.06
Uganda	1.47	0.02	0.03	0.04
Zimbabwe	1.90	0.02	0.03	0.04
Other Africa	31.46	0.48	0.75	1.02
Total – Africa	78.00	1.1	1.7	2.3
Asia (Ref 16–29)				
Bangladesh	7.99	0.08	0.13	0.17
China	134.50	0.64	0.99	1.35
India	66.79	0.74	1.15	1.56
Iran	10.76	0.16	0.25	0.34
Iraq	4.21	0.06	0.10	0.13
Israel	1.20	0.01	0.02	0.03
Japan	41.00	0.24	0.38	0.51
Kuwait	0.59	0.01	0.01	0.02
Malaysia	2.01	0.03	0.05	0.07
Mongolia	0.18	0.00	0.00	0.01
Myanmar	3.11	0.03	0.05	0.07
North Korea	3.74	0.06	0.09	0.12
Pakistan	10.34	0.11	0.17	0.22
Philippines	7.90	0.08	0.13	0.17
Saudi Arabia	3.54	0.05	0.08	0.11
South Korea	28.11	0.04	0.07	0.09
Sri Lanka	2.39	0.02	0.04	0.05
Thailand	7.04	0.09	0.15	0.20
Turkey	9.58	0.18	0.28	0.38
United Arab Emirates	0.41	0.01	0.01	0.01
Vietnam	6.29	0.09	0.14	0.20
Other Asia	34.00	0.60	0.94	1.29
Total – Asia	390.00	3.3	5.2	7.1

TABLE 3. Continued

Country	Waste generated (Tg/yr)	EPA/AEERL's regression model		
		Lower bound (Tg/yr)	Midpoint (Tg/yr)	Upper bound (Tg/yr)
Europe (Ref 30–45)				
Albania	0.37	0.01	0.01	0.02
Austria	2.60	0.05	0.08	0.11
Belgium	3.10	0.04	0.06	0.08
Bulgaria	2.20	0.02	0.03	0.04
Czechoslovakia	2.83	0.05	0.09	0.12
Denmark	2.35	0.02	0.03	0.04
Finland	2.50	0.09	0.13	0.18
France	34.00	0.41	0.64	0.87
Germany	33.94	0.48	0.75	1.02
Greece	1.78	0.05	0.08	0.10
Hungary	3.20	0.06	0.09	0.12
Ireland	1.10	0.03	0.05	0.06
Italy	17.30	0.34	0.53	0.72
Netherlands	8.50	0.12	0.19	0.26
Norway	2.00	0.03	0.05	0.06
Poland	7.90	0.11	0.17	0.23
Romania	4.50	0.04	0.06	0.08
Spain	11.00	0.22	0.35	0.48
Sweden	2.30	0.03	0.04	0.05
Switzerland/ Liechtenstein	5.80	0.03	0.05	0.07
United Kingdom	32.00	0.75	1.18	1.60
USSR (former)	40.84	0.83	1.29	1.76
Yugoslavia (former)	3.26	0.06	0.10	0.13
Other Europe	3.20	0.06	0.10	0.13
Total – Europe	230.00	3.9	6.2	8.3
North and South America (Ref 46–48, 51–56)				
Canada	21.00	0.57	0.89	1.21
United States of America	281.20	10.90	17.00	23.10
Argentina	5.67	0.10	0.15	0.20
Brazil	31.00	0.66	1.03	1.40
Colombia	6.80	0.15	0.24	0.33
Venezuela	5.23	0.08	0.12	0.17
Other N&S America	38.35	0.53	0.83	1.13
Total – N&S America	390.00	13.00	20.00	28.00
Australia and Oceania (Ref 49–50)				
Australia	11.00	0.23	0.37	0.50

TABLE 3. Continued

Country	Waste generated (Tg/yr)	EPA/AEERL's regression model		
		Lower bound (Tg/yr)	Midpoint (Tg/yr)	Upper bound (Tg/yr)
New Zealand	2.10	0.05	0.08	0.11
Other Oceania	0.54	0.01	0.01	0.02
Total – Oceania	14.00	0.3	0.5	0.6
Total Global	1102	22	34	46

Note: Decimals in country-specific estimates do not indicate precision. Estimates are considered precise to within 2 significant figures. Total may not equal sum of individual numbers due to rounding.
[a](1) Bartone, (1990b); (2) El-Halwagi et al., (1988); (3) El-Halwagi et al., (1986); (4) Kaltwasser, (1986); (5) United Nations Development Programme (UNDP) et al., (1987); (6) Holmes, (1984); (7) Monney, (1986); (8) Cointreau, (1984); (9) Cointreau, (1987); (10) World Bank, (1985); (11) Mwiraria et al., (1991); (12) United Republic of Tanzania, (1989); (13) Verrier 1990; (14) World Resources Institute, (1990); (15) Rettenberger and Weiner, (1986); (16) Bhide and Sundaresan, (1990); (17) Bhide et al., (1990); (18) United Nations, (1989); (19) Maniatis et al., (1987); (20) Lohani and Thanh, (1980); (21) Ahmed, (1986); (22) Pairoj-Boriboon, (1986); (23) Gadi, (1986); (24) Mei-Chan, (1986); (25) Kaldjian, (1990); (26) Diaz and Goulueke, (1987); (27) Cossu, (1990a); (28) Hayakawa, (1990); (29) Swartz, (1989); (30) World Resources Institute, (1990); (31) Carra and Cossu, (1990); (32) Ettala, 1990; (33) Stegmann, (1990); (34) Ernst, (1990); (35) Cossu and Urbini, (1990); (36) Beker, (1990); (37) Gandolla, (1990); (38) Cossu, (1990b); (39) Swartz, (1989); (40) Richards, (1989); (41) Kaldjian, (1990); (42) Scheepers, (1990); (43) Bartone and Haley, (1990); (44) Bartone, (1990a,b,c,d); (45) Bingemer and Crutzen, (1987); (46) USEPA, (1988); (47) Kaldjian, (1990); (48) El Rayes and Edwards, (1991); (49) Bateman, (1988); (50) Richards, (1989); (51) Kessler, (1990); (52) Kaldjian, (1990); (53) World Resources Institute, (1990); (54) Diaz and Golueke, (1987); (55) Bartone et al., (1991); (56) Yepes and Campbell, (1990).

Ongoing research by the USEPA will result in gas potential data that will provide factors for adjusting for the type of waste being landfilled (Barlaz, 1991). This is considered important because there are definite differences in geographical regions as to the types of waste being landfilled.

There are changes occurring in waste management practices worldwide. For example, industrialized countries are adopting recycling programmes resulting in less paper, food and yard waste being landfilled. The effect of this on future landfill emissions is presently not known. Developing countries are adapting 'sanitary' landfills, resulting in increased methane emissions. Other factors important to the accurate characterization of landfill methane must also be considered, such as trends in landfill technology and implementation of regulations requiring control of landfill air emissions.

The data presented in Table 3 indicate that global methane emissions from landfills amount to 22–46 million t per year (Tg/y) (with a midpoint of 34 Tg/y) or between 6 and 13% of the global methane emissions.

The geographical distribution of the emissions indicates that about half of the emissions originate from North America. Landfills have been targeted as potential sources for control because they are amenable to cost-effective control through the utilization of the methane. By utilizing the gas, as opposed to flaring, additional environmental benefits result from an offset in power plant emissions and the conservation of global fossil fuel resources. The USA regulations for air emissions for municipal solid waste landfills are expected to result in requiring about 10–15% of the landfills to collect and control methane. These regulations are expected to result in a reduction of 7–10 million t per year (Tg/y) by 2000 or of 40–45% of the methane emissions from US municipal solid waste landfills.

ACKNOWLEDGEMENT

The research that is described in this chapter was funded through the US EPA's Global Climate Change Research Program. This chapter has been reviewed in accordance with EPA's peer and administrative review policies and approved for presentation and publication.

REFERENCES

Ahmed, M. F. (1986). Recycling of solid wastes in Dhaka, in *Waste Management in Developing Countries*, Vol. 1 (ed. K. J. Thome-Kozmiensky), EF-Verlag fur Energie und Umwelttechnik GmbH, Berlin, pp. 169–73.

Barlaz, M. A. (1991). Landfill gas research in the United States: previous research and future directions, in *Proceedings of the Landfill Microbiology*

Research and Development Workshop, United Kingdom Department of Energy, London, England, November 1991.

Bartone, C. R. (1990a). Economic and policy issues in resource recovery from municipal solid wastes. *Resources, Conservation and Recycling*, **4**, 7–23.

Bartone, C. R. (1990b). Urban wastewater disposal and pollution control: emerging issues for sub-Saharan Africa, in *Proceedings of the African Infrastructure Symposium*, The World Bank, Baltimore, MD, 01/08–09/90. p. 6.

Bartone, C. R. (1990c). Investing in environmental improvements through municipal solid waste management. Paper presented at the WHO/PEPAS Regional Workshop on National Solid Waste Action Planning, Kuala Lumpur, 26 February–2 March 1990.

Bartone, C. R. and Haley, C. (1990). The Bled Symposium: Introduction. *Resources, Conservation and Recycling*, **4**, 1–6.

Bartone, C. R., Leite, L., Triche, T. and Schertenleib, R. (1991). Private sector participation in municipal solid waste service: experiences in Latin America. *Waste Management and Research*, **9**, 495–509.

Bateman, C. S. (1988). Landfill gas development in Australia, in *Proceedings of the International Conference on Landfill Gas and Anaerobic Digestion of Solid Waste*, 4–7 October, Harwell Laboratory, Oxfordshire, UK (eds Y. R. Alston and G. E. Richards), pp. 156–61.

Beker, D. (1990). Sanitary landfilling in the Netherlands, in *International Perspectives on Municipal Solid Wastes and Sanitary Landfilling* (eds J. S. Carra and R. Cossu), Academic Press, New York, pp. 139–55.

Bhide, A. D., Gaikwad, S. A. and Alone, B. Z. (1990). Methane from land disposal sites in India, in *Proceedings of the International Workshop on CH_4 Emissions from Natural Gas Systems, Coal Mining and Waste Management Systems*, Environment Agency of Japan, the US Agency for International Development, and the US Environmental Protection Agency, Washington, DC, 9–13 April 1990.

Bhide, A. D. and Sundaresan, B. B. (1990). *Solid Waste Management in Developing Countries*, Indian National Scientific Documentation Centre, New Delhi, India, pp. 14–21.

Bingemer, H. G. and Crutzen, P. J. (1987). the production of methane from solid wastes. *Journal of Geophysical Research*, **92**(D2), 2181–7.

Carra, J. S. and Cossu, R. (eds) (1990). *International Perspectives on Municipal Solid Wastes and Sanitary Landfilling*, Academic Press, New York, pp. 1–14.

Cointreau, S. J. (1984). *Solid Waste Collection Practice and Planning in Developing Countries*, John Wiley and Sons, Chichester, pp. 151–82.

Cointreau, S. J. (1987). *Solid Waste Management Study for the Greater Banjul Area, The Gambia*, Ministry of Economic Planning and Industrial Development, Banjul, The Gambia.

Cossu, R. (1990a). Sanitary landfilling in Japan, in *International Perspectives on Municipal Solid Wastes and Sanitary Landfilling* (eds J. S. Carra and R. Cossu), Academic Press, New York, pp. 110–38.

Cossu, R. (1990b). Sanitary landfilling in the United Kingdom, in *International Perspectives on Municipal Solid Wastes and Sanitary Landfilling* (eds J. S. Carra and R. Cossu), Academic Press, New York, pp. 199–220.

Cossu, R. and Urbini, G. (1990). Sanitary landfilling in Italy, in *International Perspectives on Municipal Solid Wastes and Sanitary Landfilling* (eds J. S. Carra and R. Cossu), Academic Press, New York, pp. 94–109.

Diaz, L. F. and Golueke, C. G. (1987). Solid waste management in developing countries. *Biocycle*, **28**(6), 50–55.

El-Halwagi, M. M. *et al.* (1986). Municipal solid waste management in Egypt: practices and trends, in *Waste Management in Developing Countries*, Vol. 1 (ed. K. J. Thome-Kozmiensky), EF-Verlag fur Energie und Umwelttechnik GmbH, Berlin, pp. 283–8.

El-Halwagi, M. M. *et al.* (1988). Municipal solid waste management in Egypt, in *Proceedings of the Fifth International Solid Waste Conference*, September, International Solid Waste and Public Cleansing Association, Copenhagen, Denmark, pp. 415–24.

El Rayes, H. and Edwards, W. C. (1991). Inventory of CH4 emissions from landfills in Canada. Prepared for *Environment Canada*, Hull, Quebec, pp. 25–69.

Ernst, A. (1990). A review of solid waste management by composting in Europe, *Resources, Conservation and Recycling*, **4**, 135–49.

Ettala, M. O. (1990). Sanitary landfilling in Finland, in *International Perspectives on Municipal Solid Wastes and Sanitary Landfilling* (eds J. S. Carra and R. Cossu), Academic Press, New York, pp. 67–77.

Gadi, M. T. (1986). in *Waste Management in Developing Countries*, Vol. 1, (ed. K. J. Thome-Kozmiensky), EF-Verlag fur Energie und Umwelttechnik GmbH, Berlin, pp. 188–94.

Gandolla, M. (1990). Sanitary landfilling in Switzerland, in *International Perspectives on Municipal Solid Wastes and Sanitary Landfilling* (eds J. S. Carra and R. Cossu), Academic Press, New York, pp. 190–98.

Hayakawa, T. (1990). The Status Report on waste management in Japan – Special focus on methane emission prevention, in *Proceedings of the International Workshop on Methane Emissions from Natural Gas Systems, Coal Mining and Waste Management Systems*, Environmental Agency of Japan, US Agency for International Development, and US Environmental Protection Agency, 9–13 April, pp. 509–23.

Holmes, J. R. (1984). Solid waste management decisions in developing countries, in *Managing Solid Wastes in Developing Countries* (ed. J. R. Holmes), John Wiley & Sons, Chichester, pp. 1–17.

Intergovernmental Panel on Climate Change (IPCC) (1992). *Climate Change 1992. The Supplementary Report to the IPCC Scientific Assessment*, Cambridge University Press, UK.

Kaldjian, P. (1990). *Characterization of Municipal Solid Waste in the United States: 1990 Update*, Prepared for US Environmental Protection Agency, Office of Solid Waste and Emergency Response, EPA-530-SW-90-042 (NTIS PB90-215112), June 1990.

Kaltwasser, B. J. (1986). Solid waste management in medium sized towns in the Sahel area, in *Waste Management in Developing Countries*, Vol. 1 (ed. K. J. Thome-Kozmiensky), EF-Verlag fur Energie und Umwelttechnik GmbH, Berlin, pp. 299–307.

Kessler, T. (1990). *Brazilian Trends in Landfill Gas Exploitation*, STATEC Consultores s/c Ltda, São Paulo, Brazil.

Lohani, B. N. and Thanh, N. C. (1980). Problems and practices of solid waste management in Asia. *Journal of Environmental Science*, (May–June), pp. 29–33.

Maniatis, K. *et al.* (1987). Solid waste management in Indonesia: status and potential. *Resources Conservation Recycle*, **15**(87), 277–90.

Mei-Chan, L. (1986). Waste management in the Taiwan area, in *Waste Management in Developing Countries*, Vol. 1 (ed. K. J. Thome-Kozmiensky), EF-Verlag fur Energie und Umwelttechnik GmbH, Berlin, pp. 299-307.

Monney, J. G. (1986). Municipal solid waste management – Ghana's experience, in *Waste Management in Developing Countries*, Vol. 1 (ed. K. J. Thome-Kozmiensky), EF-Verlag fur Energie und Umwelttechnik GmbH, Berlin.

Mwiraria, M. *et al.* (1991). *Municipal Solid Waste Management in Uganda and Zimbabwe*, Draft report of the United Nations Development Programme and The World Bank, 18 May.

National Academy of Sciences, (1991). *Policy Implications of Greenhouse Warming*, National Academy Press, Washington, DC.

Pairoj-Boriboon, S. (1986). State-of-the-art of waste management in Thailand, in *Waste Management in Developing Countries*, Vol 1 (ed. K. J. Thome-Kozmiensky), EF-Verlag fur Energie und Umwelttechnik GmbH, Berlin.

Peer, R. L., Eppeson, D. L., Campbell, D. C. and van Brook, P. (1992). *Development of an Empirical Model of Methane Emissions from Landfills, Final Report*, Prepared for US Environmental Protection Agency, Office of Research and Development, Air and Energy Engineering Research Laboratory, EPA 600/R-92-037 (NTIS PB92-152875), March.

Peer, R. L., Thorneloe, S. A. and Epperson, D. L. (1993). A comparison of methods for estimating global methane emissions from landfills. *Chemosphere*, **26**, (1–4), 387–400.

Ramanathan, V., Cicerone, R. J., Singh, H. B. and Kiehl, J. T. (1985). Trace gas trends and their potential role in climate change. *Journal of Geophysical Research*, **90**, 5547–66.

Rettenberger, G. and Weiner, K. (1986). Measures to improve the situation in the field of sanitation and solid waste management in Juba, Sudan, in *Waste Management in Developing Countries*, Vol. 1 (ed. K. J. Thome-Kozmiensky), EF-Verlag fur Energie und Umwelttechnik GmbH, Berlin, pp. 289–98.

Richards, K. M. (1989). Landfill gas: working with Gaia. *Biodeterioration Abstracts*, **3**(4), 317–331; and *CAB International*, **3**(4), December 1989.

Sandelli, G. J. (1992). *Demonstration of Fuel Cells to Recover Energy from Landfill Gas, Phase I Final Report: Conceptual Study*. Prepared for US Environmental Protection Agency, Office of Research and Development, Air and Energy Engineering Research Laboratory, EPA-600/R-92-007 (NTIS PB92-137520), January 1992.

Scheepers, M. J. J. (1990). Landfill gas in the Dutch perspective, *International Conference on Landfill Gas: Energy and Environment '90*, Session 3.2. United Kingdom Department of Energy and Department of the Environment, Bournemouth, UK, October.

Stegmann, R. (1990). Sanitary landfilling in the Federal Republic of Germany, in *International Perspectives on Municipal Solid Wastes and Sanitary Landfilling* (eds J. S. Carra and R. Cossu), Academic Press, New York, pp. 51–66.

Swartz, A. (1989). *Overview of International Solid Waste Management Methods*, State Government Technical Brief 98-89-MI-2, American Society of Mechanical Engineers, Washington, DC.

Thorneloe, S. A. *et al.* (1993). Global methane emissions from waste management, *The Global Methane Cycle: Its Sources, Sinks, Distributions and Role in Global Change*, NATO.

United Nations (1989). *City Profiles*, Prepared by the United Nations Centre for Regional Development and the Kitakyushu City Government, Kitakyushu, Japan.

United Nations Development Programme (UNDP), The World Bank, and the Canadian International Development Agency, (1987). *Master Plan for Resource Recovery and Waste Disposal, City of Abidjan. Final Report*, Prepared by Roche Ltd. Consulting Group, Sainte-Foy, Quebec, Canada, February.

United Republic of Tanzania (1989). *Masterplan of Solid Waste Management for Dar es Salaam. Volume II: Annexes. Ministry of Water, Department of Sewerage and Sanitation*, Prepared by HASKONING, Royal Dutch Consulting Engineers and Architects, Nijmegen, The Netherlands, and M-Konsult Ltd, Consulting Engineers, Dar es Salaam, Tanzania, pp. 1–39, 47–9, 74–94, 109–24, 142–5.

US Environmental Protection Agency (1988). *Report to Congress, Solid Waste Disposal in the United States*, Volume 1, EPA/530-SW-88-011 (NTIS PB 89-110381), October.

US Environmental Protection Agency (1993). *Report to Congress, Anthropogenic Methane Emissions in the United States: Estimates for 1990*, EPA/430-R-93-003, April.

Verrier, S. J. (1990). Urban waste generation, composition and disposal in South Africa, in *International Perspectives on Municipal Solid Wastes and Sanitary Landfilling*, (eds J. S. Carra and R. Cossu), Academic Press, New York, pp. 161–76.

World Bank (1985). *Metropolitan Area of Douala. Study of Waste Management and Resource Recovery. Part A. Phase I*, Prepared by Motor Columbus, Consulting Engineers, Inc., CH-5401 Baden, Switzerland.

World Resources Institute (1990). *World Resources 1990–91*, Oxford University Press, New York.

Yepes, G. and Campbell, T. (1990). *Assessment of Municipal Solid Waste Services in Latin America*, Report in progress prepared for The World Bank, Technical Department, Infrastructure and Energy Division, Urban Water Unit, Latin America and the Caribbean Region, pp. 1–6, 20–26.

3.7 Landfill Gas Odours

FRANZ-BERND FRECHEN

Aqua System Consalt, Am Hackenbruch 47, 40231 Düsseldorf, Germany

INTRODUCTION

Emissions of odours at landfills, although usually considered as secondary emissions, are now gaining increasing attention due to the annoyance they may cause in the area surrounding the landfill.

This chapter presents some of the basic aspects of odours, odour measurement and evaluation at landfills, and also describes a case study at a German landfill.

BASICS OF ODOUR RECEPTION AND MEASUREMENT METHODS

Odour Reception

Odour is a human sensation caused by the presence of so-called 'odorants', and is based upon a two-step process:

ODORANT → reception (physiological) → interpretation (psychological) → ODOUR IMPRESSION

The physiological part of this process is not yet known completely. In addition, psychological interpretation is an individual process that cannot be described in terms of universally valid relations or equations. The intensity of the impression reported by various test persons differs, and up to now it has not been possible

Landfilling of Waste: Biogas. Edited by T. H Christensen, R. Cossu and R. Stegmann. Published in 1996 by E & FN Spon, London. ISBN 0 419 19400 2.

to give equations that are able to predict the odour impression in relation to the presence of different air compounds. All kinds of effects, from antagonistic through additive to synergistic effects, have been reported so far. Thus, although the presence of odorants leads to an odour impression, these substances are not the main point of interest, as will be explained later on.

Measurement Methods

Figure 1 gives a general view of the situation when landfill air must be analysed and assessed.

The left part of Fig. 1 deals with the chemical aspect of the air to be analysed. Different methods, ranging from tube tests to gas chromatography in connection with various detectors, are applicable and give information about specific gas compounds. If the composition is unknown, usually mass spectrography is used to detect the compounds present in the air. But as it is not possible to evaluate odour from these substance-related values, this part of the figure is headed 'no odour'.

The relevant problem in the case of odour is not the presence

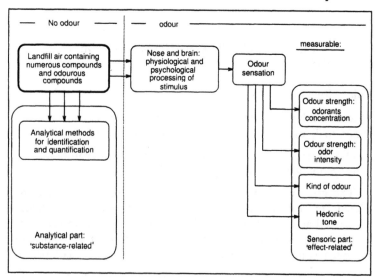

Figure 1. Analytical and sensory measurement methods for landfill air.

or absence of specific air compounds, which can be determined by means of analytical measurement techniques in terms of **compound-related** measurements, but to obtain knowledge about the effects of the air compounds: that is, to get **effect-related** measurements. Therefore measurement techniques must be based upon the judgement of test persons. The right part of Fig. 1 deals with odour. But as can be seen, different questions lead to different measurements.

The most important parameter is the 'strength of odour – **odorant concentration**'. The basic concept of measuring odorant concentration is to dilute an air sample with clean, odourless air until half of the test panel does not smell anything while the other half of the test panel still smells something: this is the odour perception threshold. The number of dilutions necessary to reach the odour perception threshold is called the 'odorants concentration (o.c.)' and is given units of 'odour unit per cubic metre (o.u./m^3)', although in fact it has no dimensions. The concept behind this is that a certain amount of 'odorant' is present, leading to the resulting odour impression. The connection between dilution and one cubic metre of air is of course arbitrary. As the device that is necessary to prepare the mixture of sample air and odourless air is called an 'olfactometer', this measurement technique is called 'olfactometry'. The measurements presented in this chapter – as far as they concern odour – are olfactometric measurement results, and thus are given as odorant concentrations. They give quite comparable and reproducible results. The VDI guideline 3881 is currently under revision, and a new issue is expected in 1995.

However, when talking about emissions, it must be kept in mind that the odorant concentration is just that – a concentration. In order to determine the source strength, it is necessary to know the odorant mass flow emitted, which is the emission concentration multiplied by the emission airflow. In cases where the sources do not have a measurable airflow, as is usually the case for landfills, it is difficult to determine the emitted odorant mass flow. In these cases special sampling has to be used, as described later.

The **intensity of odour** has to be assessed by the test persons. In order to maintain comparability, the semantic concepts used must be identical. In Germany today the following concept applies:

0 No odour perceivable.
1 Barely perceivable (perception threshold).

2   Faintly perceivable (recognition threshold).
3   Clearly perceivable.
4   Strong.
5   Very strong.
6   (Stronger than 5).

The association with intensity numbers is arbitrary at first but helps in the judging. It becomes more important when describing the above-threshold intensity behaviour of different airs in accordance with different theories of psychometry, such as the Weber–Fechner law. The German VDI guideline 3882 (VDI 3882/1) gives details of this.

The **kind of odour**, given by direct judgement, spontaneously described or according to different semantic concepts, is important when it is necessary to detect specific sources of odours.

The **hedonic tone** is highly important, as it gives the location of the odour impression on a scale representing the dichotomy of 'pleasant – unpleasant'. The German VDI guideline 3882/2 covers this question and gives advice for the conduction of the test. Considering landfill site odours, a simple answer may be given by saying that the odour emitted from these facilities is generally 'unpleasant'. For practical purposes, this statement may currently be sufficient, although the extent of unpleasantness will have to be considered in the future.

It must be understood that there is a big difference in exactness between analytical and sensory measurement techniques. This is important when evaluating measurement results, immission prognoses or any other kind of results presented in this concern.

ODOUR EMISSION/TRANSMISSION AND IMMISSION

Like other gaseous air pollutants, odorants are dispersed in the atmosphere, and thus cause immissions in the vicinity of landfill sites. Therefore in order to comprehend the whole process correctly, it is necessary to work on the three parts of the odour occurrence which are emission, transmission, immission.

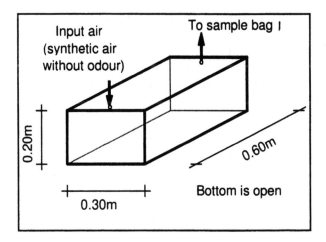

Figure 2. Sampling box for emission measurement.

Emission

It is not sufficient to measure the emission concentration, but also the emission mass flow, in order to carry out transmission calculations. The problem can be solved easily provided that there is a source of odours with measurable airflow. Usually with landfills it is not possible to measure an emitted airflow. Important sources, such as the landfill surface, are area sources without defined or measurable airflow. Thus it is necessary to determine the source odorant mass flow by using a special sampling box with an open base (Fig. 2). While the sample air is sucked into the sample bag, odourless air can flow into the sample box. As the base area of the box is known – 0.18 m^2 in this case – and as the flow of sample air in m^3/h can be chosen, it is possible to determine the specific sample air flow in m^3/(m^2 h). Determining the odorant concentration in the sample in o.u./m^3 leads to the specific odorant mass flow in o.u./(m^2 h) and thus also to the total odorant mass flow in o.u./h, which is necessary as input for the dispersion calculation.

Another approach is to conduct field inspections at different meteorological situations while at the same time determining the emission concentration via olfactometric measurement. Then a recalculation taking account of the actual meteorological circumstances can be made to determine the total emitted odorant mass flow of the source. This gives a theoretical value for the emitted air flow, which then can be used for further calculations. In general, this method is not recommended, as two sensory methods – determination of the odorant concentration at the emission source and field inspection – have to be used, and a correct description of the dispersion parameters is necessary, leading to the fact that the results are more uncertain than when using the special sampling method mentioned above. However, under certain circumstances it is necessary to use this method, especially when measurements with the sample box cannot be carried out. This is the case, for example, during turnover of composting windrows, as during this activity strong odours are emitted that cannot be connected directly with a specific surface emission.

Transmission

Transmission of odorants is usually described according to the transmission of gaseous compounds in air: thus atmospheric dispersion calculations are applicable. In Germany, the use of the Gaussian dispersion model is prescribed by law, and on behalf of the federal authorities a computer program was designed to be used for this task. Knowledge of the meteorological situation is necessary and is introduced by a three-parameter statistic consisting of wind direction, wind speed and dispersion class. Usually, these statistical meteorological data represent a ten-year period and are provided by the German Weather Agency (DWD).

The problem with this calculation is that it produces hourly means of immission concentration, whereas people smell odours within much shorter periods than 1 h. Thus, depending upon the variation of concentrations, it may often occur that the hourly mean is below the odour threshold (or any other given limit value), but there are several time periods within the hour where the concentration is above the limit value. This was first pointed out by Högström (1972), and is explained in Fig. 3. In order to assess more correctly the extent of odorous immission to people it is stipulated that an hour may be recognized as exceeding the limit value if the limit value is exceeded during 10% of 1 h: i.e. during 6 min. In practice, it is assumed to be sufficient to multiply the hourly mean by a factor of 10.

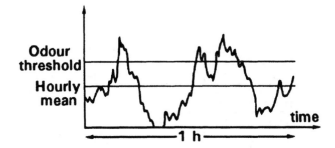

Figure 3. Schematic diagram showing how the concentration of odorous matter can exceed the odour threshold several times during a period when the hourly mean is much below the odour threshold.

The result of a dispersion calculation is the cumulated frequency of odorant concentration for every receptor point (immission point). Figure 4 presents the results of an atmospheric dispersion calculation for a landfill site under operation for one of the relevant receptor points (point no. 50). If, for example, the limit value was 1 o.u./m^3, it can be seen from the diagram that this value was exceeded for 7.5% of the time. For 5% of the time an immission concentration of 1.3 o.u./m^3 was exceeded.

Immission (Impact on Surrounding Area)

Very few immission standards exist with respect to odour. In Germany, the regulations demand that

- no unacceptable annoyance is present if only during *less than* 3% of the hours of one year odours are present that are above the *perception threshold*; and that
- unacceptable annoyance is present if during *more than* 5% of the hours of one year odours are *clearly perceivable*.

From this it can be seen that the standards to be met consist of two parameters that are connected to each other: concentration and percentage of time. The two limits given are an upper and a lower limit; in fact there is a wide range of immission situations that fall between those two limits and which have to be assessed regarding individual circumstances. Figure 4 shows that the immission situation at this receptor point falls between the two limits stipulated, so

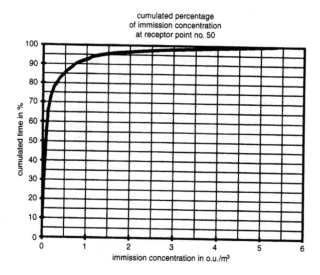

Figure 4. Cumulated frequency of immission concentrations at a receptor point.

in this case further knowledge concerning actual circumstances, for instance the land use at the receptor point, is necessary in order to judge if the immission at this point is acceptable or not.

Knowledge about odour immissions in the vicinity of odour-emitting facilities can be gained using two different methods:

- atmospheric dispersion calculations, providing data about the odour sources and meteorology; or
- field evaluation.

The method of using dispersion calculations has already been discussed. It can be applied if all relevant sources of odours are known and the emission parameters are determined correctly. If the facility with the interesting odour sources is under design, it is necessary first to perform an **emission prognosis** to obtain a set of source parameters for the dispersion calculation.

Field evaluations can be done using two different approaches, depending on the kind of information that has to be obtained:

- field inspection carried out by a test person panel according to VDI guideline 3940, duration one year or, in special cases, ½ year; or

- questionnaires to the people living in the respective area in terms of long-term questionnaires or repeated short-term questionnaires (VDI 3883, 1991).

There are some remarkable differences between these methods. The most important one is that in the case of field inspection the result is an immission parameter comparable with the result of dispersion calculations, given in time per year during which a certain odour level (in terms of odorant concentration) is exceeded, and thus also compatible with the kind of regulatory units mentioned above, whereas questionnaires will give results in terms of annoyance parameters of the people living in the immission area. Today, no regulations concerning these annoyance parameters exist, although, regarding the main task of immission prevention, precisely these standards would be most important and helpful.

A CASE STUDY

At a properly equipped and operated landfill many sources of odorous air are found. Frechen (1989) discussed the parts of the landfill that have to be considered as relevant sources of emissions, such as actual waste tipping front, site surface, slopes, leachate wells, leachate storage facilities and passive degassing installations. In fact, in most of these cases the origin of odour is biogas formed inside the landfill body, except the odour emitted during tipping of the waste. The situation may be different when other facilities are also located on the site, such as composting facilities, as will be seen in the presented case study.

Emissions

Within a study concerning a landfill site odour emissions should be measured, possible measures for emission reduction should be developed, and the effectiveness of these measures should be predicted.

Figure 5 shows the different kinds of waste deposited in the landfill, which had a surface area of about 50 000 m². In addition to the landfill a biowaste-composting facility, composting green wastes, was present within the landfill boundaries.

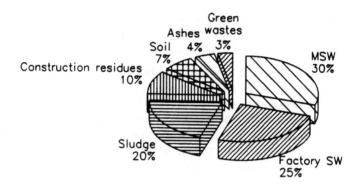

Figure 5. Waste composition in weight percent at the studied landfill.

A measurement programme was carried out at the landfill in order to determine the actual odour emissions. The main odorous emissions were:

- composting area, 45×10^6 o.u./h;
- young section of the landfill, 22×10^6 o.u./h;
- older section of the landfill (> ½ year), 15×10^6 o.u./h.

Based upon measurements at the landfill during actual operation, a specific emission rate of 500–1000 o.u./$(m^2 h)$ was determined.

Odour Controls

Frechen *et al.* (1991) described a variety of possible counter-measures against odour emissions, such as:

- location and design of landfills (with special regard to installation or enhancement of active degassing facilities);
- operation of landfills;
- use of new landfill technologies (with special regard to waste pretreatment);
- measures for reducing emitted gas flow;
- measures for reducing odorant concentration of emitted gas.

The listed measures are not all applicable in every case, but must be selected according to local conditions.

In the case presented here it seemed most effective to remove the composting facility to a place far away from housing areas. This is, however, not a removal of the source in general, but considering odour immissions it is the most effective way to reduce significantly emissions from the site.

Furthermore, an immission prognosis was needed for the future operation of the landfill, considering a mean height of the landfilled waste of about 30 m and a total area of 200 000 m^2. Studies on possible operation led to a maximum/minimum estimation, depending upon several factors such as

- the kind of waste to be landfilled in the future;
- the pre-treatment of waste;
- the different specific biogas production rates resulting from the above-mentioned factors;
- the kind of operation during placement of waste;
- the covering strategy during operation;
- the efficiency of the exhausting system.

The quantity of biogas produced as well as the composition of biogas will have a major influence on odour emissions. The main compounds of biogas are not very odorous, but the trace compounds are. Investigations concerning the composition of biogas and the effects of several operating factors on those very odorous trace compounds that are responsible for high odorant concentrations of the biogas are necessary, and current knowledge about this is very poor. Thus estimations had to be made on former olfactometric measurements showing 40 000–20 000 o.u./m^3. Assuming a specific emission potential of 0.01 m^3/(m^2 h) without degassing and an odorant concentration of 200 000 o.u./m^3, the total emission would amount to 400×10^6 o.u./h for the plant. Emission control measures including a high vacuum, sufficient amount of wells and pipes, appropriate well depth interval and others will lead to an overall gas extraction efficiency of about 90%. In addition, it must be recognized that designing and using gas extraction with the aim of emission control leads to a reduction of gas production, as more air is introduced in the perimeter zones, providing aerobic conditions in this part of the landfill body, which is most important for the emission characteristics of the site. Another effect of this type of degassing is that in the perimeter zones a reduction of odorant concentration can be found owing to dilution with fresh air – which is of no influence on total emitted odorant mass flow – and also due to aerobic biological degradation of the most odorous biogas compounds inside this zone, which affects the total emitted odorant mass flow. Estimations indicate that the emission flow rate could be reduced to 0.001 $m^3/(m^2 h)$ and the odorant concentration to about 100 000 o.u./m^3, yielding a total emitted odorant mass flow of around 10×10^6 o.u./h for the landfill surface itself.

Immission Prognosis

Different levels of control measures were refined and dispersion calculations were carried out to evaluate the efficiency of the respective measures. The comparison cannot be based only on the resulting total emitted odorant mass flows, as landfills are huge area sources and the local distribution of emissions is of importance, especially concerning receptor points near the landfill site borders. Figure 6 gives an example of the result of one of the dispersion calculations. The figure gives the percentage of time during which an immission concentration of 1 o.u./m^3 is exceeded for the respective receptor point. In addition, the curves of equal percentages for 3%, 5% and 10% of time are indicated.

It must, however, be kept in mind that the question whether immissions are acceptable or not is not easy to answer. As mentioned above, the limit value of 1 o.u./m^3 today is not the only stipulated value. Regarding illegal immissions, the stipulation is 'clearly perceivable odour', which of course is not connected with the odour threshold, but may be associated with a value of 3–5 o.u./m^3. Assuming this value, the areas in which the percentages are 3%, 4% and 5% respectively would be much smaller than those indicated in Fig. 6.

PROBLEMS AND FUTURE NEEDS

Concerning odour emissions and control at landfill sites a lot of problems and unanswered questions still exist, although the results of immission prognoses may look quite acceptable and reliable: which, as explained, is not true. Questions and thus future research needs can be divided into three major sections.

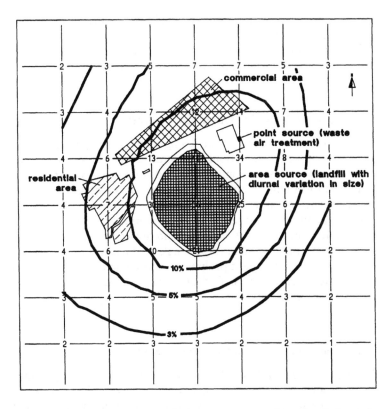

Figure 6. Results of a dispersion calculation: percentage of time that the immission concentration exceeds 1 o.u./m^3.

Basics of Odours and Measurement Improvement

Olfactometric measurements still vary too much between different institutes carrying out the measurement. Thus there is a need for more standardized conduction of sampling and measurement and for standardizing the test person panel by means of reference measurements. Reference measurement will probably have to be carried out with hydrogen sulphide H$_2$S as the reference odorant, and the results of test person panels should be around 3 µg/m^3. Although hydrogen sulphide has some disadvantages as a reference odorant, the introduction of daily panel tests and recalculation of all measured odorant concentrations to the standard value of 3 $\mu g/m^3$ together with quality criteria for the panel itself will undoubtedly give much more comparable and reliable results.

Biogas Quality Regarding Odour Concentration

As mentioned above, the main components of biogas are of no interest with respect to odour problems. In fact, the trace compounds are the most odorant compounds of biogas. An estimation of the quantity of these compounds, depending on the kind of waste, pretreatment, organic content, content of special substances, humidity, and other factors, is extremely difficult, which makes it impossible to evaluate the resulting odorant concentrations. Thus at present the only possible way to make estimates is to use results from former measurement programmes at comparable sites, or to conduct test programmes. But even when doing so the possible variations are very important.

Dispersion Calculations and Immission Standards

Although it may look as if dispersion calculations and the stipulation of immission standards are independent from each other, this is not the case, because in Germany stipulation of immission values is implicitly connected with the method of conducting dispersion calculations. Problems arising from the dispersion models are numerous and are well known, such as the incompatibility of long-distance and short-distance models, model simplifications for practical use, lack of exactness when describing meteorology, and the fact that most models are steady-state models, disregarding the dynamics of meteorology, influences of buildings, influences of orography and others. Of course, modelling can be very advanced today and may even include wind-tunnel tests, but the disadvantages are that these experiments are very complicated and costly and that, when introducing dynamic simulation into modelling, the main problem will arise with defining the 'critical case' of weather situation.

Additionally, the problem of the short-time effects of odour immissions has to be considered, and aspects such as 'adaptation' and 'difference between laboratory and field situation' may indicate the problems arising with this special topic. The actual advice in Germany, 'multiply calculated immission concentration by a factor of 10', is very simple and is connected to the model used.

Besides this, actual policy in Germany disregards the fact that annoyance is always connected with the above-threshold immission concentrations. Thus a pair of limit values (immission concentration and percentage of time) using an immission odorant concentration of 1 o.u./m^3 may be conservative, but of course is not satisfying under general considerations. Connection to parameters that are able to describe annoyance would be much better for the task, but may not make the job much easier.

REFERENCES

VDI (1986). VDI guideline 3881: Olfactometry – odour threshold determination, Part 1, May 1986, *VDI-Handbuch Reinhaltung der Luft*, volume 1.

VDI (1991). VDI guideline 3883: Effects and assessment of odours – determination of annoyance parameters by questioning, draft edition 1991.

VDI (1992). VDI guideline 3882/1: Olfactometry – determination of odour intensity, Part 1, October 1992, *VDI-Handbuch Reinhaltung der Luft*, volume 1.

VDI (1993). VDI guideline 3940: Determination of odourants in ambient air by field inspection, October 1993, *VDI-Handbuch Reinhaltung der Luft*, volume 1.

VDI (1994). VDI guideline 3882/2: Olfactometry – determination of hedonic odour tone, Part 2, September 1994, *VDI-Handbuch Reinhaltung der Luft*, volume 1.

Frechen, F-B. (1989). Odour Emissions and Controls, in *Sanitary Landfilling – Process, Technology and Environmental Impact* (eds T. H. Christensen, R. Cossu and R. Stegmann), Academic Press, London, pp. 425–436.

Frechen, F-B., Kettern, J. T. and Köster, W. (1991). Odorous emissions of landfill: estimating and reducing environmental impacts, in *Proceedings of Third International Landfill Symposium*, 14–18 October, Cagliari, Italy. CISA, Università di Cagliari, Cagliari, Sardinia, Italy, pp. 1139–1150.

Högström, U. (1972). A method for prediction odour frequencies from a point source. *Atmospheric Environment*, **6**, 103–121.

3.8 Emission and Dispersion Modelling of Landfill Gas

STEFANO CERNUSCHI & MICHELE GIUGLIANO

DIIAR – Environmental Section, Technical University of Milan, Piazza L. da Vinci, 32, 20133 Milan, Italy

INTRODUCTION

Biological activities, physico-chemical volatilization and gas-generating chemical reactions are the most important processes regulating landfill gas production and composition.

The dispersion into the atmosphere of landfill gas and of flue gases from landfill gas utilization can involve serious problems of atmospheric pollution, which can have an important role in the environmental impact assessment of the waste landfill. The atmospheric emissions of major concern are the trace components of toxicological significance and odorous substances released both with the gas migrating out of the landfill and with flue gases coming from point sources such as utilization plants and flares.

This chapter reviews the available methods for the estimation of the emission and the atmospheric dispersion of landfill gas. Some outlines are also reported about the approach for assessment of the risk to human health of landfill gas.

EMISSION RATE ESTIMATION

The evaluation of the areal emission rate of gaseous pollutants from landfills is a very difficult problem, owing to the high number of

Landfilling of Waste: Biogas. Edited by T. H Christensen, R. Cossu and R. Stegmann. Published in 1996 by E & FN Spon, London. ISBN 0 419 19400 2.

factors affecting the emission process. Emission rate values can be obtained, in principle, with two different approaches:

- utilization of theoretical and/or empirical models of gas generation and migration processes through the landfill;
- calculation of the emission from the measurement of the pollutant concentration immediately above and/or in the surroundings of the waste site.

Municipal Solid Waste Landfills

For municipal solid waste (MSW) landfills, the sizeable body of scientific data that has become available on the production and migration characteristics of gas has led to the proposal of some empirical interpretation schemes of the processes involved. However, a comprehensive model relating quantitatively the complex interactions between the decomposition processes and the physical environment, and also applicable for the evaluation of the emission of pollutants, has not yet been proposed.

The quantity of gas dispersed to the atmosphere through the top cover of the landfill depends, essentially, on the gas production rate, on its migration properties through the waste deposited and through the top layer of the landfill, on the collection efficiency of the gas extraction system and on the factors affecting the transfer of the gas from the exposed area to the atmosphere. Gas production rates are strongly dependent on the organic matter content of the refuse and its biodegradability and on the moisture and the temperature inside the landfill, with a significant decrease over time. Theoretical predictions and laboratory-scale studies can give only approximate values, whose comparison with data obtained from field studies is made difficult by the problems related to the physico-chemical characterization of the landfill environment and to the age of the refuse disposed in different landfill areas. Total specific gas productions reported in the literature for MSW (Shen, 1981; Hoecks, 1983; Stegmann, 1988; Gendebien *et al.*, 1992) Ehrig (1995), Willumsen (1996) cover a relatively broad range, between $120 \text{ m}^3 \text{ t}^{-1}$ and $300 \text{ m}^3 \text{ t}^{-1}$; specific production rates can be as high as $15–20 \text{ m}^3 \text{ t}^{-1} \text{ year}^{-1}$ in the early stages of the decomposition process (up to the first 5 years), with a decrease to values in the range $4–8 \text{ m}^3 \text{ t}^{-1} \text{ year}^{-1}$ in the successive 5–30 years. Mean rate values derived from full-scale landfills (Stegmann, 1988;

Willumsen, 1996), for a period of 10–20 years are between 1 and 10 m^3 t^{-1} year^{-1}.

The migration through the landfill of the gas produced, and its ultimate release to the atmosphere, are direct consequences of the pressure and concentration gradients of the gas inside the landfill. Movement of the gas is dependent on many factors, difficult to include in a simulation model, related to transport properties of the gas itself (diffusivity, viscosity), to physical and chemical character-istics of the waste (permeability, moisture content, temperature) and to the layout and efficiency of the gas collection system. Pressure is exerted by the gas in all directions inside the landfill, so its movement is not restricted to the vertically upward direction. How-ever, the gas tends generally to migrate preferentially upward to the cover layer (Shen, 1981), as the lateral movement will often be restricted by the static pressure within the landfill and the higher flow resistance of the surrounding soils and, eventually, of the bottom and lateral liners.

Release of gas to the atmosphere through the top layer of the landfill is also dependent on some of the factors that influence the migration inside the landfill, such as the transport properties of the gas and the permeability, moisture content and thickness of the cover. The release is also affected by some factors related to atmo-spheric meteorological conditions: mainly the wind speed (increased wind at the surface speeds up diffusion), barometric pressure fluctuations (pumping action from pressure fluctuations enhances diffusion) and air temperature (larger temperature differ-ences between the interior of the landfill and the outside top surface favour thermal diffusion).

As stated previously, a model including all the aspects related to gas production, migration and emission in MSW landfills is still not available. Emission rates can thus be evaluated, as a first approxi-mation, utilizing mean values for the gas production rate (m^3 of gas per tonne of refuse per year), and assuming the release to the atmosphere, through the top cover of the landfill, of all the gas that is not collected by the extraction system. This requires, in turn, knowledge of the gas collection efficiency of the extraction system, dependent on the characteristic of the system itself and on the particular landfill. The limited data available (Esposito, 1984; Tent and van den Berg, 1992) cover a relatively broad range, with lower limits of around 20% and maximum values as high as 90% for landfills with top liners and very efficient gas compression systems.

Emission rates E (m^3 m^{-2} year^{-1}) can thus be evaluated with the following equation:

$$E = (G) (\gamma_R)(1 - \eta)(L) \tag{1}$$

where G is the mean specific gas production rate (m^3 t^{-1} year^{-1}), γ_R is the refuse density (t m^{-3}), η is the gas collection efficiency of the extraction system, and L (m) is the depth of the landfill. Utilizing the mean values for the production rate (4–8 m^3 t^{-1} year^{-1}) and an extraction efficiency of 60%, equation (1) results in emissions between 25 and 50 m^3 m^{-2} year^{-1} for a refuse density of 0.8 t m^{-3} (Findikakis and Leckie, 1979), and a mean landfill depth of 20 m. The values are comparable with those reported in the literature (Frechen and Kettern, 1988). Emission rates E_u (g m^{-2} s^{-1}) for the compounds of interest can then be evaluated from the concentration C (g m^{-3}) of the compound in the gas and the emission rate E.

Although the approach outlined gives only approximate values, it should be considered conservative for atmospheric dispersion calculations: neither the gas remaining inside the landfill nor the gas that escapes from its lateral boundaries is subtracted from the values obtained.

Industrial Waste Landfills

For industrial waste landfills, the emission of gaseous compounds is primarily determined by the volatilization rate of the compound and its migration properties through the waste and the landfill top cover to the atmosphere.

The rate of the chemical waste volatilization at landfill sites is dependent upon the physical and chemical properties of the waste and the surrounding environment. The processes involved in the volatilization are essentially three (Shen and Tofflemire, 1980): volatilization of an organic liquid from a pure solution or a mixture of chemicals, volatilization from a water solution, and volatilization from soils or solids on which the compound is adsorbed. The first process is strongly influenced by the vapour pressure of the compound, and hence by the temperature, and by its molecular diffusion through the vapour phase close to the chemical or mixture of compounds. For the second process, water solubility also plays an important role: compounds with low vapour pressure but also with limited solubility in water can be vaporized in rates comparable with those of chemicals with vapour pressures several orders of magnitude higher but readily soluble in water. In the third process, the adsorption capacity of the solid, the strength of the adsorption between the compound and the solid and the effective surface area for the desorption process are all important in determining the volatilization rate.

In a chemical waste landfill, a complex combination of the three processes mentioned is involved. The rate of diffusion of the vaporized compound through the waste and the top cover, as stated previously, must also be considered. This, in turn, is affected by the permeability and water content of the waste and the cover, the gas pressure inside the landfill, the diffusivity of the compound at the air–surface interface and the wind speed. Some studies have been conducted to estimate theoretically the emission rates, taking into account the factors mentioned. Among the few relationships developed, one of the most easily applicable (Shen, 1980) was derived on the basis of Fick's law of diffusion, assuming that a given concentration of chemical waste volatilizes, resulting in a saturated vapour concentration, and diffuses through the landfill top cover at a rate influenced by the diffusion coefficient, the porosity and the cover depth. The equation is as follows:

$$E_i = D \ C_S P^{4/3} L^{-1}(W_i/W) \tag{2}$$

where E_i is the emission rate of the compound of interest (g m^{-2} s^{-1}); D is the diffusion coefficient of the compound in air (cm^2 s^{-1}); P is the top soil porosity (%); L is the depth of soil top cover (cm); W_i/W is the weight fraction of the compound in the landfill; and C_S is the saturated vapour concentration of the compound (g cm^{-3}).

The saturated vapour concentration C_S can be calculated from the ideal gas law:

$$C_S = \frac{pM}{RT} \tag{3}$$

where p is the vapour pressure (mmHg) at the temperature T (°K), M is the molecular weight (g mol^{-1}) and R, the universal gas constant, has a value of 6.23×10^4 mmHg cm^3 (mol^{-1}K^{-1}). The top soil porosity P can be derived from the soil bulk density d_s (g cm^{-3}) and the particle density d_p (g cm^{-3}) by the following equation:

$$P = \frac{1 - d_s}{d_p} \tag{4}$$

With most common values for d_s and d_p, P can range from 0.24 to 0.62 (Ehrenfeld, 1986).

Equation (2) was derived assuming a diffusion of the compound of interest through a completely dry soil, and should then be considered as a worst-case assumption for emissions evaluation: any water added to the soil would reduce the air-filled soil porosity.

More refined modelling approaches are based on the two-resistance theory of mass transfer, and include also the term related to the mass flux of the vapour from the air–soil interface to the overlying atmosphere (Ehrenfeld, 1986): however, the air-side resistance to mass transfer has been demonstrated to be negligible (Thoma *et al.*, 1992), and the latter term can thus in practice be disregarded. Other equations derived from the same mass transfer approach consider also the transport of chemical vapours towards the surface created by the eventual formation of landfill gas: the flux expression, therefore, contains both a diffusive and a convective term, and is especially applicable to co-disposal of hazardous waste with municipal or organic wastes.

FIELD MEASUREMENTS OF GAS EMISSIONS

The methods previously outlined for the evaluation of emissions from landfills are based on the utilization of gas production rates or volatilization properties of the landfilled waste, and on hypotheses related to the migration of the gaseous compounds through the landfill to the atmosphere. Emission rate values can also be obtained with a different approach, involving the field measurement of the pollutant concentration immediately above or in the surroundings of a real operating landfill. These methods include direct and indirect measurement techniques.

In direct measurement techniques, the landfill emission is properly sampled and analysed by the use of passive sampling methods or flux chambers. Passive sampling (Marrin, 1988) involves the utilization of sorbent probes buried in the subsurface soil for a specific time period, in order to trap gaseous contaminants that diffuse upwards through the landfill. With respect particularly to VOCs (volatile organic carbon compounds), the method is relatively simple but also time consuming, and does not have the capture efficiency for all the compounds of interest. Flux chamber techniques (Reinhart, 1992) involve the utilization of an open bottom chamber flushed with a known flow rate of clean air and placed over a representative point of the landfill: the emission rate of the pollutant is determined from its concentration in the outlet gas, the gas flow and the chamber volume. Flux chamber methods have been utilized to measure emissions from typical areal sources (landfills, chemical spill sites, surface impoundments): however, if the source is not homogeneous the emission values obtained cannot be considered as truly representative of the entire area.

Indirect measurement techniques involve the measurement of ambient air concentrations of the pollutant around the source. Emission rates can be evaluated by the integration of the wind velocity and concentration profiles determined over the source plume by continuous measurements at multiple points and heights downwind of the area source. The method is little affected by the heterogeneities of the surface but depends heavily on the accuracy of the wind and concentration measurements, the latter frequently at very low values, and thus requires an extensive and costly system of instruments. An alternative method to the use of a fixed network has also been proposed (Esplin, 1988), requiring the measurement of the average pollutant concentration downwind of the source at, at least, three different heights: average values are determined on composite samples taken with sampling trains hoisted, at the desired heights, by a balloon tethered to a cart, which moves crosswind with respect to the source plume. The technique has been applied successfully to the evaluation of total reduced sulphur emission rate from an aerated effluent lagoon, yielding values with a standard deviation of 15% over four different tests, conducted with significant variations in the wind speed (1.0–2.7 m s^{-1}) and direction and in the estimated plume boundary height (60–400 m). In comparison with other methods that rely on field measurements, the technique appears to be relatively simple and reasonably accurate, although it requires a careful measurement of the wind speed and direction during sampling and the availability of atmospheric stability that cannot be readily obtainable.

FLARE AND STACK EMISSIONS

Emissions arising from on-site disposal or utilization of landfill gas in flares and gas fuel engines should be evaluated from the volume flow and the pollutant concentration of flue gas produced. For the combustion process of landfill gas, specific flue gas volume G (m^{-3} $m^{-3}CH_4$) at normal temperature and pressure (0 °C, 101.6 kPa) can be estimated by

$$G = 11 + 9.5e + \frac{100 - [CH_4]}{100} \tag{5}$$

where $[CH_4]$ is the methane content (% vol.) of the gas feeder and e is the excess air utilized, evaluated from the oxygen content $[O]$ (vol.%) of the flue gas:

$$e = \frac{[O]}{21 - [O]} \tag{6}$$

Trace component concentrations in the gas emitted are highly dependent on landfill gas composition and combustion conditions: in the absence of any direct measurement, a representative range of values can be obtained from the few literature data available (see Chapter 6.2; also Muller and Hor, 1989; Young and Blakey, 1991).

ATMOSPHERIC DISPERSION MODELLING

The estimation of pollutant ground level concentrations downwind of the emitting source can be conducted with the widely utilized Guassian model of atmospheric transport and diffusion, whose basic formulation for point sources is as follows:

$$C(x,y) = Q(\pi u \sigma_y \sigma_z)^{-1} \exp\left[-y^2/\left(2\sigma_y^2\right)\right]\exp\left[-H^2/\left(2\sigma_z^2\right)\right] \tag{7}$$

where $C(x,y)$ (g m^{-3}) is the ground-level concentration at downwind distance x and crosswind distance y from the source, Q (g s^{-1}) is the mass emission rate, u (m s^{-1}) is the mean wind velocity at the effective release height H (m), and σ_y (m) and σ_z(m) are the diffusion coefficients in the horizontal and vertical directions respectively. The effective release height H is given by the sum of the geometric height of the source and of the plume rise determined by momentum and buoyancy forces: the latter, which can increase the geometric height by a factor of 2–10 times, is a very important parameter in determining maximum concentration values, and is usually evaluated with the formulations derived by Briggs (1969). The diffusion coefficients σ_y and σ_z are correlated with the potential dispersion capacity of the atmosphere, and are usually evaluated with relationships defined in terms of the Pasquill stability classification scheme (Hanna, 1982).

The basic Gaussian formulation is well suited for the evaluation of pollutant dispersion from typical landfill gas exploitation systems, in which the emission from the combustion installation is normally released into the atmosphere through a stack. Emissions from gas disposal by flares can still be considered as point sources, with some differences from conventional plumes related to the effective release height. Flare emissions are associated with a diffusion flame, which may release an appreciable amount of radiative heat: therefore, the sensible heat release rate utilized for plume rise evaluation (Briggs, 1969) should not consider this latter amount lost due to radiation, estimated to be roughly half of the total combustion heat of the flared gas (Leahey, 1984).

Dispersion calculations for diffuse emissions arising from the whole landfill should be performed considering the areal geometric configuration of the source. The basic Gaussian formulation for point sources can still be utilized, with the introduction of proper modifications in order to account for the initial spreading of the areal emission plume already at the source. The approach commonly utilized for the inclusion of this initial concentration spreading still considers the area source as a point source, but locates it at a distance upwind of the area boundary. This virtual distance is then added to the real source–receptor distance for the evaluation of the horizontal and vertical dispersion coefficients σ_y and σ_z utilized in equation (7). The area source is, in practice, approximated by a virtual upwind point source, whose distance from the real source is chosen to give to the plume of the virtual point source the lateral and vertical dispersion that it already has at the effective source location. The distance is dependent on atmospheric stability, and is evaluated from the relationships that approximately fit the Pasquill–Gifford curves for σ_y (m) and σ_z (m) (Turner, 1969):

$$\sigma_y = px^q \tag{8}$$

$$\sigma_z = ax^b \tag{9}$$

where p, q, a and b are stability-dependent coefficients and x is the distance (km). The lateral and vertical distances, x_y and x_z respectively, are then given by:

$$x_y = \left(\frac{\sigma_{yo}}{p}\right)^{1/q} \tag{10}$$

$$x_z = \left(\frac{\sigma_{zo}}{a}\right)^{1/b} \tag{11}$$

where σ_{yo} (m) and σ_{zo} (m) are the standard deviations of the initial lateral and vertical plume concentration distributions at the real source. For a surface-based source the initial lateral dispersion σ_{yo} is set equal to the length S (m) of the side of the source divided by 4.3 (USEPA, 1987):

$$\sigma_{yo} = \frac{S}{4.3} \tag{12}$$

and the initial vertical dispersion σ_{zo} is assumed to be equal to the vertical dimension z (m) of the source divided by 2.15 (USEPA, 1987):

$$\sigma_{zo} = \frac{z}{2.15} \tag{13}$$

The area source is required to be square. Sources with irregular shape, which constitute the most frequent situations, can be simulated by subdividing them into multiple squares that approximate the geometry of the total area. However, the evaluation of the ground concentration must then be conducted for every single source. A more simple, but less exact, approximation utilizes the square root of the area as the length S of the side, thus requiring the evaluation of the dispersion from just a single source.

The concept of a virtual upwind point source is also included in a dispersion equation for a surface-based area source recommended by the USEPA (Baker and Mackay, 1985) for assessments of air pollution from landfills. The equation is derived on the basis of a uniform horizontal concentration distribution in every wind-rose sector Θ, commonly utilized in long-term climatological models, with a virtual distance L, equal to 2.15 times the crosswind width of the landfill, added to the real downwind distance x from the landfill centre to the receptor:

$$C(x,\Theta) = 16Q[2^{1/2}\pi^{3/2}(x + L)\sigma_z u]^{-1} \tag{14}$$

where 16 is the number of sectors in which the wind rose is usually divided.

The initial area source plume dispersion can also be taken into account by considering the source as a finite crosswind line source (USEPA, 1987). Initial lateral plume spreading is included by considering the source as a line source, where the initial vertical spreading is simulated with the addition of a vertical virtual distance x_z, equal to the length S of the side of the area source, in the σ_z evaluation with equation (9). Ground level concentration $C(x,y)$ for a surface-based source can then be calculated with the following equation:

$$C(x,y) = QS[(2\pi)^{1/2}u\sigma_z] \; \{\mathrm{erf}[(0.5S' + y)(2^{1/2}\sigma_y)^{-1}]$$
$$+ \mathrm{erf}[(0.5S'-y)(2^{1/2}\sigma_y)^{-1}]\} \tag{15}$$

where erf is the error function (Burington and May, 1970) and S' in the effective crosswind width of the source, defined as the diameter of a circle whose area is the same as the area of the source:

$$S' = 2S(\pi)^{-1/2} \tag{16}$$

The approach also requires the area source to be square: for irregular sources, the same considerations previously outlined can still be applied.

Virtual point-source models appear to be indicated for the evaluation of ground-level air pollution from landfill diffuse emissions. A comparative study conducted on a hazardous waste landfill (Baker et al., 1985) resulted in more accurate estimations of the ambient concentrations monitored in the surrounding of the site when virtual source models were utilized. Average concentrations, obtained over a 10-day period, resulted in overestimation in the range 8–97%, compared with an overprediction of 374% obtained with the simple point-source model. Virtual source models also appeared to be more precise, with the smallest values of scatter around the mean and around the zero overprediction over the 10-day period.

As stated previously, emissions from landfills typically contain compounds characterized by short-term, acute effects (i.e. odorous substances) and by long-term cumulative effects (i.e. toxic trace substances). The assessment of the air quality impacts of the two categories of substances must then be conducted utilizing different approaches in the application of atmospheric dispersion models.

For compounds with acute effects, short-term versions of the models should be applied. These models calculate hourly concentration values, and consequently require meteorological data available on the same time basis. For toxic compounds with cumulative effects, long-term models, which calculate mean annual concentrations, must be applied. The models utilize statistical summaries of meteorological data in the form of joint frequency tables of wind speed, wind direction and atmospheric stability: each entry in the table represents the time fraction with which the considered combination wind speed–wind direction–stability occurs in a year. Mean annual concentrations at the receptor are then simply evaluated by summing up the concentrations corresponding to the particular wind speed and direction–stability combination multiplied by the corresponding frequency of occurrence of the combination.

EXAMPLE OF IMPACT ASSESSMENT OF LANDFILL GAS EMISSIONS ON THE ATMOSPHERIC ENVIRONMENT

The evaluation of the impact on air quality due to emissions arising from landfills makes use, as previously outlined, of a multidisciplinary approach.

The identification and quantification of the emission of compounds developing short-term effects (typically odorous substances such as thiols, alkylbenzenes, and limonene) is carried out from the type of waste landfilled. In the same way, the emission of trace substances with long-term effects (typically toxics such as benzene, organohalogenated compounds, and mercury vapours) should also be evaluated, particularly for industrial landfills.

Emissions of concern from gas utilization or flare systems should also be properly considered: the evaluation is normally conducted from flue gas quantity and composition, and may also include some conventional pollutants typically generated by the combustion process (i.e. SO_2, NO_x and CO).

The atmospheric transport and diffusion of pollutants with long-term effects is described with mathematical dispersion models in climatological versions, suitable for area sources. The concentrations are calculated on a long-term basis (for example, annual mean values), with the meteorological data usually required in input as the joint frequency of occurrence of the several combinations of wind velocity, wind direction and atmospheric stability. The influence of the mixing layer on the vertical diffusion can normally be neglected, as the source is located at ground level (USEPA, 1987). In Fig. 1 the mean annual concentration contours of benzene (a typical toxic substance) in the surroundings of a MSW landfill are

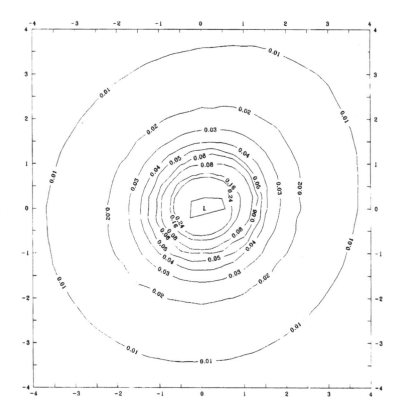

Figure 1. Contour chart showing mean annual concentration of benzene ($\mu g/m^3$) in the surroundings of a MSW landfill.

reported: the estimation of atmospheric diffusion was conducted utilizing the climatological version of the ISCLT dispersion model of the USEPA (1987). The same model was also utilized for the evaluation of concentration contours of a generic pollutant emitted by a flare, and reported in Fig. 2.

For pollutants with short-term effects, the atmospheric dispersion should be evaluated with models resulting in short-term concentration estimates (for example, mean hourly values). A screening model can be utilized in advance for the recognition of the critical combinations of wind velocity and atmospheric stability for the area of concern. The short-term dispersion model, applied with such critical meteorological data as input, gives the concentra-

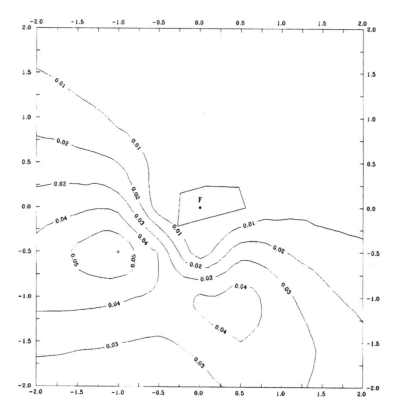

Figure 2. Contour chart showing concentration of a generic pollutant emitted by a flare.

tion contours determined by every critical combination recognized. The number of hours per year in which the estimated concentration is expected to occur can then be obtained from the joint frequency table of meteorological data, already utilized for long-term evaluations, determining the occurrence of the critical wind velocity–stability combinations for every wind direction considered. In Fig. 3, the concentration of propylbenzene (tracer of odorous compounds) around a MSW landfill for a critical combination of wind velocity and atmospheric stability is reported in terms of mean hourly concentration contours. The number appearing in each one of the eight sectors in which the wind direction has been divided gives the hours per year of occurrence of the concentration reported

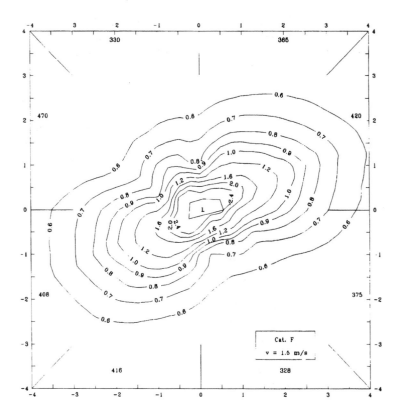

Figure 3. Mean hourly concentrations of propylbenzene ($\mu g/m^3$) around a MSW landfill.

in every sector. The atmospheric dispersion calculations were conducted with the short-term version of the ISCST model of the USEPA (1987).

The methodological approach illustrated up to now is well suited for the evaluation of the alteration of air quality in the surroundings of the landfill site. But some pollutants, owing to their high chemical stability and toxicity characteristics, are able to reach human subjects and adversely affect their health even a long time after they have been removed from the atmosphere. The exposure to these compounds, mainly toxic metals and organic polynuclear or halo-genated compounds, takes place through multiple pathways, with direct exposure from contaminated air inhalation frequently proving to be the least relevant (Travis, 1987; Stevens, 1989; Levin, 1991).

This results, both in the definition of air quality standards and in the evaluation of the environmental impact, in the necessity of a novel approach, which should also consider properly indirect exposure pathways.

A brief schematization of the approach to be adopted, for which extensive details can be found in the literature (Giugliano, 1991; Hattemer-Frey, 1991; Levin, 1991) is shown in Fig. 4. The evaluation of ground-level concentrations and depositions of the toxic contaminant constitute the fundamental basis for the subsequent estimation of direct and indirect human exposure respectively. The simulation can be performed by climatological models: the toxic action of the compounds considered is essentially related to chronic effects, so the long-term values, generally averaged on an annual basis, can conveniently be utilized. The distribution of the contaminant in the different environmental compartments through which

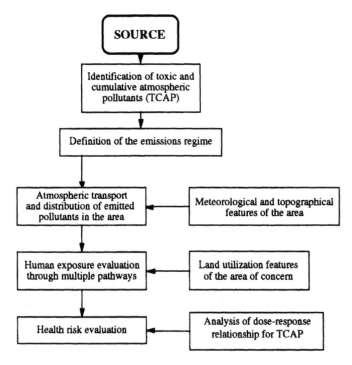

Figure 4. Generalized flow-sheet of the methodology for health risk assessment of toxic pollutants.

the main routes of exposure originate (soil, water, food) is subsequently evaluated by further environmental transport models. The efficiency of different direct (air inhalation) and indirect (dermal contact, soil and water ingestion, dietary intake) pathways in determining the total exposure of human subjects is estimated through biochemical and metabolic information, and the value obtained utilized to quantify the health-related risk. For toxic compounds without known or suspected carcinogenic effects, risk is evaluated by comparison with reference dosages with no adverse effect (ADI, admissible daily intake or NOAEL, no observed adverse effect level), whereas for carcinogenic substances, even if only suspected, risk is expressed in terms of the probability that an individual will develop cancer after exposure to the calculated dosage, over a 70-year mean assumed lifetime, derived from toxicological models of dose–response experimental data (Ricci, 1985; Colombi, 1990; Hattemer-Frey, 1991). The values obtained, even with all the uncertainties contained in the evaluation, can be utilized as a valuable index of the health impact of the emission, and represent an important contribution for the decision procedure on the acceptability of the source.

REFERENCES

Baker, L. W. and Mackay, K. P. (1985). Screening models for estimating toxic air pollution near a hazardous waste site. *JAPCA*, **35**, 1190–95.

Briggs, G. A. (1969). *Plume Rise*, USAEC Critical Review Series, TID-25075, NTIS, Springfield, VA.

Burlington, R. S. and May, D. C. (1970). *Handbook of Probability and Statistics with Tables*, 2nd edn., McGraw-Hill Book Co., New York.

Colombi, E., Maroni, A. and Foa, V. (1990). Valutazione dell'impatto ambientale-l'analisi della componente salute. *Difesa Ambientale*, **5**, 54–9.

Ehrenfeld, J. and Ong, K. (1986). *Controlling Volatile Emissions at Hazardous Waste Sites*, Noyes Publications, Park Ridge, NJ.

Esplin, G. J. (1988). Boundary layer emission monitoring. *JACPA*, **38**, 1158–61.

Esposito, A. (1984). Recupero energetico del biogas prodotto da una discarica controllata di rifiuti solidi urbani, in *Proceedings of CISPEL Course 'Il recupero energetico dai rifiuti'*, Gardone Riveria (Bs), 8–12 October.

Findikakis, A. and Leckie, J. O. (1979). Numerical simulation of gas flow in sanitary landfills. *Journal ASCE*, **105**, 927–45.

Frechen, F. B. and Kettern, J. T. (1988). Odorous emissions of domestic and toxic waste landfill sites, in *Proceedings ISWA Seminar 'Can landfill technology be improved? – Current technology and environmental aspects*', ENVIRO 88, Amsterdam (Netherlands), 19–23 September.

Gendebien, A., Pauwels, M. Constant, M., *et al.* (1991). Landfill gas: from environment to energy. State of the art in the European Community context, in *Proceedimgs of Sardinia '91, Third International Landfill Symposium,* S. Margherita di Pula, Cagliari, Italy, 14–18 October, pp. 69–75.

Giugliano, M. and Cernuschi, S. (1991). La valutazione quantitativa del rischio di esposizione ad inquinanti atmosferici tossici e persistenti. Il caso dell'incenerimento di rifiuti solidi, *Ingegneria Ambientale*, **XX**, 593–600.

Hanna, S. R., Briggs, G. A. and Hosker, R. P. (1982). *Handbook on Atmospheric Diffusion,* US Dept. of Energy, Technical Information Service, Oak Ridge, TN.

Hattemer-Frey, H. and Travis, C. C. (1991). *Health Effects of Municipal Solid Waste Incineration,* CRC Press, Boca Raton, FL.

Hoecks, J. (1983). Significance of biogas production from waste tips. *Waste Management and Research,* **1**, 323–35.

Leahey, D. M. and Davies, M. J. (1984). Observations of plume rise from sour glass flares. *Atmospheric Environment,* **18**, 917–22.

Levin, A., Fratt, D. B., Leonard, A., Bruins, J. F. and Fradkin, L. (1991). Comparative analysis of health risk assessment for municipal solid waste combustors. *Journal of Air and Waste Management Association,* **41**, 20–31.

Marrin, D. L. and Kerfoot, H. B. (1988). Soil-gas surveying techniques, *Environmental Science and Technology,* **22**, 740–45.

Muller, U. and Hor, B. (1989). Measurements of dioxins and PCBs at torches and gas engines – immission-prognosis models for PCBs and odours of landfills, in *Recycling International,* (ed. K. J. Thome-Kozmiensky, Vol. **3**, ER-Verlag für Energie und Umwelttechnik GmbH, Berlin, Germany, pp. 1766–73.

Reinhart, D. R., Cooper, D. C. and Walker, B. L. (1992). Flux chamber design and operation for the measurement of municipal solid waste landfill gas emission rates, *Journal of Air and Waste Management Association,* **42**, 1067–70.

Ricci, P. and Molton, L. S. (1985). Regulating cancer risks. *Environmental Science and Technology,* **19**, 473–79.

Shen, T. (1981). Control techniques for gas emissions from hazardous waste landfills. *JAPCA,* **31**, 132–5.

Shen, T. and Tofflemire, T. J. (1980). Air pollution aspects of land disposal of toxic wastes. *Journal ASCE,* **106**, 211–26.

Stegmann, R. (1988). Landfill gas as an energy source, in *Proceedings International ISWA Conference 'Valorization des déchets – Aspects économiques'*, Paris, France, 22–24 April, pp. 311–22.

Stevens, J. R. and Swackhamer, D. L. (1989). Environmental pollution – a multimedia approach to modelling human exposure. *Environmental Science and Technology,* **23**, 1180–86.

Tent, J. and van den Berg, J. J. (1992). Emissions and emission control at landfill sites, in *Proceedings Ninth World Clean Air Congress,* Montreal (Canada), 30 August–4 September, Vol. I, IU-4A.10.

Thoma, G. J., Hildebrand, G., Valsaraj, K. T., Thibodeaux, L. J. and Springer, C. (1992). Transport of chemical vapours through soil: a landfill cover simulation experiment. *Journal of Hazardous Materials,* **30**, 333–42.

Travis, C. C. and Hattemer-Frey, H. (1987). Human exposure to dioxin from municipal solid waste incineration. *Waste Management and Research*, **9**, 151–6.

Turner, B. (1969). *Workbook of Atmospheric Dispersion Estimates*, US Public Health Service, Cincinnati, OH.

USEPA (1981). *Evaluation guidelines for toxic air emissions from land disposal facilities*, Office of Solid Waste, Washington, DC.

USEPA (1987). *Industrial Source Complex (ISC) Dispersion Model User's Guide*, 2nd edn. (revised), USEPA 450/4-88-002a.

Willumsen, H. C. (1996). Actual landfill gas yields, in *Landfilling of Waste: Biogas*, (eds T. H. Christensen, R. Cossu and R. Stegmann), E & FN Spon, London, pp. 293–295.

Young, C. P. and Blakey, N. C. (1991). Emissions from power generation plants fuelled by landfill gas, in *Proceedings of Sardinia '91, Third International Landfill Symposium*, 14–18 October, S. Margherita di Pula, Cagliari, Italy, pp. 359–367.

4. GAS PRODUCTION

1 INTRODUCTION

4.1 Modelling Landfill Gas Production

RAFFAELLO COSSU[a], GIANNI ANDREOTTOLA[b] &
ALDO MUNTONI[a]

[a]DIGITA – Department of Geoengineering and Environmental
Technologies, Faculty of Engineering, University of Cagliari,
Piazza d'Armi, I-09123 Cagliari, Italy
[b]Department of Civil and Environmental Engineering, University of
Trento, Via Mesiano 77, 38050 Trento, Italy

INTRODUCTION

There is a pressing need for landfill gas (LFG) models able either
to forecast the yield and production rate of biogas generated, or to
evaluate the potential gas migration and related problems.

The development of landfill gas models started in the 1970s,
when several authors summarized experimental data on a rational
basis (Alpern, 1973; Boyle, 1976; Ham *et al.*, 1979). Qualitative
models were developed by Farquhar and Rovers (1973). Quantitative models were later provided in the USA (Palos Verdes, Scholl
Canyon and Sheldon Arleta models).

Depending on the approach, different classifications of models
are possible.

A general classification can be based on the availability of data
and the state of knowledge of the system, according to the diagram
shown in Fig. 1:

- *statistical analysis*, when a large number of data are available,
  but knowledge of the system is inadequate, and the data are
  collected for different purposes; this kind of model does not
  assume any cause–effect relation or deal with the temporal

Landfilling of Waste: Biogas. Edited by T. H Christensen, R. Cossu and
R. Stegmann. Published in 1996 by E & FN Spon, London.
ISBN 0 419 19400 2.

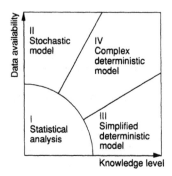

Figure 1. Model classification based on knowledge level and data availability (Marsili-Libelli, 1989).

dynamics of the system, but presents the general characteristics of the data 'population' and provides correlations;

- *stochastic model,* which describes the temporal trend of data without explaining the same; this kind of model is useful for describing the behaviour of a black-box system; it states simply which is the output related to a specific input;
- *simplified deterministic model,* which requires knowledge of the mechanisms governing the system; it is able to describe the behaviour of the system with simplified mathematical equations;
- *complex deterministic model,* which acts in a similar way to the above-mentioned model using more complex mathematical equations.

The majority of LFG models belong to the third group. The deterministic models can be further divided into static and dynamic models. In static models there is an instantaneous relation between input and output, meaning that the system has no memory of the past input and output; the state of the system is stationary, there is no time influence. In dynamic models the relation between input and output is not instantaneous, and state variables that describe the temporal evolution of the system should be introduced.

Biogas generation models can be further distinguished into different classes (Andreottola and Cossu, 1988):

- empirical models;
- stoichiometrical models;

- biochemical models;
- ecological models.

Empirical models deal with a black-box system, which is a system characterized by input and output data; generally the mathematical function that relates input to output is based on a time series of experimental data. Stoichiometric models are based on a global stoichiometric reaction, where the waste is represented by an empirical formula; generally this kind of model leads to the highest potential yield of biogas. Biochemical models consider the biodegradability of the different components of waste, and each differs in terms of kinetic expression, number of substrata and parameters. Ecological models deal with the ecosystem on which the process is based and describe the relation between the system components: these are the more complex models.

Structure of LFG Production Models

Theoretically a complete biogas model should include three sub-models:

- *Stoichiometric submodel.* This gives the maximum theoretical yield of biogas from the anaerobic degradation of the organic waste fraction. Some models proposed in the literature are simply stoichiometric and provide as a result only information on LFG yields. These models are static, according to the classification previously reported.
- *Kinetic submodel.* This is a dynamic model, which gives as a result the temporal evolution of LFG generation rates. It can be either an empirical model, based on a more or less simple equation of a defined order; or a deterministic model, based on a set of equations describing the degradation of the different biodegradable MSW fractions; or an ecological model, which describes the dynamic of microbial populations and substrata within the landfill.
- *Diffusion submodel.* This is a dynamic model, which describes the time and space variation of pressure and gas composition within the landfill body. LFG emission rates can be obtained, and the effectiveness of the gas extraction system can be verified.

In the present chapter only generation models are considered, and therefore only the aspects regarding stoichiometry and kinetics of LFG will be analysed in detail in the following sections.

ESTIMATION OF MAXIMUM LFG YIELDS

A reaction representing the overall methane fermentation process for organics in solid waste can be represented by the following equation:

$$C_aH_bO_cN_d + nH_2O \rightarrow xCH_4 + yCO_2 + wNH_3 + zC_5H_7O_2N + \text{energy} \tag{1}$$

where $C_aH_bO_cN_d$ is the empirical chemical formulation for bio-degradable organics in solid waste, and $C_5H_7O_2N$ is the chemical formulation of bacterial cells.

The biodegradable organic carbon found in MSW is transformed during anaerobic degradation into methane and carbon dioxide. The energy content of the organic matter is split into the free energy content of methane, the energy for bacterial synthesis and the enthalpy of the reaction.

The fraction of the organic matter that is converted to biomass, considering an infinite retention time in the system, is about 4% (EMCON, 1980). Therefore, for the practical evaluation of the maximum theoretical LFG yield, cell conversion of organic matter can be neglected, and equation (1) becomes

$$C_aH_bO_cN_d + \frac{4a - b - 2c + 3d}{4} \cdot H_2O$$
$$\rightarrow \frac{4a + b - 2c - 3d}{8} \cdot CH_4 + \frac{4a - b + 2c + 3d}{8} \cdot CO_2 + dNH_3 \tag{2}$$

Equation (2) has been reported several times in the literature (Tchobanoglous, 1977; Ham and Barlaz, 1987). Once the elementary composition of the waste is known, this equation permits evaluation of both the quantity and quality of the gas ($CH_4 + CO_2$) generation. Table 1 summarizes some typical data on the elementary composition of the biodegradable fraction of MSW (Tchobanoglous *et al.*, 1993).

Equation (2) states that 1 mol of organic carbon is bioconverted to 1 mol of landfill gas. Therefore, given that 1 mol of gas at 0 °C and 1 atm occupies 22.4 l:

TABLE 1. Typical Data on Elementary Composition of MSW Organic Fractions Expressed as Percentage of Dry Weight (Tchobanoglous *et al.*, 1993)

Component	Wet weight (%)	Dry weight (%)	Elementary composition (%)					
			C	H	O	N	S	Ash
Food wastes	11.4	4.6	4.7	4.7	4.4	13.0	10.0	4.0
Paper	42.8	55.0	50.8	53.0	61.3	18.5	60.0	55.2
Cardboard	7.5	9.8	9.2	9.4	11.1	3.7	10.0	8.0
Plastics	8.8	11.9	15.1	13.8	6.8	–	–	19.8
Textiles	2.5	3.1	3.6	3.3	2.4	14.8	–	1.4
Rubber	0.6	0.9	1.4	1.4	–	1.9	–	1.4
Leather	0.6	0.7	0.9	0.8	0.2	7.4	–	1.1
Yard wastes	23.3	11.2	11.4	10.8	10.8	40.7	20.0	8.3
Wood	2.5	2.8	2.9	2.8	3.0	–	–	0.6
Total	100.0	100.0	100.0	100.0	100.0	100.0	100.0	100.0

$$1 \text{ mol C in organic matter} = 22.4 \text{ l gas } (CH_4 + CO_2) \qquad (3)$$

On a weight basis:

$$1 \text{ g C in organic matter} = 1.867 \text{ l gas } (CH_4 + CO_2) \qquad (3')$$

Equation (2) allows us to estimate the maximum theoretical yield of landfill gas, starting from the general formula characterizing solid wastes ($C_aH_bO_cN_d$). The estimation can be also carried out directly according either to the specific formula of organic compound to be decomposed or to the empirical formulae introduced to represent biodegradable fractions of MSW (Table 2).

Table 2 also reports gas yield and composition deriving from the degradation of the single compounds (or classes of compounds).

The estimation of methane yield from MSW degradation can also be given in terms of COD, defined as the consumption of oxygen for the complete oxidation (biologically or thermally) of the organic matter (g O_2/g organic matter).

Oxidation of 1 mol of CH_4 requires 2 mol of O_2:

$$CH_4 + 2O_2 \rightarrow CO_2 + 2H_2O \qquad (4)$$

$$1 \text{ mol } CH_4 \rightarrow 2 \text{ mol } COD_{CH_4} \qquad (5)$$

Assuming that all C contributing to the COD will be converted to methane:

$$COD \text{ organic matter} = COD_{CH_4} \qquad (6)$$

TABLE 2. Theoretical Yields in Methane and in LFG (CH_4 + CO_2) for Some Typical Compounds or Classes of Compounds

A	*Formula*	*COD (kg/kg)*	*Gas yield (Nl/kg A)*	*Gas composition (% CH_4)*	*Reference*
Cellulose	$(C_6H_{10}O_5)_x$	1.185	830	50	Gendebien *et al.* (1992)
Protein	$(C_4H_6ON)_y$	1.619	1100	51.5	Gendebien *et al.* (1992)
Lipids	$C_{55}H_{106}O_6$	2.914	1428	71.4	Gendebien *et al.* (1992)
Total organic carbon	C	2.67	1867[a]	50[a]	Andreottola & Cossu (1988)
COD		1	700[a]	50	Andreottola & Cossu (1988)
MSW	$C_{99}H_{149}O_{59}N$	1.489	966	54	EMCON (1980)
MSW	$(CH_2O)_n$	1.067	800	50	Hoeks (1983)
No food organic waste (assimilated to paper)	$C_{203}H_{334}O_{138}$	1.398	914	54	
Rapidly decomposable fraction	$C_{68}H_{111}O_{50}N$	1.300	875	52	Tchobanoglous *et al.* (1993)
Slowly decomposable fraction	$C_{20}H_{29}O_9N$	1.710	1049	55	Tchobanoglous *et al.* (1993)
Food waste	$C_{16}H_{27}O_8N$	1.662	993	56	EMCON (1980)

[a] Ratio CH_4/CO_2 = 0.5 has been assumed.

This yields:

$$2 \text{ mol COD}_{organic\ matter} = 1 \text{ mol } CH_4 \qquad (7)$$

On a weight basis:

$$1 \text{ g COD}_{organic\ matter} = 0.25 \text{ g } CH_4 \qquad (8)$$

In terms of gas volume:

$$1 \text{ g COD}_{organic\ matter} = 0.35 \text{ l } CH_4 \qquad (8')$$

This approach does not allow estimation of CO_2 production, and therefore it is necessary either to use equation (2) or to assume a predefined ratio CH_4/CO_2 for the calculations.

In any case, the amount of CO_2 given by equation (2) is not the real quantity that can be measured in LFG, because of several causes (Ham and Barlatz, 1989):

- dissolution of CO_2 in leachate (on the contrary, CH_4 is only slightly soluble in leachate);
- solution in equilibrium with HCO_3^- and CO_3^- ions; precipitation as carbonates;
- occurrence in the landfill of aerobic decomposition (CO_2 is produced and no methane).

For this reason, in practical calculation, the evaluation of the maximum theoretical LFG yield is often carried out applying equation (3') and assuming a predefined ratio $CH_4/CO_2 = 0.55–0.6$.

Furthermore, all the previous calculations have been based on the amount of organic matter in the waste, without considering its effective biodegradability.

According to Andreottola and Cossu (1988) the organic carbon is roughly 50% on dry basis of MSW organic matter, and only 50% of this amount is biodegradable.

Other authors have estimated the total amount of organic carbon in municipal solid waste: 200 kg per ton MSW (Tabasaran, 1982); 250 kg per ton MSW (Hoeks, 1983).

As far as evaluation of the bioavailable fraction of organic carbon of each MSW organic fraction is concerned, Andreottola and Cossu (1988) have proposed the following formula to evaluate the content of biodegradable organic carbon:

$$(OC_b)_i = OC_i(f_b)i \cdot (1 - u_i) \cdot p_i \tag{9}$$

where $(OC_b)i$ is the biodegradable organic carbon in the ith component of waste (kg biodegradable carbon/kg wet MSW); OC_i is the organic carbon content in the dry ith component of waste (kg carbon/kg dry i component); $(f_b)_i$ is the biodegradable fraction of OC_i (kg biodegradable carbon/kg carbon): u_i is the moisture content of the ith component of waste (kg water/kg wet i component); and p_i is the wet weight of the ith component of waste (kg i component/kg MSW). Table 3 reports values of some of the above-mentioned parameters for the most significant MSW components.

The biodegradability of organic matter can also be estimated through the lignin content of the MSW organic fraction, using the following relationship (Tchobanoglous *et al.*, 1993):

$$(f_b)_i = 0.83–0.028 \text{ LC} \tag{10}$$

TABLE 3. Moisture Content (u_i), Organic Carbon Content (OC_i) and Biodegradable Organic Fraction (f_{bi}) in Different Waste Components (Andreottola and Cossu, 1988, modified)

Waste component	u_i (*kg H_2O/kg wet component*)	OC_i (*kg C/kg dry component*)	$(f_b)_i$ (*kg biodeg. C/kg C*)
Food waste	0.6	0.48	0.8
Yard waste	0.5	0.48	0.7
Paper and cardboard	0.08	0.44	0.5
Plastics and rubber	0.02	0.7	0.0
Textiles	0.1	0.55	0.2
Wood	0.2	0.5	0.5
Glass	0.03	0.0	0.0
Metals	0.03	0.0	0.0

where $(f_b)_i$ is the biodegradable fraction expressed on a volatile solids (VS) basis; LC is the lignin content of the volatile solids (VS) expressed as a percentage of dry weight. According to equation (10), waste with high lignin content is significantly less biodegradable.

The biodegradability of several of the organic compounds found in MSW, based on lignin content, is reported in Table 4 (Tchobanoglous *et al.*, 1993).

According to the same authors (Tabasaran, 1982), the fraction of bioconvertible carbon to landfill gas depends upon the temperature within the landfill:

$$(OCb)_i = OC_i \,(0.014 \, T + 0.28) \tag{11}$$

where T is the temperature within the landfill in °C.

Temperature, however, primarily affects generation rates,

TABLE 4. Biodegradability of Some Organic Compounds found in MSW, Based on Lignin Content (Tchobanoglous *et al.*, 1993)

Component	*Volatile solids (VS) as percent of total solids (TS)*	*Lignin content (LC) as percent of VS*	*Biodegradable fraction (BF)*
Food wastes	7–15	0.4	0.82
Paper			
Newsprint	94.0	21.9	0.22
Office paper	96.4	0.4	0.82
Cardboard	94.0	12.9	0.47
Yard wastes	50–90	4.1	0.72

which are strictly associated with the biological activity within the landfill.

In the evaluation of the amount of biodegradable organic carbon, the quantity released through leaching, especially in the acid phase when the organic content of leachate is particularly high, should be subtracted from the initial content of biodegradable C.

Finally, according to equations (3') and (9), LFG specific yield (Y_{LFG}) can be evaluated as follows:

$$Y_{LFG} = 1.867\ OC_i\ (f_b)_i\ (1 - u_i)\ p_i\ (l\ gas/kg\ MSW) \tag{12}$$

This expression represents the common theoretical basis for the majority of LFG generation models.

LFG GENERATION RATES

The main problem of modelling biogas production is not only to forecast the amount of LFG which will be produced, but also the rate and the duration of the production (Augenstein and Pacey, 1991).

The general equation that rules the biogas production is

$$\frac{dC}{dt} = f(t, C^n) \tag{13}$$

where t is time, and C is the amount of methane or of biodegradable organics. Equation (13) can express either the rate of substrate degradation or the rate of gas production.

In the majority of LFG production models, equation (13) is applied to a single batch of waste, which often corresponds to the amount of MSW disposed either in a single layer or in a year. As a consequence, LFG models are applied to each waste batch independently. The global LFG production rate is given by the sum of the single batch contributions.

The greatest absolute exponent n of the dependent variable (C in equation (13)) is called the order of the model (i.e. order of kinetics).

A zero-order kinetics means that a small increment (positive or negative) of C does not influence the rate of substrate decay or biogas production. In other words, a zero-order model indicates that the rate of methane generation is independent of the amount of substrate remaining or of the amount of biogas already produced.

According to some authors, many landfills have a biogas production that follows a zero-order kinetics, especially during the periods of highly active gas generation: probably other factors such as moisture, nutrients etc., limit the amount of methane to be formed, resulting in a relatively constant gas production independent of time (Ham and Barlatz, 1989).

The majority of LFG production models follow a first-order kinetics, which means that the limiting factor is the remaining amount of substrate or the amount of biogas already produced. In this way, other factors such as moisture or nutrient availability are not supposed to be limiting factors. Actually, in many cases the limiting factor is the water content, which plays a major role in the hydrolysis of organic matter.

Although it is clear that many factors such as moisture, temperature, availability of nutrients and presence of the necessary microorganism influence the biogas production, most authors believe that a first-order kinetics with respect to substrate is the most suitable. This choice appears to be supported by the fact that the gas production gradually declines in the long term.

Table 5 shows some examples of production equations deriving from equation (13) and the related curves of gas yield from a batch of waste (Zison, 1990). The rate constants $(k, k_1, k_2, \ldots, k_i)$ reported in Table 5 control the rate at which substrate decays and gas is produced. When time series of field data are available, these constants are usually estimated by model calibration.

Model 1 is very simple: the process of degradation of substrate is independent of the remaining amount. The gas yield is constant until all the biogasifiable matter is degraded. Obviously this is a rough schematization of the process.

In model 2 the substrate is supposed to be in excess and not to limit the gas yield rate. However, the rate declines because waste is a heterogeneous substrate composed of different classes of biodegradable matter, each of which posseses its own decay rate, so that the decay rate has to decline as the most rapidly biogasifiable class is degraded.

Model 3 is the typical, simple first-order model with respect to the substrate, bio-availability of which is the limiting factor. The rate of consumption depends on the amount of remaining substrate. It assumes that no other factors influence the process.

TABLE 5. Some Available Gas-Generation Equations (Zison, 1990).

Model	Integrated form	Order	Remarks	General yield curve
1. $\dfrac{dC_i}{dt} = -k$	$C_2 = C_1 - k(t_2 - t_1)$	0	Constant substrate consumption; methane yield rate constant.	
2. $\dfrac{dC_i}{dt} = -kt$	$C_2 = C_1 - k \cdot \ln\dfrac{t_2}{t_1}$	0	Decay rate declines over time; declining methane production rate.	
3. $\dfrac{dC_i}{dt} = -kC$	$C_2 = C_1 \cdot \exp[k(t_2 - t_1)]$	1	Exponential decay of substrate.	
4. $\dfrac{dC_i}{dt} = -\dfrac{kC}{t}$	$C_2 = C_1 \cdot \exp(k\dfrac{t_2}{t_1})$	1	Combination of models 2 and 3.	
5. $\dfrac{dG_i}{dt} = -k_1 G$	$G = \dfrac{L_0}{2} \cdot \exp[-k_1(t_h - t)]$	1	Two-stage model. Gas generation rate increases then decreases; maximal gas production rate occurs at time t_h.	
$\dfrac{dL_i}{dt} = -k_2 L$	$L = \dfrac{L_0}{2} \cdot \exp[-k_2(t - t_h)]$			

C_1 = concentration of substrate at earlier time t_1;
C_2 = concentration of substrate remaining at later time t_2; $C_2 \geq 0$;
C_i = concentration of substrate i; subscript is not shown in integrated expression for sake of clarity; it should be inferred in each case;
G = volume of gas produced prior to time t;
L = volume of gas remaining to be produced after time t; L_0 is total (ultimate) amount of gas to be produced;
t_h = time for half of total gas production to occur;
k, k_1, k_2 = decay rate constants.

Model 4, according to the Zison classification, is a combination of model 2 and model 3: this model assumes that the substrate availability is the limiting factor and that the decay rate decreases with time.

Model 5 is a two-stage model. In the first stage the rate of gas production is proportional to the amount of gas already produced, while in the second stage the rate is supposed to be proportional to the amount of gas to be produced. To avoid the consumption of substrate being zero in the first stage because the initial amount of produced gas is zero, it assumes that stage 1 starts after production of 1% of the final gas yield. This model is of the kind used by EMCON in the Palos Verdes landfill (EMCON, 1980).

As reported previously, each biogasifiable component of waste has its own degradation rate. For this reason, in many models the substrate is split into several classes characterized by a proper decay rate constant k_i. Many authors (among e.g. others EMCON, 1980; Hoeks, 1983; Andreottola and Cossu, 1988) divide the substrate into two (slowly and readily biodegradable) or three (slowly, moderately and readily biodegradable) classes. The readily biodegradable fraction is represented by food waste, the moderately biodegradable fraction by yard waste, and the slowly biodegradable fraction by paper, cardboard, wood and textiles.

In many models, values of the decay rate constant depend not only on the kind of biodegradable matter, but also on other factors such as moisture content, density, and size of waste particles. It is possible to take these factors into account by introducing appropriate corrective factors.

In Table 6 (Gendebien *et al.*, 1992, modified) values of LFG yields and generation rates are reported.

GENERATION TIME

An important result of LFG production models is the period over which biogas is generated, usually called the generation time. It is not easy to provide a general definition of generation time.

Andreottola and Cossu (1988) indicated a period of gas generation of 30 years. An average generation time of 20 years was considered possible by Bridgewater and Lidgren (1981); Ham (1989) and Richards (1989a) suggested a period of 10–15 years.

TABLE 6. LFG Generation Yields and Generation Rates Reported in Literature (Gendebien *et al.*, 1992, modified)

Source	LFG yields (m^3/ton of MSW)	LFG production rates (m^3/ton year)
Rhyne (1974)	437	
EMCON (1976)		1.22 (methane)
Tabasaran (1976)	60–180	
Rasch (1976)	300–500	
Augenstein *et al.* (1976)	128	
Bowermann *et al.* (1977)	40-50	
Rovers *et al.* (1977)		12–22
DeWalle and Chian (1978)	5.4 l/kg TS	0.181 l/kg TS d^{-1}
Rettenberger (1978)	200	
Stegmann (1978)	250	
Ham (1979)	440	16
	120–310	
Pacey (1981)	55–225	
EMCON (1981)		9.6
Beker (1981)	250	
Hoeks and Oosthoek (1981)	200	10 m^3/yr m^3
Bogardus (1983)		2–8
Hoeks (1983)	20	
Stegmann (1983)	120–150	
Campbell (1985)	100–400	
Stegmann (1986)	186–235 (USA)	12.8 (Palos Verdes)
	120–150 (Germany)	18.6 (Fresh Kills)
Wise *et al.* (1987)	100–400	3.9–130
AFRC (1988)	400	
Rae (1988)	400	
Willumsen (1988)		2–5
Richards (1989b)	135–400	

Satisfactory information concerning generation time is provided by the half time ($t_{1/2}$), the time over which the gas generation equals half of the estimated yield (Fig. 2). By definition the $t_{1/2}$ is such that the area under the production curve is the same on both sides. The range of values proposed for $t_{1/2}$ is very wide. Augenstein and Pacey (1991) reported values ranging from 2–5 years for wet to 10–25 years for dry climates in the USA.

The half time can be also calculated in first-order kinetic models by the following expression:

$$t_{1/2,i} = \ln \frac{2}{k_i} \qquad (14)$$

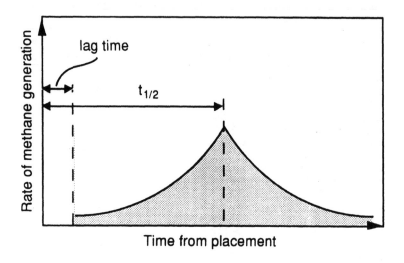

Figure 2. General gas-generation curve with lag time and $t_{1/2}$ shown.

where k_i is the decay rate constant of the ith component of organic waste.

To evaluate $t_{1/2}$ experiences carried out with lysimeters do not seem to be very useful, as generation time values appear to be an order of magnitude shorter than typical landfill values (Augenstein and Pacey, 1991).

This discrepancy could be explained by taking into account that the gas production in lysimeters is due to a single batch of waste, while the gas production of a landfill is the sum of contributions from many batches disposed in different periods.

Another parameter useful to define the generation time is the lag time: the time that passes from the placement of waste to the beginning of significant gas production, (Fig. 2). Given a typical placement period of 1 year for batch units of waste usually considered in a gas-generation model, a zero lag time corresponds to an average lag of six months between placement and start of gas generation. In practice, lag time can vary from a few weeks and months (Richards, 1989a), to 1 year (Scheithauer, 1984) and more (Andreottola and Cossu, 1988).

SELECTION OF MODELS AND PARAMETERS VALUES

The output of a model is closely linked to the selection of the parameter values.

Zison (1990) affirmed that the kinetic order is not very important. Referring to the kinetics reported in Table 5, the author made a comparison between models 1, 2, 3 and 4. The comparison is based on the application of model 3 to data from a Californian site and on the estimation of the parameters value of the three other models obtained by minimization of the sum of squares of the differences between the forecast made by each model and that of model 3. The result of the comparison is that model 1 tracks model 3 very well, and the same happens between models 2 and 4. Because models 1 and 2 are of zero order, while models 3 and 4 are of first order, it follows that the kinetic order does not have a great relevance. This is explained by the fact that long-term curves are not based on the degradation of a single batch of waste, but several batches contribute, so that the production curve of each batch is masked by those of the other batches. For this reason other factors could be considered, such as the choice of an appropriate value for the decay rate constant and the correct evaluation of the amount of degradable organic carbon in the waste (Zison, 1990).

Augenstein and Pacey (1991) dealt with the sensitivity of models to generation time and model parameters. As shown in Fig. 3 (comparison between EMCON MGM model and a constant-rate model), the sensitivity to model parameters is appreciable in the short term, while they are quite without influence in the long term. The conclusion of Augenstein and Pacey, supported also by Zison (1990), is that for many purposes, including prediction over the long term, the shape assumed for a unit batch generation curve can be relatively important.

Figure 4 shows the significance of the generation time expressed as half time ($t_{1/2}$). The curves in Fig. 4 result from the application of the EMCON MGM model to a landfill characterized by a 10 years' constant rate of waste placement. Half-time values of 5, 10 and 20 years are used, maintaining other factors constant.

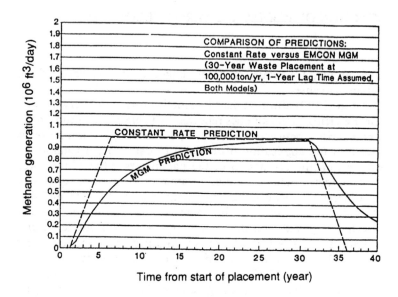

Figure 3. Long-term methane generation predictions comparing a constant-rate model and the EMCO MGM model (Augenstein and Pacey, 1991).

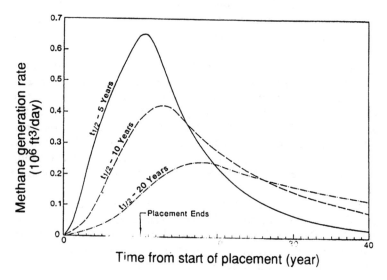

Figure 4. Effect of varying generation time on gas generation profile (Augenstein and Pacey, 1991).

EXAMPLES OF DEVELOPED MODELS

Palos Verdes Model

A two-stage first-order model was applied at the Palos Verdes landfill, CA. This model is known as the Palos Verdes model, and is of the same kind as model 5 reported in Table 5.

The first stage is described by the following equation (EMCON, 1980):

$$\frac{dG}{dt} = k_1\, G \tag{15}$$

where t is time, G is the LFG production prior to time t, and k_1 is the gas production rate constant. Equation (15) shows that, during the first stage, the gas production rate is proportional to the volume of gas already produced: as a consequence, the production rate increases exponentially with respect to time.

The second stage is described by the following equation:

$$\frac{dL}{dt} = -k_2\, L \tag{16}$$

where L is the volume of gas to be produced after time t, and k_2 is the gas production rate constant. Therefore, during the second stage, the gas production rate decreases exponentially with respect to time.

The maximum gas production rate and the transition from first stage to second stage are supposed to occur at the half time, which is the time when half of the ultimate gas production is reached.

The organic substrate is divided into three classes: readily biodegradable (food), moderately biodegradable (paper, cardboard, textiles), and refractory organic waste (plastic, rubber). For each class the model allows calculation of the gas volume produced in time t, while derivatives give production rates. Total gas volume production can be estimated as the sum of productions from each waste class.

It can be observed that, as in the first stage, the gas production rate is proportional to the gas volume already produced, it is not possible to apply equation (15) at the beginning of waste disposal. Usually the equation can be applied when at least 1% of total production is reached.

Sheldon-Arleta Model

The Sheldon-Arleta model is also based on a two-stage, first-order kinetics of gas production.

It derives from the application of a gas production curve for anaerobic digestion of sewage sludge (Fair and Moore curve; EMCON, 1980) to LFG production.

The model generates a dimensionless curve, which describes the time dependence of gas production for sequential addition of waste. The sum of production due to single increments of waste gives the total biogas amount.

The model is based on a number of hypotheses concerning waste carbon content and its biodegradability: in particular, it is assumed that waste is 26% carbon by weight, 31 % of this carbon is readily biodegradable and 66% slowly biodegradable. In this way biodegradable organic waste is divided into classes. For each class a half time is defined, and the total production time is calculated as in the Fair and Moore curve by the following expression:

$$t_{total} = t_{1/2} \cdot 0.35 \tag{17}$$

The maximum production rate is reached at half time.

The graph in Fig. 5a gives an example of Sheldon-Arleta model.

Scholl Canyon Model

The single-stage Scholl Canyon model assumes that biogas production starts at maximum rate after a lag time necessary to establish anaerobic condition and develop the microbial population.

Thereafter, the gas production rate decreases according to a first-order kinetic with respect to organic degradable substrate measured as remaining producible methane, as reported in the following equation (EMCON, 1980):

$$-\frac{dL}{dt} = kL \tag{18}$$

where t is time, L is the volume of remaining producible methane, and k is the gas production rate constant. The model considers the mass of waste as a sequence of submasses disposed in different periods.

It should be noticed that factors other than substrate availability,

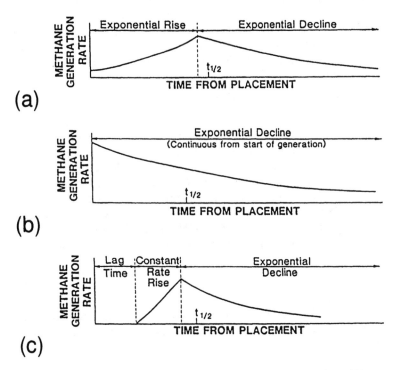

Figure 5. Examples of model predictions for unit batches (Augenstein and Pacey, 1991): (a) Sheldon–Arleta model; (b) Scholl Canyon model; (c) Emcon MGM model.

such as water content, should be considered as limiting parameters.
Figure 5b shows an example of Scholl Canyon model output.

MGM EMCON Model

A well-known computerized gas-generation model is that of MGM EMCON. This model is a very useful tool in estimating the potential production of biogas of a landfill. The main inputs are amount of waste, composition and moisture content, lag time and conversion time (time necessary for the conversion of degradable matter to biogas). The model assumes three classes of degradable matter: readily, moderately and slowly biodegradable. For each of the classes, the model calculates a methane production curve, whose sum gives the methane production curve for the composite waste. The model calculates the maximum amount of methane deriving from the optimal anaerobic degradation of component i for a certain weight of wet waste according to the following relation (Gendebien et al., 1992):

$$C_i = k \; k' \; W_t P_i \; (1 - M_i) \; V_i \; E_i \tag{19}$$

where W_t is the total weight of wet waste (kg wet waste); P_i is the fraction of component i in total waste (kg wet i component/kg wet waste); M_i is the fractional moisture content of component i (kg water/kg wet component); V_i is the fractional volatile solids content of component i (kg VS/kg dry i component); E_i is the biodegradable fraction of dry volatile solids in component i (kg biodegradable VS/kg VS in i component); C_i is the volume of methane produced from component i; $k = 351$ l CH_4/kg COD; and $k' = 1.5$ kg COD/kg l/s. The gas production for the composite waste is:

$$C_t = \sum_{i=1}^{n} C_i \tag{20}$$

where C_t is the total methane production due to the degradation of n waste components. Figure 5c reports the production curve for this model.

Other Models

Hoeks (1983) estimates the degradation rate of organic waste using a simple first-order kinetic described by the following equation:

$$\frac{dP_t}{dt} = - kP_t \tag{21}$$

where P_t is the concentration of degradable organic matter at time t, k is the degradation rate constant, and t is time. Integration of equation (21) leads to

$$P_t = P_0 e^{kt} \tag{21'}$$

where P_0 is the concentration of degradable organic matter at $t = 0$.

Values of P_0 and k are estimated for three classes of degradable organic waste:

- readily degradable, $P_0 = 120$ kg/t, $k = 0.693$ y^{-1};
- moderately degradable, $P_0 = 120$ kg/t, $k = 0.139$ y^{-1};
- slowly degradable, $P_0 = 160$ kg/t, k = 0.046 y^{-1}.

The first class (food) is assumed to be 30% by weight of waste organic fraction, while the second class (garden waste) and third class (paper, textiles) are respectively 30% and 40%.

In this model, a first stage of increasing production rate is considered to be very short and therefore negligible.

Findikakis *et al.* (1988) developed a model for calculation of LFG production and transportation rates. Referring to production rate, the model is based on an approximation of biochemical processes controlling gas generation. The function used consists of a rising hyperbolic branch and a decaying exponential branch. The use of a hyperbolic function was suggested by experimental data from the Mountain View landfill, which showed that a simple first-order kinetic is not adequate to simulate gas production during the first year of waste disposal.

The function, whose output is shown in Fig. 6, for the *k*th component of waste can be defined as follows:

$$G_k\ (t) = 0 \qquad\qquad\qquad \text{for } t \leq t_{0k};$$

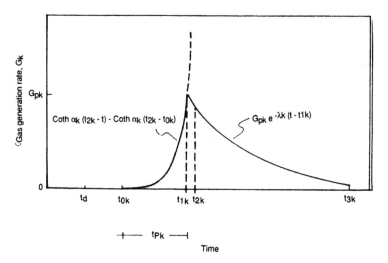

Figure 6. Gas-generation function for methane and carbon dioxide (Findikakis *et al.*, 1988).

$$G_k(t) = \coth \alpha_k \, (t_{2k} - t) - \coth \alpha_k \, (t_{2k} - t_{0k}) \qquad \text{for } t_{0k} \leq t \leq t_{1k}$$

$$G_k(t) = G_{pk} \exp(-\lambda k(t - t_{1k})) \qquad\qquad \text{for } t > t_{1k}$$

where G_k is the generation rate; t_{0k} is the time when gas production starts; α_k is a constant that determines the shape of the rising branch; t_{1k} is the time of peak gas production rate; t_{2k} is the time when the hyperbolic branch approaches infinity asymptotically; G_{pk} is the peak production rate; λ_k is a production rate constant; and t_d is the time at which the refuse is deposited.

Another model was developed using equations that describe the dynamics of the microbial landfill ecosystem (El-Fadel *et al.*, 1989). The model is described as a microbial food chain, as shown in Fig. 7.

The hypotheses of the model are as follows:

- Refuse consists of three components characterized by different biodegradation rates.

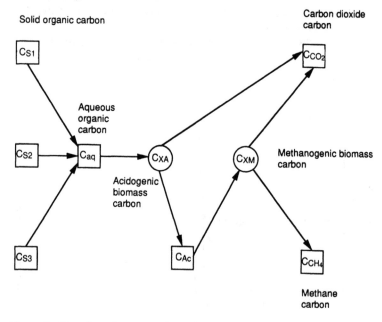

Figure 7. Organic carbon reservoirs and pathways for a batch solid substratum bioreactor (El-Fadel *et al.*, 1989). Carbon may exist in seven different forms: C_{s1}, C_{s2}, C_s, solid carbon for three different classes of waste characterized by different biodegradation rates; C_{aq}, aqueous carbon; C_{XA}, acidogenic biomass carbon; C_{XM}, methanogenic biomass carbon; C_{Ac}, acetate carbon; C_{CO_2}, CO_2 carbon; C_{CH_4}, CH_4 carbon.

- Carbon is always present in all or some of the following forms: solid carbon (C_s), aqueous carbon (C_{aq}), acidogenic biomass carbon (C_{XA}), methanogenic biomass carbon (C_{XM}), acetate carbon (C_{Ac}), CO_2 (C_{CO_2}) and CH_4 (C_{CH_4}) carbon.
- Equal volumes of methane and carbon dioxide are produced per mole of biodegradable carbon.
- Landfill cells can be considered as batch reactors.
- The hydrolysis phase follows a first-order kinetic with respect to solid carbon: $- dC/dt = k_{hi} (C_{(s)i})$ (k_{hi} is the hydrolysis rate of the ith refuse component).
- The hydrolysis products can be summed and follow the same subsequent biochemical process.
- Any void volume is occupied initially by nitrogen (80%) and carbon dioxide (20%).
- Methane generation from carbon dioxide reduction with hydrogen is negligible. For this reason a value of the methane yield coefficient in the range 60–70% is appropriate for model application, as some CO_2 is removed as HCO_3.

Concerning the calculation, the model is based on mass balances for each carbon form.

MODEL DATA VERSUS EXPERIMENTAL DATA

The reliability of LFG generation models is considered good; for instance, EMCON prefers to use its MGM model instead of field extraction tests for sizing extraction systems (Augenstein and Pacey, 1991).

Uncalibrated models, when used by an experienced analyst, can estimate gas production within a range of 30%, while models that are well calibrated on the base of reliable field data can give estimates even within a range of 10% (Zison, 1990).

Because of the poor availability of experimental data, especially in the long term, it is not easy to compare these data with the provisional data given by a model. Zison (1990) reported four cases of comparison between a first-order model (as model 3 of Table 5) and experimental data from four landfills. Only in one case were the estimated values very close to the measured ones; in the other cases the difference was always greater than 25%.

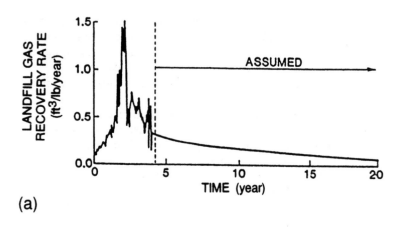

(a)

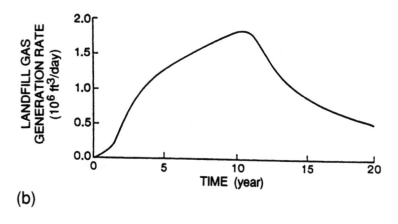

(b)

Figure 8. Prediction of field results based on Mountain View batch unit generation profile and 10-year constant refuse inflow (Augenstein and Pacey, 1991): (a) Cell 'F' unit basin results; (b) composite curve (100 000 t/yr for 10 years).

Augenstein and Pacey (1991) (see Fig. 8a, b and 9a, b) compared two long-term generation curves (Figs 8b and 9b), one based on test cell results from Mountain View landfill plus some extrapolated long-term data (obtained supposing landfilling of the same amount of waste for consecutive years with the same disposal rate), and the other based on the application of the EMCON MGM model. The results showed good correspondence between the long-term gener-

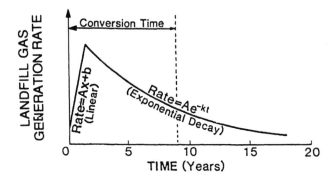

(a)

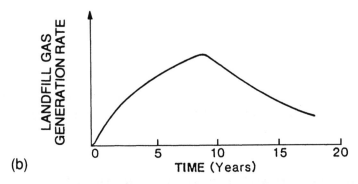

(b)

Figure 9. Prediction of field results based on MGM batch unit projection and 10-year constant refuse inflow (Augenstein and Pacey, 1991): (a) MGM incremental batch curve; (b) MGM composite curve (10 yr refuse inflow – 100 000 t/yr for 10 years).

ation curve deriving from the test cell data and that deriving from the model application.

Better correspondence can be achieved with a more accurate choice of the model parameters, although these are not simple to set.

El-Fadel *et al.* (1989) reported the results of application of the deterministic model based on a physical, chemical and biological characterization of the landfill previously described. The output of the model was compared with experimental data deriving from an

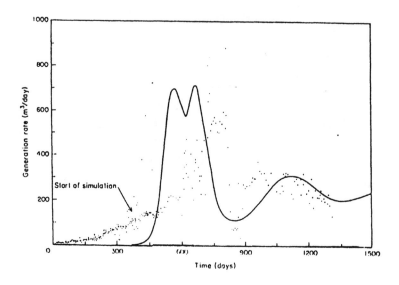

Figure 10. Measured and simulated methane production rate: –, model data; . . ., field data (El-Fadel *et al.*, 1989).

on-site control cell. As shown in Figs 10 and 11, there is a good correspondence between model output and experimental data after a lag time of 400 days.

UNCERTAINTIES AFFECTING MODELS, AND FINAL REMARKS

Some authors believe that the prediction curves are not reliable, in view of the very simple schematization of the model, the inadequate input data, the lack of sufficient long-term field data to calibrate the curves, and the fact that some models are based on ad hoc and empirical assumptions.

Actually, the main factors influencing model reliability are the degree of accuracy expected, the reliability of input data, the experience of the model users and the similarity of the site to others previously modelled (Zison, 1990).

A large number of uncertainties deriving from many sources surely affect the prediction curves, as reported in Table 7.

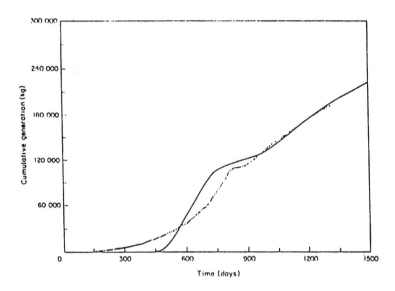

Figure 11. Measured and simulated methane cumulative production: –, model data; . . ., field data (El-Fadel *et al.*, 1989).

The importance and influence of parameters such as landfill history and location, waste composition, pH, nutrient level, temperature, moisture content, diffusion of methanogenic bacteria etc. is well known. The uncertainties due to difficulties encountered in taking these factors into account are made worse by their spatial and time variations. Other uncertainties may derive from the landfill management procedures: for example, the extraction processes may result in air infiltrations, which inhibit methanogenesis and alter the composition of the produced gas.

The differences observed between landfills as to design, construction and management are surely sources of uncertainties. For instance, the cover system influences the moisture content of the waste, and pretreatments such as shredding may influence the density and specific area of the waste.

In conclusion, it can be said that a model aiming to describe the processes occurring in a landfill will always be affected by uncertainties due to the intrinsic impossibility to control strictly what occurs in a landfill. For this reason model output is better expressed in terms of probable ranges instead of punctual values.

TABLE 7. Typical Uncertainties in Variables Affecting Methane Generation Models (Augenstein and Pacey, 1991, modified)

	Parameters	*Uncertainties*
Waste parameters	Placement History Location Composition	Difficult to trace, especially for old landfills
Biochemical parameters	Nutrients Temperature pH	Difficult to measure; varying spatially and over time;
Collection system parameters	Collection efficiency	Ranging between 40 and 90%; difficult to deduce generation from recovery
Other parameters	Moisture content	Difficulty to measure or estimate; varying spatially and over time; great influence on methane generation

In any case, given reasonable expectations, production curves probably represent a good investment in terms of cost – benefit (Zison, 1990). In fact, the cost to develop a model is surely lower than the cost of most of the field tests.

Because of the clear importance of gas production models, it is necessary to make efforts to:

- increase the knowledge of LFG generation processes;
- improve the number and the methodologies of field measurements;
- properly apply LFG production models.

Regarding the first point, the principles that rule the moisture migration, the distribution of nutrients and bacteria, the loss of biodegradable matter by leaching etc. should be studied. The degradation processes themselves need to be furthermore investigated: for instance during field tests in landfills, fractions known to be biodegradable were found to be unaltered after many years of landfilling (Augenstein and Pacey, 1991).

Regarding the second point, the lack of long-term experimental data is well known: this fact makes calibration of models very difficult. For this reason it is necessary to improve the collection of experimental data from well-managed sanitary landfills. Moreover, the gas-generation data should be better evaluated by accurately taking into account gas escaping.

As far as the last point is concerned, the choice of the order of kinetics of the model appears not to be of major importance, but mostly the selection of appropriate values for the decay rate constants and the quantification of biodegradable waste substrate.

Uncalibrated models should only be applied in the first planning stage of landfill gas exploitation programmes.

The application of empirical curves, which are not mechanistically based, should be avoided (Zison, 1990), unless these do not derive from the extrapolation of results of more complex models (El-Fadel *et al.*, 1989).

REFERENCES

AFRC (1988). *A basic study of landfill microbiology and biochemistry*, AFRC Institute Food Res UK, Contract no ETSU b 1159, Published by Department of Energy, UK.

Alpern, R. (1973). *Decomposition rates of garbage in existing Los Angeles landfills.* Master thesis, California State University, Long Beach, USA.

Andreottola, G. and Cossu, R. (1988). Modello matematico di produzione del biogas in uno scarico controllato. *RS-Rifiuti Solidi*, 2(6), 473–83.

Augenstein, D. and Pacey, J. (1991). Modeling landfill methane generation, in *Proceedings of Sardinia 91*, Third International Landfill Symposium, CISA, Cagliari, Italy, pp. 115–48.

Augenstein, D. C., Wise, D. L., Wentworth, R. L. and Cooney, C. L. (1976). Fuel gas recovery from controlled landfilling of municipal wastes. *Resource Recovery Conservation*, 2, 103–17 (cited in Gendebien *et al.*, 1992).

Beker, D (1981). Development of gas in landfill. *Rept. SVA 3747 or IVA 48*, Inst. Afvalstoffen Onderzoek, Amersfoort, The Netherlands (cited in Gendebien *et al.*, 1992).

Bogardus, E. R. (1983). Growing landfill gas recovery activity in the East. Energy and efficiency in buildings and industry, in *Proceedings Second Mid-Atlantic International Conference*, December, Baltimore, USA, pp. 259–81 (cited in Gendebien *et al.*, 1992).

Bowerman, F. R., Rohatgi, N. K., Chen, K. Y. and Lockwood, R. A. (1977). A case study of the Los Angeles County. Palos Verdes landfill gas development project. *Rept. no EPA 600/3-77-047*, US EPA, Cincinnati, OH, (cited in Gendebien *et al.*, 1992).

Boyle, W. C. (1976). Energy recovery from sanitary landfill – a review, in *Microbial Energy Conversion*, (eds H. G. Schlegel, and Barney), Erich Golze KG, Gottingen, Germany.

Bridgewater, A. and Lidgren, K. (1981). *Household Waste Management in Europe. Economics and Technics*, Van Nostrand Reinhold, Wokingham, UK.

Campbell, D. J. V. (1985). Landfilling. An environmentally acceptable method of waste disposal and an economic source of energy. Agricultural, industrial and municipal waste management in today's environment, in *Conference Proceedings*, April, Warwick, UK. Conference Publication 1985–4, Institution of Mechanical Engineers, London, UK, pp. 9–14 (cited in Gendebien *et al.*, 1992).

DeWalle, F. B. and Chian, E. S. K. (1978). Gas production from solid waste in landfills. *Journal Environmental Engineering Division Proceedings ASCE*, **104**, 415–32. (cited in Gendebien *et al.*, 1992).

El-Fadel, M., Findikakis, A. N. and Leckie, J. O. (1989). A numerical model for methane production in managed sanitary landfills. *Waste Management and Research*, **7**, 31–42.

EMCON (1976). *A feasibility study of recovery of methane from parcel 1 of the Scholl canyon sanitary landfill for the city of Glendale*, EMCON Associates, San José, CA, and Jacobs Engineering Co., Pasadena, CA, USA (cited in Gendebien *et al.*, 1992).

EMCON (1980). *Methane generation and recovery from landfills*, EMCON Associates, San José, CA. Ann Arbor Science Publishers, Ann Arbor, MI, pp. 44–51.

EMCON (1981). *Feasibility study. Utilization of landfill gas for a vehicle fuel system. Rossman's landfill, Clackamas County, Oregon*, EMCON Associates, San José, CA (cited in Gendebien *et al.*, 1992).

Farquhar, G. J. and Rovers, S. A. (1973). Gas production from landfill decomposition. *Water, Soil and Air Pollution* **2**, 493.

Findikakis, A. N., Papelis, C., Halvadakis, C. P. and Leckie, J. O. (1988). Modelling gas production in managed sanitary landfills. *Waste Management and Research*, **6**, 115–23.

Gendebien, A., Pauwels, M., Constant, M., *et al.* (1992). *Landfill Gas: From Environment to Energy*, Report No. EUR 14017/1 EN, Commission of the European Communities, Luxembourg.

Ham, R. K. (1979). Method for testing a landfill for its methane potential, US Patent No 4 159 893. Appl.: US 857574 (Dec. 5, 1977), to Reserve Synthetic Fuels, USA (cited in Gendebien *et al.*, 1992).

Ham, R. K. and Barlatz, M. A. (1989). Measurement and prediction of landfill gas quality and quantity, in *Sanitary Landfilling: Process, Technology and Environmental Impact* (eds T. H. Christensen, R. Cossu, and R. Stegmann), Academic Press, London, pp. 155–66.

Ham, R. K., Hekemian, K. K., Katten, S. L., Lockman, W. J., Lofy, R. J. *et al.* (1979). Recovery, processing and utilization of gas from sanitary landfills. *Report No. EPA 600/279-001*, US Environmental Protection Agency, Cincinnati, OH.

Hoeks, J. (1983). Significance of biogas production in waste tips. *Waste Management and Research* **1**, 323–35 (cited in Gendebien *et al.*, 1992).

Hoeks, J. and Oosthoek, J. (1981). Gas production from landfills. *Gas* (The Netherlands) **101**, 563–8 (cited in Gendebien *et al.*, 1992).

Marsili-Libelli, S. (1989). *Modelli matematici per l'ecologia*, Pitagora Editrice, Bologna, Italy.

Pacey, J. G. (1981). Prediction of landfill gas production and recovery, in *Gas*

Research, Proceedings International Conference, Los Angeles, September–October, CA, USA. Government Institutes Inc., Rockeville, MD, pp. 896–915 (cited in Gendebien *et al.*, 1992).

Rae, G. W. (1988). The control of landfill gas. HM inspectorate of pollution's waste management paper, in *Landfill Gas and Anaerobic Digestion of Solid Waste*, Proceedings Conference, Chester, UK, October, Harwell Laboratories, UK, pp. 92–9. (cited in Gendebien *et al.*, 1992).

Rasch, R. (1976). Methane generation from waste, in *Proceedings First International Symposium 'Materials and Energy from Refuse'*, Antwerp, Belgium, Van Mantgen and De Does, Leiden, The Netherlands, pp. 31–6. (cited in Gendebien *et al.*, 1992).

Rettenberger, G. (1978). Origin, consequences, collection and valorization of landfill gas, in *Progress in Landfill Technology*, 38 pp., Stuttgart berichte zur Abfallwirtschaft, Band 9. ed: Inst. Siedlungswasserbau, Wassergüte- und Abfallwirtschaft der Universität Stuttgart, Germany. ISBN: 3 503 01353 9. (cited in Gendebien *et al.*, 1992).

Rhyne, C. W. (1974). *Landfill Gas, Office Solid Waste Management Programs*, US Environmental Protection Agency (EPA), Cincinnati, OH.

Richards, K. M. (1989a). All gas and garbage. *New Scientist* (June 3) 4 pp. (cited in Gendebien *et al.*, 1992).

Richards, K. M. (1989b). Landfill gas: working with Gaia. *Biodeterioration Abstracts*, **3**, 317–31 (cited in Gendebien *et al.*, 1992).

Rovers, F. A., Tremblay, J. J. and Mooij, H. (1978). Procedures for landfill gas-monitoring and control. *Waste Management Rept EPS 4-EC-77-4*, Montreal, Quebec, Canada. Minister of Supply and Services, Canada (cited in Gendebien *et al.*, 1992).

Scheithauer, H. (1984). Deponiegasnutzung. Planungen, Erfahrungen, und Entwicklungstendenzen (Utilization of landfill gas. Design, experiences and development tendencies), in *Proceedings of Symposium of Essen* (Germany), September 1983. Bundesminister für Forschung und Technologie (BMFT), Bonn, Germany (cited in Gendebien *et al.*, 1992).

Stegmann, R. (1978). Gases from sanitary landfills. *ISWA Journal*, **26/27**, 11–24 (cited in Gendebien *et al.*, 1992).

Stegmann, R. (1983). Emission at the domestic waste landfills, in *Proceedings Seventh Mülltechnisches Seminar*, München, Germany. Published as *Ber. Wassergütewirtschaft und Gesundheitsingenieurwesen*, Techn. Univ. Munchen, pp. 75–98 (cited in Gendebien *et al.*, 1992).

Stegmann, R. (1986). Landfill gas as an energy source, in *Preparing now for Tomorrow's Needs*, Proceedings Conference Chicago, IL, USA. National Solid Waste Management Association and *Waste Age* Magazine, Washington, DC 20036, USA, pp. 307–33 (cited in Gendebien *et al.*, 1992).

Tabasaran, O. (1976). Considerations on the problem landfill gas. *Müll Abfall* **7**, 204–10 (cited in Gendebien *et al.*, 1992).

Tabasaran, O. (1982). Obtention et valorisation du methane a partir de dechets urbains. *Tribune de Cebedeau*, **35**, 483–8.

Tchobanoglous, G., Theisen, H. and Eliassen, R. (1977). *Solid Wastes*, McGraw-Hill, New York.

Tchobanoglous, G., Theisen, H. and Vigil, S. (1993). *Integrated Solid Waste Management. Engineering, Principles and Management Issues*, McGraw-Hill, New York.

Willumsen, H. C. (1988). Extraction and utilization of biogas from small and medium size landfills. *Unpublished paper*, 2 pp. (cited in Gendebien *et al.*, 1992).

Wise, D. L., Leuschner, A. P., Levy, P. F., Sharaf, M. A. and Wentworth, R. L. (1987). Low-capital-cost fuel-gas production from combined organic residues. The global potential. *Resource Conserv.* **15,** 163–90 (cited in Gendebien *et al.*, 1992).

Zison, S. (1990). Landfill gas production curves: myths vs. reality, in *Proceedings of GRCDA/SWANA Landfill Meeting*, Vancouver, Canada.

4.2 Prediction of Gas Production from Laboratory-Scale Tests

HANS-JÜRGEN EHRIG

Bergische Universität Gesamthochschule Wuppertal
Pauluskirschstr. 7, 42285 Wuppertal, Germany

INTRODUCTION

Knowledge of gas quantity and quality and their time dependance is fundamental for assessing a correct gas management strategy. There are few data available on the total amount of gas produced and even fewer on gas production with time. Most of the available data originate from lysimeter studies or laboratory-scale experiments. It is in practice not possible to measure the total gas production rate of a sanitary landfill, either at one moment or over several years. As a result no or very poor data are available from operating landfills. For most purposes data from small-scale experiments cannot be used directly, as degradation conditions are quite different from those of operating landfills.

During our own laboratory-scale experiments the liquid and gaseous emissions are measured. Leachate quality is one indicator of the kind of biological processes that take place in small-scale lysimeters as well as in full-scale landfills (Stegmann and Spendlin, 1989). Based on the knowledge that the biological processes in the lysimeters and in full-scale landfills correspond, the gas production with time can be estimated from the lysimeter results for a full-scale landfill.

After verifying these interrelationships it is also possible to estimate the effect of varying municipal solid waste (MSW) composition – e.g. reduction of paper or other organics – for future landfills.

Landfilling of Waste: Biogas. Edited by T. H Christensen, R. Cossu and R. Stegmann. Published in 1996 by E & FN Spon, London. ISBN 0 419 19400 2.

ESTIMATION OF GAS PRODUCTION

Gas Production

Quite often the total amount of gas production is just calculated from measured or estimated refuse analyses. Generally the carbon content of refuse is used. In this case the total gas production rate could be calculated by the following equation, under the assumption that the carbon is totally biologically converted into biogas (CH_4 and CO_2):

$$G_e = 1.868\ C \tag{1}$$

where G_e is the total gas production (m^3 (STP)/t MSW), C is the carbon content (kg/t), 1.868 is a conversion factor from solid carbon to gaseous carbon, and STP is standard temperature and pressure.

The relationship between methane and carbon dioxide depends on the composition of the degraded organics. Equation (1) was modified by Tabasaran (1976) in order to reflect the incorporation of carbon into biomass depending on the temperature:

$$C_a = C\ (0.014\ T + 0.28) \tag{2}$$

where C_a is the degradable carbon content, C is the total carbon content, and T is the temperature in °C. In contrast to anaerobic degradation processes in the liquid phase, the biomass is not removed from the reactor. The biomass produced in landfills can therefore be used as a new carbon source. For this reason the modification made in equation (2) does not seem to be adequate. On the other hand a significant part of carbon – e.g. from plastics and lignin – is non-degradable or very slowly degradable, so that these carbon compounds do not contribute to gas production during the first 50–100 years.

As another possibility for predicting gas production, the calorific value of MSW can be used. The idea behind this is that the methane produced contains the total energy of the refuse. However, it should be remembered that part of the energy content is used by the biological processes. In addition, as has been mentioned above, the energy of plastics etc. cannot be converted into biogas energy. Therefore this calculation method is not more realistic than that using the carbon content.

If the composition of refuse is used to predict the total biogas production rate, only the biodegradable components, such as fat,

TABLE 1. Gas Production of MSW (m^3/t wet MSW) Calculated from the Gas Production of the MSW Fractions Fat, Protein and Carbohydrate Including Cellulose

Fraction	Part of wet MSW (%)	Methane (m^3/t fraction)	Methane (m^3/t wet MSW)
Fat	2.5	1021	25.5
Protein	4.6	509	23.4
Carbohydrate (including cellulose)	18.6	453	77.2
Sum			126.1
Total gas amount (55% methane)			229.3

protein, and cellulose, should be considered (Rovers *et al.*, 1977; Ham *et al.*, 1979). The calculation of the total gas production rate from fat, protein, and carbohydrate including cellulose using analyses from our own experiments is presented in Table 1. The calculated values are higher than the values actually measured under different conditions (see Table 2). In this calculation the carbon loss by leachate could not be considered. Moreover, reproducible analyses of fat, protein, and cellulose in refuse are difficult to achieve and are very expensive.

Most of the total gas production rates measured originate from lysimeters ranging in size from a few litres to several cubic metres of refuse. As already mentioned, it is possible to measure the total gas production in gas-tight lysimeters, but the environmental conditions differ widely from those of a full-scale landfill.

TABLE 2. Published Data of Gas Production from MSW

Source	Gas production (m^3/t wet MSW)
Ham *et al.* (1979)[a]	190–240
Tabasaran (1976)[b]	60–180
	Gas production (m^3/t dry MSW)
Doedens (1985)	50–114
Chakraverty *et al.* (1981)[c]	91–413
Cooney and Wise (1975)[c]	224–328
Pfeffer (1976) [c]	131–182

[a] Different cited values.
[b] Measured at landfills.
[c] Calculated with own analysis of MSW volatiles.

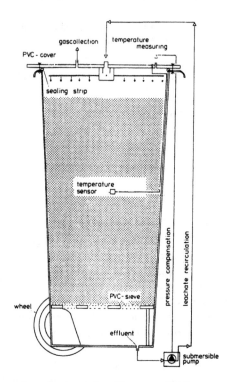

Figure 1.   Cross-section of test container (Stegmann, 1981).

Experimental Results

During the last ten years a laboratory-scale system has been developed to simulate the anaerobic degradation of refuse (Stegmann, 1981).

The laboratory-scale reactors were constructed of standard MSW containers of 120 l volume (Fig. 1). The reactors were placed in a thermostatic room at a temperature of 30 °C. Only some experiments were run at room temperature (18–20 °C). The containers were covered with gas-tight PVC plates. Leachate was collected in a small tank (2.5 l) outside the containers. Eight times a day the leachate was pumped back to the top of the container through a distribution system. Twice a week 1.5 l of liquid was exchanged. The produced gas was measured with a self-constructed gas meter. Gas samples were collected in aluminium gas bags, coated with plastic.

For the experiments, material from a compost plant was used. Both materials, fresh MSW and composted MSW, were shredded (approximately 5 cm in diameter). The two fractions were mixed with a ratio of 2 parts fresh MSW and 1 part composted MSW. Water was added until a water content of 65% wwt was reached. The added components, such as paper, grass, etc., were also mixed with both fractions. Each 120 l container was filled with approximately 80 l of mixed material. Before inserting and after removing the material it was analysed for the following components: volatile solids, COD, organic carbon, organic nitrogen, and ammonium.

Typical results of an experiment are presented in Fig. 2. From this figure the following anaerobic phases can be observed: from 0 to 50 days the acetic phase, from 50 to approximately 80 days the change from the acetic phase to the methanogenic phase, followed by the methanogenic phase. As compared with the timescale of a full-scale landfill, the acetic phase in our experiments was very short, though in principle the same processes took place. Owing to the optimum conditions in the laboratory-scale experiments the processes are – compared with full-scale plants – significantly accelerated.

During these experiments the refuse was anaerobically degraded over a period of 200–400 days. After more than 300–400 days the gas production was still measurable, but too low for gas utilization. If these results are compared with data from actual landfills it can be shown that the short-term simulation of biological and leaching processes in landfills is possible. The author estimates that one year for a full-scale landfill corresponds to less than 10 days for the biological processes and approximately 1 day for the leaching processes taking place in the test containers.

From each series the input and output of solids were measured. These values varied over a wide range. Nevertheless there was a strong relationship between volatiles and COD as well as carbon content of MSW (see Fig. 3):

$$COD \ (g/kg) = 78.4 + 1.04 \ VS \ (g/kg) \tag{3}$$

$$C \quad (g/kg) = 89.8 + 0.60 \ VS \ (g/kg)$$

The average content of volatiles (VS) was 438 g/kg dry solids with a standard deviation of 122 g/kg. The correlation was similar when the input and output material was analysed. The average carbon content amounted to 353 g/kg dry solids. The analytical variation

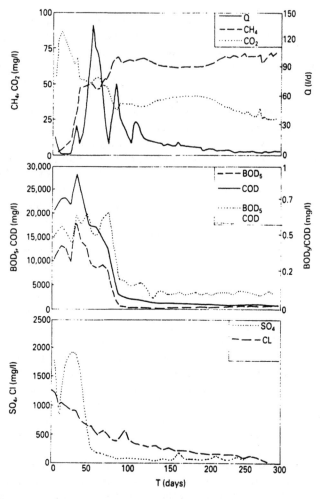

Figure 2. Typical gas and leachate data versus time from laboratory-scale experiments (Stegmann and Spendlin, 1989).

of stabilized MSW was much greater than of fresh MSW. As a consequence the degradation rates show a wide range:

- VS = 1–43%;
- COD = up to 53%;
- C = up to 47%.

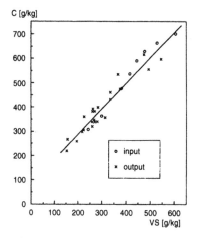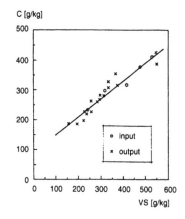

Figure 3. Relationship between (a) volatile solids (VS) and COD of the solids (left) and VS and (b) carbon content of the solids (right).

Respecting all three values the average degradation rate was between 38 and 40%: i.e. a reduction of volatiles between 92 and 242 g/kg dry solids (average = 171 g/kg) and for carbon between 89 and 187 g/kg dry solids (average = 138 g/kg).

TABLE 3. Results of Methane and Total Gas Production from Laboratory-Scale MSW Experiments

	Gas production (Nm³/t) mean	Gas production (Nm³/t) range
Production per t dry MSW[a]		
Methane	102	92–133
Total gas	186	151–271
Production per t dry MSW		
Methane	87	78–113
Total gas	158	128–230
Production per t wet MSW (30% water content)		
Methane	61	55–79
Total gas	111	90–161

[a] The degradable solids of the used MSW were reduced by approx. 15% (due to pretreatment).

TABLE 4. Carbon Balance of Laboratory-Scale Experiments

	kg carbon/ t dry refuse mean	kg carbon/ t dry refuse range
Input	354	235–468
Reduction	138	89–187
Transport by leachate as organics, soluble carbon dioxide and alkalinity	23	
Gas	100	
Sum	123	
Recovery as percentage of reduction	89%	

The total gas volumes produced during these experiments are presented in Table 3. Table 4 shows the average carbon balances of the input materials of the numerous lysimeter tests. The average carbon recovery rate of 89% is quite high, given the inhomogeneity of MSW. Compared with full-scale landfills the carbon transfer into the leachate is much higher in the lysimeter experiments, as the leachate flow over time is nearly 350 times higher. If less carbon is transported by means of the leachate the gas production is higher. On the other hand, compared with the gas-tight lysimeters, more carbon will be reduced by aerobic processes in full-scale landfills. The gas production values in Table 3 should be increased by up to 10% for actual landfills (based on wet weight):

- gas production (55% CH_4 + 45% CO_2):
- mean value 120 m^3 (STP)/t MSW
- range 100–180 m^3(STP)/t MSW

RATE AND DURATION OF GAS GENERATION

Gas Production Rates

The prediction of gas generation rates with time is even more difficult than that of total gas production. This is due to the different environmental conditions and the resulting biological processes in laboratory test containers and full-scale landfills.

The principle of gas production is given by

$$G_t = G_e \, (1\text{-}e^{-kt}) \qquad (4)$$

where G_t is the gas sum at time t (m^3 (STP)/t MSW), G_e is the total gas production rate (m^3 (STP)/t MSW), t is time in years, k is the (degradation) coefficient $= -\ln(0.5)/t_{0.5}$, ln is natural logarithm, and $t_{0.5}$ is the half-life in years. The variable coefficients of this equation are the total gas production rate and the degradation coefficient k or $t_{0.5}$. The half-life is that time interval after which 50% of the total amount of gas has been produced.

Data for the total gas production rates are given in Tables 2 and 3. Published values for the coefficient k and the half-life are presented in Table 5. The range of these data is very wide. In some publications the biodegradability of the organic MSW components is characterized as being good, moderate and low. Unfortunately, the different components are not categorized in the same way by the various authors. In order to predict gas production rates of future refuse composition it is necessary to know the the biodegradability of the different refuse components.

Equation (4) should be modified, as it is based on the assumption that the maximum gas production occurs immediately after disposal at time zero; actually this is not the case. For this reason equation (4) is not valid until some years after the refuse has been landfilled. An example of such a modified equation is as follows (Anon, 1982):

$$G_t = A \, (1 - e^{-(t/k_1)}) \, e^{-(t/k_2)} \qquad (5)$$

where G_t is the gas production rate at time t (m^3 (STP)/t MSW × year), A is the maximum gas production rate = 13.61 (m^3 (STP)/t MSW × year), and k_1 (1.1 years), k_2 (15.72 years) are coefficients.

In Fig. 4 equations (4) and (5) are compared. The decreasing part of both equations is nearly the same. Equation (5) also indicates an increasing part of the curve, but the shape of the increasing curve does not reflect the basic biological processes of such a system (see Fig. 2).

Experimental Results

If the initial gas production phase is not respected, equation (4) describes the laboratory-scale experiments in principle. Owing to the optimum environmental conditions in the laboratory-scale systems, the half-lives are not valid for actual landfills. But comparing the principal biological processes of both systems, a transfer of results looks promising.

In many cases the organic leachate content is used to describe the character of biological processes in waste deposits. The possibility and the degree of gas production are concluded from the types of process that take place. During the first phase of biological

TABLE 5. Published Values for the Coefficient k and/or the Half-Life of Anaerobic Biodegradation

Source	Coefficient k	half-life (years)	Remarks[a]
Tabasaran (1976)	0.07	10	
Rettenberger (1978)	0.288	2.4	
Tabasaran and Rettenberger (1987)	0.058–0.115	12–6	General values
Tabasaran and Rettenberger (1987)	0.081–0.092	8.6–7.5	Data from landfills
Hoeks (1983)	0.0365	19	cit. Rovers *et al.*
Hoeks (1983)	0.693	1	G. biod. (food waste)
Hoeks (1983)	0.139	5	M. biod. (garden waste)
Hoeks (1983)	0.046	15	L. biod. (paper, cardboard, wood, textiles)
Hoeks and Harmsen (1980)	0.1	7	Data from landfills
Stegmann (1978/79)		1.5	G. biod. (food and garden waste)
Stegmann (1978/79)		25	M. biod. (paper)
Moolenaar (1981)		1–5	G. biod. (food waste, paper)
Moolenaar (1981)		5–25	M. biod. (garden waste, cardboard)
Moolenaar (1981)		20–100	L. biod. (wood)
Bowermann (1976)	1.84	0.4	G. biod. (food waste, gras)
Bowermann (1976)	1.15	0.6	M. biod. (cardboard, paper, textiles, wood)
Bowermann (1976)	0.115	6	L. biod. (plastic)
Ham *et al.* (1979)		1	G. biod. (food waste)
Ham *et al.* (1979)		15	M. biod. (paper, wood, garden waste)
Ham *et al.* (1979)		0.5–10	G biod. (cited values)
Ham *et al.* (1979)		2–25	M. biod. (cited values)

[a]G. biod. = good biodegradability; M. biod. = moderate biodegradability; L. biod. = low biodegradability.

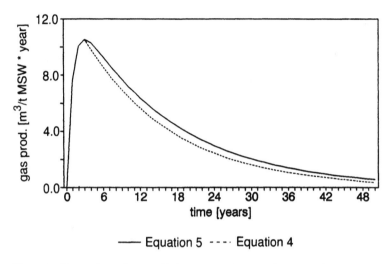

Figure 4. Comparison of gas prediction curves using equations (4) and (5).

processes in landfills the high organic leachate pollution represents the acetic phase, where the production of volatile fatty acids is predominant and no or low gas production takes place. Low organic concentrations in leachate are an indicator of the methanogenic phase, with the conversion of organic material into mainly CH_4 and CO_2. The change from one phase to the other can be seen from the BOD_5/COD ratio in the leachate. This dimensionless parameter is not influenced by dilution. The acetic phase is characterized by a BOD_5/COD ratio > 0.1. Usually the start of the methanogenic phase is defined as the point when the BOD_5/COD ratio decreases below 0.1.

Evaluating the results available from various laboratory-scale experiments (see also Funk, 1982; Wolffson, 1985; Ehrig, 1987; Wolffson, 1987; Wolffson, 1988; Stegmann and Spendlin, 1989; Spendlin, 1990) leads to a different conclusion: The increase of methane production starts much earlier, with the effect that when the BOD_5/COD ratio decreases below 0.1 the amount of methane already produced increases to 62.4–96.8% of the total gas production rate. An example of these findings is presented in Fig. 5. As a mean value of all evaluated experiments, 82.7% of methane was produced reaching BOD_5/COD ratio ≤ 0.1.

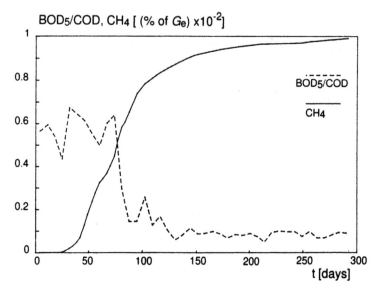

Figure 5. Comparison of the methane production rates and the decrease of BOD$_5$/COD ratios (laboratory-scale experiment). G_e: total gas production.

As a result of the evaluation of the above-mentioned experiments, the MSW biodegradation may be divided into four phases (see also Fig. 6):

- *Phase 1*:
  high organic leachate pollution;
  BOD$_5$/COD ratio > 0.4;
  low pH values (< 6.5);
  low methane production;
  increase of methane concentration.
- *Phase 2*: *beginning*:
  pH starts to increase;
  methane production starts to increase.
  phase:
     high organic leachate pollution;
     BOD$_5$/COD ratio > 0.4;
     high methane concentration (> 40%).
  end:
     see phase 3.

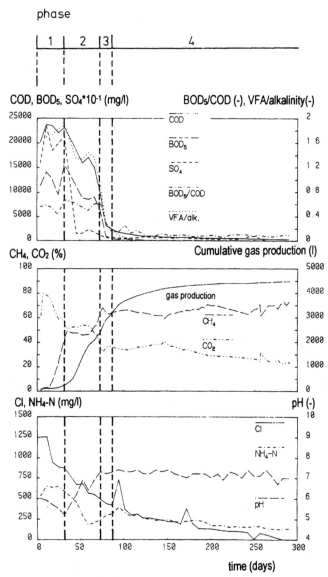

Figure 6. Description of different phases during anaerobic degradation of MSW (based on laboratory-scale experiment).

- *Phase 3: beginning:*
    pH increase is terminated (pH stabilizes);
    BOD_5/COD ratio starts to decrease;
    point of inflexion of methane production curve.
  phase:
    high methane production;
    high methane concentration (sometimes combined with a slight increase).
  end:
    BOD_5/COD ratio < 0.1;
    low organic leachate pollution;
    maximum of methane concentration.
- *Phase 4:*
    constant pH values (~ 7.5);
    low organic leachate pollution;
    BOD_5/COD ratio < 0.1;
    low methane production;
    constant methane concentrations (eventually with slight decrease).

The main methane production starts much earlier than estimated up to now from the leachate quality values.

It seems to be possible to transfer the principal results of these laboratory-scale experiments to full-scale landfills or, better, at least to the first lift of a landfill. In Fig. 6, phase 2 describes the methane production curve from the start to the point of inflexion at a methane concentration of about 50% of the total methane production rate. This part of the methane production curve reflects nearly 50% of the total gas production (half-life). During this period the pH also increases (see Fig. 6).

Figure 7 shows the organic leachate concentrations and pH values of an actual landfill in operation. In order to transfer the results from laboratory to full scale it may be possible to use the pH increase as an indicator of the half-life of the biodegradation of a full-scale landfill (in this case 2–3 years). But this value for the half-life was only valid for the first lift of a landfill. Figure 8 presents the organic leachate concentrations of the first and second lift of two lysimeters (diameter = 5 m, height = 4 m). Leachate collected from the bottom of the first lift shows an acetic phase with moderate pollution over a relatively short time. The leachate pollution of the second lift is much higher over a significantly longer period of time.

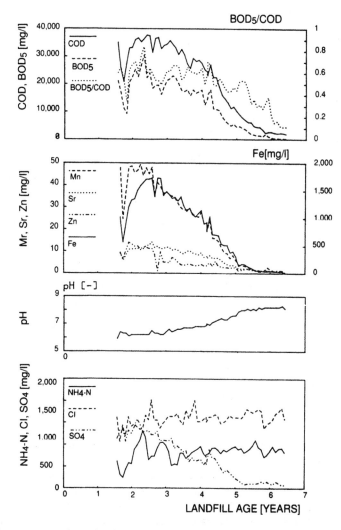

Figure 7. Time-dependent leachate concentrations of a full-scale landfill.

The reason for this phenomenon could be that the first layers are less compacted and more influenced by aerobic processes due to air intrusion. Therefore the half-life of biological degradation for the whole landfill body is longer than that calculated from measurements at the bottom of a landfill. A factor of approximately 2–2.5

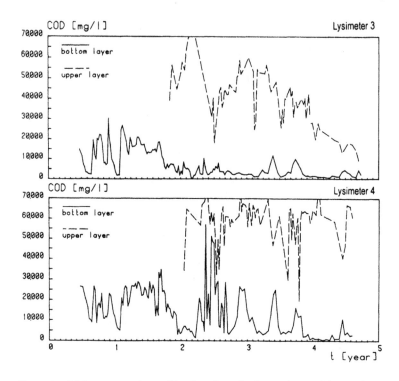

Figure 8. COD concentrations of leachate from the first and second layers of two half-technical scale lysimeters (diameter 5 m, height 4 m). Lysimeter 3: 0.40 m layers, highly compacted, emplaced every 6 weeks. Lysimeter 4: 2 m layers, highly compacted.

may describe this interrelationship, which results in a half-life of 4–7.5 years.

The evaluation of the laboratory-scale experiments shows a lag phase until gas production starts (Fig. 9). In addition, the measured and calculated gas production curves show differences in the starting phase. In reality the increase in gas production is a continuous process.

Consideration should be given to using two different equations for the prediction of gas production: The first equation describes the time from emplacement of the waste until the maximum gas production rate is reached, and the second equation simulates the subsequent gas production. The two functions are presented in equations (6) and (7):

Cumulative gas production (l)

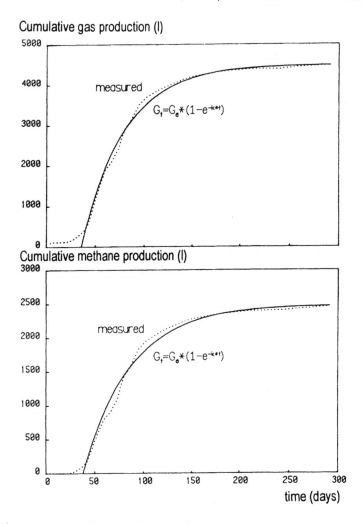

Cumulative methane production (l)

time (days)

Figure 9. Measured and calculated (non-linear regression) methane and total gas production (results from laboratory-scale experiments).

increasing gas production:
(from $t = 0$ until maximum gas production $t = t_1$)
$$G_t = G_{t\,max}\, e^{-k1(t1-t)} \qquad (6)$$
where G_t is the gas production at time t (m^3 (STP)/t MSW × year),

$G_{t \, max}$ is the maximum gas production $= G_e \, k$ (m^3 (STP)/t MSW), t is the time since dumping, G_e is total gas production, k is a coefficient $= -\ln(0.5)/t_{0.5}$ (see also equation (4)), t_1 is the lag phase from dumping until maximum gas production, $k_1 = (\ln G_{t \, max} - \ln A)/t_1$, and A is gas production at $t = 0$ years (m^3 (STP)/t MSW $\times$ year).

(The value $A = 0$ is irrelevant; the proposal is: $A = 0.01$.)

Decreasing gas production (after t_1):
$$G_t = G_{t \, max} \cdot e^{-k(t - t_1)} \tag{7}$$

In order to predict the gas production of an actual landfill it is necessary to multiply the results of the above equations with the amount of MSW put in place per time. Such exponential functions should only be used if the refuse quantities emplaced per year or shorter periods of time are known. Figure 10 gives an example how the gas production can be predicted by means of equations (6) and (7).

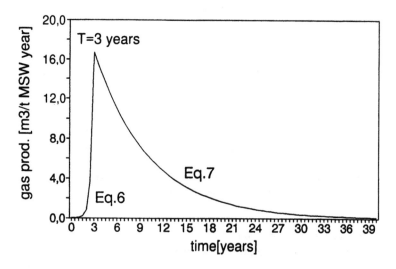

Figure 10. Gas production curves in principle for a full-scale landfill estimated from laboratory-scale experiments.

FACTORS INFLUENCING GAS PRODUCTION

Variation of MSW composition (lysimeter tests)

Owing to waste avoidance and recycling activities it will be necessary in the future to predict gas production of MSW of varying composition.

In order to gain some values for the gas production rates of specific MSW fractions, additional experiments were run. In some of these experiments specific fractions were added to the mixed refuse. In other cases only the gas production of a single fraction or an exactly mixed fraction was measured.

Environmental Conditions in Landfills

Anaerobic processes may be influenced by different landfilling techniques and/or operating conditions, but it is difficult or impossible to measure these effects on the biodegradation activity. Owing to these factors of influence the half-life of the bioactivity changes; this is not true for total gas production.

Aerobic conditions may take place if the MSW is compacted to a low degree ($\gamma < 0.5$ t/m^3). Depending upon the thickness of the lift and the frequency of new lift addition, aerobic processes may take place also in highly compacted landfills ($\gamma \leq 0.8$ t/m^3). Figure 11 shows a theoretical calculation for the aerobic degradation of MSW in such highly compacted landfills. About 40–60% of the degraded organics are used for biomass production, which may be further degraded under anaerobic conditions.

At two landfills in Germany MSW is composted over a period of 6–11 months. The composted MSW is emplaced and highly compacted. The remaining gas production potential is very low (about 13 m^3 STP/t MSW). From these data (see Table 6) it can be concluded that high aerobic degradation rates (i.e. long composting periods) are necessary to reduce the gas potential significantly.

The factor most influencing the anaerobic biological degradation processes is the water content. Differences of water content result in wide variations of half-lives. For example, wet paper is relatively easy to degrade, but dry paper (e.g. in a telephone book) is nearly undegradable. These factors of influence could be eliminated by using adequate pretreatment and/or operating techniques, such as crushing and mixing and/or prewetting of the refuse.

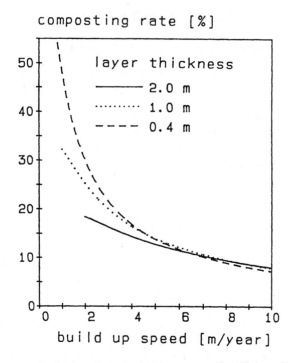

Figure 11. Aerobic biodegradation in a highly compacted landfill depending on layer thickness and build-up speed of the landfill (theoretical calculation).

In many cases in Germany the first lift in a landfill consists of composted refuse to reduce organic leachate pollution (Stegmann and Spendlin, 1989). Using such a technique it is possible to reduce organic leachate concentrations without reducing the gas potential. The half-life of anaerobic degradation in landfills with or without a first lift of composted MSW is similar, except for the first lift.

Leachate recirculation may also result in an enhancement of biological processes in landfills (see also Stegmann and Spendlin, 1989). This is mainly due to increasing and equalizing the moisture content in the landfill and optimizing the conditions for anaerobic degradation. Comparing organic leachate concentrations and pH values of landfills, where leachate recirculation is practised, with the results from laboratory-scale experiments, half-lives of gas production of up to 0.5–1.5 years for a full-scale landfill may be possible.

TABLE 6. Gas Production from Different MSW Fractions

Refuse fraction	methane (m³/t dry fraction)	gas (m³/t dry fraction)
Food waste	105–189	191–344
Food waste (with high bread content)	74	135
Grass	97	176
Leaves	33	60
Food and garden waste with a high wood content (separate collected)	22–36	40–65
Newspaper	72	120
Magazines	60–135	100–225
Cardboard	190	317
Mixture of 42% newspaper + 38% magazines + 20% cardboard	39–145	65–242
Sawdust	18	30
MSW input (in compost plant)	88	160
Composted MSW	73	133
Sieve retention	143	260
MSW input after separation of glas, paper etc. (in compost plant)	60	109
Composted material	97	176
Sieve retention	78	142

CONCLUSIONS

Prediction of gas production with time is the basis for every design of a gas extraction and utilization system. The basic data on gas production potential and time-related gas production can still be only roughly estimated.

Using controlled laboratory-scale experiments it may be possible to gain adequate data to assess the total rate and duration of gas production. The time-dependent gas production may be estimated by comparing the biological conditions (see phases 1–4 as described above) of laboratory and full-scale systems. In this way time factors may be developed allowing one to transfer the values from laboratory-scale to full-scale landfills.

In contrast to published data, no differences of biodegradability could be found in our own laboratory tests dependent upon different refuse fractions. There are indications that after the period of main gas production additional organic components are anaerobically degraded into biogas over very long periods of time (decades to centuries).

The composition of refuse influences the total gas potential and to some extent the degradation rate. However, these parameters are much more influenced by landfill operation techniques and landfill conditions, such as water content, degree of compaction, and thickness of lifts.

REFERENCES

Anon (1982). *Gas in Deponien*, Schriftenreihe Umweltschutz Nr. 3, Bundesamt für Umweltschutz, Bern.

Bowermann, F. R. (1976). *A case study of the Los Angeles County Sanitation Districts – Palo Verdes Landfill Gas Development Project*, Environmental Protection agency.

Chakraverty, C., Franklin, A. G. and Quon, J. E. (1981). *Time-Settlement Behavior of Processed Refuse, Part III*, EPA-600/2-81-135, Environmental Protection agency.

Cooney, C. L. and Wise, D. L. (1975). Thermophilic anaerobic digestion of solid waste for fuel gas production. *Biotechnology and Bioengineering*, 17, 1119.

Doedens, H. (1985). Massnahmen zur Verbesserung der Ausbeute des Gaspotentials an Deponien. *Stuttgarter Berichte zur Abfallwirtschaft*, 19, 231.

Ehrig, H.-J. (1984). Laboratory scale tests for anaerobic degradation of municipal solid waste, in *ISWA/APWA-Congress*, Philadelphia.

Ehrig, H.-J. (1987). Anaerobic degradation of municipal solid waste – laboratory scale tests, in *Global Bioconversions*, (ed. D. L. Wise, Vol. II, CRC Press, p. 121.

Funk, C. (1982). Gasproduktion aus Hausmüll. Diplomarbeit am Institut für Stadtbauwesen, Abt. Siedlungswasserwirtschaft TU Braunschweig.

Greiner, B. (1983). *Chemisch-physikalische Analyse von Hausmüll*, Berichte 7/83, Erich Schmidt Verlag.

Ham, R. K. (1979). Gasentstehung durch geordnete Deponien, Aktuelle Deponietechnik. *Abfallwirtschaft an der TU Berlin*, 5, 186.

Ham, R. K., Hekimian, K. KI., Katten, S. L. *et al.* (1979). *Processing and Utilization of Gas from Sanitary Landfills*, EPA-600/2-79-001, Environmental Protection agency.

Hoeks, J. (1983). Significance of biogas production in waste tips. *Waste Management and Research*, 1, 323.

Hoeks, J. and Harmsen, J. (1980). Methane gas and leachate from sanitary landfills. *Institute for Land and Water Management, Research Technical Bulletin*, 117, 132.

Moolenaar, W. (1981). Bioconversion of waste, a source of energy. *Resources and Conservation*, 321.

Pfeffer, J. T. (1976). Methane from urban wastes – process requirements, in *Microbiol. Energy Conversion*, UNITAR-Seminar, 139.

Rettenberger, G. (1978). Entstehung, Folgen, Erfassung und Verwertung von Deponiegas. *Stuttgarter Berichte zur Abfallwirtschaft*, 9.

Rovers, F. A., Tremblag, J. J. and Mooij, H. (1977). *Procedures for Landfill Gas Monitoring and Control*, Report EPS 4-EC-77-4, Fisheries and Environment Canada.

Spendlin, H.-H. (1990). Untersuchungen zur frühzeitigen Initierung der Methanbildung bei festen Abfallstoffen. Dissertation, TU Hamburg-Harburg.

Spillmann, P. (ed.) (1986). *Wasser- und Stoffhaushalt von Abfalldeponien und deren Wirkungen auf Gewässer*, VCH Verlagsgesellschaft mgH.

Stegmann, R. (1978/79). Gase aus geordneten Deponien. *ISWA Journal* (26/27), 11.

Stegmann, R. (1981). Beschreibung eines Verfahrens zur Untersuchung anaerober Umsetzungsprozesse von festen Abfallstoffen im Labormassstab. *Müll und Abfall*, 2.

Stegmann, R. and Spendlin, H.-H. (1989). Enhancement of degradation: German experiences, in *Sanitary Landfilling: Process, Technology and Environmental Impact*, (eds T. H. Christensen, R. Cossu and R. Stegmann), Academic Press, London, p. 61.

Tabasaran, O. (1976). Überlegungen zum Problem Deponiegas. *Müll und Abfall*, 7, 204.

Tabasaran, O. and Rettenberger, G. (1987). Grundlagen zur Planung von Entgasungsanlagen, *Müllhandbuch no. 4547*.

Wolffson, C. (1985). Untersuchungen über den Einfluss der Hausmüllzusammensetzung auf die Sickerwasser- und Gasemissionen, *Institut für Siedlungswasserwirtschaft, TU Braunschweig*, 39, 119.

Wolffson, C. (1987). Untersuchungen über den Einfluss der Hausmüllzusammensetzung auf Sickerwasser- und Gasemissionen. Institut für Siedlungswasserwirtschaft, TU Braunschweig, unpublished report.

Wolffson, C. (1988). Laboratory scale test for anaerobic degradation of municipal solid waste, in *ISWA Proceedings*, 2, 159.

4.3 Actual Landfill Gas Yields

HANS C. WILLUMSEN

Hedeselskabet, Klostermarken 12, DK-8800 Viborg, Denmark

INTRODUCTION

The methane production of a landfill is a key parameter in evaluation of the energy utilization potential and in design of the utilization system. The theoretical methane yield per weight unit of waste and the basic factors affecting the gas production is discussed in Chapter 2.1. Experiences from laboratory-scale experiments are presented in Chapter 4.2.

Many landfills already have gas utilization systems (Chapter 1.3), and some of these provide data on actual methane generation rates at full scale. This chapter presents a short summary of the data available as of 1990 in terms of specific methane production rates as a function of waste age and as a function of landfill size.

SPECIFIC METHANE PRODUCTION RATE

The specific methane production rate expressed as m^3 CH_4 per tonne of waste per year may vary substantially from landfill to landfill owing to variations in waste composition, moisture content, microbial activity in the waste and supposedly also according to the age of the waste as a measure of the stability of the organic matter present in the waste. In addition to this, observed methane production rates at actual landfills may vary owing to different efficiencies of the gas

Landfilling of Waste: Biogas. Edited by T. H Christensen, R. Cossu and R. Stegmann. Published in 1996 by E & FN Spon, London. ISBN 0 419 19400 2.

extraction systems, which again may be influenced by the shape of the landfill, the top cover, the type of extraction system and the pumping rate applied.

Most of the existing landfill utilization systems have been installed after the landfilling had started, and this may influence the observed methane production rates in a negative sense. If a landfill has been designed and operated for optimum gas recovery this might potentially lead to higher gas production rates.

Figure 1 presents observed specific methane production rates from 86 landfills as a function of the average age of the landfilled waste. The data have been extracted from Bevenyi and Goula (1989), Richards (1989), Richard and Alston (1990) and Gendebien *et al.* (1992). The data represent landfills in primarily the USA (58), Germany (6), and the UK (6). The specific methane production rates range from 0.5 to 10 m^3 CH$_4$ per tonne of waste per year with typical values around 2.5 m^3 CH$_4$ per tonne of waste per year. From the field data, no relationship between gas production rate and landfill age can be identified.

For the same data, Fig. 2 shows the specific methane production rate as a function of landfill size in terms of tonnes of waste landfilled. The largest variation is observed for the relatively small

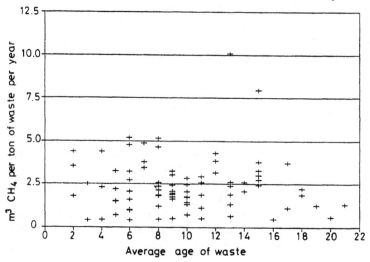

Figure 1. Specific methane production rates for 86 worldwide landfills as a function of waste age.

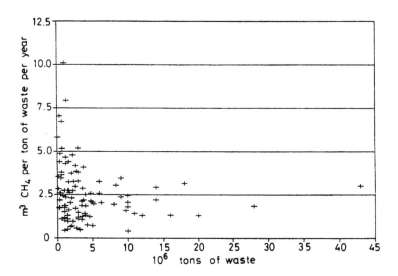

Figure 2. Specific methane production rates for 86 worldwide landfills as a function of landfill size.

landfills (less than 3×10^6 t), while the large landfills (more than 10×10^6 t) apparently yield 1.5–3 m^3 CH_4 per tonne of waste per year. It should be noted that these landfills typically are younger than 20 years, and the available data do not provide a basis for evaluating how long the methane production rate will continue.

REFERENCES

Berenyi, E. and Gould, R. (1989). *1988–89, Methane Recovery from Landfill Yearbook. Directory & Guide*, Governmental Advisory Associates, Inc., New York.

Gendebien, A., Pauwels, M., Constant, M., Ledrut-Damanet, M. J., Willumsen, H.-C., Butson, J., Fabey, R., Ferrero, G.-L. and Nyns, E.-J. (1992). *Landfill Gas. From Environment to Energy*, Office for Official Publications of the European Communities, Luxembourg.

Richards, G.-E. and Alston, Y.-R. (1990). *Proceedings of International Conference Landfill Gas: Energy and Environment '90*, 16–19 October, Bournemouth, England.

Richards, K.-M. (1989). Landfill gas: working with Gaia. *Biodeterioration Abstracts*, **3**(4), 317–331; and additions (1990) presented to the Methane Emissions Workshop, Washington, March 1990.

4.4 Field Testing for Evaluation of Landfill Gas Yields

ROBERT K. HAM

Department of Civil and Environmental Engineering, University of Wisconsin-Madison, 3232 Engineering Building, 1415 Johnson Drive, Madison, WI 53706, USA

INTRODUCTION

The procedures discussed in this chapter are designed to determine the composition and the rate of gas generation, which would then give rise to design requirements and capacity requirements for a landfill gas utilization scheme. This chapter does not include investigations regarding migration of gas off the landfill or the control of any migration of landfill gas.

The concept is to measure gas as generated by any of several techniques, each of which will be discussed in some detail in this chapter. One technique is to try to simulate the landfill by enclosing portions of the landfill or a simulated landfill in a test volume in order to measure the gas generated. The container is often called a **lysimeter**. A more common testing procedure is to pump gas from the landfill, measuring the amount and composition but also attempting to delineate the volume of the landfill giving rise to the gas being collected. Yet another concept is to measure gas passing into the atmosphere, which is termed **flex testing**, or to measure gas flowing through the cover soil. Lastly, it is possible to measure gas generated by samples of refuse taken directly from the landfill in question and placed in containers. This last technique will be called **anaerobic sampling**.

Another technique, which is totally different from those mentioned above, is to measure refuse composition changes and to relate

Landfilling of Waste: Biogas. Edited by T. H Christensen, R. Cossu and R. Stegmann. Published in 1996 by E & FN Spon, London. ISBN 0 419 19400 2.

the composition change to constituents in leachate or landfill gas. In other words this is a mass balance approach to determine gas generation.

It is clear that no one method works perfectly; this is why there are so many methods available, several of which are in common use. The best approach appears to be a combination of methods, especially if accuracy in test results is important. Note that it is very difficult to measure gas generation for a full-size landfill, given the heterogeneous nature of the waste, the non-uniformity of cover, the non-uniformity of moisture flow and availability, etc. It is even more difficult to measure landfill gas over an extended period of time in order to get total gas production. For this reason, total gas production figures vary widely, and are largely dependent upon either mass balance or laboratory work.

GAS GENERATION MEASUREMENT

Lysimeter Testing

Even though it is sometimes not considered part of field testing, it is common to simulate a landfill, or to construct a test landfill, by placing refuse in an enclosure to measure the amount of gas generated. The enclosure is usually closed except for a pipe system allowing the passage of water in and leachate out of the container, and for gas venting through the measuring device. This is the only method to give accurate gas rates; the accuracy is limited only by the accuracy of the gas-measuring device and the absence of any leaks. The containers or lysimeters range in size from holding a few kilograms of refuse to very large field test cells holding 1000 or even up to 10 000 t of refuse. This technique is widely practised, as many lysimeters and test cells have been built and tested over the last 50 or so years.

Laboratory scale. Typically, fresh refuse is placed in bottles containing up to 20 l in volume, in columns or specially built containers, in containers as used for refuse storage at the home (perhaps a volume of 100 l), or in 210 l drums. The container is fitted with a drain to allow periodical leachate removal. This has to be done in a way to avoid air leaking into, or gas out of, the container. The temperature of the container is controlled by enclosing it in a temperature-controlled heat-source device, or by placing it in a controlled-temperature room.

Water is added periodically, or leachate will be recirculated, to provide adequate moisture and moisture flow to promote decomposition. It is common to de-aerate the water first to avoid any oxygen entrance to the test container. Water can be added by a distribution system to try to wet the refuse uniformly. It is important to try to avoid or minimize any water channelling along the walls of the container.

Tubing is used to direct gas from the container to any of several types of gas measuring device, including water displacement devices, wet test meters, or other containers for gas storage and accumulation, followed by periodic measurement (for example, a tedlar bag). A sampling port is provided so that gas samples can be taken for composition analysis, typically by gas chromatography.

Gas leakage and errors in gas measurement are the major problems with this technique. Generally, however, the data are useful regarding gas generation rates over the period of interest, given careful experimental design and operation.

Field scale. The concept is to build test cells of virtually any size manageable. Such cells have been cited ranging from 100 to 10 000 t of refuse. The cell is typically lined with clay and probably plastic film, and it may be placed in groundwater or in clay soils to further minimize any leakage of gas out of the sides or bottom of the test cell. The temperature in this case will be set by natural conditions, and is virtually impossible to control independently. The test cell is covered with a plastic film and/or compacted wet clay to avoid gas flow direct to the atmosphere.

A leachate collection system is provided by an underdrain and pipe network system, with provision for leachate removal either by gravity or by pumping. A water trap must be used to avoid any possibility of air leaks into or gas leaks out of the system through the leachate removal line. Water is added or leachate is recirculated using a pipe network located at the top of the test cell but under the cover. This network may include a water distribution layer of sand, gravel or other material in order to facilitate uniform water distribution. It is important to minimize the channelling of moisture along the walls of the lysimeter or test cell by uniform waste placement and compaction, and by using water diverters along the walls as much as possible. Note, however, that as refuse decomposes it will often pull away from the walls, and channelling can result.

Gas is measured by venting the lysimeter only through a gas-measuring device. Typically a gas meter is used as is commonly used in the natural gas industry. Sampling ports are provided to allow for periodic sampling and gas chromatography analysis of the gas.

A major problem in this system is leaks. All sides of the cell have to be sealed to minimize any leaks; however, these seals must be continually repaired and maintained. One must also continually maintain and recalibrate gas meters, as corrosion, condensation and other problems will occur and have lead to erroneous data in the past.

The main advantage of the lysimeter approach to measuring gas generation is that, in theory, 100% of the gas can be collected and measured by either laboratory or field test cells. Another advantage is that this process gives data regarding gas generation rates over a period of time, so that it is possible to follow decomposition patterns and gas generation changes over as a long a period as desired. In some cases, data have been gathered over much of the decompositional life of the refuse, to give an indication of total gas generation rates. It is possible to use different conditions and different refuse compositions to determine the effects on gas generation. Variables can include temperature, moisture flow and moisture content, and water can be used to carry nutrients, pH-adjusting chemicals, buffers or other additives. Further, refuse can be specially mixed to examine the effects of different compositions and different additives to the refuse.

There are several disadvantages associated with lysimeter testing. The most important is that it is very difficult to mimic real-world landfill conditions, and therefore it is very difficult to apply the results to full-scale landfills. The reasons for this stem mostly from water addition, which is not typical of precipitation events, and from the fact that refuse tends to be placed over a short time in the test cell and so has tendencies to progress through the different stages of decomposition as a unit. The test cell does not provide the intermix of different decomposition stages and states of decomposition as is the case in the real-world landfill; therefore, for example, it is common to observe acid souring, which can reduce, postpone, or even virtually eliminate methane generation. It is also possible to get a great variety of results, many of which are not observed in real-world landfills. For example, acid souring, significant amounts of hydrogen generation, and the inability to reach relatively low concentrations of leachate typical of methanogenic decomposition are all problems in test cell experiments. In order to obtain results somewhat like real-world observations, designers have had to resort to unusual and perhaps unimplementable actions under real-world conditions. Such manipulations include the addition of seed organisms, the use of large amounts of moisture and moisture flow to promote decomposition, neutralization of leachate or the refuse, pre-aeration of the refuse, etc. For the above reasons, the results of lysimeter experiments tend to be largely of theoretical or mechanistic interest, and help to improve our understanding of refuse decomposition and gas generation and the effects of various variables and additives. It is very difficult to apply the results of such an experiment to estimate gas generation rates at full-scale landfills.

Landfill Gas Pump Tests

The most common method used to determine gas generation rates in full-sized landfills is to perform landfill gas pump tests. In this process, one basically sets up an entire gas withdrawal system or portion of a system such as would be used to withdraw gas directly from the landfill for gas control or gas productive use purposes. This system can incorporate either trenches or vertical wells for withdrawing the gas, of which wells are the most common. The use of trenches implies that a significant amount of waste is placed after the trenches are placed, because the trenches have to be at lower elevation within the landfill. Wells are used for gas withdrawal after a significant portion or all of the landfill has been placed. In either case, the concepts to be described will apply.

The overall concept is to withdraw gas from the landfill using the gas withdrawal system, and then to determine the volume of landfill from which the measured gas is withdrawn so one can project these rates to other portions of the site or to other landfills. Devices used in this concept are gas wells to withdraw the gas, and probes to measure the volume of landfill from which gas is withdrawn.

Wells are typically placed in boreholes 30–90 cm in diameter excavated directly into the landfill. Some borings extend to within 3 m or so of the bottom of the landfill or to groundwater. Other designers extend the boring only to a portion of the total depth. The depth should depend on the contour of the landfill. If the boring is near to a side slope of the landfill, it is common to extend the boring further down to keep the active (vacuum) portion of the well away from the surface. If the landfill has good impermeable cover of wet clay or plastic, the borehole does not have to extend as far down because one can use relatively high vacuum to remove gas from lower depths without pulling large amounts of atmospheric air into the landfill.

A pipe is placed in the borehole. This pipe is typically made of PVC and is of diameter 10–30 cm. The larger pipe is more likely to be used if the well will be used for gas withdrawal over a long period in the ultimate gas use system.

The bottom portion of the pipe should be slotted or have holes drilled, typically 6 mm or less in diameter or width, according to the gravel size, so as to avoid gravel plugging the holes or slots. The typical perforated interval will be over the lowest ⅓ to ⅔ of the pipe length. Once again, if cover is impermeable the perforated length can run closer to the surface without pulling in a large amount of atmospheric air, and vice versa. Slots or holes are placed around the circumference of the pipe, spaced so as not to unnecessarily weaken the pipe. Typically the holes or slots are 5 cm or so apart.

The annular space over the slotted interval is backfilled with coarse gravel or stone, and from the top of the slotted interval to the surface a good seal should be used. It is best to use slurry or a semi-liquid seal to get the least flow and hence the best seal possible along the solid pipe length. It may be unreasonable to use slurry sealant all the way to the surface in some cases, in which case slurry layers along with natural available fill material can be used. The point is to get the best possible seal to avoid air flowing down the pipe or boring wall into the gas well.

In addition to wells, probes must also be placed. The borehole for probes is typically 10–15 cm in diameter. The depth will vary. Some operators extend the depth of the bore hole to 1 m or so beyond the centre of the perforated length of the well or wells nearest the probe. It is unusual to go deeper than this, but some operators will go to a shallower depth depending on the desired information to be gained. The shallowest depth will typically be about 4 m from the surface of the landfill. The slotted or perforated interval for the probe pipe will be approximately 2 m or less, with the remainder of the pipe being solid to the surface of the landfill. The annular space over the perforated interval is backfilled with gravel, above which the annular space is sealed to the surface with a slurry seal either all the way to the surface or in layers. Some operators try to have multiple probes at different depths in the same borehole. This is somewhat risky practice, as it has been shown that it is difficult to effectively seal the different perforated intervals so there is no communication of gas pressure or flow between the different probes within a given borehole.

A pump is placed on each well or combination of wells. The pump should be capable in general of at least a 15 cm water gauge vacuum as a first estimate and should have a capacity of at least 100 l per minute per metre of well depth. Vacuum and pumping requirements will sometimes be greater than these values, but will most often be lower. A flame arrester should be installed after the pump, as for example between the pump and the flare. A flare is typically used, as opposed to free venting of the gas to the atmosphere, to control odours and to minimize or avoid other potential problems associated with direct release of landfill gas. The pump is typically electrically driven, using the local power line or a portable generator. Some operators have driven the pump with internal combustion engines, adjusted to operate on landfill gas. Typically these engines are started with propane and are switched to landfill gas when sufficient gas is available. If this is done, the gas must be pumped to a sufficient pressure to operate the engine, which will require more pumping capacity than would be required for simple flaring.

Measurements to be taken are gas pressure, flow and composition at each well, and gas pressure at each of the probes.

The simplest configuration is to use one well surrounded by three or more probe lines extending radially outward from the well in different directions. This is to try to accommodate the variability within the landfill with respect to refuse composition, voids, density, compaction, moisture content and other factors affecting gas permeability. Probes are then located at various distances from the wells. A typical probe pattern might call for three probes perhaps 10 m from the well, three additional probes 20 m from the well, etc. on out to perhaps 50–75 m. Seldom will probes be placed beyond 75 m, unless the landfill is unusually dry or poorly compacted and so has high gas permeability.

In performing a pump test, it is necessary to determine background pressures with the well capped or not being pumped. When the background pressures become stabilized, which may take several days, the average pressures are recorded at each probe. The well is then pumped at perhaps 1.5 l per minute per metre of well depth. Once the pumping is initiated probe pressures are noted, and when probe pressures stabilize, all data including probe pressures are recorded. The pumping rate is then increased by perhaps 50% of the starting rate, and when probe pressures have stabilized a complete data set is recorded, including probe pressures. This entire process is repeated at each of several increasing pumping rates.

The gas composition in the well is monitored frequently or continuously. If nitrogen is obtained in the landfill gas being pumped, and the nitrogen content doesn't decrease to basically zero with continued pumping as nitrogen is purged out of the system, the pump is operating too fast or there is too much vacuum being applied. Air is being pulled in through leaks in the piping, down the borehole if the bore is improperly sealed, or air is being pulled in through the cover soil of the landfill. Some operators measure oxygen for this purpose, but as oxygen can be utilized by decomposing refuse, nitrogen is the more sensitive indicator of air infiltration. If the nitrogen-to-oxygen ratio is approximately 4 and doesn't change even at low well flow rates, air is most likely leaking directly into the pipe system or into the bore hole. If the ratio is 5 or more at least some of the leakage is attributed to air being pulled in through the cover and refuse, and the test can be considered to be a valid indicator of landfill characteristics and hence maximum pumping rates. If leaks are in the piping, it may be possible to use the data by subtracting off the direct air leak, but it is still best to find the leak and fix it.

Depending on gas use and quality requirements, the maximum pumping vacuum (and hence gas flow rate) allowed for a given well can range from producing a steady 1% nitrogen (for the highest-quality gas or most sensitive gas use) to perhaps 10% nitrogen (for low-quality gas requirements, such as for flaring). It is undesirable to pump at rates beyond perhaps 10% nitrogen, to avoid pulling too much air into the landfill, thereby inhibiting or poisoning the methane generation process. Once methane generation is inhibited it can take many months to return gas generation rates to normal, so it is important to avoid over-pumping a landfill for either testing or gas withdrawal purposes.

The above method is used to determine the amount of gas and the quality of gas that can be pulled from a given well. It will also provide vacuum and pumping requirements. This information is relatively easy to obtain, and is good design information for long-term gas withdrawal and use. The more difficult part in gas testing is to determine the specific gas generation rate (in other words the volume of gas or methane produced per tonne of refuse per day) or the radius or volume of influence of a well. This information is important for well field design and to determine the total extractable gas flow for full-scale gas withdrawal programmes. The probes are necessary to determine this information.

The concept is, if the vacuum applied to a given well affects the pressure at a given probe, the refuse in the vicinity of the probe is under the influence of the well, and gas generated at that point will flow to the well. If no pressure changes are noted upon pumping, that probe is unaffected and therefore is not influenced by the well. In concept, the change in pressure at a probe caused by pumping compared with not pumping at all will decrease with increasing distance from the well. The no-flow pressures are called **static pressures**. This change in pressure from static pressures should decrease similarly for all three probe lines, so the radius of influence or influenced volume can be established at each well flow rate. The radius or volume of influence at the maximum well flow rate will be the radius or volume to use in well field design. The specific generation rate can then be calculated at each well flow rate, but certainly would be calculated at the maximum flow rate by assuming the refuse in the influenced volume is in a cylinder of length equal to the average depth of refuse at each well and of a radius equal to the average radius of influence in the three probe trains. The specific gas generation rate (for example, litres of gas generated per cubic metre of landfill) can be calculated. If the refuse density is known or can be assumed, the generation rate per unit weight of refuse can also be estimated.

In practice it is much more difficult than the concept suggests. The refuse is not homogeneous, so seldom will all three probe sets give the same result. Consequently, averages are typically used. This still does not account for the large differences in permeability that typically occur, with an apparent radius of influence perhaps 60–70 m in one probe set and 20–30 m in another.

A second difficulty arises from the fact that gas flows vertically as well as horizontally, so a cylinder is unlikely to be a true description of the refuse mass affected by a given well. Horizontal influence, as a rule of thumb, is often ten times that of vertical influence, reflecting the much greater horizontal permeability than vertical. This is clearly influenced by cover layers within the refuse mass, perched water, the method and extent of compaction, etc. These permeability differences suggest that the influenced volume will be shaped more like a flattened sphere if the perforated interval of a well does not extend to the bottom of the landfill or like a cone if it does. The shape gets even more complex, however, because over the perforated interval, the radius of influence will not be constant but will probably decrease with depth, because the overburden of waste compresses lower layers more than upper layers, reducing the void space and therefore the permeability in lower layers. Some gas testers use different probe depths in an attempt to describe the volume of influence more accurately and hence the shape of the refuse mass affected by a well. Others have used models to project the shape, using a few probe readings to give the overall size of the affected volume of refuse. No one method is satisfactory, however, given the non-homogeneity of the landfill, so it often simply assumed that a cylinder does on the average approximate the affected refuse volume. Careful observations during drilling can aid in determining whether special conditions exist favouring a certain volume of influence configuration. For example, if there is a moist refuse layer, ponded perched water, or intermediate cover layer, etc., one would modify the interpretation of probe pressures accordingly. One logical modification is to discount the top 5–10 m of the cylinder if the cover is permeable or sparse, because this waste may not be producing methane owing to air infiltration, and any gas generated in this area would probably escape to the atmosphere (at least at some of the greater distances from the well).

Another problem is to determine when to take readings at each pumping rate. How long does one pump the well before readings are assumed to be consistent and an 'equilibrium' condition is reached? Theoretically, the radius of influence is infinite or at least extends to the boundaries of the refuse mass. Using the groundwater drawdown model, groundwater pumping lowers the level of groundwater most next to the well, with the amount of drawdown decreasing with distance from the well, gradually approaching the prepumping level. Similarly, regarding gas flow to a well, if pumping continues indefinitely, pressure influence will extend continuously outward from the well, eventually encompassing the entire landfill This is especially true if the cover is relatively impermeable. For a permeable cover, gas generated near the cover will flow through the cover into the atmosphere irregardless of the well, unless the well reduces pressure below atmospheric pressure, in which case at least convective flow will be to the well. Clearly it is impossible to pump the well at a given rate for an infinite time period before taking readings, so in practice operators wait from a few hours to a few days and then take several readings to be sure no trends to lower probe pressures persist.

The next problem follows from the concept that the radius of influence will extend continuously outward with time for a landfill with 'impermeable' cover. The radius of influence depends on the sensitivity of the device measuring the change in pressure at a probe from the static (no pumping) condition. If one had an extremely sensitive instrument, a very small change in pressure could be measured and the observed radius of influence would increase, and vice versa.

Finally, all of the above assume stable conditions regarding gas generation and flow as well as atmospheric pressure, permeability of the landfill cover, and short-term effects. These factors greatly complicate the approach to stable conditions at any pumping rate, including no pumping or static conditions. It is well known that under static conditions, pressures measured by probes will change throughout the day. Some of these changes are reasonably predictable on a daily basis, the so called diurnal effect; others cannot be tied to a 24 h cycle. Various theories have been proposed regarding the basis for diurnal variations, often dealing with the moisture content and hence permeability of cover soil, or daily atmospheric pressure changes, but the mechanism is not yet proven.

Atmospheric pressure changes will affect probe pressures within the landfill according to the permeability of cover soil. If a landfill is relatively permeable, with direct communication between atmospheric and probe pressures, it is logical to subtract atmospheric pressure and use gauge pressure to describe gas flow. On the other hand, if there is a lag period between atmospheric pressure changes and change in pressure at a given probe, or if the landfill is not so permeable that probe pressure is only partially affected by atmospheric pressure changes, this solution is not so simple. If the cover on a landfill is very impermeable, only pressure changes within the landfill are important and there should be little or no effect of atmospheric pressure on probe readings.

Changes in permeability of cover soil will affect the pressures within a landfill. Precipitation, drying conditions, freezing conditions, etc. will directly affect gas flow permeability, making it difficult to compare and use gas-testing data properly before and after a rainfall event, for example.

Finally, even if atmospheric pressure, precipitation events, cover soil permeability, diurnal effects, etc. are accounted for, pressure changes within a landfill occur from point to point. These changes may stem from minute moisture content changes, changes in refuse temperature, or some other factor or combination of factors.

The net result of all of the above complications is variations in pressure readings at the probes beyond those caused by pumping alone. Typically, a test programme will call for readings after a set number of days at each pumping rate, at set times of the day, plus normalization of readings by dealing only with gauge pressures (subtracting atmospheric pressure) to account for at least some of the factors causing variability. Still, however, variability remains a significant problem for gas testing. Note that these problems have their greatest impacts on the use of probes, which means the determination of radius or volumes of influence and/or specific generation rates. Gas quantity and composition at a given well are relatively unaffected by these complications.

To remove some of the difficulty associated with use of probes and defining influence zones, and to account for different refuse ages and conditions in different parts of a landfill, it is necessary to use more than one well. Wells should be placed at different parts of the landfill so the locations are representative of the entire landfill or at least of the portions of the landfill considered for gas withdrawal. Characteristics to be considered in defining representative test areas include general shape (for example near or far from a side slope), different refuse ages, different cover characteristics, different moisture conditions (for example near or away from groundwater, perched water, etc.), different depths, and so forth.

Wells may be placed in clusters to allow probes to be used for more than just one well. This also gains information about variability in a given area of the landfill and can lessen the reliance on probes to determine volumes of influence. If a very large number of wells were installed covering the entire landfill, no probes would be necessary because the entire landfill would become the volume influenced. Specific generation rates can be readily established. Decreasing the number of wells and concentrating them in specific portions of the landfill lowers the cost of testing, and if the area is reasonably isolated, probes are not as critical in defining the influenced volume.

Given the above complications with pump testing, if it is likely that gas will be withdrawn from a landfill for gas use or for gas control purposes, many designers will simply install a well field designed from experience and use it at first for testing purposes to determine gas composition and quantity and vacuum requirements, and to indicate any refinements necessary from the initial design. For example, if the nitrogen content of the gas increases sharply before increased vacuums result in marginal or no increases in methane, more wells or a better cover are called for to collect the most gas.

The same concepts apply to using trenches for gas withdrawal as was described above for use of wells. In this case, the probe pattern has to incorporate different distances from a trench in order to fully describe the volume of the landfill influenced by a trench. Also, in this case, the perforated interval of any pipes established in the trench cannot be closer than a certain distance from the side of the landfill (typically 50–100 m) because of the likelihood of drawing in atmospheric air. A solid pipe and impermeable backfill will be used to extend the pipe to the surface of the landfill.

MEASUREMENT OF GAS EXITING A LANDFILL

Given the difficulties of measuring gas generation within a landfill, there has been periodic interest in measuring gas leaving or exiting a landfill. The concept is to measure gas flow through the cover for the portion of the site above ground, and gas flow into the surrounding soils for the portion of the site below ground. Gas flow below ground uses the same estimation techniques used for gas migration testing and modelling and so will not be discussed in detail here.

By measuring gas composition and pressure changes across a known thickness of soil, the pressure and concentration gradients driving convection and diffusion flow respectively can be determined. Testing of the soils within the layer for hydraulic permeability and diffusivity allows calculation of convective and diffusive flow rates. Different layers of soil can be used, representing trends and changes in soil characteristics in different directions from the landfill, or as a check on estimates, and to estimate flow in all directions.

Estimation of flow through cover to the atmosphere involves similar estimation of convective and diffusive flow. Concentration and pressure gradients must be measured over or within cover soil layers, whereupon measurement of diffusion coefficients and hydraulic permeability corrected for gaseous flow allows calculation of flow by both mechanisms. The problem with such projections, whether for below-ground migration or through cover soil, is the need to have uniform soils or to at least be able to account for all the different soil characteristics surrounding the landfill. Differences in soils, moisture content, density of soil, compaction etc., will affect flow rates for a given pressure or concentration gradient, and it is to be expected that there will be such differences in the actual landfill. Some variability can be corrected by measuring grain size distribution, organic content etc. for the different soils. If a landfill is lined such that little or no flow occurs below ground, the problem is simplified, but even then one has to deal with surface irregularities such as cracks, whether by settlement or freeze/thaw, erosion irregularities, different vegetation types and development, different moisture contents across the cover, differences caused by the shape or contour of the surface, worm/insect/animal effects, and a host of other irregularities. A surface crack can clearly vent much more gas than would otherwise be vented: any estimate must take such a crack into account. Such complications have limited the use of this technique.

Another method of estimating gas flow to the atmosphere is simply to collect the gas as it passes out of the cover. In concept, if one could completely cover the landfill with a balloon-like membrane or plastic sheet, all gas leaving the landfill would be collected and so could be measured. In effect, the gas is forced to flow to the atmosphere through measuring devices. If no gas leaves the site underground (for example a clay or plastic liner), conceptually all gas generated would be measurable by this technique. It is unrealistic to completely enclose a full-size landfill, so it is necessary to place such a gas collector over representative portions of the landfill cover, projecting the measured rate of gas flow for the area test over the entire surface of the landfill. Gas collectors have ranged in size from plastic bags covering perhaps 3000 cm^2 to plastic sheets of perhaps 25 m^2 or so. The device must be dug into the soil or sealed in the soil in such a way as to collect all gas flowing from the soil within the test area. Sealing devices tested include metal rings embedded several centimetres into the soil, to which plastic bags are attached, and the use of excavated trenches bounding the test area into which the edges of the plastic sheet are placed and backfilled with water. Gas can be collected and stored until the collector is evacuated, perhaps by pumping gas through a gas meter.

Even though the concept of flow measurement into the atmosphere sounds simple and easy to accomplish, carrying out such measurements in reality is very difficult and subject to error. First, there can be no pressure differential between the measurement or test area and adjacent areas: otherwise, measured gas flow per unit cover area will not be representative. Any back pressure, caused for example by the weight of a heavy plastic membrane used for the collection device, will cause gas to flow to some degree into the surrounding area for venting to the atmosphere. Consequently, very light plastics have been used to minimize the back pressure.

Another problem is caused by weather effects during measurement. Condensing vapour, freezing or wet conditions, and temperature all have an influence on measuring devices and the accuracy of the resulting data. Finally, surface irregularities, changes in vegetation, changes in organic content, moisture content, compaction or basic soil characteristics, cracks or holes caused by settlement or living creatures etc., all have to be accounted for to project properly results for the measured areas to the entire landfill.

Largely because of the lack of uniformity of flow characteristics of cover soils and soils surrounding the landfill, surface flow and below-ground flow techniques are rarely used to estimate gas generation at a full-sized landfill. At best these techniques can be used with other techniques for mutual benefit, but it is seldom that one could use only such techniques for landfill gas testing. Even if one obtains good data for the test area measured, which in itself is difficult, the results must be projected over the entire landfill. It must be assumed that the areas tested are representative of the entire landfill area or the test values must be adjusted accordingly.

GAS GENERATION BY ANAEROBIC LANDFILL SAMPLES

This method is related to the concept of enclosing a landfill perhaps as in a balloon and channelling the gas through a measuring device. In the application discussed here, samples of refuse are taken directly from a portion of the landfill for which is it desired to obtain gas generation rates and placed in a container that simulates conditions within the landfill. The waste will continue to decompose, supposedly as though it continued to be in the landfill, and gas generated can be measured and related to the weight of refuse to give the specific generation rate. Clearly the bigger the sample, and the larger the number of them, the more closely will the results relate to the actual landfill. Samples should be replicated, and many sampling points will be necessary simply to bracket the different gas generation rates of refuse. Experience has indicated that at least two but preferably more samples should be taken at each sampling location, and as many locations should be tested as have different refuse ages, moisture availability, temperature, compaction, etc. as are found in the landfill. Results from this technique have been shown to relate to gas generation rates in full size landfills; however, variability from sample to sample suggests that testing many samples over much of a landfill is necessary to reasonably represent the actual landfill.

The technique involves excavating refuse from each area and depth of the landfill to be tested and placing these samples quickly, to minimize oxygen contact, into incubation containers fitted with gas collection and measuring devices. Samples are typically brought to the surface of the landfill by augers from a boring device, although direct excavation has been used for shallower samples. Sample size should be such as to fill the container totally, and will generally range from 1 to 10 kg. The auger or bucket can be used to mix and grind the samples by advancing the boreholes slowly at the point at which samples are to be taken. Representative samples of refuse should be packed quickly into the incubating containers, and the containers should be capped and sealant applied quickly to minimize air leaks into or gas leaks out of the container except through the gas-measuring device. The container is then placed in a room or water bath with the temperature controlled to that noted within the landfill.

Gas generation rates will mimic rates of generation within the landfill to the extent that the same refuse, micro-organisms, nutrients, moisture etc. that set decomposition patterns and rates within the landfill will be operable within the container. There will be an increased rate of gas generation resulting from the mixing that inevitably occurs in obtaining the sample and packing it into the container, but gas rates have been shown to return to gas rates similar to rates obtained by other test methods at the same landfill. Once gas rates have reduced to a level that is held with relatively little additional drop over a period of several weeks, the rate has been found to be similar to that of the landfill. The technique has been used to test for the effects of changes in temperature, moisture content, moisture flow, nutrients added, buffer added etc. on, gas generation rates at various landfills.

Problems with this technique include the inability to sample any landfill in a truly representative way. Many samples should be taken to represent the heterogeneous nature of the landfill. Another problem is knowing whether the data obtained truly describe what would have happened had the sample continued decomposing in the landfill. Even though results suggest similarities, there is no proof that the laboratory results represent adequately the full-scale conditions. Again, this technique should be used in conjunction with other techniques to obtain, hopefully, mutually reinforcing results.

MEASUREMENT OF REFUSE COMPOSITION CHANGES OVER TIME

This last technique is also the most indirect, but it is the only method providing estimates of gas generation rates over extended periods of time without the continuous monitoring of gas generation. The concept is to consider the loss of organic matter from the refuse to be equal to gas generated plus the loss of organic matter via leachate flow. Mass balances have been performed in enclosed laboratory-scale reactors simulating the methanogenic decomposition of MSW in a landfill. Except for experimental error, the amount of gas can be directly related to the loss of organic matter from the refuse plus whatever organic matter was removed, if any, in leachate. The approach has also been utilized at full-scale landfills, but without long-term gas and leachate monitoring data the validity of the mass balance approach in this situation has not been proven.

Methanogenic degradation of organic matter of general composition below is related to the theoretical amount of methane and CO_2 by the following equation:

$$C_a H_b O_c N_d + \left(\frac{4a - b - 2c + 3d}{4} \right) H_2O \rightarrow$$

$$\rightarrow \left(\frac{4a + b - 2c - 3d}{8} \right) CH_4 + \left(\frac{4a - b + 2c + 3d}{8} \right) CO_2 + dNH_3$$

Clearly real microbiological systems cannot provide 100% turnover to methane and CO_2; there must be some loss of energy and some production of cell mass. This loss is generally held to be 15%. Therefore it is reasonable to expect that approximately 85% of the theoretical gas calculated from the above equation will be produced in a mature biological system. To the extent that little organic matter is lost in the leachate in comparison with the amount of gas produced, loss of partially degraded organic matter in leachate can be neglected, and the loss of organic matter from the refuse can be related directly to the gas produced. More specifically, as cellulose and hemicellulose are the major chemically identifiable components in MSW that can decompose methanogenically, representing more than 80% of the methane potential in MSW, measurement of these specific compounds over time, or relating present results to initial or projected final cellulose and hemicellulose contents, can provide information about the state of decomposition, the potential for future gas generation, and average generation rates to date.

Sampling a landfill to give reliable chemical composition ranges from difficult to impossible. Many samples must be taken at sufficient locations throughout the landfill so as to represent the overall landfill. If only a portion of the landfill is under investigation, the problem can be simplified accordingly. In general, three to five replicate samples must be taken at each sampling point, and sampling points must be distributed both vertically and laterally to represent fully the refuse composition, age, moisture content, access to moisture, depth, and other factors affecting the decompositional state of each microenvironment within the landfill. If composition data are available describing the average composition of the MSW as placed, or such information can be estimated, the difference in organic content is related to the gas generated. Conversely, if sampling of old landfills or other evidence is available to describe the realistic final composition of decomposed waste, the potential for future gas generation can be estimated.

Potential errors in this concept are obvious and relate largely to sampling problems, loss of organic matter to products leaving the landfill in leachate, and the conversion efficiency of lost organic matter and MSW to methane and CO_2; however, these uncertainties are in the range of uncertainties associated with other methods of landfill testing to determine present gas generation rates. The difficulty of estimating present gas rates is compounded when it is necessary to perform tests over an extended period of time to obtain long-term gas rates. The mass balance approach is probably the most useful method for long-term estimation of gas generation potential.

CONCLUSIONS

Field testing of landfills to determine present gas generation rates or long-term gas production potential is difficult and inexact, subject to problems in obtaining reliable data but also to problems stemming largely from the inherent heterogeneity of a landfill, which makes data interpretation and application difficult. The fact that several different techniques are used as outlined above, plus specific procedures unique to the organization or individuals carrying out each technique based on individual experience and preference, suggests that no one technique or procedure has emerged as being the best. The best approach, if the most exact estimates of gas generation rates are to be obtained, is to use several of these techniques in accordance with time and budget constraints. In this case, the results will hopefully reinforce each other and there can be more confidence in the outcome.

Because of the cost, complexity and difficulty of obtaining reliable results, many operators will not do independent landfill gas testing; rather they build a gas withdrawal system and run it initially on a test basis. This initial operation then allows refinement of the withdrawal system and allows determination of actual gas quantity and quality as a function of vacuum and pump capacity. If it is reasonably certain that gas will be withdrawn from a given landfill for emission control purposes or for use as an energy source, so a withdrawal system will be required anyway, the concept of using the initial withdrawal system for test purposes is logical.

5. EXTRACTION AND TRANSPORTATION

5.1 Zones of Vacuum Influence Surrounding Gas Extraction Wells

RONALD J. LOFY

Lofy Engineering, PO Box 5335, Pasadena, CA 91117, USA

INTRODUCTION

There is very little known about gas movement in landfills. Gas extraction well design and operation is mostly done on the basis of experience and not on scientific knowledge. Owing to this situation, in those cases where the experience is relatively low, mistakes have been made that were costly and/or dangerous. Although gas prediction models are in use and fairly reliable, they are always linked to the LFG recovery rate. The latter is difficult or impossible to measure under field conditions. If landfills are covered with soil or liners, it is very difficult to predict the effect of the cover on the gas recovery rate.

In the following three chapters (5.1, 5.2, 5.3), results from scientific investigations are presented in order to understand better what happens when LFG is extracted or passively vents into the atmosphere. Based on these results models are developed in Chapters 5.2 (active gas extraction) and 5.3 (passive gas venting) that help to understand and design better LFG extraction systems.

Landfilling of Waste: Biogas. Edited by T. H Christensen, R. Cossu and R. Stegmann. Published in 1996 by E & FN Spon, London. ISBN 0 419 19400 2.

RESEARCH OBJECTIVES AND BACKGROUND

The primary objective of this very basic research was to determine the limits of the vacuum sphere of influence and pressure profile for the mechanically induced vacuum zone around an extraction well or system of extraction wells. The shape and extent of this vacuum zone was thought to be heavily influenced by the strength of the exerted vacuum, the gas extraction rate, the permeability of the refuse, the length and depth of perforated screen(s), and number of perforations per screen. It was believed that once 'steady state' extraction had been reached, the zone of influence did not change.

Another interest was to gain a better appreciation and understanding of the phenomenon of vacuum breakthrough to the surface, sometimes called 'daylighting'. Daylighting of the vacuum was thought to be simply a matter of applying too great a vacuum, resulting in withdrawing too much gas per unit time from a given landfill volume. Some in the industry felt that breakthrough could be mitigated by simply determining the critical maximum extraction rate and then throttling back on the wellhead vacuum until breakthrough was eliminated. The theoretical calculations presented in the next chapter relating extraction rate to permeability and radial distance to zero isopleth line as a function of barometric pressure provide some credence to this approach.

At the time, there was considerable controversy about how long and deep a well screen needed to be in relationship to the total depth of the landfill. One camp argued that as the gas rose to the surface to eventually vent to the atmosphere anyway, only a relatively short well close to the surface was necessary to capture the gas before it was lost to the atmosphere. The other camp argued that the much more expensive deep wells were necessary to maintain higher gas quality and capture the gas from the deeper, otherwise unaffected, portion of the landfill before it permeated out of the landfill sides. Neither side was able to prove its point because of the considerable 'site-specific differences in the landfills'.

A widely held and important assumption in the industry was that the zero vacuum contour line (zero isopleth), dividing positive pressure zones from negative pressure zones in the landfill, represented the limits of the area of vacuum influence within which a gas molecule will be drawn to the well. This somewhat arbitrarily established dividing point took into consideration the recognized accuracy of the pressure gauges then commonly in use and the fact that the typically dry California landfills did not have an ambient *in situ* pressure much above zero gauge. This logical, but now recognized as erroneous, assumption stemmed from the belief that a vacuum in a landfill is an abnormal phenomenon that only occurs when a mechanically induced vacuum is applied. It was always assumed that under ambient conditions the generated landfill gas could not escape at a fast enough rate because of the permeability and porosity of the landfill, and therefore would always maintain a slight positive pressure within the landfill. The measurement of negative pressures within reasonable proximity of a well was considered presumptive evidence of gas withdrawal by the mechanically applied vacuum. This assumption was also used in the subsequent evaluation of the data. However, even before beginning this research, there was some question concerning the validity of this assumption.

An extensive number of pressure or vacuum gauge measurements were to be taken to delineate those areas of the landfill that were experiencing a vacuum. By plotting the pressure isopleths and calculating the pressure and gas concentration gradients, it was hoped to determine the extent of the vacuum and/or its effects as a function of distance from the well and its relative shape within the landfill.

LAYOUT OF SYSTEM

Building on prior industry experience and an intuitive understanding of the gas extraction process, a landfill test facility was constructed comprising 37 monitoring-probes in a three-dimensional grid pattern, which covered the presumed area of vacuum influence surrounding three wells.

In laying out the probe locations shown in Fig. 1, it was desired to have more probes close to the well in order more precisely to identify the shape of the vacuum zone as it developed and possibly detect any apparent daylighting of the vacuum in the immediate vicinity of the well. At the same time, the maximum extent of the anticipated vacuum, based on prior engineering experience, suggested that the probes should not reach out more than about 2 ½ to 3 times the depth of the well. For a 25.9 m (85 ft) deep well, this was approximately 64.6 m (212 ft), and for a 32 m (105 ft) deep well, it was 80.2 m (263 ft): thus the selection of monitoring-probes out to 91.4 m (300 ft) maximum radial distance from the well to detect the maximum anticipated radius of influence of well vacuum.

The multiple depth probe layout shown provided a three-dimensional grid system within the landfill, which was used in a qualitative manner to define the approximate shape of the vacuum zone of influence. The random, radially occurring distances from the well allowed a greater opportunity and probability of picking up any unusual phenomena occurring with regard to the transitory development of the vacuum zone of influence.

Within the constraints of the allowable budget and physical configuration of the site, this particular layout provided a high degree of research flexibility and ability to use one portion of the system as a control for another part of the system, i.e. the probe configuration surrounding Well 1 is to a large degree a mirror image of the Well 3 configuration. Each of the individual well-probe arrangements was to complement and reinforce the information obtained from other well-probe systems.

The numbering system for each multiple-depth monitoring-probe station consists of three components that uniquely identify each probe in relationship to an extraction well. The first digit of the three-component station label identifies a grid line passing through the well; the second component signifies the direction away from the well, i.e. west (W), east (E), and north (N); and the third component indicates the station, in ascending order, in a particular direction away from the well. For example, 1W3 means that the probe station is on the east – west trending line passing through Well 1, and it is the third probe along the grid line west of Well 1.

The test facility site plan is shown in Fig. 2. The original contour elevations were compiled from an old topographic map of the site. Because of an almost complete lack of interim topographic maps for this area, there was no way of determining where the original crests of the canyons and backside hills had been excavated to provide cover material and additional inventory. The final buried contours of the canyon over which the well and probe system was placed were painstakingly determined from the soil-boring logs, discussions with old site operators, and reliance on the few records that were available.

As can be seen by a careful perusal of the monitoring-probe – extraction well system layout (Fig. 1) that has been superimposed

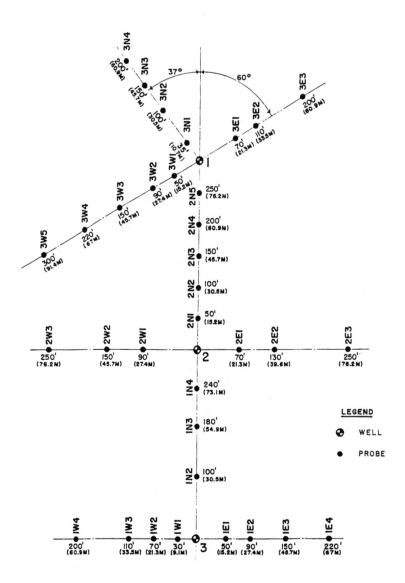

Figure 1. Location plan of multiple-depth monitoring probes and extraction wells.

on the site plan (Fig. 2), the three wells fairly well straddled the deepest part of the buried canyon. The probe spacings and centring of the east – west trending grid lines are staggered over the width of the canyon so as to distinguish any geometry effects due to the presence of the canyon sidewalls.

As will be seen later, the N–NE trending buried ridge separating the test facility area from the main portion of the landfill to the east is believed to play an important role in the peculiar phenomena observed. Based on bore logs and the other available data, it is believed that the crest of the ridge had been 'shaved' down some-what from the 183 m (600 ft) elevation. Note how the crest of the ridge separates the test canyon from the rest of the landfill to the east, and particularly how it acts as a barrier at the east end of grid 2W3–2E3, but drops down to create a backside 'tunnel' connection between the test canyon and the larger portion of the landfill near grid 1W4–1E4.

The multiple-depth monitoring-probe grid line identified as 3W5–3E3 is roughly parallel to the top of the slope of the canyon wall. Line 3N4–Well 3 is perpendicular to the edge of the slope. The three wells are exactly 91.46 m (300 ft) on centres along a straight line. As shown in Fig. 2, Well 3 is on a bench that is approximately 4.9 m (16 ft) lower than the upper terrace on which Wells 1 and 2 are located.

The particular site area selected for the three-test-well facility affords an excellent chance to observe the effects of extraction in a typical buried landfill as well as one with exposed exterior sidewall. The series of three wells installed for this project traverse the approximate centre of the slightly curved canyon. Well 3 lies approximately 76.2 m (250 ft) in from a perpendicular to the exposed outer sidewall top of slope. The three-dimensional multi-probe depth gas-monitoring grid pattern surrounding Well 3 is quite similar to that placed in the area of Well 1. This affords an excellent opportunity to compare the well radius of influence for a typical flat or buried type landfill compared with Well 3, which exemplifies that for an exposed sidewall condition. The system of probes was installed in a random non-symmetrical pattern to provide greater chance for picking up any idiosyncrasies or addi-tional information that might otherwise have been lost with a symmetrical system.

While the effects of an individual well were of some importance in terms of identifying the development of a transitory zone of

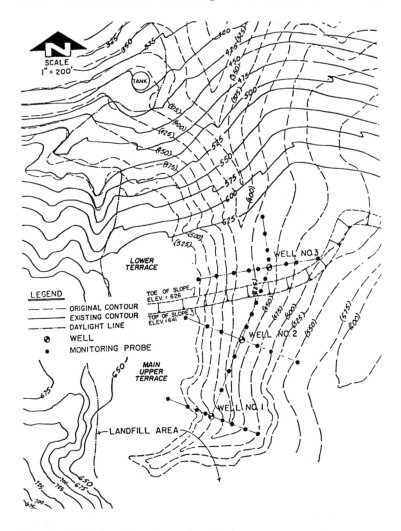

Figure 2. Facility layout superimposed on site plan.

vacuum influence, they did not provide the more realistic picture of what happens when several wells are operated simultaneously. Thus the series of three wells in line provided a practical opportunity to observe the effects of different well interactions under different operating conditions. The three wells were connected to the centrifugal extraction blower by means of 15 cm (6 in) diameter, schedule 40 PVC pipe between Wells 2 and 3, and 20 cm (8 in) diameter, schedule 80 PVC pipe between the blower (near Well 1) and Well 2.

Well Design

Each pipe casing had its own 7.6 m (25 ft) long well screen, shown in Fig. 3. The design of the wells consisted of two annular well pipes. The inner 10 cm (4 in) diameter pipe was sealed off at the 16.8 m (55 ft) level such that the gas taken in through the lower screen was conveyed up the inner 10 cm (4 in) diameter pipe to the wellhead take-off/collection pipe manifold. The gas taken in through the upper screen moved up the annular space between the 10 and 20 cm (4 and 8 in) casing walls and was drawn off at the wellhead take-offs as shown. The special wellhead design allowed for three modes of operation:

- withdrawal of gas from only the upper well screen (screen 1) at an elevation of –9.1 to –16.8 m (–30 to –55 ft);
- withdrawal of gas from only the lower well screen (screen 2) at elevations between –18.3 and –25.9 m (–60 to –85 ft); and
- withdrawal of gas from both well screens (screen 1 + 2) simultaneously.

As shown in Fig. 3, each well screen section was isolated from the other well screen section in the borehole by an impermeable barrier of bentonite clay of approximately 2 ft (60 cm) thickness followed by an additional 3 ft (90 cm) of relatively impermeable native soil.

Upon completion of phased filling in an area, and knowing that they might not return for over a year, the Los Angeles County Sanitation Districts (LACSD) often added significant amounts of cover soil in order to provide a better than normal interim cover, as well as provide appropriate grade for drainage of storm waters. Typically, this consisted of a relatively impermeable layer of silty clay measuring 1.2–1.8 m (4–6 ft) in thickness. This occurred at least twice: at a base elevation of approximately 176.8 m (580 ft) and again at about 193.6 m (635 ft). Normally the interim cover layer is only 15–30 cm (6–12 in) thick. As it turned out, the thick cover layer at elevation 176.8 m (580 ft) coincided, as best as could be determined, with the 1.5 m (5 ft) plug separating the upper well

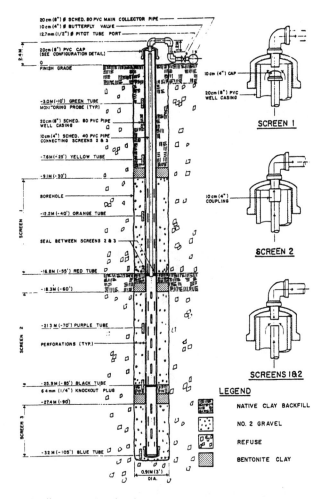

Figure 3. Well construction detail.

screen from the lower well screen. At the time that this research was performed, the landfill surface cover was also a minimum of 1.5 m (5 ft) thick. As can be shown later by careful study of the graphical plots, these two layers created a relatively impermeable barrier to the vacuum, which appeared to force it to move more horizontally than vertically.

At the insistence of the LACSD, an additional 6 m (20 ft) section of pipe was added to the proposed three wells. The lower 4.6 m

(15 ft) of this pipe section was perforated (screen 3) to correspond to an elevation below the surface of –27.4 to –32 m (–90 to –105 ft). A 0.64 cm (¼ in) scored PVC plastic plate was inserted between the two pipe sections at the couple separating screens 2 and 3. This effectively prevented any vacuum to well screen 3. These plugs were later knocked out for the last two experimental runs prior to turning the well system over to the LACSD. Experimental runs 36 and 37, the last two weeks of the experimental field research programme, extracted gas from screens 2 and 3, which correspond to a zone of the landfill between –25.9 and –32 m (–60 and –105 ft).

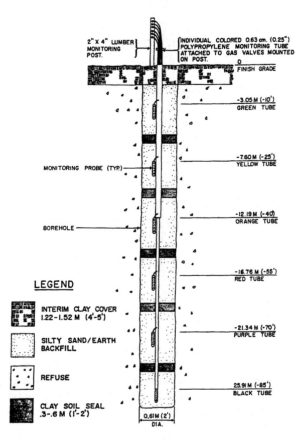

Figure 4. Typical 25.9 m (85 ft) multiple-depth monitoring probe installation.

Monitoring Probe Design

Because drilling was one of the largest cost items in the construction of the test facility, and as it was desired to maximize the number of monitoring-probes surrounding the series of wells, the multiple-depth monitoring-probe and sampling system in a common bore hold shown in Fig. 4 was developed.

The individual monitoring-probes were 30 cm (12 in) long, 12.7 mm (½ in) diameter, schedule 40 PVC pipe which had a number of 3.2 mm (⅛ in) holes drilled in it. This short plastic pipe was capped at both ends, and a continuous polypropylene tube, of 4.7 mm (³⁄₁₆ in) O.D. and 1 mm (0.04 in) sidewall thickness, reaching to the surface was inserted in a small drill hole through the cap into the probe. The individual probes were wrapped with 6.3 mm (¼ in) thick foam rubber padding to prevent particulate matter and debris from plugging the gas inlets. The polypropylene tubing connecting each of the individual probes was placed inside a 12.7 mm (½ in) and a 19 mm (¾ in) schedule 40 PVC protective pipe to the surface.

The 25.9 m (85 ft) long multiple depth monitoring-probe was installed in 33 of the 0.61 m (2 ft) diameter boreholes. Sandy soil available on site was used for backfill around the probes. At pre-scribed intervals between individual probes, a 0.3–0.6 m (1–2 ft) thick bentonite clay plug was installed to isolate each probe from an adjoining probe and to prevent a chimney effect in the backfilled borehole. The uppermost 1.2–2 m (4–5 ft) was backfilled with the relatively impermeable clay available on site and water flooded to provide a compacted backfill.

The individual probes were installed at depths of −3.05 m (−10 ft green), −7.60 m (−25 ft yellow), −12.19 m (−40 ft orange), −16.76 m (−55 ft, red), −21.34 m (−70 ft purple) and −25.91 m (−85 ft, black), respectively. Each probe, corresponding to a specific elevation level, was identified by its own specific colour-coded polypropylene monitoring tube to the surface. The colours used in descending order of depth were: green, yellow, orange, red, purple and black. Using the same design, a seventh, light blue coloured, monitoring-probe was added at a depth of 32 m (105 ft) at Well 1, Well 2, Well 3, and Probe Station 3N1.

The six or seven monitoring tubes from each borehole were fastened to a 5 cm × 10 cm (2 in × 4 in) lumber monitoring post at the surface. Each of the monitoring tubes was connected to a plastic aquarium-type valve to close off the tube. The tubes were fastened to the post in descending order of depth corresponding to the colour code of green, yellow, orange, red, purple, black and blue.

The portable instrument case was carried from one station to another along a monitoring line. Starting with the uppermost probe tube mounted to a post, the pressure transducer would be connected to the monitoring tube sticking out of the ground. After opening the valve and reaching a steady-state condition, the pressure would be read and recorded. The valve would be closed and the instrument disconnected. It would then be connected to the next lower tube and the process repeated until all of the six or seven monitoring-probe tubes had been read and recorded. A station could be read in about 5 min or less.

The typical work week for the monitoring crews was from 8:00 Monday morning until 1:00 Saturday morning. For those experimental runs of more than one day, monitoring was performed over two shifts (17 h total) between 8:00 and 1:00 the next morning. The extraction equipment would operate unattended between the hours of approximately 1:00 and 8:00.

Portable Gas Rig Power Source

A Hauck centrifugal blower, driven by a Waukesha dual-fuel naturally aspirated gas engine, was used to extract the gas from the landfill for the entire series of experiments. The blower was capable of pulling a maximum of about 34 m^3/min (1200 cfm) of landfill gas. The desired gas flow rate was obtained by regulating engine rpm. Total dynamic head of the blower was approximately 122 cm (48 in) of water column with a maximum suction of about 51 cm (20 in) of water column.

ANALYSIS OF DATA

Evaluation and Interpretation of Zone of Vacuum Influence Plots

A total of 37 experiments or 'Runs' were performed. A list of these 37 experiments indicating dates, basic parameter setting, baseline measurement period(s), and/or actual duration of run is indicated in Table 1.

TABLE 1. Tabulation of Experiments Run

Experiment no.	Date(s) of run	Well(s)	Screen(s)	Approx vacuum (kPa) (in.)	Approx. av. flow (m³/min) (scfm)	Cumulative baseline (h)	Time run (h)
1	24 Jun	1	2	2.77 (11.10)	6.17 (218)	0:07	1:21
2	25 Jun	3	1	2.09 (8.40)	2.78 (98)	0:52	4:53
3	27 Jun	3	2	3.66 (14.70)	3.37 (119)	6:34	1:10
4	30 Jun	2	2	2.99 (12.00)	4.79 (169)	2:09	5:40
5	2-3 Jul	1	2	1.52 (6.10)	5.52 (195)	7:07	26:52
6	7 Jul	2	1+2	1.04 (4.19)	4.76 (168)	2:04	8:44
7	8 Jul	2	1+2	1.84 (7.40)	5.58 (197)	—	7:55
8	9 Jul	2	1+2	1.74 (7.00)	5.21 (184)	1:34	6:59
9	10 Jul	2	1+2	1.74 (7.00)	5.21 (184)	3:05	4:21
10	11 Jul	1	1	1.12 (4.50)	4.79 (169)	—	1:30
11	14-15 Jul	1	1	0.62 (2.47)	2.04 (72)	5:45	32:49
12	18-21 Jul	3	2	1.72 (6.91)	4.70 (166)	11:13	8:47
13	23-24 Jul	3	2	1.72 (6.91)	4.70 (166)	—	37:07
14	25 Jul	2	2	1.41 (5.68)	3.29 (116)	—	15:20
15	26 Jul	2	2	1.32 (5.29)	3.09 (109)	—	16:45
16	29 Jul-1 Aug	1+2+3	2	1.39 (5.59)	10.65 (376)	28:12	19:58
17	6-8 Aug	1+2+3	2	1.05 (4.20)	7.99 (282)	5:14	57:51
18	11-13 Aug	1+2+3	1	0.87 (3.50)	12.23 (432)	11:54	37:39
19	14-15 Aug	1+2+3	1	0.91 (3.67)	9.20 (325)	23:01	1:26
20	18-19 Aug	1	2	0.65 (2.59)	3.77 (133)	—	35:09
21	20-21 Aug	3	1	2.23 (8.97)	5.89 (208)	2:23	20:30
22	21-22 Aug	2	1+2	1.76 (7.08)	6.97 (246)	—	38:21
23	25-26 Aug	1+2+3	1+2	2.22 (8.91)	22.54 (796)	—	33:34
24	27-29 Aug	1	1	1.76 (7.06)	6.83 (241)	2:46	50:14
25	2-5 Sep	1+2+3	1+2	0.84 (3.38)	17.47 (617)	2:30	84:06
26	8-9 Sep	1	2	1.98 (7.93)	6.26 (221)	1:24	24:07
27	9-12 Sep	1+2+3	1+2	1.56 (6.25)	20.11 (710)	—	96:31
28	15-17 Sep	1+2+3	1+2	1.62 (6.51)	19.82 (700)	—	62:53
29	18 Sep	1+2+3	1+2	1.58 (6.33)	20.02 (707)	7:11	13:03
30	19 Sep	1	1+2	—	—	15:00	—
31	22-26 Sep	3	1	1.29 (5.19)	4.19 (148)	9:47	100:02
32	29 Sep-3 Oct	1+2+3	1 + 2	0.98 (3.92)	14.92 (527)	5:58	103.17
33	6-8 Oct	1+2+3	2	—	—	17.17	—
34	13-16 Oct	1+2+3	2	3.15 (12.65)	10.31 (364)	—	81:00
35	20-31 Oct	1+2+3	2	2.56 (10.29)	—	18:38	171:39
36	3-7 Nov	3	(2+3)	3.46 (13.89)	2.04 (72)	7:03	100:52
37	10-14 Nov	1	(2+3)	1.76 (7.08)	5.75 (203)	16:16	76:57

Exceeded range of instrument.

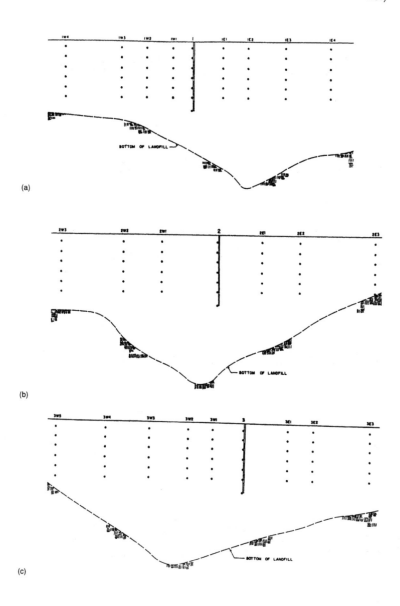

Figure 5. Location of the probes (see also Fig. 1): (a) East–west cross-section through well 1; (b) east–west cross-section through well 2, and (c) east–west cross-section through well 3.

In order to 'visualize' the large amount of data obtained, and to adequately analyse and evaluate the results of the individual experiments, some of the results were plotted on a two-dimensional grid system, which corresponded to an imaginary vertical cross-section of the landfill as defined by the series of multi-depth monitoring-probes and wells. By plotting the data in the same chronological sequence as they were obtained, it was possible to follow the continuously changing, dynamic and synergistic effects of the vacuum from time of start-up until it reached 'steady-state' conditions. A widely held assumption within the industry was that the radius of vacuum influence was all of that negative pressure between the well and the zero pressure contour line. As a result, the contour plot is divided into areas of positive and negative pressure. Areas of negative pressure are shown slightly shaded to differentiate them from the positive areas. The contour intervals selected vary from one graphic presentation to another. This was done so as not to crowd the plot with many contour lines, while still showing all of the important information and delineation of interesting phenomena. The contours typically employed were the zero contour line, the 125 Pa (0.5 in) and negative 125 Pa (−0.5 in) contour lines, and the positive or negative integer contour lines. The locations of the probes in relation to radial distances from the wells are shown for the respective grids in Fig. 5 and 6.

Vertical cross-section sheets showing the locations of the probes in relationship to the well were drawn to a horizontal and vertical scale of 1 cm = 6 m (1 in = 50 ft). The vertical distance between all probes is 4.6 m (15 ft) while the horizontal distance between probes is between 9.1 and 30.5 m (30 and 100 ft).

The data shown in these plots do not represent an 'instantaneous snapshot' of the entire vertical cross-section of the plane of interest. Instead the results are analogous to a movie camera slowly panning a landscape, starting at one side and moving across the landscape to the other side. This effect occurs as a result of the time lapse involved in connecting, reading and recording the results of a single portable pressure transducer. The technician would start at a monitoring station at one end of the grid, connect the instrument to the uppermost monitoring tube, and proceed to take measurements at all six multiple-depth probes before moving on to the next station in line. This procedure was continued until all of the monitoring-probes were read. Depending on the technician doing the work, the

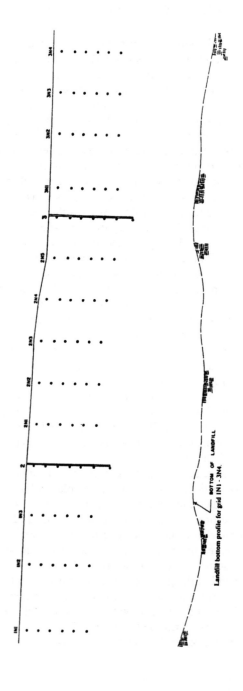

Figure 6. Location of the probes (see also Figs 1 and 5).

number of monitoring stations involved, technical problems with keeping the instrument clean, and time required to obtain a stable pressure reading, the time to complete a traverse of the monitoring stations in a grid ranged from 20 to 60 min. Thus considerable change could have occurred in the shape of the zone of vacuum influence in the time that it took to complete the traverse, particularly during the transitory condition following start-up and until 'steady-state' conditions were reached.

By a process of 'logical contouring', contour lines of equal pressure/vacuum were drawn in. Logical contouring assumes that the scalar distance is proportional to the difference in numerical value: i.e. if the numerical values for two adjacent points were 125 and 375 Pa (0.5 and 1.5 in) of positive water column, then the 250 Pa (1 in) contour line would be exactly halfway between the two points. Conversely, if the numerical values of the two points were 125 and 625 Pa (0.5 and 2.5 in) then the 250 Pa (1 in) positive pressure contour line would be one-quarter of the scalar distance from the 125 Pa (0.5 in) point.

The range of values observed for a particular traverse determined the density of contour lines and the contour interval(s) to be used. Typically, the zero contour line and at least one or more contour intervals were used in all of the plots.

The following series of plots have been culled from those prepared because they illustrate specific points that need to be discussed.

Experimental run 3. This, one of the first experimental runs, was conducted on 27 June 1980, between 9:00 and 16:46. Between 2.83 and 3.68 m³/min (100–130 cfm) of gas were intermittently extracted from screen 2 (−18.3 to −25.9 m or −60 to −80 ft below the surface) of Well 3 when the blower was operating. Figures 7–11 provide a good example of the information that can be gleaned from an experiment that does not always go according to plan.

Between 9:00 and 9:26 (Fig. 7) baseline measurements (the engine was not operating) were taken starting at Station 3E3 and moving westward to 3E1. The engine was started at 9:26 before the technician measured the fourth probe at Station 3E1 (3E1-55′). The maximum vacuum at the top of the well screen was 3.31 kPa (13.3 in of water column). The engine ran until 9:45 when it unexpectedly stopped. From 9:45 until 10:20, baseline measurements were again taken. These transitions from baseline to run measurements and

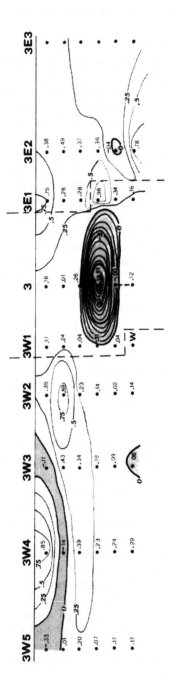

Figure 7. Run 3, 27 June 1980, 9:00–10:20 (all values in inches of water column). Shaded areas: negative pressure.

back again are indicated by the dashed lines on both sides of the shaded (vacuum) area around the well. While the engine had been running, an elliptical zone of vacuum influence had developed in 19 min as determined by the process of logical contouring.

Note that the vacuum did not penetrate all the way to the bottom of the well screen because the well was flooded to −23.2 m (−76 ft) during the heavy 1979–1980 winter rains. The vacuum instead moved out over the top of this perched water zone.

For future comparisons, note the small vacuum zones at 3E2–70′, at 3W3–85′, and the band of vacuum in the upper horizon stretching between 3W3 and 3W5.

Baseline measurements were again taken approximately 1 h later (Fig. 8). There is now a large zone of positive pressure in the approximate area where the vacuum extraction zone used to be. The maximum positive pressure is 457 Pa (1.83 in), almost double the largest positive pressure that existed an hour earlier. The negative bubble previously at 3E2–70′ has enlarged and moved upward to 3E2–40′. The positive pressure bubbles at 3W2–25′ and 3W2–40′ appear to be slowly moving to the landfill surface. The negative zones to the west of the extraction well have completely disappeared.

When first observed in 1980, this phenomenon was not immediately understood. However, petroleum engineers were very cognizant of this phenomenon and have developed a number of useful diagnostic and evaluation techniques based on it. Transferring this high-pressure oil and gas concept to the analogous but much lower-pressure, near-surface situation (which is affected by many other significant variables and interactions), the process can be described as follows.

Once gas molecules are placed in motion, the gas continues to flow to the shut-in well even though the blower itself has stopped. This phenomenon appears to continue for half an hour to several hours depending on the vacuum and the permeability and porosity of the landfill. The situation is perhaps analogous to planetary gravitational physics in that the distant gas molecules feel the tug of the negative pressure gradient, which causes the molecules to begin their flight or trajectory toward the well. As they move closer to the well, they accelerate owing to the exponentially increasing negative pressure gradient that existed prior to shut-off of the mechanical extraction blower. Even though the blower may have been shut-off for some time, the gas molecule has 'memory' in the form of kinetic energy or momentum that has been imparted to it. It cannot slow

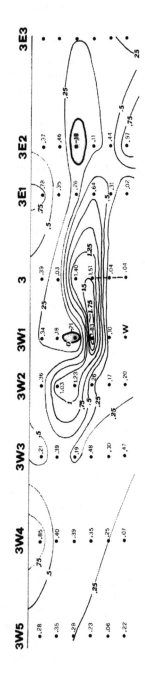

Figure 8. Run 3, 27 June 1980, baseline 10:29–11:18 (all values in inches of water column). Shaded areas: negative pressure.

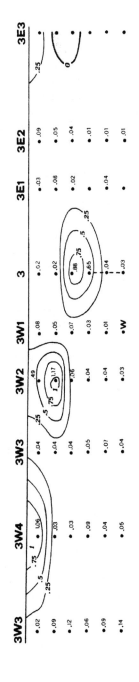

Figure 9. Run 3, 27 June 1980, baseline 11:36–14:28. Shaded area: negative pressure.

down or reverse this process until the positive pressure building around the well is large enough to counteract the momentum and kinetic energy that has been given to the gas molecule. The apparent reason for the higher than ambient positive pressure in the immediate vicinity of a well is that gas molecules are being pulled in from a large, presumably circular, area and are both converging and accelerating into an ever-smaller volume surrounding the well. The compression and molecular collisions of all the gas molecules near the well create the positive pressure measured.

The third set of readings of the grid (Fig. 9) was started at 11:36 and finished at 14:28, following a break for lunch. The negative pressure 'bubble' at 3E2–40′ has now disappeared or moved and there is the same or a new 'vacuum bubble' at 3E3–25′ and 40′. The large positive pressure zone in the vicinity of the well has now broken up into two positive pressure zones, each of less pressure. The gas at 3W2 appears to be venting to the surface. There is an increased positive pressure zone at 3W4–10′.

Baseline measurements continued from 15:32 to 15:54, after which the engine ran for approximately 22 min (Fig. 10). The vacuum zone of influence reached about the same radial distance (slightly more on the west side) even though the maximum vacuum was only 2.54 kPa (10.2 in) compared with the earlier 3.3 kPa (13.3 in) in 19 min shown in Fig. 9.

By 16:30 (Fig. 11), the vacuum zone of influence has thickened vertically somewhat and expanded an additional 12 m (40 ft) to the west. There is a positive pressure area at the east end of the vacuum zone, which has a maximum pressure of 219 Pa (0.88 in), and a very high positive pressure zone between 3W3 and Well 3, which encroaches upon and merges into the negative vacuum zones, above and below. The large negative pressure zone at the surface on both sides of Well 3 suggests that breakthrough has occurred.

This first monitoring/evaluation effort revealed some rather surprising results, as follows.

- Slight negative, as well as positive, pressure areas exist within a landfill during static (non-extraction) periods.
- These zones of negative and positive pressure change location, size, and pressure, and are apparently in a dynamic state of flux.
- After 'shut-in' of a well, gas continues to move into the high-vacuum area, and can increase pressure to above ambient

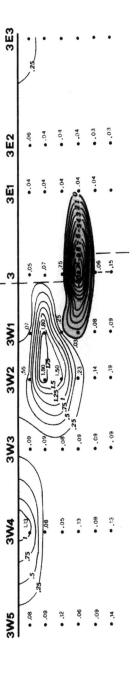

Figure 10. Run 3, 27 June 1980, 15:33–16:17. Shaded area: negative pressure.

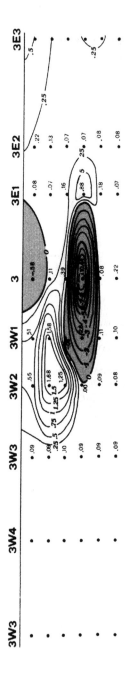

Figure 11. Run 3, 27 June 1980, 16:27–16:46. Shaded area: negative pressure.

positive pressure levels. This has some interesting implications. Every time an extraction operation is stopped (equivalent to individual well 'shut-ins') for more than a few minutes, a high positive-pressure zone is created around each well (which is a function of refuse permeability, gas flow rate etc.), which by virtue of its greater-than-ambient pressure will rather quickly vent to the surface and cause perhaps exacerbated odour conditions in the vicinity of the well. It also implies that operating a landfill as a (peaking power) gas reservoir may be far less beneficial or feasible than originally thought.

- During extraction, even in locations somewhat distant from the well, large positive pressure zones can be created.

Experimental run 8.   The next series of vacuum influence diagrams reveal:

- the apparent process by which breakthrough occurs;
- continual shifts in the presumed zone of vacuum withdrawal;
- the non-symmetrical nature of the fluctuating zone of vacuum influence as shown by the two perpendicular grids passing through Well No. 2, and;
- the apparent effects of landfill geometry as gas comes over the buried ridge from the larger portion of the landfill.

Experimental Run No. 8 was conducted on 9 July between 8:49 and 17:32. Total cumulative run time was 6 h and 59 min. Baseline measurements were taken between 13:36 and 15:10. Wellhead vacuum was approximately 1.75 kPa (7 in) and the withdrawal rate was 3.68 m^3/min (130 cfm) from the combined upper (−30 to −55 ft) and lower (−60 to −85 ft) well screens of Well 2.

Figures 12–15 represent the east – west cross-sections 2W3–2E3 and the ridge dividing the research test area from the larger landfill area to the east. A zone of positive pressure as high as 750 Pa (3 in) over the crest of the buried ridge indicates a positive flow of gas from the larger portion of the landfill toward the extraction well. The zone of vacuum influence surrounding the well has the appearance of a concave dish. All of the monitoring-probes along the entire surface in Fig. 12 had negative pressure, which suggests vacuum breakthrough to the surface. In Fig. 13, between 10:10 and 10:34, the outer perimeter of the vacuum receded somewhat and created a positive pressure area between 2W1 and 2W3.

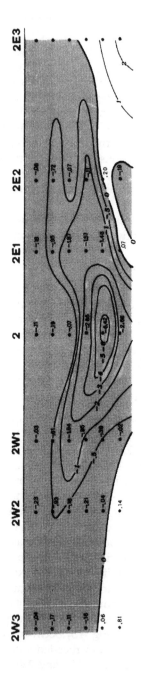

Figure 12. Run 8, 9 July 1980, 9:13–9:30. Shaded area: negative pressure.

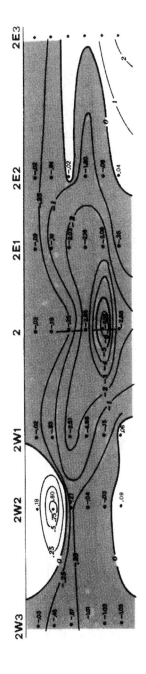

Figure 13. Run 8, 9 July 1980, 10:10–10:34. Shaded area: negative pressure.

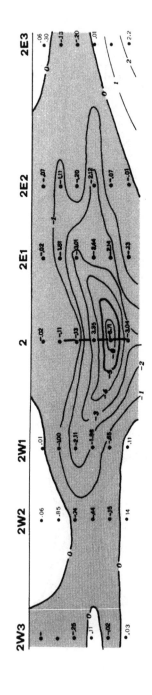

Figure 14. Run 8, 9 July 1980, 11:09–11:41. Shaded area: negative pressure.

In Fig. 14, between 11:09 and 11:41, the breakthrough to the surface had diminished in a broad area halfway between Well 2 and 2W3, as well as at 2E3.

By 14:51–15:09 (Fig. 15) the breakthrough to the surface had diminished even further and the entire vacuum zone had receded toward the well. The top of the –250 Pa (–1 in) contour line had sunk between 2.4 and 4.6 m (8 and 15 ft) deeper into the landfill. Bands of alternating negative and positive pressure at different strata also appear to be forming within the landfill.

The non-symmetrical nature of the zone of vacuum influence and the pronounced fluctuations in the shape of this zone are clearly shown in Figs 16–20, which are at right angles to the previous four views shown in Figs 12–15. The dynamic process by which breakthrough occurs is shown rather dramatically in Figs 16–20. In Fig. 16, there is a pronounced vacuum bulge at 2N2–28' and 2N2–40', while the remainder of the –250 Pa (–1 in) vacuum isopleth is below the –12.2 m (–40 ft) line.

An hour later, as shown in Fig. 17, a large negative pressure area had developed at 1N3–25', indicating a breakthrough of the vacuum from below the thick interim cover at roughly –16.9 m (–55 ft) into the cell above. The vacuum in this cell appears to be increasing, analogous to a vacuum flask being evacuated. The zone of vacuum influence around probe 1N3 had also begun to expand. A positive pressure zone had also emerged between the two vacuum bulges, breaking the continuity of the vacuum breakthrough to the surface in the area between Well 2 and 2N2.

Between 11:45 and 14:38 (Fig. 18), the vacuum breakthrough along the entire surface area north of Well 2 to 2N4 stopped, corresponding to the complete disappearance of the vacuum bulge at 2N2–25'/40'. However, the vacuum at 1N3 increased and caused discernible breakthrough to the surface in the area between 1N1 and 1N3.

Figure 19 shows that the strength of the vacuum deep within the landfill to the north of Well 2 has increased and has caused a gradual expansion of the zone of vacuum on that side. Figure 20 shows only minor changes.

Some of the interesting things learned from this series of plots were the following.

- The process of vacuum breakthrough occurred in stages within the landfill. In this particular illustration, the process

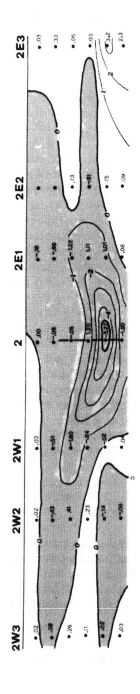

Figure 15. Run 8, 9 July 1980, 14:51–15:09. Shaded area: negative pressure.

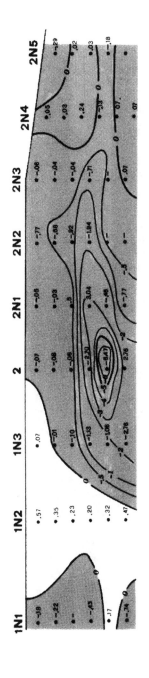

Figure 16. Run 8, 9 July 1980, 9:37–10:05. Shaded area: negative pressure.

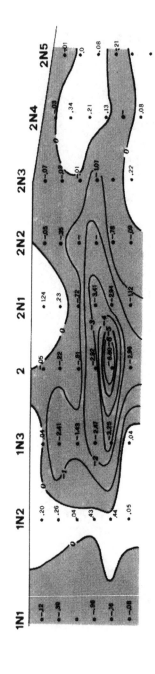

Figure 17. Run 8, 9 July 1980, 10:36—11:03. Shaded area: negative pressure.

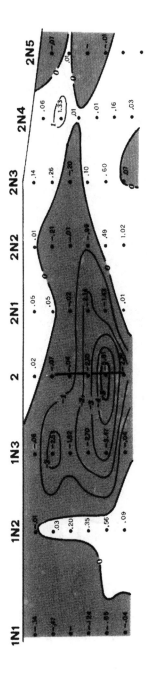

Figure 18. Run 8, 9 July 1980, 11:45–14:38. Shaded area: negative pressure.

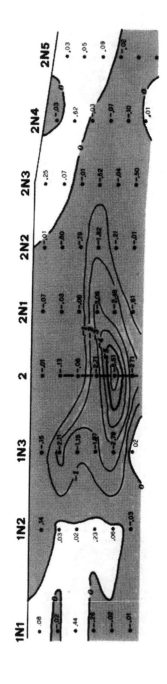

Figure 19. Run 8, 9 July 1980, 15:17–15:53. Shaded area: negative pressure.

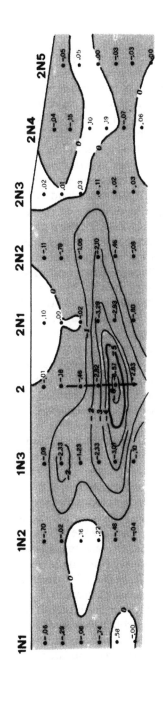

Figure 20. Run 8, 9 July 1980, 16:25–16:44. Shaded area: negative pressure.

was slow enough that each step in the layer-by-layer vacuum evacuation and progressive breakthrough to the surface was clearly evident. In other instances, breakthrough occurred before the second monitoring was even begun ½ to 1 h later. Fluctuations in vacuum within a cell spread to adjoining areas and were responsible for the growth or corresponding decline in the vacuum in that region of the landfill. As was observed during this particular experiment, one or more vacuum 'bulges' can occur, while another vacuum bulge 'collapses' or declines in another part of the landfill.

- The vacuum zone of influence around a well is not symmetrical. When comparing the east–west grid with the north–south grid passing through Well 2, it is apparent that the shape of the zone of vacuum influence, using both the −250 Pa (−1 in) and the 0 Pa (0 in) contour isopleths, varies significantly, indicating a non-symmetrical vacuum draw around the well. This has rather important implications for those gas generation/gas extraction models that rely on symmetry, and axiometric, isotropic, homogeneous conditions in the development of the model.

- Given that the centrifugal blower was operating at a constant rpm as controlled by the rpm governor-regulated engine, then the zone of vacuum influence should be fairly constant once steady-state conditions were reached. However, the zone of vacuum influence was observed to be in a continual state of dynamic change. This was the first indication that there are powerful external variables that had not been addressed, which were influencing the mechanical extraction process and the zone of gas withdrawal.

- The zone of almost continuous positive pressure over the crest of the canyon ridge separating the test areas from the larger landfill suggests that the zone of vacuum influence is not described by the zero vacuum contour line but actually extends beyond that to some undefined degree. This hypothesis is supported by the 27 June data plots showing the unusual positive pressure areas that merged into the high negative pressures close to the well. It has a further ramification in that landfill geometry appears to influence the dynamics of gas withdrawal. It is important to note that throughout the data that have been assembled, there was never a corresponding high-pressure area on the west side (edge of the landfill) of our test plot. The high-positive-pressure areas always existed at or over the crest of the canyon ridge separating the two landfill areas.

Experimental run 13.   The next series of vacuum influence isopleths are of experimental run 13 data collected on 23 July and 24 July, 1980. The average wellhead vacuum was 1.72 kPa (6.9 in) on screens 1 and 2 of Well 3. Average gas extraction rate was approximately 4.66 m³/min (165 cfm).

The pronounced elliptical areas of high vacuum intensity surrounding each well screen are shown in Fig. 21. Although the vacuum isopleths are relatively symmetrical out to the −250 Pa (−1 in) contour line, the shape of the zone of vacuum influence out to the zero line is quite amorphous. The positive pressure area extending the full depth of the landfill between 3W4 and 3W2 separates another large negative pressure area between 3W5 and 3W4: that is, at the edge of the landfill. By the time represented by the plot in Fig. 22, the shape of the vacuum zone of influence has shifted and expanded to the east, causing an additional breakthrough between 3E2 and 3E3. The large positive pressure area between 3W5 and 3W4 and the well would normally be considered the end of the well vacuum zone of influence, yet separated from this area by a distance of 21–46 m (70–150 ft) was another solid negative pressure area at the side of the fill, which has to be created by the vacuum emanating from Well 3. The presence of these two areas strongly contradicts the previously held belief that the zero line constitutes the edge of vacuum influence.

By 23:33 (Fig. 23), the shape and areal extent of the vacuum had changed significantly. By 10:06 (Fig. 24), the vacuum was almost entirely on the west side of the well, with only the concentrated vacuum around the well evident on the east side.

Figures 25–28 of the north–south profile 2N1 through 3N4 reveal the non-symmetrical nature of the vacuum zone of influence and the fluctuations that occur with time. The possible effect of the relatively thick interim cover at −16.8 to −18.3 m (−55 to −60 ft) is shown in the separation between the two well screens' zones of vacuum influence. There appear to be alternating negative and positive pressure layers the length of the grid. The positive pressure bubble at 2N3-85′ has moved upward in Fig. 26. By the time represented by the plot in Fig. 27, the single positive pressure bubble has elongated into two prominent bubbles, which are sandwiched

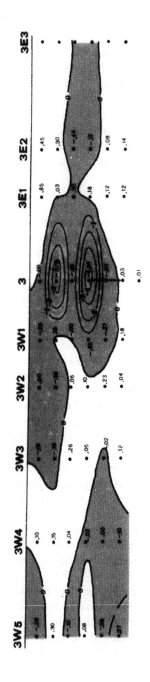

Figure 21. Run 13, 24 July 1980, 20:49–21:03. Shaded area: negative pressure.

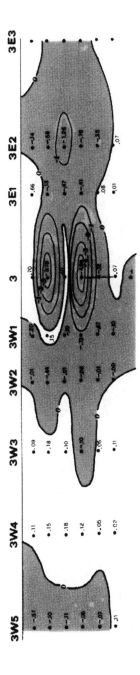

Figure 22. Run 13, 24 July 1980, 22:01–22:31. Shaded area: negative pressure.

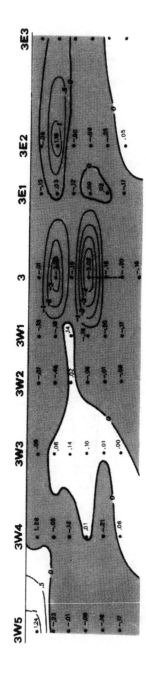

Figure 23. Run 13, 24 July 1980, 23:33–24:02. Shaded area: negative pressure.

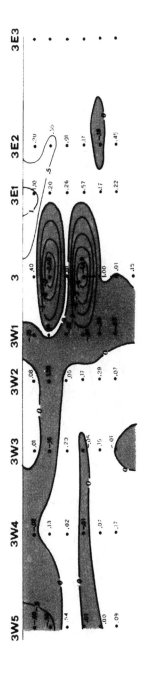

Figure 24. Run 13, 25 July 1980, 10:06–11:05. Shaded area: negative pressure.

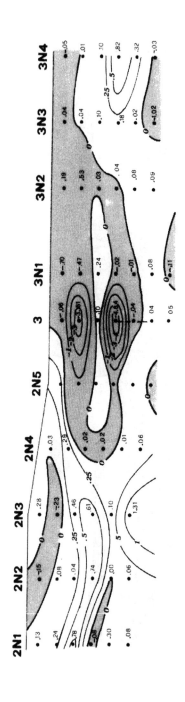

Figure 25. Run 13, 24 July 1980, 20:07–20:22. Shaded area: negative pressure.

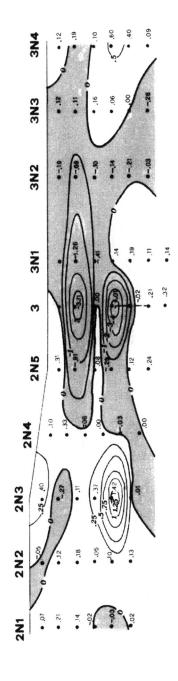

Figure 26. Run 13, 24 July 1980, 21:26–21:59. Shaded area: negative pressure.

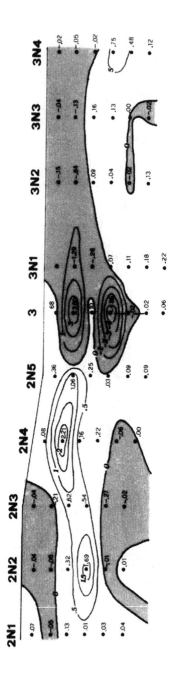

Figure 27. Run 13, 24 July 1980, 23:02–23:25. Shaded area: negative pressure.

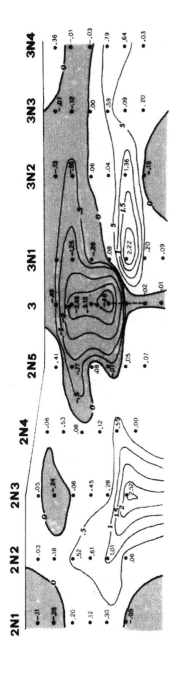

Figure 28. Run 13, 25 July 1980, 11:10–11:54. Shaded area: negative pressure.

between two negative pressure areas leading towards the upper well screen. The apparent vacuum zone of influence immediately around the well is predominantly on the north side of the well.

In Fig. 28, a new positive pressure bubble centred around 2N3–85′ has appeared. The two distinct ellipsoids around the respective well screens have merged into one large negative pressure zone, which has an almost square shape. Another positive pressure zone coincides somewhat with the known wet area that occurred during the winter rains of 1979–80.

Positive pressure zones that can move both vertically and horizontally (dependent on a number of as yet unexplained factors) are apparently created within the landfill. One of the possible causes of positive pressure build-up is that gas flowing towards the well cannot pass through some refuse areas (perhaps less permeable or more saturated refuse) fast enough, and accumulates until such time as the vacuum increases to a point in front of it where 'end runs', piping or breakthrough bleeds off the gas. The positive pressure zones(s) appears to rise to the surface owing to barometric venting if the gas cannot find its way to the well first.

Experimental run 27.   The data for experimental run 27 were taken between 9 September and 12 September 1980, using well screens 1 and 2 of all three wells. The approximate average vacuum was between 1.49 and 1.74 kPa (6–7 in). The flow rates varied between 5.65 and 8.5 m^3/min (200–300 scfm).

This experimental run and the one following it are of extreme interest because of the unusual positive pressure phenomenon observed on the east side of the field. This condition existed for a period of about 2 weeks (experimental runs 27 and 28). Owing to presumed atmospheric conditions and/or a combination of other unknown variables, the pressures on the east side of all three wells evidenced tear-shaped or elliptical positive pressure areas, which merged with the zone of vacuum surrounding a well. In Fig. 29–32, the pressures reached as high as 2.1 kPa (8.44 in), and resembled a thick wedge that ebbs and flows according to the time of the day. The vacuum seems to envelop these positive pressure zones as they approach within 16.8 m (55 ft) of the well. These wedges are fairly well aligned with the height of the upper well screen. They also coincide quite well with the presumed crest of the buried ridge, and seem to sweep and flow over the ridge. The negative areas are all west of the well and remain fairly constant over this period of time.

The constancy of the negative areas suggests a stable atmospheric condition. This condition was most pronounced for grid 1W4–1E4.

Figures 33–36 for grid 3W5 through 3E3 reflect a similar phenomenon on the east side of Well 3 but not to the same areal extent. The pressures, however, are typically from 1.49 to as high as 2.7 kPa (6–11 in) of water column positive pressure. It was somewhat surprising to find thin, elongated bands of negative pressure sandwiched between these high-positive-pressure areas. At both Well 2 and Well 3, the typical zone of vacuum surrounding each well screen does not exist as it usually does; instead the two vacuum envelopes are merged together to produce more of a 'sphere' of influence.

Figures 37–40 for grid 1N2–2N5 reveal that the amorphous zone of vacuum influence, as measured out to the zero contour line at least for this series of vacuum isopleths and grid direction, does not overlap that of the adjoining wells. Although it may not be correct (based on interpretations of prior results) to use the zero contour line as the true limit of the vacuum zone of influence, these plots suggest that the wells should have been constructed on less than about 61–73 m (200–240 ft) centres for optimum overlap. The ratios of major axis to minor axis for the core of the admittedly amorphous zone of vacuum ranges from about 3.2:1 to 3.8:1.

Analysis and Interpretation of Barometric Pressure Induced Changes in Gas Flow Rate, Static Pressures, and Subterranean Gas Pore Pressures at Different Depths

It was perceived shortly after the research began that the experimental results were wholly different from what had been expected and that one or more external variables were most likely responsible for the pronounced fluctuations in (a) the zone of vacuum influence (as measured by the zero pressure contour line), (b) the wellhead vacuum, (c) the flow from the well, and (d) the in-refuse gas pore pressures. The following analyses use all of the applicable available data on wellhead vacuum, velocity head, barometric pressure, and pore gas pressures immediately surrounding the well to chronicle the transitory nature of supposedly 'steady state' operation and hopefully identify the causative factor(s).

Experimental run 11. The data for experimental run 11 were obtained on 14 and 15 July 1980, using only screen 1 of Well 1. The

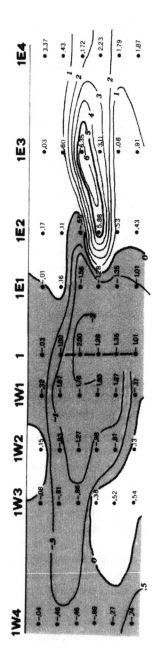

Figure 29. Run 27, 10 September 1980, 9:21–9:50. Shaded area: negative pressure.

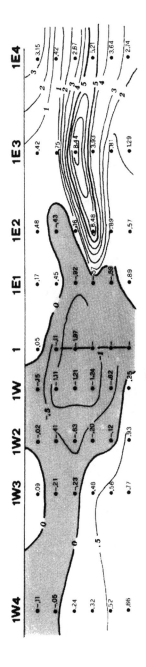

Figure 30. Run 27, 10 September 1980, 14:32–14:58. Shaded area: negative pressure.

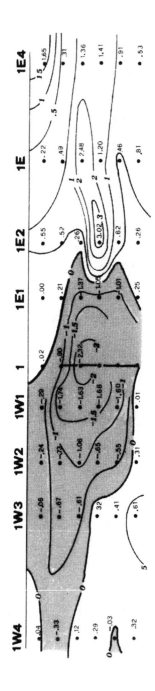

Figure 31. Run 27, 10 September 1980, 19:45–20:05. Shaded area: negative pressure.

369

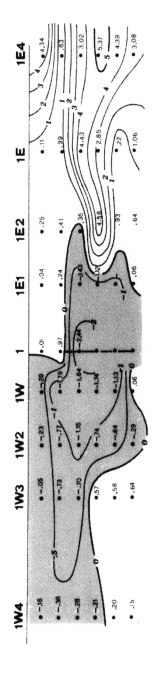

Figure 32. Run 27, 10 September 1980, 22:35–22:53. Shaded area: negative pressure.

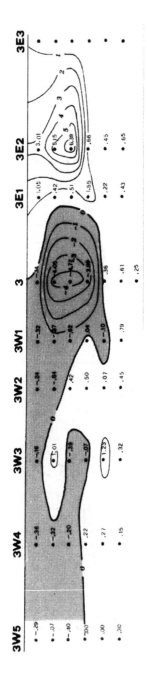

Figure 33. Run 27, 10 September 1980, 10:51–11:26. Shaded area: negative pressure.

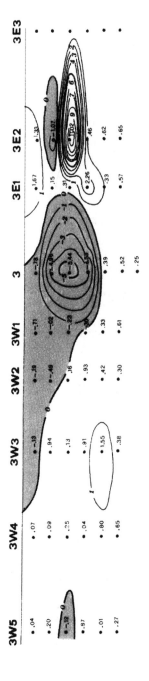

Figure 34. Run 27, 10 September 1980, 15:43–16:10. Shaded area: negative pressure.

372

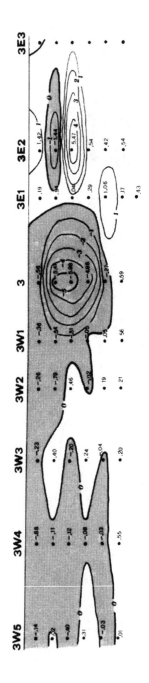

Figure 35. Run 27, 10 September 1980, 18:31–19:05. Shaded area: negative pressure.

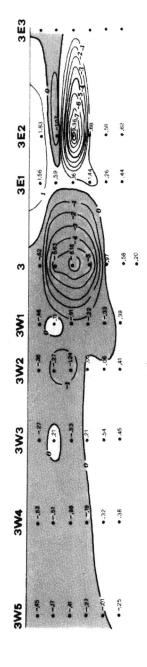

Figure 36. Run 27, 10 September 1980, 21:57–22:16. Shaded area: negative pressure.

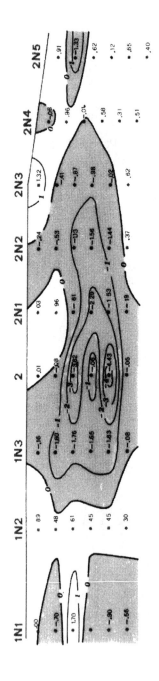

Figure 37. Run 27, 10 September 1980, 9:55–10:31. Shaded area: negative pressure.

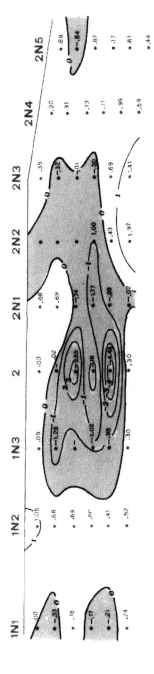

Figure 38. Run 27, 10 September 1980, 15:02–15:28. Shaded area: negative pressure.

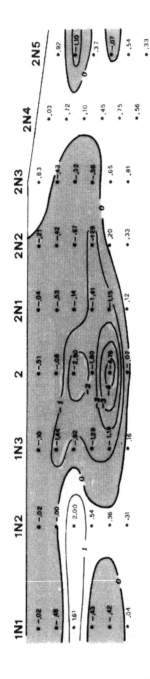

Figure 39. Run 27, 10 September 1980, 17:52–18:16. Shaded area: negative pressure.

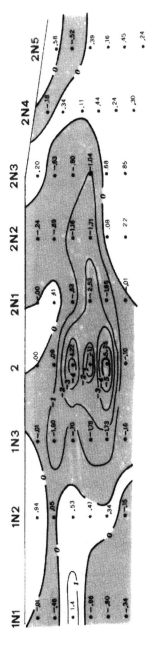

Figure 40. Run 27, 10 September 1980, 20:25–20:45. Shaded area: negative pressure.

wellhead vacuum varied between 0.6 and 1.15 kPa (2.4–4.6 in), and the flow also varied over the course of the experiment. This plot (Fig. 41) provided the first indication that there is not necessarily a direct constant correlation between the magnitude of the wellhead vacuum and the magnitude of the gas flow from a well. This contradicted a then prevalent assumption in the industry. The velocity head, which has a direct correlation with flow rate, was varying between 0.58 and 1.09 cm (0.23–0.43 in) of water, equivalent to 4.11–5.7 m³/min (145–203 cfm) over the latter half of 14 July as a result of a wellhead static vacuum that varied between 1.05 and 1.15 kPa (4.2–4.6 in) of water column. What was startling was that the static wellhead vacuum had diminished to approximately

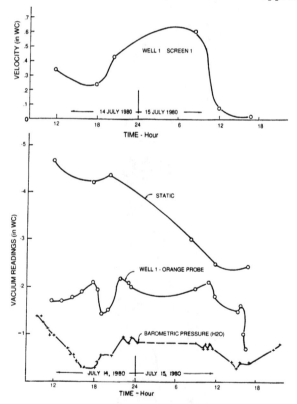

Figure 41. Experimental run 11, 14–15 July 1980. Fluctuations in flow, wellhead vacuum and in refuse gas pore pressure with changing barometric pressure. 1 in WC ≡ 2.54 cm WC.

0.75 kPa (3 in) of water column by the morning of 15 July and was still declining while the velocity head measured in excess of 1.52 cm (0.6 in) of water column, which is equivalent to a flow of 6.88 m³/min (243 cfm). As a result of this observation, more frequent readings of flow rate, static pressure and barometric pressure were taken during subsequent experimental runs. It was only some months later that plots of barometric pressure were superimposed upon this data, which showed a distinct correlation between the diurnal variations in barometric pressure and the resultant variations in static wellhead vacuum and flow rate.

Of equal interest over the approximately 30 h of monitoring, was

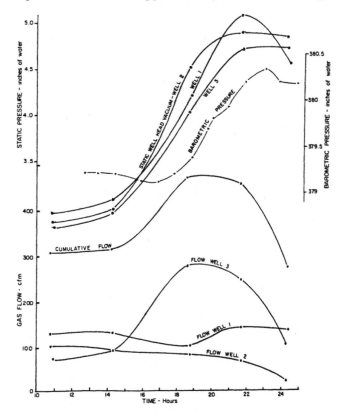

Figure 42. Experimental run 18, 12 August 1980, screen 1 of Wells 1, 2 and 3. Fluctuations in static wellhead vacuum and flow of gas pore pressure with changing barometric pressure. 1 in WC ≡ 2.54 cm WC; 1 cfm = 1.7 m³/h.

that the internal landfill gas pressure, as indicated by the orange probe at a depth of −12.2 m (−40 ft), remained fairly constant even though the wellhead static pressure was declining fairly uniformly over that time period until the very end, when both flow and internal refuse gas pore pressure dropped off.

Experimental run 18. Experimental run 18 was conducted between 11 and 12 August 1980, using screen 1 of Wells 1, 2 and 3. The data presented in Fig. 42 cover the time period between 11:00 a.m. and midnight of 12 August, 1980, and are interesting because of the considerable insights they provide into what occurs inside a landfill.

Perusal of the plots show a direct and almost immediate response correlation between barometric pressure and static wellhead vacuum at all of the wells. It should be noted that: (a) the static wellhead vacuums at the individual wells were almost identical for all intents and purposes, which is remarkable in that this condition was typically quite difficult to maintain over the course of an experimental run with wells in series; and (b) the magnitude of change in the static wellhead vacuum is somewhat greater than that of the magnitude of the barometric pressure change.

Prior to the increase in barometric pressure which occurred between about 15:00 and 23:00 hours, the flows from the three wells were fairly constant. Because of the limited number of data points, it is difficult to tell exactly when the change in flow occurred, but it appears to coincide with the change in barometric pressure. However, the increase in flow rate was not uniform from the individual wells but appeared to occur almost exclusively from Well 3.

The analogy that can be made is that a landfill is like a gas-filled sponge. As the barometric pressure rises, it is analogous to squeezing the sponge and forcing out the gas from the sponge, or in this case, the gas from the landfill. As barometric pressure decreases, it is equivalent to the hand being removed and the sponge rebounding. Perception of this cause-and-effect relationship can be masked by the often observed delayed response or lag in equilibration of soils and refuse pressures after a change in barometric pressure. Even with dry soils and refuse, which is the simpler case, the magnitude and rate of response is perceived to be a function of depth and length of well screen, operating vacuum, magnitude and rate of change of barometric pressure, quality of cover cap, refuse permeability and variation in air and landfill temperatures. The situation is complicated immeasurably by random periods of precipitation and/or spring melt, which affect the surface and subterranean layers as the moisture percolates through the landfill. In the presence of these different layers and degrees of saturated soil and refuse, the synergisms become so complex that the response relationships are almost indecipherable. These complex interactions are currently being researched.

Another analogy can be made between a landfill and a balloon as regards the phenomenon of a flow change occurring at only the one well. When squeezing an inflated balloon, it is not known where the redistributed pressure will find a point of weakness on the surface of the balloon and cause a bulge, the bulge in this case being the point of least resistance and the point at which the maximum gas will be withdrawn.

Experimental run 31, using screen 1 of Well 3. For the four day run, conducted between 23 and 26 September 1980, distinct diurnal barometric pressure fluctuations occurred, which varied in magnitude between 962.7 and 969 cm (379 and 381.5 in) of water column (Fig. 43 and 44). The flow varied between 3.80 and 5.58 m^3/min (99–197 cfm), with an average flow of 4.19 m^3/min (148 cfm) for the average wellhead vacuum condition of −1.29 kPa (−5.29 in).

The static pressure at the wellhead appeared to follow the changes in barometric pressure almost immediately and in almost direct proportion to the measured changes in barometric pressure. On many plots from other experimental runs, the magnitude of the static pressure change was slightly greater than the corresponding change in barometric pressure. However, these plots show that it can be either more or less and does not necessarily coincide identically with the order of magnitude change in barometric pressure.

The maximum interior pore pressures (vacuums) were measured at the orange probe −12.2 m (−40 ft) and at the yellow probe −7.6 m (−25 ft). Well screen 1 is a 7.6 m (25 ft) long screen between −9.1 and −16.8 m (−30 and −55 ft). There was an approximately 2.5 cm (1 in) of water column decrease in vacuum between the static vacuum measured at the top of the well and that measured at the orange probe. There was an additional approximately 1 cm (0.4 in) of water column head loss at the yellow probe over and above that measured at the orange probe.

The data confirm an earlier hypothesis that the vacuum apparently can dissipate quickly at the top portion of the well screen and reach zero vacuum within 7.6 m (25 ft), as shown by the measured

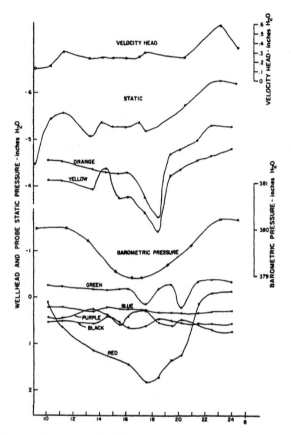

Figure 43. Experimental run 31, 25 September 1980. Barometric pressure induced changes in flow, and static wellhead and multiple depth probe pore pressures. 1 in WC ≡ 2.54 cm WC.

response at the red probe, which is at −16.8 m (−55 ft) and at the bottom of the well screen. The red probe showed a slight positive pressure in almost all cases throughout the course of the experimentation. For the first two days, it was fairly constant and showed minor fluctuations. On the third day of the run (25 September 1980), the red probe experienced a perceptible increase in positive pressure. On 26 September, the increase in positive pressure was due to shut-in of the well.

While the data for the red probe suggest that the vacuum is completely dissipated before it reaches the bottom of the well screen,

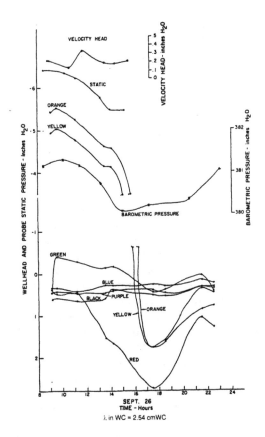

Figure 44. Experimental run 31, 26 September 1980. Barometric pressure induced changes in flow, and static wellhead and multiple depth probe pore pressures. 1 in WC ≡ 2.54 cm WC.

this does not mean that there is not a flow of gas into the lower end of the well screen, as there needs to be a pressure gradient in order for gas to flow. This leads to the inescapable conclusion that the exclusive use of vacuum pressure measurements for the determination of the radius of vacuum influence is erroneous, and other complementary means of determining the radial distance of influence and other rheological characteristics of gas flow to a landfill gas well are required.

The run was terminated at approximately 15:20 hours on 26 September. The pressure built up for the next 2 h, owing to the momentum of the gas still travelling to the well. This increase in positive pressure was approximately 6.2 cm (2.5 in) over and above what was normally observed. Approximately 2 h after shutdown of the blower, the pressure began to return to normal in response either to the shift in barometric pressure or to the dissolution of the positive pressure 'bubble' surrounding the well. The dissipation of the high positive pressure bubble continued for approximately 4 h after it reached its maximum pressure at about 17:30 hours.

Analysis of Gas Sample Results

One of the research objectives was to determine whether gas concentrations change in direct correspondence with the measured variations in internal gas pore-pressure. While an extensive amount of data on internal pore-pressures were taken as part of this research project, only a limited amount of gas concentration data were taken for correlation with the corresponding pressure measurements within the landfill because of the considerable cost of gas chromatographic analyses. The plotting and evaluation of gas concentration isopleths alongside available pressure isopleths was for the purpose of delineating the extent of air intrusion.

Based on a review of the available plots of gas pressure isopleths for various times of the day and various conditions of gas extraction, air intrusion appeared to be quite dependent on the rate of extraction, barometric pressure, time of day, and other such dependent and independent variables.

A simple saline water displacement process was used to obtain the gas samples in the field. Because of the cost, no more than 12 samples were generally taken at a time near the end of an experimental run, when air intrusion was presumed to be at its maximum. Glass burettes used for collecting the samples were pre-evacuated with a vacuum pump in the laboratory and transferred to the field in a large shipping container.

The collection system consisted simply of two glass burettes, connected in series, the uppermost one filled with a saline solution to prevent absorption of carbon dioxide and other soluble gases. The system of flasks was connected with rubber surgical tubing to the buried probe and suspended high in the air. As the saline solution flowed to the lower burette, it created, in combination with the vacuum in the lower burette, a suction on the probe, which drew the sample. After the saline solution had been completely transferred from one flask to another and the sample obtained, the sample burette was shut off at both ends and placed in the shipping box. Back at the laboratory, it was analysed on a gas chromatograph for methane, carbon dioxide, oxygen, and nitrogen.

Because of the considerable vacuum exerted by the mechanical extraction system under certain test conditions, not all of the desired probes were sampled because of the limitations of the water displacement procedure in creating the necessary suction. As a result, a number of samples could not be obtained and a number of discontinuities occurred in the sample grid pattern. Only two of the five monitoring efforts showed any perceptible continuity in probe results surrounding a well sufficient to provide useful information.

Experimental run 37. Plots of oxygen, nitrogen, methane and carbon dioxide concentration levels, as determined by gas chromatographic analysis for the 12 probes in the immediate vicinity of Well 1, are presented in Fig. 45. The samples were taken on 13 November 1980, while gas was being extracted from screens 2 and 3 of Well 1 during experimental run 37. The average wellhead vacuum was about -1.5 kPa (-6.0 in) of water with an average flow of approximately 4.3 m^3/min (152 cfm).

Based on the oxygen and nitrogen values, it would appear that there is significant intrusion of air into the upper well screen gravel pack, as evidenced by the high air concentrations at the -7.6 m (-25 ft level) at Well 1 and probe station 1W1. As well screen 1 was closed off and gas was being extracted only from screens 2 and 3, air presumably was being drawn down in the immediate vicinity of the well somewhere between 1W1 and 1E1. Conversely, it was thought that there could be short-circuiting of the vacuum around the clay seal from the lower well screen gravel pack through the refuse, into the upper well screen gravel pack. However, clear evidence of a chimney effect is not evident because of the low values at the -12.2 m (40 ft) depth. Air intrusion is extremely pronounced immediately below the clay layer which is presumed to be between about the -15.2 and -16.8 m (-50 and -55 ft) elevation. The air appears to be coming in from beyond 1W3 and penetrating ever deeper as it proceeds toward well screens 2 and 3 after passing 1W2 and 1W1.

The methane and carbon dioxide gas analyses basically confirm the interpretation derived from the oxygen and nitrogen gas plots.

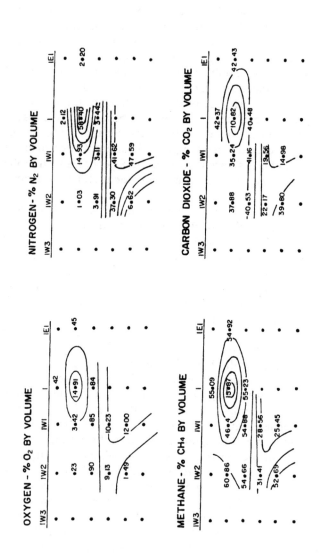

Figure 45. Experimental run 37, 13 November 1980. Gas concentration isopleths for extraction from screens 2 and 3 of Well 1

However, because of the limited database, it is risky to place too much confidence in a single data point or a single set of data, particularly for the −7.6 m probe data at the well. Nevertheless, it is indicative of what may be occurring within the landfill under this particular site-specific set of conditions. It is interesting to note that the air intrusion may not be entering the area immediately surrounding the well borehole, but may be coming in at a considerable distance from the well and moving through more permeable horizons deep within the landfill.

Experimental run 32. The oxygen, nitrogen, methane and carbon dioxide gas concentrations, as determined by gas chromatographic analysis, have been plotted in Figs 46 and 47 for experimental run 32, taken on 2 October 1980. The approximate average flow from Well 3 was 3.34 m³/min (118 cfm), with an average wellhead vacuum of −770 Pa (−3.20 in). Gas was being extracted from screens 1 and 2 of Wells 1, 2 and 3. This is the only gas analysis plot that purports to delineate the effect of sidewall air intrusion. Probe station 3N4 is 60.96 m (200 ft) on a perpendicular from the well and 15.24 m (50 ft) from the top edge of the outside canyon wall. From that point on, the canyon face drops off in a series of terraced benches on an approximately 2:1 slope.

Perusal of the plots for oxygen and nitrogen show the anticipated air intrusion from the surface at Stations 3N1 and 3N2, with the maximum air intrusion occurring near the upper well screen of Well 3. What is surprising is the distinct horizontal intrusion of air at the −25.9 m (−85 ft) level at probe station 3N4. Because there is only one data point, the isopleths of equal percentage concentration show a rather horizontal intrusion at this depth, which would imply that air is coming through the sidewall and penetrating approximately 67 m (220 ft) at that point and a total of about 117 m (385 ft) into the interior of the landfill. This interpretation is predicated on the assumption that this single point is a true and representative sample. Based on similar results obtained with the previous plots, it would appear that air intrusion at such depths is possible, and therefore it does have some credibility.

Based on the limited amount of data, it does not appear that there is necessarily a direct correlation between gas concentration isopleths and pressure isopleths. However, both types of plot indicate that air intrusion is occurring and indicate the general areas and depths to which air intrusion does occur. At present it is not believed

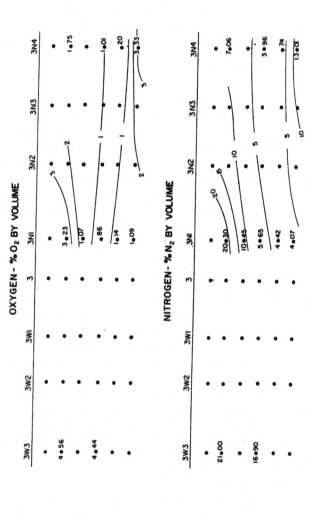

Figure 46. Experimental run 32, 2 October 1980. Oxygen and nitrogen concentration isopleths for extraction from screens 1 and 2 of Well 3.

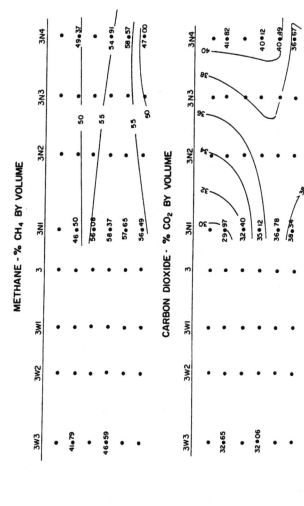

Figure 47. Experimental run 32, 2 October 1980. Methane and carbon dioxide concentration isopleths for extraction from screens 1 and 2 of Well. 3.

that the gas extraction process alone causes a reduction in the methane concentration values in or near the intensive zones of vacuum extraction close to the well. Any reduction in the methane and carbon dioxide concentration isopleths is presumed to be due entirely to the forced intrusion of air, which dilutes the existing ambient gas concentrations.

CONCLUSIONS

Negative as well as positive pressure areas can exist within a landfill under static, non-extraction conditions. These zones of negative and positive pressure change location, size and pressure, and appear to be in a continual state of change in apparent response to one or more external variables.

The 'apparent' zone of vacuum influence around a well is continually changing in perceived response to a number of synergistic variables, such as barometric pressure, availability of gas, ratio of gas drawn to gas generation rate, transport of liquids to the well induced by the applied vacuum, climate, preferential pathways for gas movement, precipitation and percolation into the landfill, and other such variables.

The zone of vacuum influence around a particular well does not appear to be necessarily symmetrical or consistent in its shape from hour to hour or day to day. This has crucial implications for those landfill gas-extraction models that assume symmetrical, isotropic, and steady-state conditions.

The zone of negative pressure surrounding a well out to the zero contour line is not a true indicator of a well's actual 'sphere' of vacuum influence. As was shown on one occasion, the zone of vacuum influence appears to extend far beyond the zero contour line.

The vacuum zone of influence around each deep well screen was fairly reproducible for a given wellhead vacuum out to the -250 Pa (-1 in) of water column contour line. It was less consistent and less reproducible around the upper well screen. A possible explanation is that the lower well screen was immediately below a thick, compacted layer of interim cover, which confined the vacuum and minimized barometric-pressure-induced fluctuations and random breakthrough of the vacuum to the surface.

For the well design and range of operating levels used in this research programme, changes in the available vacuum did not appear to significantly increase or decrease the horizontal radius of vacuum influence as measured by the −250 Pa (−1 in) contour line.

For the experimental data plotted, there was never an overlapping of the −250 Pa (−1 in) contour lines between wells. On a few occasions, there was overlapping of the zero contour lines between wells. This suggests that the 91 m (300 ft) on-centre well spacing was excessive and did not provide adequate zones of vacuum coverage of the landfill mass.

Based on the few available probes and the perceived location of the zero contour line, the vacuum did not appear to go deeper than about 9 m (30 ft) below the bottom of the perforated well casing.

The approximate maximum 'reach' of the vacuum as measured by the −250 Pa (−1 in) contour lines was typically achieved in less than 20 min.

During extraction of the gas, large positive pressure zones can be created at distances of up to 76 m (250 ft) or more from the well.

There were significant differences in the yield characteristics of the three wells even though each was only separated by a distance of 91.4 m (300 ft), perhaps reflective of the unique individual characters of the refuse making up each of the cells surrounding each of the wells.

Although the probes were spaced closer vertically than horizontally, the patterns of the equal pressure contours (isopleths) obtained by the process of logical contouring suggest greater migration of the vacuum in the horizontal direction than in the vertical. The shape of the zone of vacuum influence at least to the −250 Pa (−1 in) of water column contour line resembles an ellipsoid in the vertical plane. The ratio of major horizontal axis to minor vertical axis for a single well screen ranged from about 2 to 1 to 3.8 to 1. It would appear that screen length, in combination with other variables, dictates the shape of the ellipsoid and the ratio of horizontal to vertical axes.

The measurement of a negative pressure in a probe near the surface near an extraction well is not conclusive proof of vacuum breakthrough, as fluctuations in barometric pressure and/or other as yet unidentified variables also induce negative pressure zones in the near-surface horizon, as was observed during periods of non-extraction. Thus the use and interpretation of data obtained from shallow probes should be weighed carefully, through a more statistical approach than was used in this study.

The process of vacuum breakthrough occurs as a series of vacuum evacuations of areas leading to the surface with a corresponding decrease in pressure surrounding these areas. Dependent on the applied vacuum, porosity, and other landfill related physical characteristics, the process can occur in a matter of minutes or take hours to accomplish.

Air intrusion did occur through the exposed sidewall to Well 3 in a distance in excess of 76 m (250 ft) on a normal from the top of slope.

Gas flow rates under supposedly constant mechanical extraction conditions appeared to increase or decrease in response to changes in barometric pressure.

Landfill geometry (distance to buried or exposed sidewalls, and height of buried ridges or divides in relation to well screen) appears to have a pronounced effect on creation of vacuum areas, induced flow patterns, and gas-flow dynamics within a landfill.

After a well is shut-in, the gas molecules continue to move into the high vacuum areas and can increase the pressure in these areas to levels considerably above ambient pressure. These positive pressure areas break up and dissipate, typically in a matter of hours.

ACKNOWLEDGEMENT

This research was funded by Argonne National Laboratory under contract number 31-109-38-5265, Mr Michael L. Wilkey, Project Officer. The author also wishes to thank: the Los Angeles County Sanitation Districts for use of the Puente Hills Landfill for this research and for providing both equipment and personnel during the installation of the test facility; and Lockman & Associates staff personnel Brian Whitaker, John Chin, Paula Durham, Mark Himel, Michael Lofy, Rogelio Paniagua and Greg Scofield, who, day after day, conscientiously and untiringly collected over two million bits of data, which are the basis of this study.

REFERENCES

Lofy, R. J. (1983). The study of zones of vacuum influence surrounding landfill gas extraction wells, Argonne National Laboratory Methane From Landfill Program, Contract No. 31-109-38-5265 (1982), ANL/CNSV-TM-113 (1983).

5.2 Predicted Effectiveness of Active Gas Extraction

RONALD J. LOFY

Lofy Engineering, PO Box 5335, Pasadena, CA 91117, USA

INTRODUCTION

This chapter on introductory concepts and theory is an attempt to explain, with mathematical models where possible, the preceding chapter, which summarized the empirical results of an illustrative array of actual landfill gas vacuum well extraction experiments conducted in 1980, some 13 years earlier. These two chapters tie together the theory and the reality and, in the process, point out the many areas of weakness in our understanding and comprehension of the subject.

The design and evaluation of landfill gas extraction systems is still more of an art than a science. The task of modelling the complex synergistic interactions that occur within a landfill is complicated by the fact that a landfill is by its very creation a non-homogeneous, non-symmetrical, non-isotropic, non-constant entity. In addition to the many intrinsic internal variables, it also appears to be subject to external variables, such as barometric pressure, frost and snow cover (where applicable) for varying periods of the year, infiltrating precipitation, and atmospheric conditions, all of which appear to influence the performance of a vacuum gas collection system.

Because of the acknowledged complexity, an adequate mathematical model of the in-refuse gas extraction operation has not yet been developed. Thus practitioners and students of this emerging

Landfilling of Waste: Biogas. Edited by T. H Christensen, R. Cossu and R. Stegmann. Published in 1996 by E & FN Spon, London. ISBN 0 419 19400 2.

science are constrained to observing field examples and trying to correctly interpret the importance of observed changes in response to different modes of landfill construction and operation.

In order to develop the reader's comprehension of landfill gas vacuum extraction dynamics and the considerable synergisms involved, a conceptual and, where possible, theoretical foundation will be laid by presenting some of the more easily perceived cause-and-effect relationships that occur: (a) during the construction of a landfill and, again, (b) when a gas extraction well is installed and operated under different conditions.

This chapter attempts to develop a fundamental, basic understanding of the complex processes and variables that influence the dynamics of vacuum gas extraction, to the degree that they are presently known; and introduce a mathematical gas reservoir pressure model and some of the insights it provides.

INTRODUCTORY CONCEPTS

Some of the implicit or postulated bases to begin this process are as follows.

- The design, construction, operational management, composition of refuse placed, landfill geometry, depths of individual cells, original and intruded moisture content, initial landfill compaction density, and nature and thickness of interim and final cover cap of a landfill, for example, create a unique entity with a host of 'fixed' conditions, which will predetermine the performance of a gas extraction system for a long period of time into the future.
- A landfill typically comprises a sequentially placed series of refuse layers. Each layer, which may vary from 10 to 30 ft (3.05–9.15 m) in thickness, dependent on the design and operational philosophy at the landfill, may have an initial compacted density of typically between 800 and 1500 lbs per cubic yard (475–890 kg/m^3), dependent on the type of landfilling equipment used.
- Each landfill may have different thicknesses and composition of interim and final cover with their own unique porosity, permeability and moisture retention capacity, depending on

the type and nature of available soil or composite materials used.

- Dependent on the unique time sequencing and operation of the individual landfill, each layer is thus exposed to the elements for different amounts of time, which influences: the time period during which aerobic, facultative aerobic, facultative anaerobic, or strict anaerobic conditions exist in that particular layer; the amount of time for which the layer is exposed to seasonal ambient temperature conditions; and the time during which a certain amount of percolating moisture can enter that particular zone or layer.

- Successive incremental layers of refuse apply a surcharge to lower layers of refuse, which, coupled with the chemical, biological, and physical decomposition of the organic refuse material, eventually causes that and adjacent layers to slowly settle and compact to an increasing density and subsequent lesser permeability.

This concept is shown in Fig. 1, wherein a quarry or pit type landfill is sequentially filled. F_1 represents the first fill area or cell. The first layer comprises fills F_1 to F_3. The second layer comprises fills F_4 to F_6, and the third layer consists of fill areas F_7–F_{10}, etc., until completion. Thus the age, presumed state of decomposition, compaction density etc., of each cell is assumed to be: $F_1 > F_2 > F_3 > F_4 > F_5 > F_6 \ldots F_n$.

Assuming that the initial compaction density etc. of each layer is approximately the same, the age therefore presumably correlates reasonably well with the stage of decomposition. It is recognized that there may be random variations between different cells in that the amount of percolated precipitation and nature of the refuse etc. may vary significantly from one cell to another because of seasonal and other variations in loads coming from different parts of the community. Thus it can be assumed that the permeability of each cell as the refuse decays leads to the following general relationship: $K_1 < K_2 < K_3 < K_4 < K_5 < K_6 \ldots K_n$.

Assuming the indicated time progression and sequential age of refuse cells with the concomitant changes in other variables, it follows that the permeability of layer 1 in Fig. 1 is less than overlying layer 2, which is correspondingly less than layer 3, etc. Looking at this relationship in more detail for extraction well 1, which penetrates fill area F_{15}, F_{11}, F_7, F_4 and F_1, the permeabilities of the

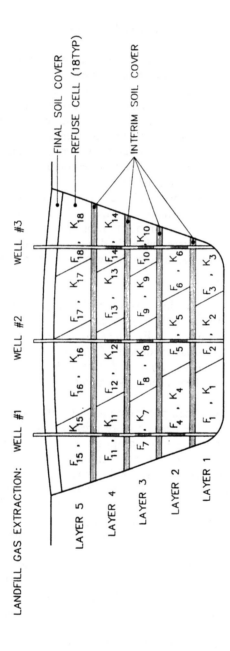

Figure 1. Fill sequence effects on refuse characteristics, cell age, state of decomposition, and refuse permeability.

respective refuse cells increase as one approaches the surface. Similarly, adjacent extraction well 2, which penetrates fill areas F_{16}, F_{12}, F_8, F_5 and F_2, may well have much different moisture contents and permeabilities from different depths than the preceding well because of their different ages and unique cell histories. Thus it is easy to comprehend that each extraction well across the face of a landfill may be uniquely different. These unique levels of synergistically and antagonistically interacting variables lead to the pretty much confirmed observation that the vacuum sphere of influence surrounding a well is non-symmetrical, non-isotropic, and therefore consequently quite difficult to predict.

Figure 2 illustrates the simplest, most idealized depiction of pressure isopleths around a small gas extraction well inlet screen deep within a large volume of totally homogeneous refuse (same porosity and permeability throughout the refuse mass). This figure illustrates the calculated pressure gradient with isocontours or isopleths (contours of equal pressure in two-dimensional space) for an idealized, hypothetical 110 ft (33.5 m) deep well with a 3 ft (0.9 m) long perforated screen section in homogeneous material. The steady-state well vacuum and gas flow is such that the vacuum sphere of influence around the well screen, as identified by the progressively decreasing negative pressure contour lines out to the zero pressure line, is still well within the homogeneous refuse mass, and has never broken the plane of the landfill surface to the atmosphere. Under these idealized conditions, the vacuum sphere of influence is perceived of as an idealized sphere of constant radius.

Figure 3 shows the idealized effect of decreasing the distance to the surface without changing the operating conditions. Instead of having the entire sphere of vacuum influence contained within the homogeneous refuse mass, the contours of equal vacuum now transcend beyond the landfill surface boundary. The distance across the landfill surface between the intercepting outer extremities of the vacuum sphere of influence (point a to point b) represents the surface area through which atmospheric air can move through the landfill surface boundary and through the refuse mass to the inlet well screen. (Lesser and ever-decreasing amounts of air could be expected to penetrate the surface circle defined by points a and b as one moved away from the centre and beyond points a and b.) This process of the applied vacuum intercepting the surface boundary is known as **vacuum breakthrough** or

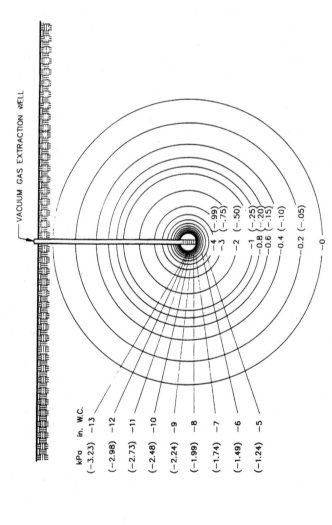

Figure 2. Idealized sphere of vacuum influence totally contained in homogeneous refuse out to zero gauge pressure.

atmospheric air intrusion. In reality, as soon as this sphere of vacuum influence begins to intercept the surface, the idealized condition (illustrated by the perfect spherical vacuum isocontour lines) begins to collapse and dynamically change in response to this violation of the idealized boundary condition. The rate at which this change comes about is a distinct function of the size of the surface area(s) intercepted, the magnitude of the vacuum at the surface boundary, and the concomitant resultant flow rate of intruded air through these area(s). The lumped average flow rate is controlled by the permeabilities of both successive cells and/or soil layers as the intruded air moves progressively downwards to the well screen. A comparison of Figs 2 and 3 clearly demonstrates certain intuitively evident and mathematically verifiable truths about landfill gas and vacuum extraction, in that the closer that a well screen is to the surface, all other operating and landfill conditions being equal, the greater is the vacuum at the surface, the greater is the averaged incremental layer permeabilities to the surface, the greater is the potential induced air intrusion flow rate, and the greater is the potential for rapid and progressively more severe air intrusion.

If a similar deep well casing were to be placed in the same homogeneous refuse material, except with a longer perforated screen section of perhaps 40–60 ft (12.2 or 18.3 m), the idealized plot of isopressure contours might look something like that in Fig. 4, due to the extended perforated screen length.

In the previous chapter, there were some pressure profile plots that resembled Fig. 4, but there were many that appeared to be elongated ellipsoids with the major axis being horizontal. The ratio of the major axis to the minor axis for these ellipsoids ranged from about 2.5:1 to about 3.8:1. One possible explanation for this deviation from theory is the suspected 'channelling' of the vacuum by the confining upper and lower interim soil layers due to the much larger gas transmissivity of the refuse as compared with that for the interim soil layers.

Figures 1–4 introduced some basic concepts under rather idealized conditions. Additional complicating variables that are known to have appreciable influence on the operation of a landfill gas vacuum extraction are: (a) refuse and soil permeability; (b) the influence of fluctuating barometric pressure; and (c) the impact of a modern-day landfill cover cap.

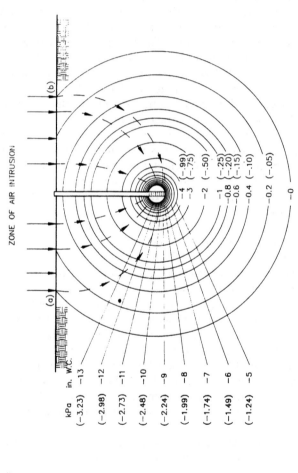

ZONE OF AIR INTRUSION

kPa   in. w.c.
(−3.23)   −13
(−2.98)   −12
(−2.73)   −11
(−2.48)   −10
(−2.24)   −9
(−1.99)   −8
(−1.74)   −7
(−1.49)   −6
(−1.24)   −5

−4 (−.99)
−3 (−.75)
−2 (−.50)
−1 (−.25)
−0.8 (−.20)
−0.6 (−.15)
−0.4 (−.10)
−0.2 (−.05)
−0 (−.05)

(a)   (b)

Figure 3. Same idealized sphere of vacuum influence as Fig. 2 with shorter well casing depth such that breakthrough occurs (shown prior to collapse/equilibration of vacuum).

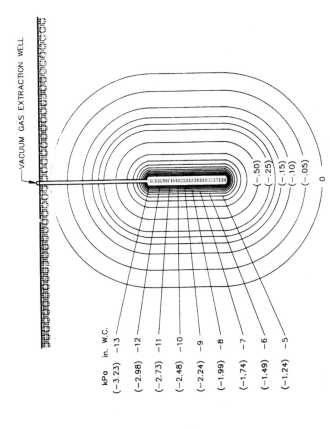

VACUUM GAS EXTRACTION WELL

kPa	in. W.C.
(−3.23)	−13
(−2.98)	−12
(−2.73)	−11
(−2.48)	−10
(−2.24)	−9
(−1.99)	−8
(−1.74)	−7
(−1.49)	−6
(−1.24)	−5
(−.50)	
(−.25)	
(−.15)	
(−.10)	
(−.05)	
0	

Figure 4. Idealized sphere of vacuum influence for long well screen totally contained in homogeneous refuse out to zero gauge pressure.

Permeability

Figure 5 depicts some assumed sequential changes in cell or layer thickness and associated permeabilities over time for a relatively dry landfill. In this illustrative example, each layer was initially constructed at a constant height of 25 ft (7.62 m) with an interim soil cover layer thickness of 1 ft (0.3 m). Four such layers were placed with a final cover cap of 4 ft (1.22 m) of 10^{-5} cm/s (0.01 D) soil placed on top. Thirteen years later, some assumed values for the

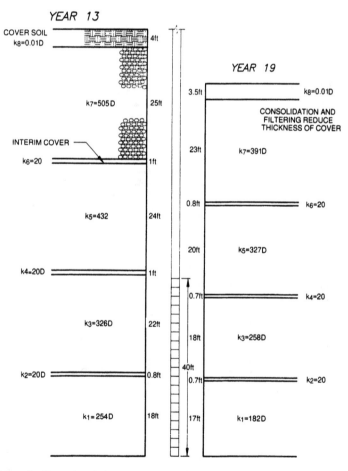

Figure 5. Example of decreasing permeability as refuse consolidates around a vacuum well. D = Darcy; 1D = 10^{-3} cm/s.

landfill refuse are: $k_1 = 254$ D, $k_2 = 20$ D, $k_3 = 326$ D, $k_4 = 20$ D, $k_5 = 432$ D, $k_6 = 20$ D, $k_7 = 505$ D, and the cover cap, $k_8 = 0.01$ D.

Six years later, the individual layers, owing to consolidation, have been reduced to the heights and permeabilities shown. Also, as depicted, the individual layers have also shifted in relationship to the 40 ft (12.2 m) well screen, originally placed to the bottom of the landfill. The average vertical permeability can be shown to be:

$$\sum_{i=1}^{n} \frac{k_i H_i}{H}$$

where H is the total depth in feet and H_i is the depth in feet of an incremental refuse or soil layer.

Substituting the values for year 13 into the above equation, the average vertical permeability for the illustrative example is 363.18 D. The same process leads to an average vertical permeability for year 19 of 278.55 D. Several things are illustrated by this example. A wellhead vacuum that permeates out into the refuse at different radial angles and directions encounters refuse of different permeabilities, i.e. at the bottom of the landfill (for year 13), the first horizontal layer permeability is 254 D, followed by the second horizontal layer permeability of 326 D, both of which are in direct continuity with the well screen, versus the overall average vertical permeability of 363.18 D. Similarly, for year 19 of the hypothetical example, the average vertical permeability of 278.55 D is much larger than the 182 D for the bottom first layer and the 258 D for the second horizontal layer. This would suggest that as the vacuum permeates out and away from the well screen, it will take the path of least resistance, and this has an overall tendency to move in an upward arc as it finds the extraction of gas to be progressively easier as it goes upward to the surface. One would expect, therefore, to see a 'banana-shaped' pressure profile of vacuum isocontours going to the surface as a result of this permeability change. Likewise, at the upper layer, the permeability is much larger, and therefore the ease and the rate at which air from the surface can be brought down to the well screen are progressively enhanced. This possible cause-and-effect relationship can be observed in a perusal of the actual field data presented in the previous chapter.

By a similar derivation process, the following proportional relationship can be derived for comparing different layers of different soils in terms of their relative permeabilities and equivalent thicknesses:

$$G_1 = k_1/L_1$$

$$G_2 = k_2/L_2$$

and by continuity

$$G_1 = G_2$$

Thus

$$k1/L1 = k2/L2$$

and rearranging

$$L_2 = k_2 L_1/k_1.$$

For a 1 ft thick interim soil cover layer with an assumed soil permeability of 10^{-3} cm/s (1 D), compared with an adjacent refuse layer of 60 D, the equivalent thickness of the interim layer as equivalent trash would be

$$L_2 = \frac{k_2 L_1}{k_1} = \frac{60 (1)}{1} = 60 \text{ ft of refuse}$$

Similarly, for a 1 ft thick cover soil cap using a material of 10^{-5} cm/s (0.01 D) adjacent to the same refuse having a permeability of 60 D, we obtain $L_2 = 60/0.01 = 6000$ equivalent ft of refuse. According to this, breakthrough and atmospheric air intrusion should theoretically not occur because the pressure gradient, if it were to be drawn out in the clay, would be a very tight succession of compressed pressure gradient contour lines. However, as is well known, breakthrough does occur, probably due to fissure, desiccation, and settlement cracks in the clay.

In justification of the theoretical analogy, detailed examination of actual three-dimensional and two-dimensional plots of vacuum isocontour lines does show a significant horizontal elongation due to this presumed effect of interim soil layers as well as the effect of the lengthier vertical well screens distorting the idealized spherical model. Some possible explanations are as follows.

- While a landfill operator can achieve some degree of recompaction in placing an interim soil cover layer, the same *in situ* permeability as the original material taken from the borrow pit is never truly achieved because of the equipment and the way in which the soil is laid down, and the often insufficient compaction effort applied at the close of the day.

- Owing to the vibration of the heavy equipment moving over the interim soil cover deck and the eroding process of soil transport caused by percolating precipitation, the interim soil cover material has a tendency to filter down into the refuse, and therefore the presumed interim soil cover thickness and concomitant impermeability are never realized. In many years of drilling and sometimes excavating into sides of refuse, the initial 6–12 in (0.15–0.3 m) interim soil cover layers are often almost indistinguishable or often cannot even be found, therefore lending some credence to the fact that, over time, they disappear dependent on their original thickness.

- The integrity of both the interim soil cover layers and the final cover cap is compromised by fissures, desiccation cracks and/or settlement cracks. These cause a concomitant significant reduction in the lumped parameter permeability of the respective layers. The actual propensity for such dilution and/or compromising of the integrity of the cover layer of soil would appear to be somewhat a function of its original thickness. Thicknesses of 6 in (15 cm) or less could be expected to be much more vulnerable to waterborne soil transport and/or filtering of the soil material, and the origination of either settlement cracks or desiccation cracks, which could essentially go through the entire layer.

Barometric pressure

Atmospheric barometric pressure is known to fluctuate in several ways, one being the rather modest diurnal (daily) pressure variations due to the heating and cooling of the earth's atmosphere, as well as the more pronounced and potentially more extreme fluctuations due to seasonal and/or violent weather fronts passing through a region. It is possible in different parts of the world to see atmospheric barometric pressure fluctuate anywhere from 2 to 5 or more inches of mercury (27–68 in of water column or 6.77–16.93 kPa) in a matter of hours. Several landfill operators in Alaska have indicated that the barometric pressure has fluctuated by as much as 6 in of mercury (20.3 kPa) within 30 min. The importance of barometric pressure was at first not recognized. But, with hindsight, every change in barometric pressure forces the interstitial gas pressures within incremental layers of the landfill, starting from the surface and moving progressively downwards, to come into equilibrium with the atmospheric pressure. As one can imagine, the extent of the equilibration process is first affected by the rate and magnitude of the atmospheric pressure change and then by the intrinsic properties of the landfill cover and underlying refuse, as well as the superimposed landfill gas extraction system design and operation.

There is also a perceived double hammer effect, in that a centrifugal blower vacuum system, which is in direct and immediate contact with the atmosphere, and whose flow characteristics fluctuate in direct response to a change in the flow-pressure system curve for that particular blower impeller configuration, is working from the other side of the equilibration process deep in the landfill, and is perceived to be responsible for the compounding of the observed pressure-flow oscillations.

These atmospheric fluctuations are often several times greater than the typical range of operating landfill gas vacuums. For example, a typical landfill gas extraction system may usually employ a vacuum at the wellhead of between about 2 and 15 in of water column (-0.5 to -3.74 kPa). (Other systems, because of considerable saturation of the refuse or other operational considerations governing the vacuum application rate, may go as high as several inches of mercury.) The significance of this is that the magnitude and rate of some barometric pressure changes in relationship to the presumably steady-state mechanically induced vacuum can be overwhelming, and typically can swamp out the puny mechanically applied vacuum. Thus it can be perceived that even small shifts in barometric pressure can cause or accelerate vacuum breakthrough during such periods of change when a landfill gas system is designed or operated with little tolerance built into the system. This will be illustrated by three examples presented later in Tables 1–3.

As those systems that utilize centrifugal blowers are in continuity with the atmosphere, the pressure difference (ΔP) between the atmospheric pressure at the surface is measured in relationship to an average of the slowly changing refuse gas pressures at progressively and incrementally deeper depths within the landfill, which are slowly trying to change and equilibrate to the change at the surface. This can significantly increase and/or decrease the ΔP across the centrifugal gas extraction system and therefore cause a significant change in gas flow and pressure gradient for the system.

Figure 6, which is a simplified introduction to the sequential process of refuse pore pressure equilibration to a changing barometric pressure, illustrates two different hypothetical responses to a

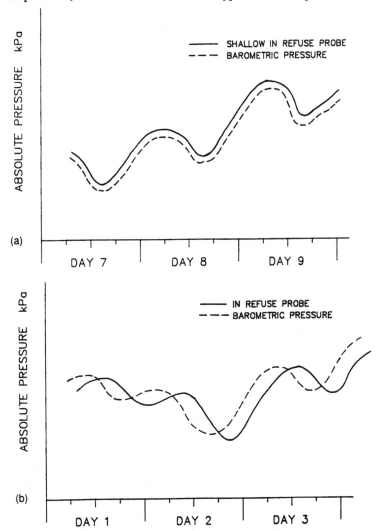

Figure 6. (a) Shallow *in situ* probe response to fluctuating barometric pressure under quite permeable cover soil cap (assumes relatively dry refuse and cover soil); (b) shallow *in situ* probe lagged response to fluctuating barometric pressure under less permeable cover soil cap (assumes relatively dry refuse and cover soil).

change in barometric pressure on subterranean *in situ* refuse probe gas pressures dependent on the nature and condition of the cover cap. Figure 6a illustrates a cover cap comprising extremely permeable sandy soil and a shallow *in situ* probe, which responds almost immediately and in direct proportion to a change in barometric pressure. Figure 6b indicates a condition where the soil is more compact and/or the *in situ* gas-monitoring-probe is at a deeper elevation, with a resultant increase in lag time or delayed response. Under such conditions, the *in situ* probe gas pressure and the barometric pressure curves can actually cross, leading to a situation where sometimes the barometric pressure is confining the release of gas and at other times it is accelerating the escape of gas.

Figure 7 illustrates a more realistic and complicated picture of barometric pressure changes and the resultant effects within the near-surface cover cap and underlying refuse, based on the data obtained by Bogner *et al.* (1987) for a time period when the cover cap soil and underlying refuse were relatively dry (Fig. 7a and 7b), and for a corresponding period of time, when the cover cap and near-surface refuse were partially saturated owing to wet climatic conditions (1–3 October 1985) (Fig. 7c).

As one might expect, the pressure probe closest to the surface tracks the barometric pressure over the course of the cycles much better than the deeper probe, which reflects the longer lag or delayed response to the change in atmospheric pressure at the surface.

While the changes experienced at progressively increasing depth are somewhat rational for the drier soils, the situation is extremely complex and not readily understandable in terms of the lag time response to changing barometric pressure when the cover cap and underlying refuse are saturated or partially saturated. In the latter case, there is a significant build-up in pressure, which occurs at a faster and higher rate than the apparent change in barometric pressure. There is apparent sensitivity to slight downturns in barometric pressure and, apparently even in anticipation of a downturn in the atmospheric barometric pressure, the soil gas pore-pressure within the landfill begins to drop suddenly at a much faster rate than the barometric pressure and, for some unexplained reason, actually goes significantly negative, far below the bottoming-out point for barometric pressure. The cycle then repeats into the next day.

These research data clearly demonstrate the inordinate importance of the soil cover cap and the variable degrees of saturation of this soil over time on the performance, effectiveness, and degree of

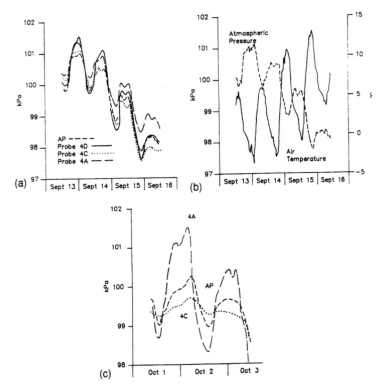

Figure 7. (a) Data from probe nest 4 for dry-weather period 13–16 September 1985. Atmospheric pressure (AP) and soil gas pressures for probe 4A (deep), probe 4C (intermediate) and probe 4D (shallow). (b) Data from probe nest 4 for dry-weather period 13–16 September 1985. Atmospheric pressure (AP) and air temperature. (c) Monitoring data from probe nest 4 for wet-weather period in early October 1985. Atmospheric pressure (AP) and soil gas pressures for probe 4A (in refuse at 1.7 m) and probe 4C (in cover at 1.3 m) (Bogner, 1987).

air intrusion of a typical landfill gas extraction system. This surface boundary flux condition appears to be an extremely erratic and highly variable condition, which is neither reproducible from day to day nor easily predicted.

Cover Cap

The modern-day landfill cover cap is one of the most important variables, besides barometric pressure, influencing the performance of a vacuum gas extraction system. As will be introduced in the following sections on mathematical modelling, the cover cap is the critical and often porous 'boundary condition' that totally dictates and controls the effective radial sphere of influence and the potential for barometric pressure fluctuations or flux through this boundary layer. It is the interfacial area through which all or most of the atmospheric air intrusion occurs with the resultant decrease in gas quality and collapse of vacuum sphere of influence.

While landfilled refuse, dependent on age and degree of compaction, may have a gas permeability estimated to range from a few darcies to several hundred darcies, modern landfill cover caps may have hydraulic permeabilities in the range 10^{-3} to 10^{-6} cm/s (1 to 0.001 D). However, the existence of fissure cracks, desiccation cracks, settlement cracks, or refuse-native soil interface cracks can significantly degrade the relative impermeability of the cover cap and allow unpredictable and extensive air intrusion, with the effect of significantly altering and truncating the spheres of influence for individual wells. This has the additional resultant effect of drawing in excessive amounts of air, which dilute and thereby decrease the percentages of combustible methane gas, cause short-circuiting and subsequent collapse of the sphere of influence, and increase the potential for underground fires due to continuous air (oxygen) intrusion.

THEORETICAL–MATHEMETICAL MODELLING

The dynamics of landfill gas movement and the mathematical models that would be used to describe the process(es) vary, dependent on the situation. For example, the mathematical model to describe the fluctuations in vacuum gas extraction within a landfill is quite different from the equation(s) for offsite lateral landfill gas migration under a pressure and/or diffusional gradient. The modelling of linear, two-dimensional diffusional transport of migrating landfill gas outside a landfill was one of the first processes that was mathematically modelled (Metcalfe *et al.*, 1987) to what some consider to be an acceptable level of realism. The same is not true regarding the process of modelling vacuum gas extraction within the landfill.

Gas Reservoir Pressure Model

A great deal of investigative research and development of mathematical models for pressurized oil and gas formations has occurred within the US petroleum industry. These gas well and well field equations are typically for very deep oil and gas formations, which may consist of a single stratum of fairly uniform permeability or a series of strata of different permeabilities under quite high pressures and temperatures. These oil and gas formations are usually confined by an overlaying and underlying confining zone, which creates ideal boundary conditions such that the pressure on the outer perimeter of the formation (outer sphere of influence of an operating well) is the controlling variable.

The mathematical model used herein to predict the performance of a vacuum gas extraction well under a host of combinatorial landfill conditions was a modification of the basic petroleum industry hemispherical gas pressure model (Muskat, 1937):

$$G = \frac{0.703k \, (P_e^2 - P_W^2)^n}{\mu Tz \, (1/R_W - 1/R_e)}$$

which has historically been used in the petroleum industry to estimate extraction of natural gas from oil- and gas-producing geological formations. It is important to note that the model was developed for confined, subsurface strata, which are generally found at significant depths and typically operate at much higher natural gas pressures and temperatures than found in sanitary landfills. This important discrepancy, which could potentially affect the veracity of the predicted vacuum gas extraction well results, is the reality of some air leakage through the landfill surface, as depicted in Fig. 3, versus the model's inherent assumptions of confined, impermeable boundary conditions for the confining strata assumed to exist both above and below the permeable gas recovery zone. Thus these mathematical models, indiscriminately applied to landfill situations, are subject to error in that they do not address the very real-world landfill surface boundary condition. Modelling is further complicated by the fact that refuse density decreases, and the relative permeability of each cell increases, as one proceeds from the lowest layers to the surface.

Because the landfill surface is a 'porous' boundary, which does not rigorously satisfy the mathematically necessary confined condition used in the development of the petroleum engineering reservoir gas models, the use of the petroleum engineering hemispherical model can only approximate or provide some indication of how the system might perform under varying refuse permeability and cover-cap integrity conditions. Obviously, the closer the cover cap comes to achieving a confined condition, as might be obtained through the use of a gas-impermeable thin-film barrier integrated into the cover cap, the closer the model results will be to reality.

The petroleum engineering reservoir equation transformed from dimensions of reservoir barrels per day into suitable landfill gas industry cubic feet per minute (cfm) dimensions is:

$$G = \frac{0.4882 \, k \, (P_e^2 - P_w^2)^n}{\mu Tz \, (1/R_w - 1/R_e)}$$

where G is the gas flow (cfm) (1 cfm $\equiv$ 1.7 m^3/h), k is the permeability (D) (1 D $\equiv$ 10^{-3} cm/s), n is a coefficient, to be determined for each individual well field (assumed equal to 1 for all modelling done), P_e is the presumed pressure at maximum radial distance from well (psia), typically assumed to be atmospheric pressure (1 atm = 407.5 in water column = 29.921 in Hg = 101.325 kPa) P_w is the pressure at well casing (psia), typically assumed to be wellhead vacuum (1 psia = 27.23 in water column = 6.77 kPa), R_e is the radial distance out to the presumed maximum pressure point Pe (ft) (1 ft $\equiv$ 0.305 m), R_w is the well casing radius (ft), T is the *in situ* refuse temperature °R (°F + 460 °), μ is the viscosity, (cP) (1 cP = 0.001 Pa s), z is the gas compressibility factor.

The validity of the model results are directly related to the assumed degree of gas impermeability of the landfill cover cap. For the purpose of illustrating the considerable insights provided by the model, and the interrelationships that exist between the respective variables, it was assumed that the cover cap was contiguous, extended beyond the zone of calculated influence of the well and was essentially impervious.

The parameter settings used in these initial calculations were:

- a refuse permeability (k) of 100 D;
- an atmospheric pressure of 1 atm (i.e. 29.921 in of mercury, 407.5 in of water column or 101.325 kPa);
- ambient landfill pressure (P_e) typically assumed to occur at a radial distance (R_e) of 100 ft (30.5 m);

- a landfill refuse-gas temperature of 100 °F (37.77 °C) and the concomitant gas viscosity (μ) of 0.018 cP;
- gas compressibility factor (z) equal to 1.0;
- landfill well field coefficient (n) assumed equal to 1.0;
- gas pressure (vacuum) at the well (P_w) equal to the wellhead vacuum;
- landfill surface considered to be gas impermeable to beyond the calculated zone of influence defined by radial distance R_e; and
- refuse considered to be completely homogeneous and of constant uniform permeability in the area defined by radial distance R_e.

This modified hemispherical landfill gas model was used to calculate the following scenario conditions described in the remainder of this section.

Sphere of Influence for Different Barometric Pressures

Table 1 summarizes the results of mathematical calculations for the idealized sphere of vacuum influence out to and beyond the 0 psig (0 kPa) pressure isocontour for different levels of barometric pressure. It is assumed for these calculations that a positive displacement blower was capable of maintaining a constant flow at the well of 128.27 scfm (3.63 m³/min) with a constant well screen vacuum of 13 in of water column (−3.23 kPa). The permeability of the homogeneous refuse volume was 100 D. The table shows that for fluctuating barometric pressures ranging between 380 and 415 in of water column (94.49–103.2 kPa), the radial distance from the small inlet screen to the 0 psig isopressure contour line varies from 23.08 ft (7.04 m) at a barometric pressure of 380 in (94.5 kPa) to 103.35 ft (31.55 m) for 407.52 in (101.33 kPa) barometric pressure. It is further shown that the 0 psig isocontour line for the same condition at 415 in (103.2 kPa) extends far beyond 1500 ft (457.2 m), exemplifying the considerable sensitivity and the seemingly considerable propensity for atmospheric air intrusion under this particular condition.

Another way of interpreting Table 1 is that for a well screen inlet at approximately 53 ft (16.15 m) below the surface and a barometric pressure of 400 in of water column (99.47 kPa) under idealized

TABLE 1. Pressure (Vacuum) Under Idealized Conditions vs. Radial Distance: Different Levels of Barometric Pressure[a]; Refuse Permeability of 100 D [b]

Barometric pressure (in WC)	380	385	390	395	400	407.5	415
Flow (scfm)	128.28	128.28	128.28	128.28	128.28	128.28	128.28
Permeability[c] (D)	100	100	100	100	100	100	100
Well screen vacuum (in WC)	–13	–13	–13	–13	–13	–13	–13
Pressure (vacuum) (in WC) at radial distance							
10 ft	(1.589)	(1.738)	(1.883)	(2.024)	(2.162)	(2.363)	(2.556)
20	(0.187)	(0.353)	(0.516)	(0.674)	(0.828)	(1.053)	(1.269)
23.08	(0.000)						
26.87		(0.000)					
30	0.279	0.107	(0.061)	(0.225)	(0.385)	(0.618)	(0.841)
32.18			(0.000)				
40	0.512	0.337	0.166	(0.000)	(0.163)	(0.400)	(0.627)
50	0.652	0.475	0.302	0.134	(0.030)	(0.269)	(0.499)
52.99					(0.000)		
60	0.745	0.567	0.393	0.224	0.058	(0.182)	(0.414)
70	0.812	0.633	0.458	0.288	0.122	(0.120)	(0.353)
80	0.861	0.682	0.507	0.336	0.169	(0.074)	(0.307)
90	0.900	0.720	0.544	0.373	0.206	(0.037)	(0.271)
100	0.931[d]	0.781	0.575	0.403	0.236	(0.008)	(0.243)
103.5						(0.000)	
300				0.582	0.412	0.165	(0.072)
500				0.618	0.448	0.200	(0.038)
1000				0.645	0.475	0.226	(0.012)
1500					0.484	0.235	(0.0035)
1889							(0.000)

[a] Assumes a constant extraction flow rate and a constant well screen vacuum.

[b] 1 darcy = 10^{-3} cm/s; 1 scfm = 1.7 m³/h; 1 ft = 0.305 m; 1 in WC = 0.248 kPa.

[c] Assumes a perfectly homogenous soil of equal grain size, constant porosity and permeability so as to provide perfectly isotropic and symmetrical conditions.

[d] Note that, for the purpose of this example, there was no purpose served in carrying the calculation beyond one inch of water column positive pressure.

conditions, vacuum breakthrough is imminent. Conversely, the well screen would have to be at 103.35 ft (31.55 m) below the surface at a barometric pressure of 407.52 in (101.33 kPa) to have the same imminent vacuum breakthrough under idealized conditions. This latter idealized condition is shown in Fig. 2. At a barometric pressure of 415 in (103.2 kPa) the well screen inlet would have to be in excess of 1500 ft (457.2 m) below the ground in order to prevent atmospheric breakthrough for these particular idealized conditions.

Several things should be noted about this hypothetical example. For the low barometric pressure of 380 in (4.49 kPa), the sphere of influence of the well extends out to only 23.08 ft (7.04 m). In reality, this relatively small spherical volume of refuse is incapable of providing the 128.27 scfm (3.63 m³/min) being demanded by the positive displacement blower (see later Fig. 18). Any well field/system imbalance would most likely cause a new equilibrium steady state to occur for the new conditions.

If, by design, the top of the well screen was placed at 50 ft (15.24 m) below the surface and it was desired that the 0 psig isocontour line never got closer than 10 ft (3.05 m) to the surface, then the modelling would suggest that the well could be operated safely only at a barometric pressure less than or equal to 395 in of water column (98.22 kPa). Below that level, vacuum breakthrough would theoretically never occur. However, for a barometric pressure between 395 and 400 in of water, vacuum breakthrough would presumably occur, with its resultant decrease in gas Btu quality, because of intruded air as well as an almost instantaneous collapse of the vacuum sphere of influence as soon as the vacuum started to break through the surface, causing a readjustment of the entire vacuum flow relationship. This particular consideration caused the calculation of a slightly different condition, shown in Table 2.

Flow as a Function of Barometric Pressure

Table 2 assumes that, instead of a positive displacement blower, the system is somehow able to float along the blower system operation curve, much as one might with a centrifugal blower while maintaining the idealized design conditions of −13 in water column (−3.23 kPa) at a well screen placed 110 ft (33.53 m) below the surface. Other vacuum isocontour design points would be a vacuum of −4 in water column (−0.099 kPa) at 40 ft (12.2 m) from the

TABLE 2. Extraction Flow Rates In Adjusted Proportion to Changes in Barometric Pressure so as to Maintain Well Field Vacuums and Radial Sphere of Influence at Constant Levels[a]

Barometric pressure (in WC)[b]	380	385	390	395	400	407.5	415
Flow (scfm)	119.47	121.07	122.67	124.27	125.87	128.28	130.67
Permeability[c] (D)	100	100	100	100	100	100	100
Well Screen Vacuum (in WC)	-13	-13	-13	-13	-13	-13	-13
Pressure (vacuum) (in WC) at radial distance							
40 ft	(0.400)	(0.400)	(0.400)	(0.400)	(0.400)	(0.400)	(0.400)
50	(0.269)	(0.269)	(0.269)	(0.269)	(0.269)	(0.269)	(0.269)
60	(0.182)	(0.182)	(0.182)	(0.182)	(0.182)	(0.182)	(0.182)
70	(0.120)	(0.120)	(0.120)	(0.120)	(0.120)	(0.120)	(0.120)
80	(0.074)	(0.074)	(0.074)	(0.074)	(0.074)	(0.074)	(0.074)
90	(0.037)	(0.037)	(0.037)	(0.037)	(0.037)	(0.037)	(0.037)
100	(0.008)	(0.008)	(0.008)	(0.008)	(0.008)	(0.008)	(0.008)
103.4	(0.000)	(0.000)	(0.000)	(0.000)	(0.000)	(0.000)	(0.000)
110	0.015	0.015	0.015	0.015	0.015	0.015	0.015

[a] Assumes a constant extraction flow rate and a constant well screen vacuum.

[b] 1 darcy = 10^{-3} cm/s; 1 scfm = 1.7 m^3/h; 1 ft = 0.305 m; 1 in WC = 0.248 kPa.

[c] Assumes a perfectly homogenous soil of equal grainsize, constant porosity and permeability so as to provide perfectly isotopic and symmetrical conditions.

screen and −0.12 in water column (−0.03 kPa) at a radial distance
of 70 ft (21.34 m) from the screen. In order to maintain this
constant pressure gradient under the varying assumed barometric
pressure conditions, the flow rate would have to be adjusted to the
values shown of 119.5 scfm (3.383 m³/min) at 380 in water column
(94.49 kPa) barometric pressure to 130.67 scfm (3.701 m³/min) at
415 in water column (103.2 kPa) barometric pressure. This hypothetical calculation suggests a very useful way, in theory, of controlling vacuum breakthrough for a known operating condition by
simply adjusting the flow upwards or downwards proportional to
the change in barometric pressure.

Influence of Permeability

Table 3 examines the same scenario conditions as illustrated in
Table 1 only using a different permeability, k, of 300 D. Table 3
utilizes the hemispherical gas model to predict the vacuum readings
at different spherical radii for different levels of barometric pressure
under the previously postulated idealized conditions. It assumes a
constant extraction flow rate of 384.3 scfm (10.898 m³/min), a
constant well screen vacuum of −13 in water column (−3.23 kpa)
and a permeability of 300 D. Somewhat surprisingly, comparison
of Table 3 with Table 1 reveals that the vacuum readings for the
range of barometric pressure fluctuations at different distances from
the well screen are identical to Table 1, with the only difference
being a threefold larger flow rate commensurate with the tripling of
the well field permeability from 100 to 300 D. This proportional
increase in gas flow rate can be expressed by the relationship $G_2 = G_1 k_2/k_1$. The flow rate theoretically tripled, but the radial distances
out to the respective vacuum isocontour lines are the same as for
the examples with a refuse permeability of 100 D.

Thus, according to the model, one could expect that as a landfill
decomposes and consolidates, the well field radial sphere of vacuum
influence can remain unchanged if the wellhead vacuum is maintained constant at the original design/operational levels and the flow
is allowed to decrease proportional to the decline in permeability
according to the above relationship, $G_2 = G_1 k_2/k_1$.

The model results imply that a landfill gas designer could create
a landfill gas extraction system on some kind of fixed diamond or
grid pattern and withdraw gas at a rate proportional to the refuse

TABLE 3. Pressure (Vacuum) Under Idealized Conditions vs. Radial Distance: Different Levels of Barometric Pressure[a]; Refuse Permeability of 300 D[b]

Barometric Pressure (in WC)	380	385	390	395	400	407.5	415
Flow (scfm)	384.83	384.83	384.83	384.83	384.83	384.83	384.83
Permeability[c] (D)	300	300	300	300	300	300	300
Well Screen Vacuum (in WC)	−13	−13	−13	−13	−13	−13	−13
Pressure (vacuum) at radial distance				(in WC)			
10 ft	(1.589)	(1.738)	(1.883)	(2.024)	(2.162)	(2.363)	(2.556)
20	(0.187)	(0.353)	(0.516)	(0.674)	(0.828)	(1.053)	(1.269)
30	0.279	0.107	(0.061)	(0.225)	(0.385)	(0.618)	(0.841)
40	0.512	0.337	0.166	(0.000)	(0.163)	(0.400)	(0.627)
50	0.652	0.475	0.302	0.134	(0.030)	(0.269)	(0.499)
60	0.745	0.567	0.393	0.224	0.058	(0.182)	(0.414)
70	0.812	0.633	0.458	0.288	0.122	(0.120)	(0.353)
80	0.861	0.682	0.507	0.336	0.169	(0.074)	(0.307)
90	0.900	0.720	0.544	0.373	0.206	(0.037)	(0.271)
100	0.931	0.781	0.575	0.403	0.236	(0.008)	(0.243)
300				0.582	0.412	0.165	(0.072)
500				0.618	0.448	0.200	(0.038)
1000				0.645	0.475	0.226	(0.012)
1500					0.484	0.235	(0.0035)

[a] Assumes a constant extraction flow rate and a constant well screen vacuum.

[b] 1 D = 10^{-3} cm/s; 1 scfm = 1.7 m^3/h; 1 ft = 0.305 m; 1 in WC = 0.248 kPa.

[c] Assumes a perfectly homogenous soil of equal grain size, constant porosity and permeability so as to provide perfectly isotropic and symmetrical conditions.

permeability. Over time, the only change to the operation would be to proportionally reduce the quantity of recovered gas commensurate with the decline or reduction in permeability. However, perhaps because this relationship has been masked by all of the other variations, the industry approach over the years has been to maintain the overall volume of gas recovered by progressively increasing the applied mechanical vacuum to the well field to compensate for the reduction in permeability. Intuitively, the ramifications and/or consequences of this would be a shift to a more unstable operating condition, which leads to increased and more rapid atmospheric air intrusion. This explanation seems to fit nicely with some of the emerging long-term (5–10 or more years) operational data that are beginning to emerge for landfill plant extraction operations. Another concomitant ramification of this would be an increased probability of underground fires, long-term decrease in recoverable gas quality, and shortening of operational design life by shifting from more of an anaerobic environment to perhaps more of a facultative anaerobic or facultative aerobic landfill.

The inexplicable or solitary rebuttal to this line of argument may be that the gas generation design life curve, may be totally different from the porosity/permeability curve, and thus the gas may continue to be generated at a fairly constant rate for a number of years in total contrast to what is occurring with landfill gas permeability.

Driving Force for Gas Movement for Gas Extraction Systems and Passive Vent Systems

Figure 8 graphically illustrates the actual physical upper and lower pressure limits that define the driving force for inducement of flow for a vacuum gas extraction well and a passive landfill gas vent well. It shows that the true driving force for inducement of flow to a vacuum gas extraction well is the difference between the applied vacuum at the well screen (P_w) and the ambient *in situ* landfill gas pressure (P_e), whether that be positive or negative. In most cases, the magnitude of an applied wellhead vacuum is much larger than that which can be achieved with just the passive landfill gas-venting option. The passive venting range of operation is merely the difference in positive *in situ* ambient refuse soil pressure (P_e) and the atmospheric (0 gauge) pressure, which is the pressure at the point of discharge.

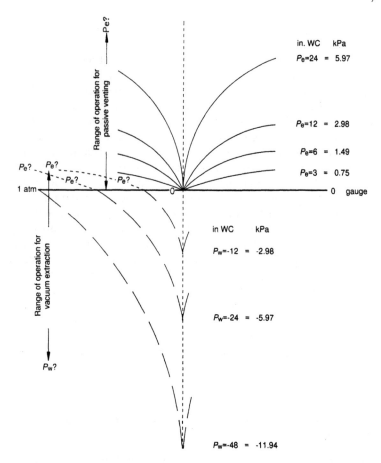

Figure 8. Schematic of driving force for inducement of flow for vacuum gas extraction well as function of internal land fill gas pressure (P_w-P_e) vs range for passive vent well (P_e-atm).

The plot reveals two things: (a) the true sphere of influence is not merely out to the widely accepted zero in water column (0 kPa) isocontour line, but to an isocontour line defined by the value P_e, which is typically slightly positive (more correctly, P_e should perhaps be considered as an averaged value over depth); and (b) as was illustrated in Table 1, the small additional extra pressure that + P_e (P_e – 0 gauge) represents, effects a significant incremental increase in the sphere of influence, with all of its subsequent ramifications on gas yield or recovery values per unit mass of refuse.

Size of Well Casing and Borehole on Effective Sphere of Influence and Flow Rates

The vacuum profiles for various applied wellhead vacuums, ranging from 2 to 100 in water column (0.5–24.9 kPa) at different radial distances from the well were calculated using the gas pressure model for different borehole diameters of 12, 24, 36 and 48 in (30.5, 61, 91.4 and 121.9 cm) respectively (Figures 9–12). The landfill conditions modelled were an assumed atmospheric pressure (P_e) of 1 atm, a refuse permeability (k) of 100 D, an assumed maximum ambient radial distance (R_e) of 100 ft (30.48 m), and a landfill gas temperature of 100 °F (37.7 °C).

One of the most obvious realizations from a study of the four plots is that most of the potential work and/or value of large applied wellhead vacuums is essentially wasted with small-diameter boreholes. From a practical standpoint, the larger the well borehole, the more effective the utilization of the applied wellhead vacuum, particularly for the more extreme applied vacuums. It also suggests that shallow wells of small diameter would be subjected to much more rapid and premature daylighting of high vacuums, and sealing

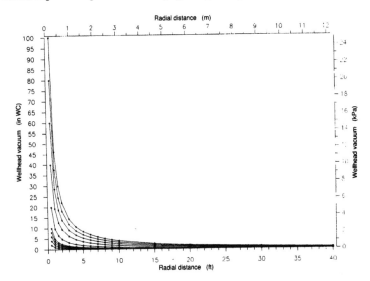

Figure 9. Theoretical vacuum profiles vs radial distance for 12 in (30.5 cm) diameter vacuum gas extraction well. $k = 100$ D; $P_w = 1$ atm; $R_e = 100$ ft (30.48 m); $T = 100$ °F (37.7 °C).

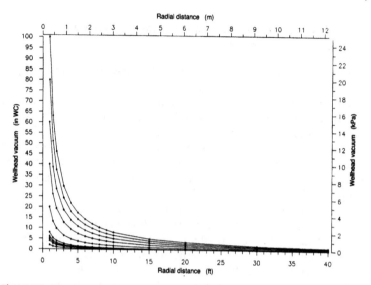

Figure 10. Theoretical vacuum profiles vs radial distance for 24 in (61 cm) diameter vacuum gas extraction well. $k = 100$ D; $P_w = 1$ atm; $R_e = 100$ ft (30.48 m); $T = 100$ °F (37.77 °C).

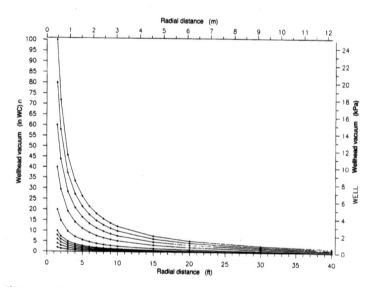

Figure 11. Theoretical vacuum profiles vs radial distance for 36 in (91.4 cm) diameter vacuum gas extraction well. $k = 100$ D; $P_w = 1$ atm; $R_e = 100$ ft (30.48 m); $T = 100$ °F (37.77 °C).

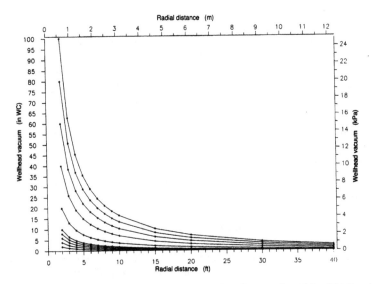

Figure 12. Theoretical vacuum profiles vs radial distance for 48 in (121.9 cm) diameter vacuum gas extraction well. $k = 100$ D; $P_w = 1$ atm; $R_e = 100$ ft (30.48 m); $T = 100$ °F (37.77 °C).

of the backfilled borehole would be much more critical because of the chimney-like effect of vacuum breakthrough up the borehole. Taking the extreme case of a 100 in water column (24.9 kPa) vacuum applied to a 12 in (30.5 cm) diameter borehole, 89.3% of the available vacuum is dissipated in the first 4 ft (1.22 m) radial distance from the centreline of the well. For a 24 in (61 cm) diameter well, it is 78.2% of the available vacuum; for a 36 in (91.4 cm) diameter well, it is 66.6%; and for a 48 in (122 cm) diameter well, it is 54.5% of the available 100 in water column vacuum. The plots suggest that while the application of high vacuums in order to extract gas may be necessary under certain circumstances, the arbitrary and indiscriminate use of high vacuums is actually counterproductive.

The corresponding gas extraction flow rates for the vacuum and other conditions presented in Figs 9–12 are plotted in Fig. 13. This figure shows the theoretical quantity of gas that could be recovered at different wellhead vacuums ranging from 2 in water column (0.5 kPa) up to 100 in water column (24.9 kPa) for seven different borehole diameters of 12 in (30.5 cm), 18 in (45.72 cm), 24 in (61 cm), 30 in (76.2 cm), 36 in (91.4 cm), 42 in (106.7 cm), and

48 in (121.9 cm) at an assumed refuse permeability (k) of 100 D. Conventional practice today in the USA employs well boreholes that range typically between 24 and 36 in in diameter and occasionally 48 in in diameter. However, earlier in various parts of the USA, equipment availability often constrained borehole diameters to 12 or 18 in in diameter. Doubling or tripling the permeability (k) will double or triple the gas flow, respectively. Likewise, the results of the modelling reveal that flow is also directly proportional to borehole diameter: i.e. doubling or quadrupling the borehole diameter doubles or quadruples the theoretical expected flow rate. Normal experience would say that one would indeed be fortunate to get maybe 200–300 cfm (5.7–8.5 m³/min) for a reasonable applied wellhead vacuum of up to perhaps 20 or 30 in of water column (5–7.5 kPa). In those applications where much higher wellhead vacuums are used, it is usually in quite saturated refuse where the permeability is assumed to be less. Figures 9–12 clearly prove the considerable long-term operational advantages of larger-diameter boreholes for increased theoretical gas recovery effectiveness and the concurrent increase in effective sphere of influence. Also, as was pointed out in the discussion of Table 3, the sphere of influence and pressure profiles, as depicted in Figs 9–12, are constant irrespective of permeability, and thus the plots can be used for a wide range of landfill applications as long as the other variables remain fixed.

Figure 13 provokes an interesting question about the relationship between the predicted gas extraction rate and the corresponding hemispherical volume of refuse required to generate that amount of gas on a continuous steady-state basis assuming no vacuum breakthrough or air intrusion. Figure 14 via Fig. 13 allows translation of a specified gas extraction rate to an equivalent theoretical hemispherical, radial sphere of influence when an average gas generation rate expressed in cubic feet per minute per million tons in place (cfm/MMt) is either known or assumed for the given landfill along with an assumed or known refuse density expressed in pounds per cubic yard (lb/yd³). The eight curves shown cover a range from 200 000 to 960 000 (cfm/MMt × lb/yd³) (3704 to 17 779 m³/min MMt$_m$ × kg/m³). The value 200 000 can represent any combination of numerical values for gas generation rate in cfm per MMt times in-place refuse density in lb/yd³.

For example, it could be 200 cfm/MMt (6.25 m³/min per million metric tons (MMt$_m$) × 1000 lb/yd³ (593.3 kg/m³) or 500 cfm/MMt (15.61 m³/min per MMt) × 400 lbs/yd³ (237.3 kg/m³). For the

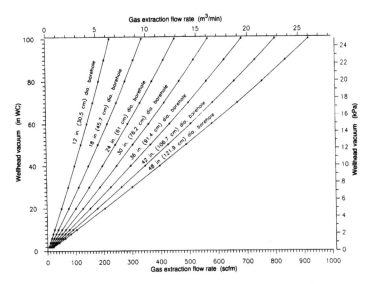

Figure 13. Theoretical gas extraction rates as function of wellhead vacuum and borehole diameter. $K = 100$ D; $P_w = 1$ atm; $R_e = 100$ ft (30.48 m); $T = 100$ °F (37.77 °C).

illustrative example cited for an extraction rate of 150 cfm (4.25 m³/min), the equivalent hemispherical radial sphere of influence is 160 ft (48.76 m), if the assumed gas generation rate is 600 cfm/MMt (18.73 m³/min per MMt_m) and the refuse density is 1600 lb/yd³ (949.2 kg/m³). The equivalent hemispherical radial distance for 150 cfm is 265 ft (80.8 m), if the assumed gas generation rate is 200 cfm/MMt and the assumed in-place refuse density is 1000 lb/yd³.

Figure 14 reveals that the effective sphere of influence, assuming no air intrusion, varies widely dependent on the values assumed for gas generation rate and in-place refuse density. Because the curves were calculated for a hemispherical volume of refuse, so as to be compatible with the gas pressure model, there may be occasions, dependent on the geometry of the landfill, where the landfill is not 265 ft (80.8 m) deep (for the latter example provided), and the equivalent radial sphere of influence would have to be recalculated for a truncated hemispherical or perhaps a cylindrical volume, if the depth to radial sphere of influence is too small. Figure 14 also provides a useful cross-check on the correctness of the initial value assumed for the value R_e used in any modelling.

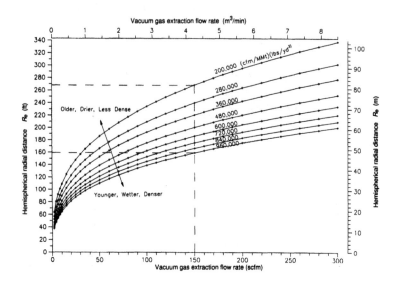

Figure 14. Vacuum extraction well gas flow rate vs radial sphere of influence for assumed gas generation rate and refuse density. $k = 100$ D; $P_w = 1$ atm; $R_e = 100$ ft (30.48 m); $T = 100$ °F (37.77 °C).

Well Casing Head Losses

Figure 15 is a plot of head loss for a range of gas flows up to 900 cfm (25.49 m³/min) through a 2 in (5 cm), 3 in (7.6 cm), 4 in (10 cm), 6 in (15.2 cm) and an 8 in (20.3 cm) well casing of 100 ft (30.5 m) length. It confirms the prevalent use in the industry of either 4 in or, preferably, a 6 in diameter well casing for most flow rates customarily expected in a good producing well field.

Well System Curves

Figure 15, used in conjunction with Fig. 13, allows one to calculate the actual system head loss or pressure drop for any combination

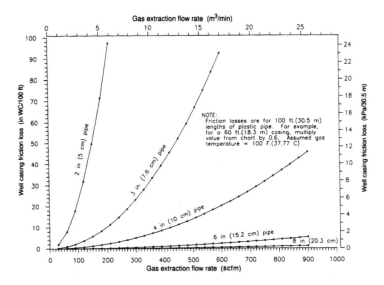

Figure 15. Well casing friction losses for different sizes of plastic pipe. Friction losses are for 100 ft (30.5 m) lengths of plastic pipe. For example, for a 60 ft (18.3 m) casing, multiply value from chart by 0.6. Assumed gas temperature = 100 °F (37.77 °C).

of well casing and borehole diameter and thus determine the most cost-effective combination. More importantly, it allows a well system operating curve to be created for any condition of well casing and well borehole diameter (for refuse with a reference base permeability of 100 D) by adding the wellhead vacuum from Fig. 13 and the well casing friction loss for the correct length casing from Fig. 15 to create a combined system pressure drop versus flow curve, as shown in Fig. 16. This is shown for the illustrative example of a 2 in (5 cm) diameter well casing in a 12 in (30.5 cm) diameter borehole, a 4 in (10 cm) diameter well casing in a 24 in (61 cm) diameter borehole, and for a 6 in (15.2 cm) diameter well casing in a 36 in (91.4 cm) diameter borehole. It is important to note that the borehole flows vs vacuum relationship was calculated using a refuse permeability k of 100D. A different k of say 200 would reduce the vacuum by one-half for the borehole curve. The optimization trend is definitely towards a minimum 6 in diameter

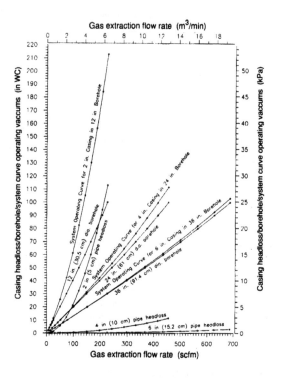

Figure 16. Gas extraction system curves for the following: 2 in well casing in 12 in borehole diameter; 4 in well casing in 24 in borehole diameter; 6 in well casing in 36 in borehole diameter. (Borehole curves based on $K = 100D$.)

well casing in the largest borehole that one can cost-effectively drill.

REFERENCES

Bogner, J., Vogt, M., Moore, C. and Gartman, D. (1987) Gas pressure and concentration gradients at the top of a landfill, in *Proceedings GRCDA Tenth International Landfill Gas Symposium.*

Metcalfe, D. E. and Farquhar, G. J. (1987). Modeling gas migration through unsaturated soils from waste disposal sites. *Water, Air and Soil Pollution*, **32,** 249–259.

Muskat, M. (1937). *The Flow of Homogenous Fluids through Porous Media*, McGraw-Hill Book Company, New York.

5.3 Predicted Effectiveness of Passive Gas Vent Wells

RONALD J. LOFY

Lofy Engineering, PO Box 5335, Pasadena, CA 91117, USA

INTRODUCTION

The installation of passive landfill gas vent wells is increasingly being promoted or even mandated by regulatory agencies in the belief that they are a cost-effective and practical technical solution for stopping landfill gas migration and/or relief of landfill gas pressure build-ups under completed cover caps. There are those who even argue that passive gas vent wells achieve almost the same results as mechanically induced vacuum gas extraction wells without the costly, extensive landfill gas-collection pipe system and the continuous power-consuming vacuum pumps. As a result, they are being trumpeted as one of the least expensive ways of meeting present-day regulatory requirements.

 However, as more of these systems are installed and operated, there is the growing perception that they are not achieving the desired levels of offsite facility protection originally anticipated. When their advocates are pressed to explain the often glaring discrepancies, and to provide reliable quantitative data to support their assertions, there is none. The problem appears to be that few, if any, have used the proper low flow/low pressure gas meters necessary to accurately measure gas flow from the vent well stacks; few, if any, have properly instrumented the landfill in the immediate vicinity of the well to determine the radial sphere of influence, to profile the pressure drop as one proceeds towards the well, to obtain

Landfilling of Waste: Biogas. Edited by T. H Christensen, R. Cossu and R. Stegmann. Published in 1996 by E & FN Spon, London. ISBN 0 419 19400 2.

statistically representative ambient *in situ* landfill pressures or to determine the relative permeability of the refuse surrounding the vent well so as to prove how effective these passive vent systems truly are. All too often, landfill operators, either lacking an understanding of the physics of passive vent flow or constrained to using the only measuring equipment and resources available to them, simply perform easy routine measurements and observations such as measuring the methane gas content, the pressure and the gas flow rate at the top of the vent pipe. Predictably, the results are usually inconclusive as well as unimpressive.

Because of this perceived inability to ascertain the performance of passive vent wells via conventional field measurements, this chapter takes a theoretical approach to predicting their performance based on a simple pressure (convective) flow model for an individual passive gas vent well; it does not take into consideration diffusive (concentration based) gas flow or the interaction of multiple wells. It attempts to address the relatively important influences of refuse permeability, as well as the additive effects of moisture penetration and/or snow cover of the landfill surface.

TYPICAL DESIGN

A passive landfill gas vent well typically consists of a 2–8 in (5.1–20.3 cm) plastic or metal pipe inserted through the cover cap of a landfill, perhaps to a depth of a few feet to as much as 15 ft (4.57 m) or more below the cover cap. Only a relatively small percentage of the wells goes to one-half or almost the full depth of the landfill. Usually, the entire length of well casing below the cover cap is perforated. The vent well casing may also be immersed in a larger borehole, filled with coarse gravel, of perhaps 12–30 in (30.5–76.2 cm) in diameter.

PHYSICS OF FLOW

The common conception is that the pressure of the landfill gas beneath the confining cover cap forces the gas to the more porous gravel-filled vent well, which is close to equilibrium with the atmosphere. The gas, under the diffusive (concentration) and convective (pressure) gradient, moves to this low-pressure, low-concentration area created by the vent well and from there dissipates harmlessly into the atmosphere. This process is therefore seemingly influenced by the following factors:

- magnitude of ambient landfill gas pressure, which is the primary driving force causing flow of gas;
- magnitude and rate of fluctuations in barometric pressure, which amplify or attenuate the internal landfill gas pressure gradient to the surface or to the well;
- permeability of the landfill refuse, which determines the rate at which gas can vent;
- variations in degree of water saturation of cover cap material and refuse as moisture percolates down through the landfill, which alters the permeability;
- degree of hermetic integrity of cover cap, which influences the distance and direction of gas movement; and
- existence of any frost and/or snow cover, which alters the permeability and/or hermetic integrity of the cover cap.

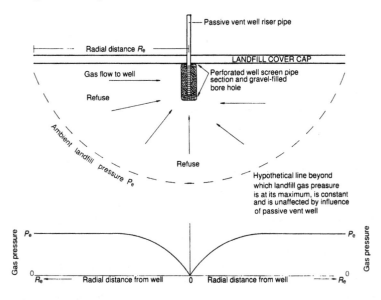

Figure 1. Illustrative passive gas vent well situation and corresponding pressure vs distance profile.

Figure 1 depicts an illustrative passive landfill gas vent well situation with a positive ambient landfill gas pressure (P_e) at some radial distance (R_e) from the well with a concomitant pressure decrease (headloss) in the immediate vicinity of the vent well to a presumed pressure of zero gauge. For the purposes of defining the situation with a mathematical model, it is assumed that beyond a certain radial distance (R_e) from the vent well, the landfill gas pressure (P_e) is constant in all directions, is at its maximum, and is no longer measurably influenced by the vent well. It is further assumed that the pressure at the discharge vent well is equivalent to zero gauge pressure. With these assumptions, the magnitude of the driving force for convective (pressure) gas flow towards the vent well is simply the positive ambient landfill pressure value (P_e). However, because of discontinuities and imperfections in the cover cap, near-surface variations in pressure due to possible barometric pressure fluctuations, random pockets of less permeable or even saturated refuse, and the influence of adjacent vent wells or other landfill gas relief operations, the sphere of influence of the well may be distorted or reduced.

PROBLEMS WITH TESTING AND EVALUATING A PASSIVE WELL

Study of Fig. 1 reveals much about the way the typical passive vent well performs and how it should be properly monitored. Because of the typically low ambient pressures and the concomitant relatively low flow rates out of a vent well, it is often difficult, if not impossible, to get a detectable measurement of velocity head using a primary measurement device such as a Pitot tube. The situation is further confounded by the fact that the static pressure is typically near or at zero gauge. The ability to get an accurate measurement is further exacerbated by the fact that even if the Pitot tube is placed into a sample port at mid-length of the vertical vent stack or at least ten diameters upstream of the vent outlet, wind velocity across the top of the pipe creates a venturi effect, which can seriously distort an already difficult measurement. Similarly, the taking of a pressure reading in this pipe is simply erroneous, as it is known that the pressure is at or near zero pressure gauge near the top of the vent well.

Another source of measurement error occurs when inserting the gas detector probe or sample tube into the top of the pipe to obtain a reading or sample. Because of wind and/or the normal atmospheric air dilution that occurs near the top of the pipe, a somewhat diminished gas concentration value is typically obtained rather than that which might be obtained at a point further or deeper into the landfill. Obviously, the top of a passive vent well is not the proper location to evaluate its performance. Other approaches are required, but the list of easy, practical and reliable methodologies seems sparse indeed.

In situ pressure probes to ascertain the magnitude of the driving force seem like a logical requirement, but how many, how deep, and at what distance from the well? Typically, only the vent well exists. In the absence of any intermediate ambient pressure measurement probes, the vent well itself could serve as an ambient pressure probe if it were capped for a period of time, presumably long enough to reach steady-state ambient conditions and a pressure measurement taken under controlled conditions. The disadvantages of this approach are as follows.

- The vent well is taken out of service for some period of time.
- Dependent on the gas-generation rate of the landfill and the effectiveness of the confining cover cap, it is difficult to ascertain when steady-state conditions are reached because of near-surface pressure variations due to constant equilibration with barometric pressure.
- Because of the considerable sensitivity of slight pressure variations in the near-surface horizon of the landfill, it could have ripple-like ramifications on adjoining vent wells or other venting opportunities.

Because of the extremely limited perceived alternatives, even if it were desired to calculate theoretical flow rates using a calibrated mathematical model, true and representative measurements of pressure as well as permeability would be essential to determine the true venting flow rate correctly. However, there are numerous potential measurement errors that could significantly alter or misrepresent the true field status:

- Because of discontinuities in the landfill surface, there could be localized pressure relief venting through fissures or cracks in the landfill surface.

- The measurement probe could be in a relatively depressed area of the landfill surface, which, because of precipitation, could be saturated and therefore give a somewhat more elevated pressure reading than occurs in other areas within the sphere of influence of the well.
- Refuse in the pathway to the well could be saturated and create a temporal restriction or blockage of gas to the well, which reduces the actual flow rate to the vent well. (Moisture movement through soil and refuse appears to have a disproportionate influence on gas movement and pressure distribution in the near surface horizons of a landfill.)

Therefore, a number of permeability measurements and pressure measurement points would lead to greater statistical confidence in the results.

THEORETICAL-MATHEMATICAL MODEL

The mathematical model used to predict the performance of a passive vent well under a host of combinatorial landfill conditions was taken from the petroleum industry. This basic hemispherical gas pressure model (Theis, 1935; Muskat, 1937; Katz *et al.*, 1959; Craft and Hawkins, 1959):

$$G = \frac{0.703 \, k \, (P_e^2 - P_w^2)^n}{\mu T z \, (1/R_W - 1/R_e)}$$

has historically been used in the petroleum industry to estimate extraction of natural gas from oil- and gas-producing geological formations. It is important to note that the model was developed for confined, subsurface strata, which are generally found at significant depths and typically operate at much higher natural gas pressures and temperatures than found in sanitary landfills. The important discrepancy, which could potentially affect the veracity of the predicted passive gas vent well results, is the reality of some air leakage through the landfill surface versus the model's inherent assumptions of confined, impermeable boundary conditions for the confining strata assumed to exist both above and below the permeable gas recovery zone.

Because the landfill surface is a 'porous' boundary, which does not rigorously satisfy the mathematically necessary confined condition used in the development of the petroleum engineering reservoir gas models, the use of the petroleum engineering hemispherical model can only approximate or provide some indication of how the system might perform under varying refuse permeability and cover cap integrity conditions. Obviously, the closer the cover cap comes to achieving a confined condition, as might be obtained through the use of a gas-impermeable thin-film barrier integrated into the cover cap, the closer the model results will be to reality.

The application of this model to the prediction of passive gas vent well flow rates was done ignoring the coinciding diffusive transport of gases in response to a concentration gradient, as it is customarily assumed that flow resulting from a pressure gradient is usually far greater in magnitude than diffusive flow. It was rationalized that the predicted results, using a convective (pressure) model, would establish the upper boundary for predicted flow, as diffusive flow is an order of magnitude slower and less important phenomenon influencing transport of landfill gases.

The petroleum engineering reservoir equation transformed from dimensions of reservoir barrels per day into suitable landfill gas industry cubic feet per minute (cfm) dimensions is

$$G = \frac{0.4882 \, k \, (P_e^2 - P_w^2)^n}{\mu Tz \, (1/R_w - 1/R_e)}$$

where G is the gas flow (cfm), k is the permeability (darcies) (1 darcy $= 10^{-3}$ cm/s $= 10^{-5}$ m/s), n is a coefficient to be determined for each individual well field (assumed equal to 1 for all modelling done), P_e is the presumed maximum pressure at maximum radial distance from the well (psia) (1 psia = 27.23 in water column = 6.77 kPa), P_w is the pressure at the well casing (psia) (1 atm = 407.5 in water column = 29.921 in Hg = 101.325 kPa), R_e is the radial distance out to the presumed maximum pressure point P_e (ft), R_w is the well casing radius (ft), T is the *in situ* refuse temperature in R° (F° + 460 °), μ is the viscosity (cP) and z is the gas compressibility factor.

Modelling Results

The situation modelled is as depicted in the cross-section profile shown earlier in Fig. 1. Essentially, a shallow vent well is placed at the centre of a presumed homogeneous section of landfilled refuse of uniform permeability. The pressure outside the dashed hypothetical reference gas line constitutes the steady-state ambient landfill pressure, which is at its maximum for that area of the landfill and constitutes the driving force to propel gas towards the lower-pressure vent well outlet.

The validity of the model results is directly related to the assumed degree of gas impermeability of the landfill cover cap. For the purpose of illustrating the considerable insights provided by the model, and the interrelationships that exist between the respective variables, it was assumed that the cover cap was contiguous, extended beyond the zone of calculated influence of the well and was essentially impervious.

The modified hemispherical landfill gas model was used to calculate a number of scenario conditions looking at different landfill gas ambient gauge pressures of 0.5, 1, 2, 3, 4, 5, 6, 8, 10, 12, 16, 20 and 24 in of water column (0.12, 0.25, 0.5, 0.75, 1, 1.24, 1.49, 1.99, 2.49, 2.98, 4.97, and 5.95 kPa) for individual well diameters of 4, 6, 8, 10, 12, 18 and 24 in (10.1, 15.2, 20.3, 25.4, 30.5, 45.7 and 61 cm). The parameter settings and assumptions used in these initial calculations were:

- a refuse permeability, k, of 100 D (10^{-3} m/s);
- an atmospheric pressure of 1 atm (i.e. 29.921 in of mercury, 407.5 in of water column or 101.325 kPa);
- the ambient landfill pressure P_e initially assumed to occur at a radial distance R_e of 400 ft (121.9 m);
- a landfill refuse-gas temperature of 100 °F (37.77 °C) and the concomitant gas viscosity, μ, of 0.018 cP;
- gas compressibility factor, z, equal to 1.0;
- landfill well field coefficient, n, assumed equal to 1.0;
- gas pressure at the vent well point of discharge, P_w, equal to 0 in (0 kPa) of water column;
- the landfill surface considered to be gas impermeable to beyond the calculated zone of influence defined by radial distance, R_e; and
- the refuse considered to be completely homogeneous and of constant uniform permeability in the area defined by radial distance R_e.

Figure 2 is a plot of the calculated predicted pressure profiles versus radial distance for a 24 in (61 cm) diameter vent well out to the assumed hypothetical boundary beyond which the respective land-

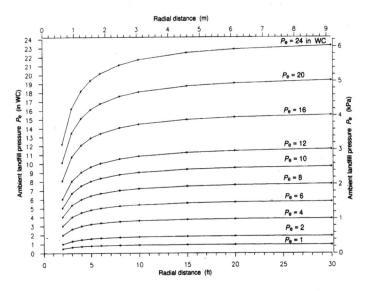

Figure 2. Pressure profiles for 24 in (61 cm) diameter passive vent well. k = 100 D; P_w = 1 atm; R_e = 400 ft (121.9 m); R_w = 1 ft (30 cm); T = 100 °F (37.77 °C).

fill ambient pressures are assumed to be constant and at their maximum. What it reveals is that the respective internal refuse pressures quickly approach almost ambient conditions within a distance of perhaps 10–30 ft (3.05–9.14 m) from the vent well. The calculated distances within which the pressure approaches ambient are less at the lower ambient pressures and increase slightly as the ambient pressure levels increase.

Within the space of 15 ft (4.57 m) and, most certainly, 30 ft (9.14 m), the pressures have already reached or are close to the respective ambient pressures. This clearly suggests that placement of a gas pressure measurement probe at about 25–30 ft (7.62–9.14 m) from the vent well should approximate the ambient landfill gas pressure under almost all scenario conditions. It likewise suggests that perhaps the relative radial sphere of influence of a passive vent well is only about 25–30 ft for a refuse permeability of 100 D. This may explain why passive vent wells, placed typically at 100 ft (30.5 m) or more on centres, seem to perform so poorly in controlling gas migration.

The projected passive gas-venting flow rates in standard cubic feet per minute (scfm), shown in Fig. 3, are plotted over the range

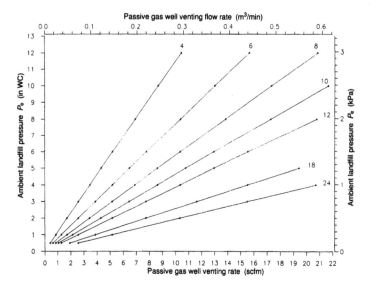

Figure 3. Passive gas well venting rate vs ambient landfill pressure for vent well diameters of 4–24 ins (10–61 cm). $k = 100$ D; $P_w = 1$ atm; $R_e = 400$ ft (121.9 m); $R_w = 1$ ft (30 cm); $T = 100$ °F (37.77 °C).

from 0.5 to 13 in of water column (1.25–32.4 kPa). This range of ambient landfill pressures is seldom exceeded. Study of Fig. 3 reveals that the curves for the respective individual well diameters are almost linear and that the predicted gas flow rates for the smaller-diameter wells at the quite high (and somewhat improbable) pressures are still quite low.

While the results presented herein have been calculated for a barometric pressure of 1 atm (101.325 kPa or 29.921 in Hg) and a refuse permeability of 100 D, it is a simple matter to calculate the corresponding flow rate for either higher- or lower-permeability refuse through the simple proportional relationship of $G_2 = K_2 G_1 / K_1$. (For example, the flow would be decreased by one-half for a refuse permeability of 50 D, and would be doubled for a permeability of 200 D or quadrupled for a permeability of 400 D).

Determination of Effective Distance

The model was used to assist in the determination of what constitutes an effective operating sphere of influence around a passive vent well. The aforementioned calculations assumed that the radial or spherical zone of influence around the well extended out to a maximum of 400 ft (121.9 m). At 400 ft, the prevailing ambient (positive) pressure was the theoretical maximum pressure or driving force to cause flow towards the passive vent well. However, upon plotting the calculated predicted results and relating field knowledge and empirical experience of:

- observed fluctuations in shallow probe pressure measurements over the course of a day or several days, which reflect the strong influence of barometric pressure,
- variations and/or discontinuities in the surface cover soil cap, and
- induced horizontal and/or vertical flow variations due to percolating precipitation and/or snow and frost cover,

it became clear that it is highly unlikely that the maximum positive driving pressure P_e for inducing flow would occur at such a considerable distance from the vent well. Rather, based on the nearly ambient levels reached within only 15–30 ft (4.57–9.14 m) from the well, the sphere of influence and actual driving force for induced flow could just as likely occur within a presumed 15–50 ft (4.57–15.25 m) radius of the well. Because of this intuitive perception, coupled with the acknowledged inability to verify this under realistic monitoring and/or testing scenario conditions, a gas flow rate sensitivity analysis was performed using the model, which assumed different radial distances from the well beyond which the positive driving force P_e was a constant and at its maximum.

For this sensitivity analysis, the radial distances at which the maximum assumed P_e values occurred were at 20, 30, 40, 50, 100, 200 and 400 ft (6.1, 9.1, 12.2, 15.2, 30.5, 61, 122 m). The results for the different size vent wells and different ambient pressure conditions are presented in Tables 1–3. As suspected, there is a slight increase in the predicted flow rate, holding P_e constant, as the radial distance from the well decreases. The percentage deviation in flow rates between the 400 ft (121.9 m) radial distance and the 20 ft (6.1 m) radial distance for each of the assumed P_e conditions is presented at the bottom of each of the tables. The percentage deviation in predicted passive vent well gas flow rates for the two extremes of 400 ft and 20 ft ranges from less than 1% for a 4 in (10 cm) diameter well to 5% for a 24 in (61 cm) diameter well.

TABLE 1. Sensitivity Analysis for 4 in Diameter Well Comparing Flows at R_e of 20, 30, 40, 50, 100, 200 and 400 ft

R_e (ft)	Flow rate (scfm)												
20	0.431	0.864	1.730	2.599	3.470	4.343	5.218	6.974	8.739	10.512	14.085	17.690	21.330
30	0.430	0.862	1.726	2.592	3.460	4.331	5.203	6.955	8.715	10.483	14.045	17.641	21.271
40	0.430	0.860	1.723	2.588	3.455	4.325	5.196	6.945	8.702	10.468	14.026	17.616	21.241
50	0.429	0.860	1.722	2.586	3.452	4.321	5.192	6.939	8.695	10.460	14.014	17.602	21.223
100	0.429	0.858	1.719	2.582	3.447	4.314	5.183	6.928	8.681	10.442	13.990	17.572	21.188
200	0.428	0.857	1.717	2.580	3.444	4.310	5.179	6.922	8.673	10.433	13.979	17.558	21.170
400	0.428	0.857	1.717	2.579	3.443	4.309	5.177	6.919	8.670	10.430	13.973	17.551	21.162
P_e (in WC)	0.5	1	2	3	4	5	6	8	10	12	16	20	24
Dev. (%)[a]	0.701	0.817	0.757	0.775	0.784	0.789	0.792	0.809	0.796	0.786	0.802	0.792	0.794

[a] Percent deviation between flows at R_e of 20 and 400 ft.

TABLE 2. Sensitivity Analysis for 12 in Diameter Well Comparing Flow at R_e of 20, 30, 40, 50, 100, 200 and 400 ft

R_e (ft)	Flow rate (scfm)												
20	1.317	2.637	5.280	7.930	10.587	13.249	15.919	21.277	26.661	32.071	42.968	53.969	65.073
30	1.306	2.614	5.235	7.863	10.497	13.137	15.784	21.097	26.435	31.799	42.604	53.511	64.521
40	1.301	2.603	5.213	7.830	10.453	13.082	15.717	21.008	26.324	31.665	42.424	53.286	64.249
50	1.297	2.597	5.200	7.810	10.426	13.049	15.678	20.955	26.257	31.585	42.317	53.151	64.087
100	1.291	2.584	5.174	7.771	10.374	12.983	15.599	20.849	26.125	31.426	42.104	52.884	63.765
200	1.288	2.577	5.161	7.751	10.348	12.951	15.560	20.797	26.060	31.347	41.999	52.751	63.605
400	1.286	2.574	5.155	7.742	10.335	12.935	15.541	20.772	26.027	31.309	41.947	52.686	63.526
P_e (in WC)	0.5	0	2	3	4	5	6	8	10	12	16	20	24
Dev. (%)[a]	2.411	2.448	2.824	2.428	2.438	2.428	2.432	2.431	2.436	2.434	2.434	2.435	2.435

[a] Percent deviation between flows at R_e of 20 and 400 ft.

TABLE 3. Sensitivity Analysis for 24 in Diameter Well Comparing Flow at R_c of 20, 30, 40, 50, 100, 200 and 400 ft

R_c (ft)	Flow rate (scfm)												
20	2.704	5.412	10.839	16.278	21.731	27.197	32.676	43.674	54.726	65.830	88.198	110.779	133.571
30	2.658	5.319	10.652	15.997	21.356	26.728	32.113	42.921	53.782	64.695	86.678	108.869	131.268
40	2.635	5.274	10.561	15.861	21.174	26.499	31.838	42.554	53.323	64.142	85.937	107.938	130.146
50	2.622	5.247	10.507	15.780	21.066	26.364	31.676	42.337	53.050	63.815	85.498	107.387	129.482
100	2.595	5.194	10.401	15.620	20.853	26.098	31.356	41.910	52.515	63.170	84.635	106.303	128.174
200	2.582	5.168	10.348	15.542	20.748	25.967	31.198	41.699	52.251	62.853	84.209	105.769	127.530
400	2.576	5.155	10.323	15.503	20.696	25.902	31.121	41.595	52.120	62.696	83.999	105.504	127.211
P_c (in WC)	0.5	1	2	3	4	5	6	8	10	12	16	20	24
Dev. (%)[a]	4.969	4.985	4.999	4.999	5.001	5.000	4.997	4.998	5.000	4.999	4.999	5.000	5.000

[a] Percent deviation between flows at R_c of 20 and 400 ft.

Based on a number of calculations using the model, beyond a distance of about 50 ft (15.25 m) from the well, the change in flow rate and pressure is relatively minuscule and does not appreciably affect the accuracy for refuse permeabilities of less than 200 D.

A rule of thumb suggested by the modelling results for locating a monitoring-probe for measuring ambient pressure(s) P_e would be at least 25 or more feet (7.62 m) from the subject vent well or approximately mid-distance between adjacent venting wells but probably no more than 50 ft (15.25 m) for ambient pressures P_e of up to 30 in of water column (74.7 kPa). This range of recommended distances would necessarily increase in direct proportion to any further increase in ambient pressure or well diameter.

The model is currently unable to distinguish between a vent well casing in direct contact with refuse and a vent well encased in a sheath of large granular material of larger total diameter. Thus, for example, a passive vent well listed nominally as 24 in (61 cm) in diameter in direct contact with refuse could in reality be a 6 in (15.2 cm) diameter vent pipe casing in a gravel-filled borehole of 24 in (61 cm). At some point, the ratio of well casing to gravel-filled nominal borehole size breaks down if the ratio becomes too small and the gas cannot effectively vent out the restricted vent pipe. Exactly where this occurs has not been determined.

Sensitivity to Barometric Pressure Changes

It is known from actual field experience with vacuum gas-extraction wells that barometric pressure fluctuations have a significant impact on extraction well performance, and the same is probably true with passive vent wells. However, the model is not structured to handle the complex transient (complex pressure lag response) conditions as the refuse pore gas pressures at progressively deeper depths in the landfill slowly equilibrate to a change in barometric pressure. Figure 4 is thus only representative of a landfill situation where there is an almost instantaneous equalization of *in situ* refuse pore gas pressure at all pertinent depths of the landfill in response to a change in barometric pressure. In the model, an increase or decrease in barometric pressure is merely seen as a relatively small increase or decrease, respectively, in the absolute pressure and does not reflect the sometimes large pressure differentials that can occur during this transient lag response period or the minimum pressures or even

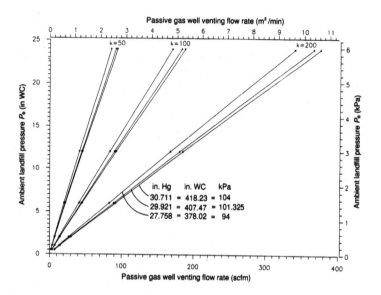

Figure 4. Passive landfill gas venting flow rates as function of refuse permeability and barometric pressure (assumes zero lag response time). $k = 100$ D; $P_w = 1$ atm; R_e = 400 ft (121.9 m); $R_w = 1$ ft (30 cm); $T = 100$ °F (37.77 °C).

negative pressures that occur during the period of pressure cross-over.

The sensitivity of gas flow to different barometric pressure levels (which are assumed to have equilibrated almost instantaneously with the *in situ* refuse gas pore pressures) ranging from 378 to 418.2 in of water column (27.76–30.71 in Hg, or 94–104 kPa), holding all other variables constant, was evaluated to produce Fig. 4. It shows gas flow both as a function of a range of assumed barometric pressures as well as at three different refuse permeabilities ($k = 50$, 100 and 200 D). The differences in the rate of flow over the range of barometric pressures are relatively small, but the effect of different permeability values is much more significant.

The effect of barometric pressure fluctuations can be approximated by measuring the absolute atmospheric pressure using a barometer and the *in situ* pressures at several depths in the landfill with a multiple depth monitoring probe(s) and then averaging the results to get the ΔP pressure difference. This pressure difference (ΔP) is the same as P_e. Entering Fig. 2 or 3 (or similar figures for the appropriate permeability or other range of parameter settings) will allow one to approximate the flow or relative sphere of influence for the perceived condition.

Sensitivity of Sphere of Influence and Flow to Well Diameter

An interesting observation pertaining to the model was that the pressure vs radial distance profile and the flow are a direct function of the well diameter and the ambient pressure P_e. But the pressure vs radial distance profile is unchanged by changes in permeability k, and relatively unchanged by changes in radial distance R_e, holding all other parameters constant. The model thus suggests that the effective sphere of influence and the flow rate are largely determined by the magnitude of the well diameter, R_w, and the absolute pressure P_e. Figure 5 shows that the effective sphere of influence for a 4 in (10 cm) diameter well is only a few feet while that for a 36 in (91.4 cm) diameter well (Fig. 6) is up to eight times greater. The corresponding flow rates for the 36 in and 4 in diameter boreholes (shown in Fig. 7) reveal the considerable advantages of utilizing larger-diameter boreholes.

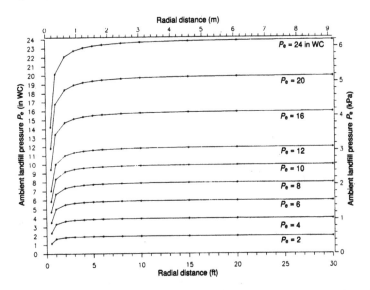

Figure 5. Pressure profiles for 4 in (10 cm) diameter passive vent well; effect of small-diameter well on sphere of influence. $k = 100$ D; $P_w = 1$ atm; $R_e = 100$ ft (30.5 m); $R_w = 0.1667$ ft (5 cm); $T = 100$ °F (37.77 °C).

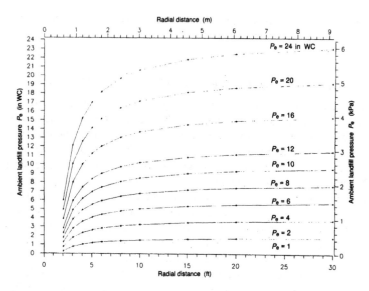

Figure 6. Pressure profiles for 36 in (91.4 cm) diameter passive vent well: comparison of large-diameter well on sphere of influence. $k = 100$ D; $P_w = 1$ atm; $R_e = 100$ ft (30.5 m); $R_w = 1.5$ ft (45.7 cm); $T = 100$ °F (37.77 °C).

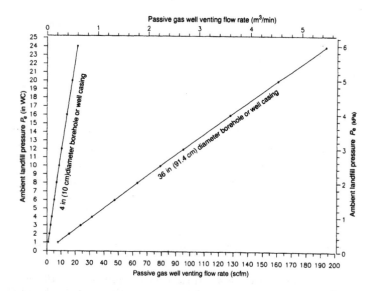

Figure 7. Passive gas well venting rate for 4 vs 36 in (10 vs 91.4 cm) diameter borehole/well casing. $k = 100$ D; $P_w = 1$ atm; $R_e = 100$ ft (30.5 m), $T = 100$ °F (37.77 °C).

SUMMARY AND CONCLUSIONS

A widely used hemispherical petroleum gas-reservoir model has been adapted, subject to some simplifying assumptions, to predict the approximate passive gas vent flow rates for a number of assumed representative landfill conditions. Subject to the validity of the simplifying assumptions, the model provides considerable insight into the interactive effect of the different variables as well as delineating, in the absence of hard-to-get field measurements, the predicted approximate performance of passive vent wells which heretofore had been notoriously hard to analyse and evaluate. The most significant perceived fallibility in the use of the model is that it assumes that the landfill surface is an impermeable boundary similar to the confined strata for which the petroleum reservoir model was originally developed. The more permeable or 'porous' the boundary surface condition is, the less reliable the predicted results will be. However, based on the empirically acknowledged low ambient gas pressures and the perceived relatively small radial sphere of influence of a passive vent well, it is likely that the effects of moderate flux of gas through this boundary condition may not be as detrimental as feared and the actual performance may at least approximate the results predicted by the mathematical equation. It can be inferred that a relatively impermeable cover cap or a gas-impermeable membrane incorporated in the cover cap would essentially satisfy the mathematical model's requirement for a confined or non-porous boundary condition.

The way in which passive gas vent wells have typically been monitored in the past is believed to be primarily responsible for the inconclusiveness and lack of reliable data upon which to judge their performance in that:

- One cannot correctly evaluate the performance of a passive vent well by merely taking pressure and flow measurements at the individual vent well stacks using the procedures and type of portable field test equipment typically used today.
- Measurement of pressure at the vent well provides little or no indication of the true driving force for flow to the well, as the reading would typically be close to zero gauge.
- Because of the almost zero gauge pressure readings within a vent well, measurement of flow using a primary velocity head measurement device, such as a Pitot tube, leads to low, if not

totally unreliable, measurements (if a measurement can even be obtained).

Because of the already low pressures, any flow reading at or close to the top of the vent well is also subject to potentially significant and serious measurement errors due to wind velocity-induced pressure fluctuations across the top of the pipe. Likewise, any measurement of gas concentration would most likely be diluted by the intrusion of atmospheric air and/or mixing near the top of the vent pipe.

- Capping the vent well for a period of time may provide some indication of the ambient landfill gas pressure within that area (conditioned upon the pressure reaching truly steady-state conditions). In the course of capping the pressure distribution and flow patterns are most likely distorted and the conditions measured may not be representative.

In summation, the top of a vent well, typically the only point of access to a passive well system on the entire landfill, is simply not the proper location to get meaningful data using present equipment and methodologies.

It would appear that there are few, if any, easy approaches to obtaining an accurate evaluation of a passive vent well's performance. Even the results of the model are only as good as the input values. While statistically representative pressure measurements could be taken from a number of probes arranged around the well at distances from about 15 to 40 ft (4.57–12.2 m), the determination of just 'average refuse permeabilities' is a significant research project fraught with many of the same measurement problems. As the permeability of refuse changes with the percolation of precipitation and is constantly decreasing as refuse decomposes and consolidates, such measurements will have to be periodically updated.

For the typical range of empirically observed ambient landfill gas pressures, the resultant gas flow rates predicted by the model, even under idealized conditions, are quite low.

The model results suggest that the effective sphere of influence around a passive vent well may only be 15–40 ft (4.6–12.2 m), dependent on absolute ambient pressure. Even these indicated distances need to be tempered by the recognition that in a landfill with conventional soil cover it may be easier and more direct for a gas molecule to travel the vertical upward distance of 5–10 ft

(1.5–3.05 m) than to travel the horizontal distance of 20 ft (6.1 m) or more to the well.

Increasing the well casing diameter (or alternatively placing the entire perforated length of a reasonably sized well casing of perhaps 4 or 6 in (10–15.2 cm) diameter in a gravel-packed borehole of perhaps 18–36 in diameter (45.7–91.4 cm) significantly increases the flow rate for the well.

A passive vent well screen that is both short in length (possibly only a couple of feet long) and is placed extremely close to or just below the cover soil cap is a relatively ineffective design. An increase in well screen length and deeper burial, as if designing for a vacuum gas extraction well, will result in greater effectiveness.

Owing to the sometimes dramatic and rapid changes in barometric pressure, which exceed the capability of the landfill to equilibrate, there are times when there may be a net inward flow of atmospheric air into the landfill and/or a capping of the venting process due to these barometric pressure vacillations, which effectively negates the effectiveness of passive venting systems for perhaps as much as 50% of the time. This is reflected in ambient soil pore pressures (P_e), particularly in the near-surface horizontal strata, which may actually be negative for a period of time during this lag-response time, and thus the resultant zone of influence and flow rates can be affected.

Major changes in barometric pressure appear to have an appreciable impact on a well's predicted radial sphere of influence, in that increasing barometric pressures will significantly expand the zone of influence of the well, reaching out sometimes for hundreds of additional feet upon a modelled increase in barometric pressure. Because of possible weaknesses, penetrations or fissures in the surface cover cap, the effect of these barometric pressure changes is difficult to evaluate truly under actual field conditions.

A snow or frost cap on a landfill and/or water saturation of the cover cap or some interim refuse layer might increase the equilibrium steady-state ambient landfill pressure and therefore increase the flow rate out of a vent well during such periods. However, it is highly unlikely that a conventional landfill would be able to sustain ambient pressures much greater than a few inches of water column except under conditions of extremely impermeable cover cap, snow/frost cover, saturated refuse or perched water in a landfill, or some combination of the above. Nevertheless, even with these higher ambient pressures, passive vent flow rates are still relatively small.

As most landfill gas ambient pressures are typically less than 5 in (12.7 cm) of water column, and the model results suggest a relatively small effective radius around the vent well, there is reason to believe that the relative flux through the annular zone surrounding a passive landfill gas vent well may be small enough that the model can still approximate the real physical situation occurring in the vicinity of the well. It is felt that the validity of the model might begin to deteriorate as near-surface pressures and/or radial distances from the vent well increase because of the many antagonistic interactions that can occur.

Many regulatory agencies are looking to passive vent wells to prevent 'ballooning' of the membrane liner in the modern-day composite cover cap. If the wells are in refuse, the modelling results suggest that nine or more wells would be required per acre (greater than 22 wells/ha).

For those assumed scenario conditions cited herein, it became evident that diffusive flow may constitute a greater percentage contribution than originally anticipated. However, the development of a model that combines both convective and diffusive flow for the purpose of better quantifying the flow rates from a passive vent well is probably a moot point, as the passive vent well model results clearly suggest that a passive landfill gas vent well is probably not a desirable or preferred methodology for achieving present-day environmental goals.

REFERENCES

Craft, B. C. and Hawkins, M. F. (1959). *Applied Petroleum Reservoir Engineering*, Prentice-Hall, Englewood Cliffs, NJ.
Katz, D. L. *et al.* (1959). *Handbook of Natural Gas Engineering*, McGraw-Hill, New York.
Muskat, M. (1937). *The Flow of Homogenous Fluids through Porous Media*, McGraw-Hill, New York.
Theis, C. V. (1935). The relationship between the lowering of piezometric surface and rate and duration of discharge of wells using ground-water storage. *Trans, AGW*, **II**, 519.

5.4 Design of Combined Landfill Gas Abstraction Systems

ANDREW LEACH

82 Bower Street, Bedford, Bedfordshire MK40 3QZ, UK

INTRODUCTION

There is a wide variation in the design of gas abstraction systems and particularly in the detail design of gas wells and the practical considerations of installation and operation.

Essentially, however, the objectives for any gas abstraction system are the same: that is, to abstract the maximum volume of gas for the lowest suction in the most cost-effective way.

This chapter examines the design, installation and operation of gas well systems from the operator's perspective. Particular attention is paid to those parameters that adversely effect the efficiency of gas wells and prevent the objectives being achieved.

GAS WELL DESIGN

Gas well design will inevitably be to some degree site specific. Designs will vary according to depth and volume of waste, operational practices, leachate level and leachate control practices, the age and make-up of waste, and the degree of settlement that is expected.

Some different well designs that have been tested are described below.

Landfilling of Waste: Biogas. Edited by T. H Christensen, R. Cossu and R. Stegmann. Published in 1996 by E & FN Spon, London. ISBN 0 419 19400 2.

Standard Vertical Well

A typical well design is as shown in Fig. 1 and consists of 120–150 mm nominal bore (NB) mineral-based polypropylene well casing inserted within a borehole within the waste, with the annulus filled with pea gravel. The top 4 m of the annulus are sealed using polyurethane foam or bentonite clay. The casing is slotted (3 mm) for the length inserted within the gravel pack. A wellhead made up of UPVC fittings enables the installation of a valve for fine-tuning the suction on each well, a sampling point to enable the well to be dipped for leachate, and a manometer to allow the wellhead suction to be measured. A length of flexible pipe is used to connect the wellhead to the collection pipework to allow for a different settlement rate between the two.

Common variations on this design include the use of polyethylene for well casing and different well casing/borehole diameter ratios.

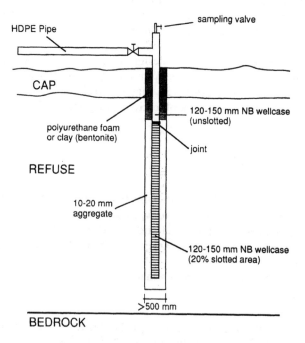

Figure 1. Scheme of a vertical well installation.

Large Diameter Vertical Well

A typical well design is shown in Fig. 2. The well consists of large-diameter perforated concrete rings (typically 0.9 m diameter) constructed during filling operations to the surface or just below. To protect against the possible deterioration of the concrete rings in an aggressive leachate environment a well casing is placed within the concrete rings. Sealing the well can be achieved either by terminating the concrete rings below the final waste level and extending the well casing to the surface or by extending the concrete rings to the surface and sealing the top 2–4 m with polyurethane foam or bentonite clay.

Again, a wellhead is required for control and monitoring purposes

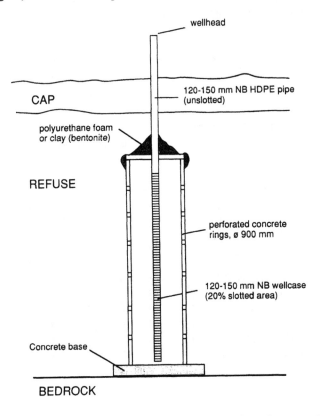

Figure 2. Scheme of a large-diameter vertical well with perforated concrete rings.

including a length of flexible pipe to connect the well to the collection pipework.

Small Bore Horizontal Well

A diagram of this well design is included in Fig. 3. This design consist of 2 in (50 mm) NB slotted drainage pipes placed in an aggregate surround within an excavated trench. The trench was between 0.5 m and 1 m deep within the refuse. Refuse and cover are then backfilled over the pipe and aggregate to try and achieve an airtight seal. Unslotted pipe joins the gas-collecting pipe to a central wellhead. The final connections to this central manifold can be sealed with polyurethane foam or bentonite.

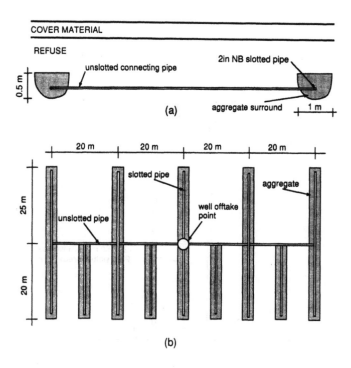

Figure 3. (a) Typical section and (b) plan of a horizontal well abstraction system.

Gabion Well

A gabion well consists generally of a metre diameter cylinder of porous material within the landfill, held in place by an openwork gabion shell. A length of MDPE pipe connects the centre to a wellhead. A cross-sectional diagram is included depicting a typical gabion well (Fig. 4). Installation of these wells is undertaken by excavation of a pit within the refuse, placing the gabion shell at the required depth and backfilling with the excavated material.

Hybrid Well

As the name suggests, the hybrid well combines both vertical and horizontal forms of gas abstraction. Slotted pipe is placed vertically

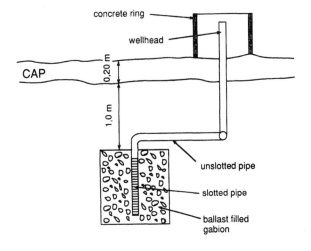

Figure 4. Cross-sectional view of a typical gabion well construction. Three gabions are connected to the wellhead at 20 m intervals.

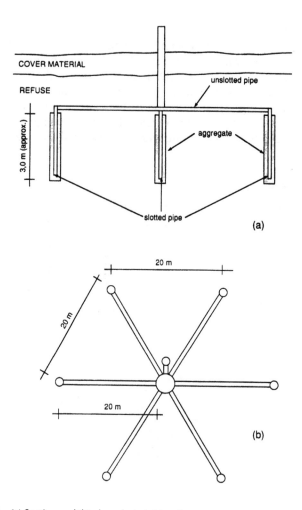

Figure 5. (a) Section and (b) plan of a hybrid well system.

into the refuse. Vertical wells are then connected to a central well take-off point using unslotted connecting pipe, (Fig. 5). Installation of hybrid wells involves excavation from the existing refuse level, placing the slotted pipe and porous fill, then backfilling.

Large Bore Horizontal Wells

This type of well consists of 150 mm NB slotted drainage pipe placed in an aggregate surround within an excavated trench. The trench is excavated to approximately 2 m depth in waste and backfilled after installation of waste, and a further 2 m of waste is added to give a final installed depth of 4 m. These are connected via polyethylene header pipes to a common wellhead for control purposes.

WELL PERFORMANCE TESTING

To determine the effectiveness of the different well types, pumping trials were carried out using a fan flare trailer. Gas flows were measured at increments with increasing wellhead suction. An interval of 5 min was left between each reading to allow flow rates and suctions to stabilize.

To make comparisons between different gas wells the quotient of the flow rate against wellhead suction is determined for the range of flows where the gas quality is steady. This constant is obtained using the units m^3/h and mm Wg for flow rate and wellhead suction respectively. This method enables a quick and easy comparison of different gas wells and gas well types to be made. The higher the constant, the more prolific the gas well, and likewise the lower the constant the lower the gas yield for an equivalent wellhead suction.

Using this method to test wells of different designs in similar sites or areas it is possible to rate the performance of different well types (Table 1).

Obviously the results in Table 1 will only be repeatable in a site of similar waste inputs, waste age, depth etc. However, the comparisons that can be made are useful, and some conclusions can be drawn from them:

- High leachate levels have a disastrous effect on vertical gas wells.
- The productivity of a well is in part related to the area of interface it has with the waste.
- Vertical wells perform better than horizontal wells in relation to the 'area of interface' with the waste.

TABLE 1. Values of the Gas Well Performance Constant (flow rate/well head suction) for Different Well Types Observed in Similar Landfill Sites. Low Leachate Level = Distance from Ground Level <10 m. High Leachate Level = Distance from Ground Level >10 m

Well type	Gas well performance constant $(m^3/h.mm\ H_2O)$	
Large-diameter vertical	5.0	Low leachate level
Standard horizontal	1.5	
Standard vertical	1.0–3.0	
Small-bore horizontal	1.0	High leachate level
Gabion	0.7	
Hybrid	0.6	
Standard vertical	0.3	
Large diameter vertical	0.3	

These conclusions can be explained as follows. First, the effect of a high leachate level on a vertical well is to greatly reduce the length of perforated well casing through which gas can be drawn, as gas cannot be sucked through saturated waste. If the leachate level rises too close to the top of the perforated section then a leachate 'foam' is often produced in the well as gas is drawn into the well with leachate.

Second, the increase in well productivity with a greater surface area exposed to the waste is as expected. Unfortunately, in general the greater the surface area the greater the cost of the well and its installation.

Finally, the better performance of the vertical well with respect to the horizontal well can be explained by consideration of the lateral and vertical permeabilities of waste. Waste is put down in layers and sandwiched every day between further layers of daily cover material. As a consequence the lateral permeability is greater than the vertical permeability by up to as much as 40:1. The vertical well penetrates all layers and is able to take full advantage of the higher lateral permeabilities.

PRACTICALITIES OF GAS WELL DESIGN (INSTALLATION, OPERATION AND COSTS)

The choice of gas well for a particular site depends as much upon the practicalities of installation as it does on the design. Some of the advantages and disadvantages of the main well designs are summarized as follows.

Standard Vertical Well

This is the most easily installed well, particularly if the waste is at final restoration levels. A good seal can be obtained around the top of the well, allowing high suctions to be applied. Vertical wells will tend to settle at a different rate to the upper layers of waste surrounding them as they are subject to reducing settlement rates with increasing depth, creating a 'flotation' effect. As a consequence the installation must allow the well casing to 'rise' above the surrounding ground level without damaging pipework and fittings or end-loading the casing. A well casing too highly loaded, owing to settlement, differential or otherwise, will buckle and fail, usually in its mid-section. For this same reason solid connections from vertical wells installed to add horizontal branches within the waste and increase the influence of a well may fail when used in sites taking a high proportion of domestic waste.

Vertical wells also allow the operator to finely control gas quality and easily identify and isolate rogue wells. When used for gas migration control they allow manipulation of the system to cover problem areas.

Large-Diameter Vertical Well

These wells are commonly used for leachate-monitoring purposes. The installation is relatively easy provided care is taken by the site operators. Uniform compaction of waste around large-diameter wells is hard to obtain, and this can lead to increased settlement around the well, poor sealing of the well or distortion and collapse of the well. During construction the odours from the plume of the well during its passive venting stage may prove unacceptable in some locations.

If the well is sited on the base of the site then the settlement of the waste surface relative to the well will be greater than that experienced by the standard vertical well. Provided a good seal can be achieved around the top of the well then it can be a very effective gas well.

Horizontal Wells

This well type presents different construction alternatives. Horizontal wells should preferably be installed to a depth of 4 m or greater to prevent air ingress through the cap. This may be achieved by trenching from final restoration levels; however, this is only at the expense of a considerable odour nuisance during installation and a loss of waste density immediately above the installed pipes allowing greater air ingress during operation.

Alternatively, the wells can be installed with the filling operation; however, this assumes that very careful machine operators are employed, and even then there is no easy way of checking that the pipe has not been crushed. A compromise solution is to shallow trench the horizontal gas well into the waste prior to the placing of the final layers of waste. Land drainage machinery has been tried but generally results in expensive machine breakages. Horizontal gas wells may be surrounded by a gravel pack; however, this could prove more expensive than the pipe.

Horizontal wells should always run uphill to allow the gas to be drawn from the highest point, thus leaving the condensate to drain back into the site. To install them alternatively may result in an effective land-drainage system that swamps the gas plant. Also, excessively long runs should be avoided, particularly where small corrugated 'drainage' pipe is used, as the corrugations severely reduce the effective diameter of the pipe.

Settlement damage is most likely to occur at the transition point where the horizontal well is connected to a vertical riser, owing to stresses brought about by different settlement rates between the horizontal and vertical.

The control of horizontal wells is difficult, as in general badly performing sections cannot be isolated; also, collapsed or blocked pipe runs cannot be easily located. Gas abstraction systems based upon horizontal wells generally produce gas of a lower quality than their vertical well counterparts.

In term of costs, a horizontal well system is equivalent to a vertical well system.

VERTICAL WELL DEVELOPMENT

The vertical well proved to be the most efficient in terms of gas flow against suction and it is the system more widely used. In this section the design and development of vertical wells are discussed in detail, on the basis also of experience at the Brogborough landfill site.

TABLE 2. Results of the Well Calibration Test Carried Out for Vertical Wells with Different Size at the Brogborough Landfill Site. Oxygen Percentage in the LFG was Always Below 1.0%.

Well diameters[a] (mm)	Flow m^3/h	Suction (mm)		Gas quality CH_4 %
		W/H^b	F/T^c	
$D_c = 150$	50	6	14	44
$D_b = 350$	100	14	36	44
	150	20	92	45
	200	27	164	45
	250	36	232	44
	280	42	350	44
$D_c = 300$	50	2	14	47
$D_b = 350$	100	4	31	47
	150	6	63	47
	200	7	165	47
	250	9	257	46
	320	12	348	46
$D_c = 75$	50	9	20	33
$D_b = 350$	100	21	46	32
	150	41	100	33
	200	70	225	32
	250	96	330	32
	300	108	373	32
$D_c = 300$	90	1	25	42
$D_b = 400$	140	1	57	43
	190	3	112	44
	240	5	230	43
	290	10	320	44
	340	12	330	43
$D_c = 75$	50	1	20	45
$D_b = 350$	100	18	50	46
	150	54	142	46
	200	75	205	46
	250	95	310	46
	300	113	390	46

[a]D_c = well casing diameter, (mm); D_b = borehole diameter.
[b]W/H = wellhead.
[c]F/T = flare trailer.

Casing Diameter

The selection of borehole and casing diameters can have a considerable influence on the efficiency and cost of a vertical well installation. In an effort to determine the best well casing size a number of different vertical wells were installed at the Brogborough landfill site. These wells were as listed in Table 2.

The wells were all drilled to the same depth (22 m) and were located 15 m apart. All wells were connected with a conventional wellhead arrangement. The waste in this area was all less than 18 months old and therefore not in a fully methanogenic state. A well calibration test, as described earlier, was carried out on each well in turn with the other two adjacent wells isolated. The results of these tests are given in Table 2. As can be seen from the table, the benefit of lower section obtained by using 300 mm casing over 150 mm casing does not warrant the additional expenditure, 300 mm casing being some 2.2 times more expensive than 150 mm casing. The difference between 75 mm casing and 150 mm casing is more marked, with considerable pressure drops being experienced in the well casing at the higher flow rates.

This suggests that well casing diameters of 100 mm and 125 mm may be adequate.

Borehole Diameter

It is generally considered that the larger the diameter of the borehole the more productive/effective the gas well is. However, large-diameter boreholes of say 1 m diameter involve expensive plant and equipment for installation purposes. The alternative of building wells up from the base is often undesirable for the reasons given earlier. Borehole sizes for gas wells installed using conventional drilling equipment typically vary between 200 mm and 400 mm.

WELL FAILURE

Well failure normally occurs as a result of casing collapse through waste movement and settlement, or through blockage and silting-up. In the case of settlement the forces experienced by the well will depend upon the age, depth and make-up of waste. Deep sites filled quickly with a high percentage of putrescible waste will obviously experience a great deal of settlement regardless of the initial compaction technique employed. The resistance of well casings to collapse can be minimized by the following measures:

- The use of polypropylene-based materials, which have a higher collapse resistance, particularly at the temperatures experienced in a landfill, than other materials such as polyethylene.
- The use of horizontally slotted casing as opposed to vertically slotted casing. The latter removes a greater proportion of the well casing hoop strength, making it more susceptible to collapse.
- A correctly placed gravel pack to provide uniform support to the casing and enhance the collapse resistance of the well casing. Here the choice of drilling technique and borehole/well casing diameter ratio will have a direct influence. Augered boreholes will often produce a roughly finished hole, which is likely to cause bridging of the gravel pack.

Well failure as a result of blockage and silting-up usually occurs over a number of years, either by a gradual reduction in depth as silt accumulates in the bottom of the well or by blockage of the gravel pack and well casing. In the case of the former the problem is easily identified and the remedy is usually to employ conventional well flushing techniques or to re-drill the well.

In the case of the latter the diagnosis is more difficult. At one site it was possible to carry out gas well pumping tests and compare these with similar tests carried out five years earlier. The deterioration in performance is quite marked, as shown in Fig. 6. The blockage on this site was possibly due to the use of sandy materials for daily cover. New gas wells were drilled, which were found to have similar production characteristics to those of the original wells.

Prevention of well casing blockage may be possible using screen filters; however, if the filter becomes blocked it may still be necessary to replace the well. The design of the well and gas collection system can increase the life expectancy of the well by minimizing the section pressures. Through choice of borehole/well casing diameters and location/spacing of wells, the probabilities of well blockage will be reduced.

LEACHATE LEVEL CONTROL

Proper control of the leachate level is vital for the efficiency of the gas abstraction system (see Table 1). Where leachate sumps and drainage facilities have not been installed in the site during filling, or where they have subsequently blocked or collapsed, it is possible to reduce leachate levels by pumping from vertical gas wells.

Four methods are commonly available for leachate removal:

- ejector pumps (pneumatic);
- shaft-driven sump pumps;
- submersible electric pumps;
- eductor pumps (hydraulic).

These also need to be considered as an integral part of the design of the gas abstraction system. The advantages and disadvantages of each system are briefly outlined.

Ejector Pumps (Pneumatic)

Pneumatic ejector pumps use an intermittent pressurized air supply to force leachate out of a vessel. The vessel is fitted with check valves and the appropriate connections to allow leachate to flow in, then

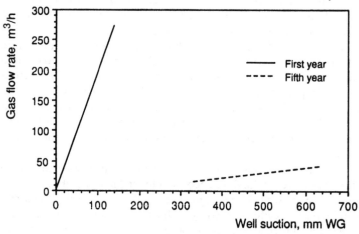

Figure 6. Deterioration of performance observed in a gas vertical well after 5 years of operation.

to allow pressurized air to force the leachate out through a riser pipe. The system can be made to fit down boreholes as small as 100 mm diameter. Some systems involve controlled leakage of air to the borehole, which makes them unsuitable for use in combined gas and leachate wells. The ejector pump has the advantage that it can pump down to within 1 m of the base of a site and is unlikely to be adversely affected by sludges and fine particles entrained in the leachate.

Shaft-Driven Sump Pumps

Shaft-driven pumps with a motor mounted at the wellhead driving a pump at depths of up to 120 m are commercially available. The drive shaft is held concentrically within the riser pipe by rubber spacers, which allows the leachate to flow up the annulus. Drive shaft and riser can be assembled in sections to give the desired depth. Flow control can be achieved by speed control or adjustment of pulley sizes or by timed operation of the unit.

The advantages of this type of pump are that, like the ejector pump, it can be installed in small-diameter casings (75 mm and above) and that it will pump down to within 1 m of the base of the site. Unlike the ejector pump there is no release of air into the gas stream, and therefore the pump is suitable for installation in combined gas and leachate extraction wells.

Submersible Electric Pumps

Experience has shown that conventional electric submersible pumps are suitable for installation in combined gas and leachate extraction wells; however, a number of design considerations need to be made.

First, the pump needs to be installed to the maximum depth possible without affecting the integrity of the site base seal and preferably within a sump or length of plan casing. This minimizes the ingress of gas drawn in at the pump suction, which can otherwise seriously affect the pumps performance.

Second, the pump should be constructed of 316 stainless steel or similar and should include a high-temperature motor suitable for temperatures experienced within a landfill.

Finally, the method of level control needs to be carefully considered, as the aggressive and difficult conditions within a gas well prohibit the use of conventional float switches and level sensors. Digital timers can be used so that pumping rates can be matched to well performance.

The advantage of this type of pump is that large volumes can be pumped.

The disadvantages are that a 150 mm diameter casing or greater is required for installation purposes, and that the pump always needs to be immersed in liquid with sufficient free board above any sludge or silt at the base of the well, which means that leachate can only be pumped down to within 3 m of the base of the site. Another disadvantage is that abrasives within the leachate can cause significant wear of moving parts and seals. Practical experience has shown that it is necessary to carry out a major overhaul every 6–12 months on each pump section.

Eductor Pumps

One pump design that is ideally suited to leachate extraction is the eductor or ejector pump. In this system leachate is circulated down a borehole and through an ejector nozzle. The increased velocity through the nozzle creates a suction effect similar to that experienced in the venturi of a carburettor. In this way additional leachate is induced into the flow for discharge at the surface.

The advantages of this system are low maintenance costs, the ability to pump a well to within 1 m of the base of the site and the ability to pump safely without need of any level control device.

The disadvantages are the low efficiency achieved using this system and the costs of maintaining a leachate-pumping circuit for inducing leachate flow.

5.5 Landfill Gas Pumping and Transportation

RAFFAELLO COSSU[a], GIOVANNI MARIA MOTZO[a] &
MATHIAS REITER[b]

[a] DIGITA – Department of Geoengineering and Environmental
Technologies, Faculty of Engineering, University of Cagliari,
Piazza D'Armi, 09123 Cagliari, Italy
[b] CISA – Environmental Sanitary Engineering Centre, Via Marengo
34, 09123 Cagliari, Italy

INTRODUCTION

To extract gas using one of the systems presented in Chapter 5.4 it
is necessary to create a depression (vacuum) by pumping the gas.
In fact, owing to the density of the LFG (which is generally higher
than that of air), a passive venting system may not be effective in
ensuring a proper degassing action. Furthermore, pumping is always
necessary when a gas utilization step is adopted.

The pumped gas has to be conveyed through pipelines to the final
disposal/utilization system. During transportation the condensate
has to be removed regularly (see Chapter 5.6), and safety control
measures have to be installed in order to prevent the introduction
of explosive mixtures into the pumping station.

Good engineering, choice of proper materials and equipment and
a careful control and maintenance programme are particularly
important, not just to ensure system efficiency but in particular to
guarantee a high safety level.

These aspects are discussed in detail in this chapter.

Landfilling of Waste: Biogas. Edited by T. H Christensen, R. Cossu and
R. Stegmann. Published in 1996 by E & FN Spon, London.
ISBN 0 419 19400 2.

PUMPING FACILITIES

The vacuum values to be adopted at the extraction wellhead can vary dramatically according to the desired quantity and quality of LFG and to the different specific situations. These are related to such factors as the height of the abstraction well, the characteristics of daily cover, the *in situ* waste quality (age, moisture, density, waste origin), and the top cover design.

A top cover with a very low permeability barrier can bear lower negative pressures than landfills with a poor top cover system.

Air penetration in the waste mass has several adverse effects:

- low-quality gas for combustion;
- the risk of formation of an explosive mixture in the transport pipelines or in the disposal facilities;
- inhibition of anaerobic processes.

Generally, the pressure values at the wellhead are maintained in the range −5 to −20 mbar for landfill with no or low lining systems, and can reach higher values when good landfill bottom and surface barrier systems are present.

Adopting the usual equations of fluid mechanics it is possible to evaluate the pressure levels necessary at the entrance of the pumping and end-use station respecting the minimum values required by the different facilities (flares, engines, etc.).

The gas flow can be adjusted by use of valves, by flow recirculation with a bypass circuit and by regulation of the number of revolutions of the engine. The latter could be done by utilizing frequency regulation of the supply current.

According to the specific use planned, it is possible to have mobile or fixed pumping facilities. The former are mainly used for survey purposes or as emergency or temporary solutions. In this case the blower is often also combined with a mobile flare. Generally, the blower is installed in a building or a metallic container together with other disposal facilities, secluded from atmospheric exposure.

For LFG pumping purposes the following types of blower are normally used (see Fig. 1 for basic types):

- radial blower;
- rotary blower;
- lateral canal blower.

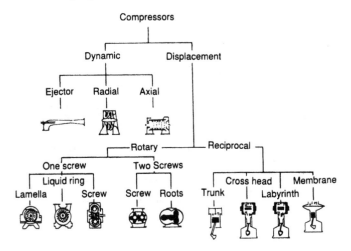

Figure 1. Basic compressor types (Coker, 1994).

These types of blowers have different costs, performance and characteristic curves, and hence are suitable for use in different situations. There are other variants, but the types mentioned above are those that are more widely used and have therefore been tested more for these applications.

Radial Blower

The most widely used blower is the undeflagrating version of the radial type (Fig. 2). Flow is adjusted by valves placed in the suction and discharge pipes or by one of the means previously mentioned, bearing in mind that this kind of blower is not able to accumulate pressure due to the slack between the impeller and the casing.

These blowers (in the case of satisfactory efficiency) afford a maximum pump head of about 300 mbar. The trend of characteristic curves of radial blowers allows a wide range of adjustment, thus maintaining more or less stable efficiency (Fig. 3).

The use of radial blowers has proved to be effective when constant values of output pressure are not required. In fact, pockets of condensate in the transport line could induce the transmission of pressure pulsation also in the discharge pipe.

The use of a radial blower is particularly convenient from an economic point of view, as the fitting of a simple double flame-

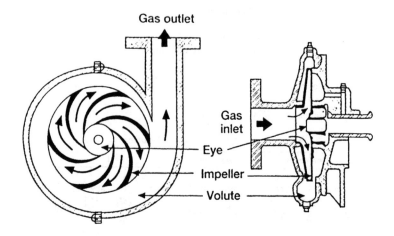

Figure 2. Section of a radial blower (Coker, 1994).

breaker filter can assure a high safety level without the need for more sophisticated and expensive measures.

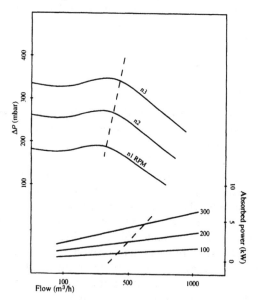

Figure 3. Characteristic curve for a radial blower: RPM = rev per minute; ΔP = blower head.

Rotary Blower

These blowers are displacement rotary machines comprising two rotors that turn inside a properly shaped casing (Fig. 4): during rotation the gas is trapped by the rotors at the suction side and is conveyed towards the discharge. The movement of the rotors is synchronized by timing gears located on the shafts: hence no friction occurs between the rotors or between the latter and the casing, thus resulting in oil-free biogas. The volume of the biogas conveyed is proportional to the speed of rotation and is almost constant at operating pressure, which depends on the head losses encountered by the gas in the network: it is therefore easy to regulate the operating pressure and to maintain it practically constant. For these reasons the characteristic curve of pressure and of flow is rectilinear (Fig. 5).

During the compression phase the gas temperature rises considerably, and as the temperature of the incoming biogas is already high, temperatures could be reached that would endanger the hold of some elements and of the electric control system. In these cases a heat exchanger is used in order to lower the temperature.

The high compression ratios of this type of blower cause a high noise level in both the suction and discharge pipes. Suitable silencers are used in both lines in order to reduce the acoustic impact to acceptable levels.

These blowers are manufactured so as to exert a pressure

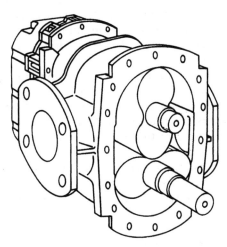

Figure 4. Exposed view of a displacement rotary blower (Anon, 1994).

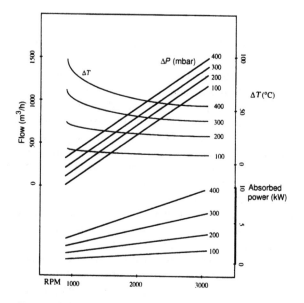

Figure 5. Characteristic curve for a displacement rotary blower: RPM = rev per minute; ΔP = blower head.

accumulation, and therefore the system is normally equipped with a bypass circuit and safety valve. During the starting phase of the blower, the bypass opening is adjusted by means of an automatic valve until the required revolution number for functioning is reached. As mentioned before, the bypass, driven by an electrovalve, may also be used to adjust flow.

Further control and adjustment of the flow is provided by frequency regulation of the supply current in order to determine the number of revolutions of the electric motor and therefore the flow.

Rotary blowers are equipped with an independent lubrication system, which is run each time a casual or programmed breakdown of functioning occurs.

The cost of these blowers is relatively high, but their functioning is very reliable and offers the best performance in terms of pressure head constancy. Furthermore, they make it possible to reach higher vacuum values.

In addition to a protection system of the engine, instruments for monitoring of pressure and temperature are used for continuous control of efficiency.

Lateral Canal Blower

This blower can be used for relatively low flow and pressure head. The price is similar to that of the radial blower. The characteristic curves do not allow wide ranges of adjustment, and therefore these blowers are often used in parallel (Fig. 6). Because of their particular functioning and construction characteristics, these machines (as for radial blower) have no contact between rotor and stator, and this affords some advantages:

- functioning without wear;
- no need for lubrication;
- minimum maintenance;
- gas conveyed without loss of quality;
- silent functioning;
- absence of pulsation.

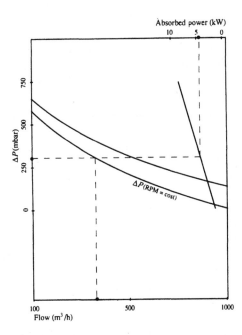

Figure 6. Characteristic curve for a lateral canal blower: RPM = rev per minute; ΔP = blower head.

Constructive Materials

The body of blowers is normally made of nodular cast iron, while the impeller of the radial and lateral canal blowers may be in aluminium and fitted directly to the motor axis.

Electric motors are sized for continuous running, IP 54 protection, constructed according to IEC standards. When required, their surfaces may be treated for better resistance to attack by the chemical agents typically found in landfill gas. A proper combination of suitable materials can reduce the potential for spark generation.

Two types of corrosion can occur: all-over superficial, and local. The first type occurs when a material with constant characteristics is adopted; the mode of action depends on the concentration of corrosive substance or other factors such as pH. When corrosion occurs in restricted zones, it is a sign that particular conditions are present (stresses, temperature) that render the material susceptible to the action of some substances.

The presence of oxygen in the gas mixture may not only induce a dangerous situation but also, owing to the fact that oxidation may occur, represents an important factor for the durability of materials.

It has been observed in practice that the presence of condensates can affect the durability of blowers. Characteristic parameters that can influence the quality of condensates and the risk of corrosion are as follows (see also Chapter 5.6):

- *pH*: hydrogen concentration assumes a particular significance for metals such as zinc or in general non-ferrous metals.
- *BOD*: this parameter indicates the degree of organic pollution of a liquid, but endangers equipment only when hydrogen sulphide is generated from decomposition of these organic compounds.
- *PAH*: polynuclear aromatic hydrocarbons can affect the characteristics of some materials (blistering).
- Cl^-: chlorine concentration exceeding 150 mg/l may lead to localized corrosion. The risk is greater the lower the carbonate hardness of the liquid.
- SO_4^{2-}: at concentrations higher than 250 mg/l, a bond transformation into cast iron occurs. In the case of steel localized corrosion may occur.
- S_3^{2-}: The presence of these ions is caused by hydrogen sulphide, and as temperature (> 70 °C) rises, the capacity of

attacking ferrous materials increases. Copper and related alloys are also attacked.

- NO_3^-: at concentrations higher than 50 mg/l, ferrous metals and zinc corrode if the liquid is not sufficiently buffered.
- NO_2^-: all easily oxidizable materials such as zinc are subjected to the corrosive action of nitrites.
- NH_4^+: non-ferrous metals are not able to resist ammonium. Ferrous metals resist up to 20 mg/l. At a higher concentration they begin to show localized signs, and 'perbunan' too is inadequate to ensure good protection.
- *phenols* (C_6H_5OH): for concentrated phenol solutions the resistance of carbon steel and 'perbunan' is inadequate.

Deposits of a different nature can often occur, compromising the functioning of the blower. This risk is reduced if the performance of the condensate separation systems is satisfactory. In fact, corrosive chemicals are transported with the condensate, which is generally characterized by low pH values (see Chapter 5.6).

TRANSPORT PIPELINES

The length of pipeline network on the landfill should be minimized to reduce negative settling effects. For this reason, the different lines are either connected to a peripheral collection line with increasing diameter, which is located at the external boundaries of the landfill, or connected to pipe collectors fitted with control and regulation systems for all the wells.

On the landfill a slope greater than 5% is adopted for the pipelines in the direction of the condensation separator or in the direction of the wells. In this latter case the flow velocity should be reduced by using larger pipe diameters. The problem of rapid removal of condensate may be avoided by using the above-mentioned collectors in an elevated position in order to allow flow of condensation towards the wells (see Chapter 5.6).

In order to prevent, in connection with settling phenomena, pipe bursting or rupture, a straight-line layout should be avoided by adopting slight bends in the pipeline.

The slope necessary for the peripheral collection line is obviously less marked as the latter is placed on solid ground not subjected to relevant settlements. Generally, a slope up to 1% is sufficient when condensate and biogas flow in the same direction.

The transport pipelines should be placed on the surface of the landfill and final top cover should be applied only after the majority of settlement has occurred. The gas network has to be located above the surface liner in any case, and sometimes new gas pipelines have to be installed.

To design the gas network pipeline, the following transportation rates could be used (Rettenberger, 1988; Raddatz, 1988):

- $v < 5$ m/s (without condensate separator and with different flow direction of gas and condensate);
- $v < 10$ m/s (with condensate separator).

A design with higher flow velocity obviously leads to higher head loss and possible damage to the network pipeline due to vibration.

Sagging of the transport lines caused by differential settling may be avoided by placing wooden boards or similar at regular intervals, securely fastened under the transport lines to stiffen them. Better control of settling movements may be obtained by connecting to the transport lines some rigid tape measure which come out from the cover soil.

Friction Losses

Various formulae can be used to calculate friction losses. For practical evaluation the chart reproduced in Fig. 7 can be used. This chart has been plotted by using the semi-empirical formula of Colebrook and White:

$$\frac{1}{\sqrt{f}} = -2 \log\left(\frac{2.51}{R_e} + \frac{k}{3.7\,D}\right)$$

where

$$f = \frac{h_f}{\left(\dfrac{L}{D}\right)\dfrac{u^2}{2g}}$$

is the friction coefficient, which could be calculated by means of successive iterations or by use of special charts; L is the pipe length (m); $D = 4A/P$ is the hydraulic diameter (m); $u^2/2g$ is the velocity head (m); $R_e = uD/v$ is the Reynolds number; k is the roughness value

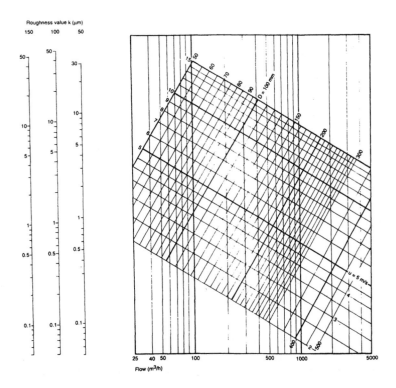

Figure 7. Chart for calculation of friction losses in landfill gas piping.

(mm) (see Fig. 7); u is the flow velocity; A is the wet area (m²); and P is the wet perimeter (m). The loss values in the chart are obtained for an LFG composition of 55% CH_4 and 45% CO_2, a temperature of 30 °C and a kinematic viscosity $v = 12.1 \times 10^{-2}$ cm²/s.

A value of 50 μm for the roughness value may be used for HDPE straight pipes with slight bends (including losses for the non-perfect alignment of joints).

COLLECTOR PIPE STATION

It is common practice to connect each single well to an individual transportation pipe. The ends of these single pipes (HDPE NP6)

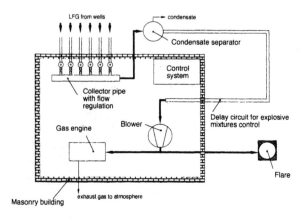

Figure 8. Schematic view of a gas end-use station.

are connected to a collector pipe placed inside a suitable sheltered small building. Here the pipes are supported by straps and are placed higher than the collector pipe, in order to discharge the condensation formed between the well and collector pipe (see Fig. 8). The pipes must be covered in order to avoid exposure to sunlight. The lines should enter the building by means of an opening in the wall equipped with a suitable gas link-seal (Fig. 8).

The following elements should be placed on every single pipe connected to the collector pipe, with the exception of special pieces such as flanges and pipe bends:

- fittings of $\frac{1}{4}$ in and $\frac{3}{4}$ in (6 and 19 mm) for sampling in order to detect the qualitative and quantitative characteristics of biogas (in particular the more important parameters such as methane content and flow inside the tube);
- a compensator (normally in stainless steel) for the slight movements of the pipe;
- a pressure regulator in order to maintain a constant flow and to remove possible pockets of condensation (optional);
- a butterfly valve (link seal) fitted with a scale and lever showing the degree of opening in order to adjust gas flow to suit each line separately. However, flow control is possible only to a certain extent.

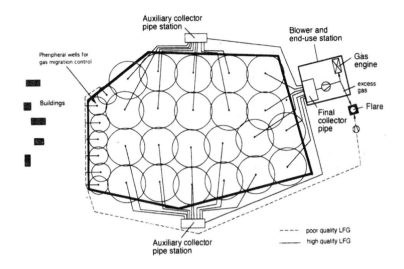

Figure 9. Schematic plan of a gas pipeline network.

The collector is normally made of cold-galvanized steel or HDPE with a maximum of eight to ten input lines. It must be placed with a slope of 1% in the direction of the condensate separator, in order to drain condensate away from the lines.

In the case of a complex network layout, additional collector pipes could be placed at the landfill boundary in order to minimize the length of transport lines (Fig. 9). Inside the buildings where the collectors are placed, a detector should be installed to reveal the presence of explosive mixtures caused by loss of biogas. This may be avoided if the building is open to atmosphere.

The use of two parallel collection pipelines placed outside the landfill area provides the possibility of separating gas flows from different wells according to the quality (percentage of CH_4, presence of halogenated or sulphonated compounds). In this way, LFG of poor quality could be directed to different end uses (Fig. 9). For example, in the case of gas utilization with electric power generation, this solution makes it possible to feed the engines with high-quality gas, while the poor-quality gas can be flared.

REFERENCES

Anon (1994). 'Positive displacement rotary blowers', *Robuschi SpA, Catalogue* Parma, Italy.

Coker, A. K. (1994). 'Selecting and sizing process compressors.' *Hydrocarbon processing*, 73(7), 39–47.

Raddatz, W. (1988). 'Deponieentgasung mit horizontalen Systemen auf der Deponie der Stadt Hannover'. *Deponiegasnutzung*, Economica Verlag, 255–282.

Rettenberger, G. (1988). 'Planerische, bauliche und sicherheitstechnische Gesichtpunkte bei der Erstellung von Entgasungsanlagen mit dem Ziel der Emissionsminimierung'. *Deponiegasnutzung*, Economica Verlag, 37–50.

5.6 Condensate from Landfill Gas: Production, Quality and Removal

RAFFAELLO COSSU[a] & MATHIAS REITER[b]

[a]*DIGITA – Department of Geoengineering and Environmental Technologies, Faculty of Engineering, University of Cagliari, Piazza d'Armi, 09123 Cagliari, Italy*
[b]*CISA – Environmental Sanitary Engineering Centre, Via Marengo 34, 09123 Cagliari, Italy*

INTRODUCTION

Landfill gas is characterized by high relative humidity, near that of the saturation point. This leads to the formation of significant quantities of condensate within the gas network, causing a series of problems that affect the functionability and durability over time of the gas extraction and utilization plant.

The principal cause of condensate formation is related to atmospheric conditions (temperature), which can be seen especially in the winter months. The relatively warm gas generated within the landfill body (where temperatures are on average around 30 °C with peaks above 50 °C) rises in the well to the top, where it comes into contact with lower-temperature bodies (pipes, compensators, valves, etc.), on the surface of which condensate forms. Other lesser factors that may influence condensate formation are the pressure and composition of the gas.

When the condensate accumulates in the lowest points of the gas network, a collection point of liquid is formed, which can create a true syphon. In this manner an obstacle to the gas passage is formed (registered as a pulse in the flux) with the probability of a total

Landfilling of Waste: Biogas. Edited by T. H Christensen, R. Cossu and R. Stegmann. Published in 1996 by E & FN Spon, London. ISBN 0 419 19400 2.

blockage of the flux and the deactivation of one or more of the gas transport pipelines.

This phenomenon, predictable depending on the layout of the gas network, is also present as an unavoidable eventuality related to the phenomena of bulk settling of the landfill and associated differential settlements. The elimination of condensate syphons can be made more difficult because the polyethylene lines must be buried to protect them from temperature variations and exposure to the sun. In the most difficult situations, a complete substitution of the affected lines is necessary.

The critical points of the network for the formation of condensate occur where transport line is connected to wellhead (here the gaseous mixture is first rapidly cooled) and where pipe discontinuities are found (reductions, enlargements, Ts, elbows, valves, collection devices and aspirators). In the latter, turbulence, cyclones, pressure and temperature gradients can be created, which favour condensate formation.

A further problem related to condensate is its elevated potential for corrosion of the normal materials used in a gas plant. Items such as valves, aspirators and measurement devices are particularly at risk unless they are made of materials specifically designed to resist the corrosive condensate.

Other problems may be found in the utilization phase of the gas. The combustion of a humid gas mixture, enriched with halogenated organic compounds, can lead to the formation of hydrochloric and hydrofluoric acids. When considering gas motors, the possible transport of these corrosive substances through the lubrification oil could affect all the mechanical parts of the system.

Without an efficient condensate separation system the functioning and durability of the entire gas plant may be compromised. Therefore condensate must be separated from the landfill gas and properly disposed of. It represents, because of its characteristics, an additional landfill emission.

THE CONDENSATE FORMATION PROCESS

The saturated landfill gas contains a large number of components (see Chapter 2.2). Apart from methane, carbon dioxide and water, one finds organic substances produced by biological degradation of the organic materials or volatile substances from the materials in the waste itself (paints, solvents, etc.) that enter the gas directly. Given their small quantity in comparison with the main LFG components, such substances cannot in any sense influence the formation processes of the condensate but may contribute to the characterization of its quality.

Therefore, as a simplification of the problem, one can consider the gaseous mixture to result from the anaerobic degradation of the waste and to be composed exclusively of methane, carbon dioxide and saturated vapour. The carbon dioxide and methane do not contribute to the formation of condensate but are present only as a dissolved gas, as can be seen by considering the critical points and vapour pressures for the two gases (see Chapter 2.4).

Therefore, at the end of the formation of the condensate, even if some trace components in the biogas are present in the liquid phase, one can refer exclusively to the liquid–vapour system of water. The liquid–vapour system of water is heterogeneous and monovariant according to the Gibbs phase rule, which means that when a single thermodynamic parameter (P, V, or T) is determined, the other remaining parameters that define the system can be calculated. Therefore, to determine the temperature, the value of the vapour partial pressure of the water in the mixture can be at most equal to the saturated vapour pressure corresponding to the given temperature.

A gas–vapour mixture can be identified through the relative humidity, defined from the relation between the partial pressure and the saturated vapour pressure. The absolute humidity is defined as the value in mass units of the water vapour contained in the mass of the dry gas mixture.

Considering that the specific weight of humid air is not significantly different from the specific weight of biogas, one can make a series of interesting observations about the formation of the condensate by examining the Mollier diagram, which displays the behaviour of air–vapour mixtures at a pressure of 1 atm. In the diagram (Fig. 1) enthalpy values through absolute humidity are represented. The axes are set up in a manner in which the isotherms are on a straight line with an angular coefficient slightly greater than zero. For every point the corresponding value of the vapour saturation pressure can be read.

Heating of the mixture does not cause any variation of the absolute humidity, while the relative humidity is reduced and the

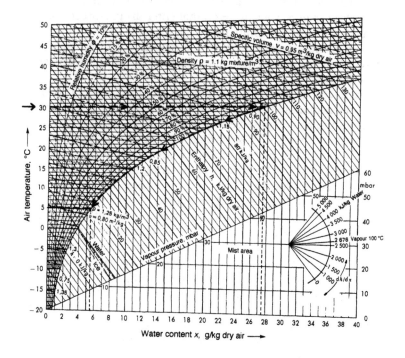

Figure 1. Mollier diagram, *h*, *x*, for humid air at 1 bar pressure (modified from Recknagel and Sprenger, 1981).

vapour pressure is increased correspondingly. In contrast, cooling of the gaseous mixture leads to an increase in the relative humidity with constant absolute humidity until the saturation curve. Further cooling follows the saturated vapour curve, with constant relative humidity (100%) and a decreasing absolute humidity. This second process leads to the formation of a determined quantity of liquid equal to the change in the absolute humidity.

For example, suppose that gas leaves the extraction well at a temperature of 30 °C and is in a condition of saturated vapour (relative humidity = 100%). At the corresponding point on the intersection of the isotherm with the saturation curve, one reads an absolute humidity of 27.5 g/kg equivalent to 35.4 g/Nm³, when taking into account a specific weight of the gas of 1.287 kg/Nm³. Consider an instantaneous cooling from 30 °C to 5 °C: following the saturation curve to meet the corresponding 5 °C isotherm, one can read an absolute humidity value of 5.5 g/kg, equivalent to 7.0 g/Nm³ (Fig. 1). This signifies that the cooling of the mixture forms a quantity of condensate equal to 22.0 g/kg or 28.3 g/m³. These values constitute the theoretical maximum on the basis of the approximations made.

In the landfill gas network, there is an absolute pressure less than atmospheric, and the Mollier diagram must be modified. On the basis of the relation between relative and absolute humidities, one can define the relation between external pressure and humidity as:

$$X = \frac{0.629 \ (U_r \ P_s)}{P - U_r \ P_s} \tag{1}$$

where X is the absolute humidity, U_r is the relative humidity, P_s is the saturated vapour pressure, P is the external pressure, and 0.629 is the ratio between the densities of the dry mix and the water vapour. Assuming that the temperature remains constant and the external pressure changes, one can determine a new value for the relative humidity. Considering no loss of mass, one can hypothesize that the absolute humidity remains constant, and therefore

$$X_1 = X_2$$

Combining the above with equation (1), one can determine the following relation:

$$\frac{U_{r1}}{P_1} = \frac{U_{r2}}{P_2} \tag{2}$$

From equation (2), one can see that the relative humidity is directly proportional to the external pressure. This means that by halving the pressure applied to a gas mixture saturated with water vapour ($U_r = 100\%$), a mixture with a relative humidity of 50% would result. To return to saturation level, the mixture is ready to absorb an equivalent quantity of water vapour supplementing its previous content.

One can apply this consideration to the real behaviour of the biogas network. For example, if within a feed line one creates an absolute pressure of 990 mbar, at atmospheric pressure the gas mixture would be able to absorb 27.0 g/kg of water vapour, or 34.7 g/Nm³. On the basis of this consideration the same saturated gas mixture is characterized by an absolute humidity of 28.2 g/kg, or 36.4 g/Nm³.

SYSTEMS FOR THE REDUCTION AND SEPARATION OF THE CONDENSATE

To eliminate the problems caused by condensate in the extraction and transportation of LFG, one must act to minimize the stagnation of the condensate inside the pipelines allowing for separation or at least natural draining-off of the condensate.

Reduction of the Presence of the Condensate in the Lines

If local topographical conditions permit, one can position the transport pipelines in a manner to direct all the condensate produced towards a low point, where a proper system of condensate separation can be installed. One must ensure, in any case, a slope of at least 5%, but even that can be inadequate in relation to the potential settlement that can occur in the landfill. The geomechanical behaviour of the landfill is in fact difficult to predict and can vary within a distance of a few metres.

A possible measure to adopt to increase the stability of the transport pipeline slope would be to install under the lines a support system (either of wood or another material) suitable to distribute the settlement load.

The dimensions of the LFG transport pipelines influence the related phenomena of the condensate. With a larger-diameter pipe, there is a reduction in the head loss, a greater rigidity and a reduced risk of pipe blockage by condensate. Small-diameter pipes, under the same conditions, carry the condensate at a higher flux velocity and as a consequence give an increased driving force to the liquid and solid particles, with less deposit on the internal surface.

To avoid changes in the pipeline geometry following settlement of the landfill body, it is convenient to place the transport pipelines on the surface of the landfill body until the greatest part of the settlement has been completed.

To control the presence of condensate within the lines, there exist automated systems to isolate and evacuate the condensate, mechanically regulated by a pressure stabilizer. Stabilization valves are installed in the gas network and calibrated for a specific pressure value. On the occurrence of a partial or total blockage of the line, caused by a pocket of liquid, one would see a variation of the pressure, and the valve would open completely to achieve the greatest pressure gradients. In such a manner, it is possible to aspirate the liquid blockage and free the line thereby stabilizing the normal operating conditions. Following this, the valve closes slowly to its initial position, regulating the gas flux according to the specified value. This stabilization valve system is also used to reduce the decompression that can be created in the landfill body following a reduction of the specific production of gas. However, when the settlement and therefore the reduction of the lines is considerable, this system can also prove to be inadequate.

Condensate Separators

Condensate separation systems can be defined as either active or passive.

Active separators are systems that make use of external energy to cool or compress the gaseous mixture to optimize the separation efficiency (Fig. 2). These systems are convenient when high qualitative standards are needed for the gas or when the length of the transport pipeline is such as to require an elevated pressure causing a large variation in the relative and absolute humidity.

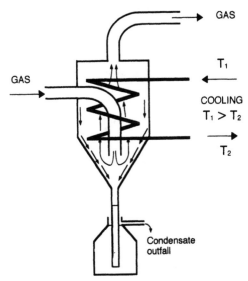

Figure 2. Active condensate separator with cooling system (modified from Damiani and Gandolla, 1992).

Passive separators use physical phenomena such as a variation in the local pressure and temperature, expanding of a section, generation of turbulence, or sharp inversion in the direction of the flux. There is no dehumidification of the mixture as such, but a separation of the liquid and solid particles along the transport pipelines. In fact the temperature and pressure variation does not reach levels that can influence relative humidity significantly. Among the different passive separators, the gravity and cyclone types are widely applied (Fig. 3 a–c). The separators are fitted with a hydraulic syphon made to avoid infiltration of air, and with different systems for condensate discharge (Fig. 3 d–f).

The efficiency of passive separators, according to different specific design solutions, can vary between 70 and 90% (with respect to the total amount of suspended particles).

Before the separator, it is advisable to insert a condensate

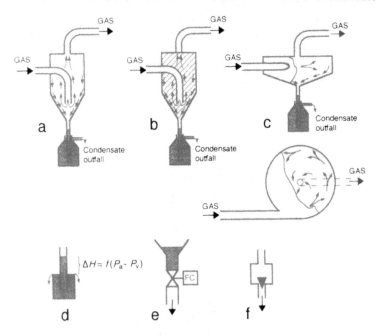

Figure 3. Different passive condensate separators and condensate outfall systems (modified from Damiani and Gandolla, 1992): (a) gravity separator; (b) gravity separator with packing; (c) cyclone separator; (d) syphon valve; (e) time-controlled valve; (f) floating valve; P_a = atmospheric pressure; P_v = vacuum.

discharge placed along the transport pipeline to avoid the formation of liquid suspensions that can bypass the separator.

The separators are generally installed within special wells, where the separated condensate also is collected. The well also houses the other accessory equipment, such as a submerged pump, level probe, and valvès. The wells containing the condensate separators are often enclosed in a confined environment under a system of passive aeration, which may not be sufficient in the case of a gas loss. It must therefore be treated as a risk zone and appropriate explosive gas probes should be used.

All parts in contact with the condensate require an appropriate choice of materials. In particular, this is important for the pump used to transfer the condensate from the well to the expected treatment phase.

An additional measure to take is to insert a series of condensate discharge points and valves in the connection lines at diverse points on the system and in particular in the connection between blower and flare and between blower and engines.

EXPERIMENTAL QUANTITATIVE DATA

The available experimental data regarding the specific production of condensate (expressed in mass per volume of extracted gas) are random in nature and incomplete with regard to the specification of the gas network (for example the slope towards the wells or towards the separators), the applied depression values (generally from -5 to -40 mbar), the meterological conditions of the possible superficial water infiltration (leachate), and the type of separation system used for the condensate.

In most cases, the specific production of condensate must not increase beyond a value of 25–26 g/Nm^3, corresponding to an increase in temperature from 30 °C to 15 °C (Martens, 1988; Sperl, 1991).

At the Huckarde landfill site (Germany) in a period of one month, nearly 20 500 l of condensate were collected; the quantity of extracted biogas in the same period equalled 1 622 590 Nm^3, with a specific production, therefore, of 13 g/Nm^3 (Schmidt, 1991).

In the Berlin-Wannsee landfill site, with a maximum capacity of the gas plant of 4500 Nm^3/h about 120 l/h of condensate were collected (Martens, 1988). In this plant, a double separation system was adopted, comprising a series of cyclones followed by an active system of refrigeration of the gaseous mixture.

In the Hannover central landfill site, in one year 798 m^3 of condensate was separated for an LFG quantity of 7.34×10^6 Nm^3. In the following year, 715 m^3 of condensate was separated, corresponding to 7.0×10^6 Nm^3 of extracted gas (Raddatz, 1988). These values correspond to specific production of 108 g/Nm^3 and 102 g/Nm^3 respectively. These values, at first glance, confirm the theoretical estimate calculated on the basis of the Mollier diagram. The temperature of the gas measured at the exit of the extraction well was, in fact, found to be between 60 °C and 55 °C. Corresponding to these values one can read on the saturation curve corresponding absolute humidity values. These are 116 and 154 g/kg respectively, or 149.3 and 198.2 g/Nm^3. At the collector pipe station, the temperature of the gaseous mixture varied throughout the year between 15 °C and 34 °C, with an average value of 25.4 °C in the first year and 23.7 °C in the second year. With these temperature values, one reads on the saturation curve of the Mollier diagram absolute humidity values of 16 and 20 g/kg. Taking into account a specific weight of the biogas of 1.287 kg/Nm^3 one obtains values of 20.6 and 25.8 g/Nm^3 respectively. Neglecting the effect of the depression on the humidity, one can calculate the following theoretical values of specific production of the condensate: 128.7 g/Nm^3 (first year) and 172.4 g/Nm^3 (second year).

It should be noted that the Hannover landfill utilizes a horizontal gas extraction system. This fact, together with the high levels of leachate found in the landfill, accounts for the greater formation of condensate.

In the Ravensburg landfill site, in one year, a total of 25 000 l of condensate were measured at the purifying plant corresponding to an average amount of 850 m^3/h (Hofstetter, 1988). From these values, one can calculate a specific production of almost 3.4 g/Nm^3.

QUALITATIVE CHARACTERISTICS OF THE CONDENSATE

The disposal of LFG condensate is a relatively new problem. For this reason, there is still not enough data to provide an overall characterization of condensate. The available data from the literature are reported in Table 1. As can be noted, the data are scarce and fragmented, and often the parameters that best characterize the condensate are not quantified: halogenated organic compounds, dissolved carbon dioxide and sulphur compounds for example. Moreover, information about the sampling method and corresponding parameters of the leachate and LFG is not reported. The sampling method is important because, as shown also by the data in Table 1, a contamination of condensate by normal leachate often occurs. Leachate can easily infiltrate the gas network, corresponding to high depression values and elevated velocity of the gas transport (> 10–12 m/s).

The chemical characteristics of condensate are very similar to those of normal leachate. In Fig. 4, the values of pH and the measured conductivity of the condensate are reported graphically for different landfills. The condensate generally is characterized by acidic pH value (pH < 7), while the pH of leachate coming from a landfill in a stable methane phase is usually higher.

In fact, there is negligible transport in the gaseous phase of salts such as carbonate, phosphate, borate and silicate, which normally buffer the leachate, while in the condensate many of the dissolved volatile substance present in LFG behave in an acidic manner in an aqueous solution. This is the case for several halogenated compounds present in the biogas (Thorneloe, 1991). The acidic character of the condensate is caused principally by the carbonic acid that is formed through the solubility of carbon dioxide. One cannot exclude a partial oxidation of the hydrogen sulphide with the formation of sulphur dioxide and sulphide, which leads to the formation of sulphuric acid in solution.

Condensates with basic pH values (pH > 7.5) are considered atypical, and are presumably caused by contact with the leachate or from elevated ammonia values in the biogas, which in solution can furnish a certain buffer power to the condensate.

The relative absence of salts in the 'pure' condensate is confirmed by the low specific conductivity values and by the low total solid content.

Organic substances (COD, BOD, TVA) show a strong variation depending above all on the separation efficiency of the leachate (Fig. 5). In general, the values associated with the pure condensate are lower than those of the older leachate. High concentrations (with respect to the corresponding leachate) of polycyclic aromatic hydrocarbons have been found in the condensate in some cases.

TABLE 1. Analytical Data for Landfill Gas Condensate Measured at Different Landfill Sites (A–N)[a]

Parameter	Unit	A	B	C	D	E	F	G	H	I	L	M	N
pH	—	6.3	6.35	7.6	6.3	8.2	7.14	5.6	8	5.2	8.3	8.7	8.7
COD	mg/l	870	—	—	135	15600	53	—	2010	570	6201	3000	7670
BOD	mg/l	490	—	—	11	130	6	—	1630	300	74	680	1020
TOC	mg/l	—	—	3000	—	—	—	—	—	—	—	—	—
TVA, as CH$_3$COOH	mg/l	—	—	—	3.6	81	0.5	—	—	1.5	—	170	313
N–NH$_4$	mg/l	<1	—	—	220	3414	34	—	4180	45	2699	7800	1010
N–NO$_3$	mg/l	—	—	—	—	—	—	—	1.4	4	—	—	—
SO$_4$	mg/l	222	—	—	—	—	—	—	—	—	—	—	—
S$^-$	mg/l	—	—	<0.05	3.2	11.2	9.6	—	—	—	—	0.4	—
Cl$^-$	mg/l	2.8	4.5	<0.1	10	567	4	—	—	15	—	1100	1200
F$^-$	mg/l	—	4.7	<0.1	—	—	—	0.04	—	—	—	—	—
Alkalinity, as CaCO$_3$	mg/l	—	—	—	—	—	—	—	—	—	—	—	—
Conductivity	µS/cm	1340	792	3300	1320	13220	190	28500	21500	251	—	28000	4200
Total solids	mg/l	—	—	—	—	—	—	—	—	193	4093	20	—
CO$_2$	mg/l	—	—	—	—	—	—	441	—	—	—	—	—
Ca	mg/l	—	—	—	—	—	—	—	—	19	—	—	—
Fe	mg/l	—	—	—	—	—	—	—	—	10.7	—	—	—
Si	mg/l	—	180	190	—	—	—	—	—	—	—	—	—
Zn	mg/l	—	—	—	0.92	0.15	0.18	0.47	—	2.32	0.58	—	—
Cu	µg/l	—	5.0	12.0	—	—	—	—	—	9.3	41.0	160	—
Cd	µg/l	—	—	—	0.3	0.2	<0.6	—	—	2	0.2	20.0	—
Hg	µg/l	—	—	—	1.0	1.0	1.0	<0.5	—	0.2	0.25	—	—
Pb	µg/l	—	—	—	5.0	1.4	8.0	—	—	5.1	46.0	40.0	50.0
As	µg/l	—	—	—	—	—	—	<1	—	—	—	—	—
Se	µg/l	—	—	—	—	—	—	<1	—	—	3350	—	100
Mg	µg/l	—	—	—	—	—	—	—	—	1550	—	—	—
Mn	µg/l	—	—	—	—	—	—	—	—	720	—	—	—

TABLE 1. Continued.

Parameter	Unit	A	B	C	D	E	F	G	H	I	L	M	N
Ni	μg/l	—	—	—	—	—	—	—	—	140	83.0	—	—
Cr	μg/l	—	—	—	—	—	—	—	—	3.0	97.0	50.0	—
AOX	mg/l	—	0.24	—	—	—	—	—	1.4	0.16	—	—	—
Trichloroethylene	μg/l	10.0	5.3	—	52.0	13.3	1.3	0.8	—	—	2.7	1.0	1.4
Tetrachloroethylene	μg/l	6.0	5.5	—	8.6	37.4	0.16	0.7	—	—	56.0	1.2	4.5
Trichloromethane	μg/l	—	<0.1	—	0.5	0.5	6.3	—	—	—	20.0	1.0	3.2
Tetrachloromethane	μg/l	—	<0.1	—	—	—	—	—	—	—	1.2	0.5	0.5
Trichloroethane	μg/l	—	1.4	—	—	—	—	—	—	—	2.5	1.0	1.5
Dichloroethane	μg/l	—	<0.5	—	—	—	—	1.1	—	—	1.7	0.8	1.0
Organic chlor	mg/l	16	—	—	—	—	—	—	—	—	—	—	—
Organic fluor	mg/l	<10	—	—	—	—	—	—	—	—	—	—	—
Toluene	mg/l	—	—	—	—	—	—	0.1	—	—	—	—	—
Benzene	mg/l	—	—	—	—	—	—	0.12	—	—	—	—	—
Xylene	mg/l	—	—	—	—	—	—	0.28	1.9	—	—	—	—
Hydrocarbons	mg/l	0.7	—	1.3	—	—	—	—	15	13.8	—	—	—

[a]References: Landfill A (Hofstetter. 1988); B,C (Cossu and Reiter, 1994); D,L,F (Bressi et al., 1991); G,H (Cossu and Reiter, 1994): I (Martens, 1988); L (Cossu and Reiter, 1991), M,N (Cossu and Reiter, 1994)

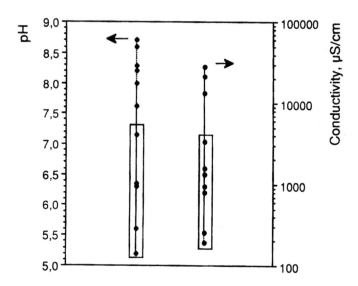

Figure 4. pH and conductivity values for landfill gas condensate measured at different landfill sites. The areas of typical values for condensates without leachate contamination are outlined. See Table 1.

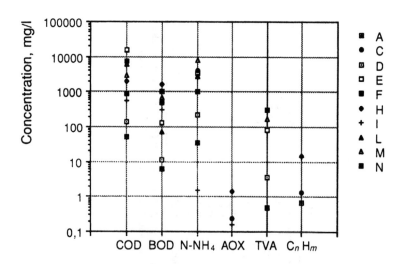

Figure 5. Concentration of some macroparameters measured for landfill gas condensate at different landfill sites (A–N). See Table 1.

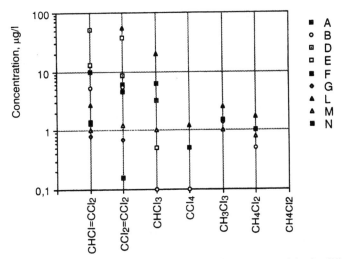

Figure 6. Concentration of halogenated hydrocarbons measured for landfill gas condensate at different landfill sites (A–N). See Table 1.

Several substances from LFG are found in the condensate in higher concentrations (Arendt, 1990). In particular, this is the case for halogenated compounds (Fig. 6), which are characterized by high boiling points and low vapour pressures at ambient temperature. These substances can be found dissolved or mixed in the condensate in more or less elevated quantities depending on their solubility in water.

Carbon tetrachloromethane ($P_s = 116$ mbar at 20 °C), trichloromethane ($P_s = 211$ mbar at 20 °C) and above all tetrachloroethylene (perchloroethylene), whose vapour pressure at 20 °C is only 18.6 mbar and which is characterized by a solubility of 0.4 g/l in water, are the halogenated compounds that can be found in large concentration in the condensate as dissolved gas or in liquid form mixed with condensate water. Many of these substances, coming from solvents, degreasers and paints, have corrosive characteristics in solution and are suspected carcinogens.

The heavy metals show concentrations comparable with that of leachate (Fig. 7). The high values encountered for arsenic in two cases are linked to specific characteristics of the corresponding leachate (Cossu and Reiter, 1994). The relatively high zinc values should be related to the constructive material of the condensate separators.

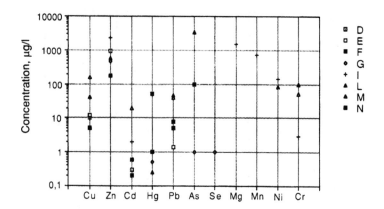

Figure 7. Concentration of heavy metals measured for landfill gas concentrate at different landfill sites (A–N). See Table 1.

Reported in Table 2 are the data that are believed to best represent the quantitative characteristics of the condensate, uncontaminated by the presence of leachate.

TABLE 2. Analytical Data that can be Considered Typical for a Condensate Without Leachate Contamination

Parameters	Unit	Value
pH	–	<7.0
COD	mg/l	<1000
BOD	mg/l	<500
N–NH$_4$	mg/l	20–40
Conductivity	μS/cm	<2000
AOX	mg/l	0.2–2.0

CONCLUSIONS

The condensate problem is strictly related to the design and performance of a LFG gas plant, and for that reason a series of precautionary measures must be considered to reduce the formation or at least neutralize the negative effects that humidity in the gas and condensate can produce.

The formation process of the condensate appears well defined in theoretical aspects; similarly the operational problems of condensate accumulation in the gas network have been well defined. However, there is a significant shortage of experimental data on the quantity of condensate collected and above all on its quality.

The latter aspect is important in defining a strategy for the treatment of these landfill emissions related to the environmental impact that they can create. In particular, the presence of condensates containing halogenated organic compounds and their influence on the treatment processes must be carefully considered. In the case of a recirculation of the condensate within the landfill, these compounds can have an inhibitive effect on the methane production (Poller, 1990).

To be able to make definitive conclusions on the qualitative and quantitative characteristics of the condensate, additional studies must be conducted, considering homogeneous parameters and considering also the leachate and LFG in the same location and time period.

Furthermore, the sampling method should always be specified as well as the waste and landfill characteristics, and the gas network and condensate separation systems should be described.

REFERENCES

Arendt, G. (1991). Schwerflüchtige Kohlenwasserstoffe im Deponiegas und Kondensat, In *Deponiegasnutzung* (eds G. Rettenberger and R. Stegmann), Trierer Berichte zur Abfallwirtschaft, Economica Verlag, Bonn.

Bressi, G., Dugnani, L., Baldi, M. and Gandolla, M. (1991). Analytical characterization of the biogas condensate, in *Sardinia '91*, Proceedings of the Third International Landfill Symposium, S. Margherita di Pula, Italy, pp. 249–55.

Cossu, R. and Reiter, M. (1991). Indagine sperimentale sulla captazione ed utilizzo del biogas presso lo scarico controllato di Serdiana. *Internal Report*, CISA, Environmental Sanitary Engineering Centre, Cagliari, Italy.

Cossu, R. and Reiter, M. (1994). Produzione di condense dal biogas: processi di formazione, metodi di rimozione, analisi qualitativa. *Technical Report, n. 4*; CISA, Environmental Sanitary Engineering Centre, Cagliari, Italy.

Damiani, A. and Gandolla, M. (1992). *Gestione del biogas da discariche controllate*, Monografia n. 8, Istituto per l'Ambiente, Milano.

Hofstetter, K. (1988). Deponiegasverwertung am Modell Ravensburg, in *Deponiegasnutzung*, (eds R. Stegmann and G. Rettenberger), Hamburger Berichte 1, Economica Verlag, Bonn, pp. 231–53.

Martens, J. (1988). Kondensat im Deponiegas-Problemstellung und Beseitigung, in *Deponiegasnutzung* (eds R. Stegmann and G. Rettenberger), Hamburger Berichte 1, Economica Verlag, Bonn, pp. 219–30.

Poller, T. (1990). *Hausmullbürtige LCKW/FCKW*, Hamburger Berichte 2, Abfallwirtschaft, Economica Verlag, Bonn.

Raddatz, W. (1988). Deponieentgasung mit horizontalen Systemen auf der Deponie der Stadt Hannover, in *Deponiegasnutzung* (eds R. Stegmann and G. Rettenberger), Hamburger Berichte 1, Economica Verlag, Bonn, pp. 255–82.

Recknagel, H. and Sprenger, E. (1981). *Taschenbuch für Heizung und Klimatechnik*, 61st edn, R. Oldenbourg, München.

Schmidt, C. (1991). Einspeisung von Deponiegas in eine Kokerei, in *Deponiegasnutzung*, (eds G. Rettenberger and R. Stegmann), Trierer Berichte zur Abfallwirtschaft, Economica Verlag, Bonn.

Sperl, J. (1991). Anforderungen an Deponiegasförderstationen, in *Deponiegasnutzung*, (eds G. Rettenberger and R. Stegmann), Trierer Berichte zur Abfallwirtschaft, Economica Verlag, Bonn.

Thorneloe, S. A. (1991). U.S. EPA's global climate change programme – landfill emissions and mitigation research, in *Sardinia 91*, Proceedings of the Third International Landfill Symposium, S. Margherita di Pula, Italy, pp. 51–67.

6. GAS TREATMENT

6.1 Combustion Processes and Emissions of Landfill Gas Flares

GERHARD RETTENBERGER[a] & WOLFGANG SCHREIER[b]

[a]*Fachhochschule Trier, Schneidershof, 54293 Trier, Germany*
[b]*Ingenieurgruppe RUK, Schockenriedstrasse 4, 70565 Stuttgart, Germany*

INTRODUCTION

The flaring of landfill gas (LFG) may give rise to several emissions. On the one hand these emissions depend on the quality of LFG, and on the other hand they are influenced by combustion techniques (shape of the combustion chamber, air supply, etc.). This chapter provides an overview of the possibilities of influencing flare emissions.

FUNDAMENTAL OF COMBUSTION PROCESSES

Suitability of Landfill Gas for Combustion

The suitability of LFG for use as fuel depends primarily on its methane content. Methane represents the main combustible component of LFG and has a high energy content. The most important physico-chemical and thermal parameters of methane are listed in Table 1 (see also Chapter 2.4).

Among the main LFG components, carbon dioxide (CO_2) and nitrogen (N_2) are not involved in the combustion process, while oxygen is the oxidizing agent. Trace components do not influence the combustion process, owing to their low concentration in LFG. Nevertheless, they are important with regard to the composition of

Landfilling of Waste: Biogas. Edited by T. H Christensen, R. Cossu and R. Stegmann. Published in 1996 by E & FN Spon, London. ISBN 0 419 19400 2.

TABLE 1. Physical Characteristics and Thermal Behaviour Parameters for Methane

Parameter	Value
Molar mass (kg/mol)	16.043
Density (kg/Nm^3)	0.7175
Upper burning value (MJ/Nm^3)	39.82
Lower burning value (MJ/Nm^3)	35.88
Upper heating value (MJ/kg)	55.50
Lower heating value (MJ/kg)	50.01
Ignition temperature[a] (°C)	645
Lowest ignition limit[b] (vol.%)	5.1
Upper ignition limit[b] (vol.%)	13.5

[a] In air at 0 °C.
[b] In air at 20 °C.

the exhaust gas in order to not exceed the limit standards of existing emission regulations.

Combustion Reactions and Thermodynamic Processes

The reactions that take place during the combustion process can be described by chemical equations showing the initial and final components of the reactions. The actual course of combustion is far more complex and consists of many individual steps. The combustion process of methane can be described by the following equation:

$$CH_4 + 2O_2 \rightarrow CO_2 + 2H_2O \tag{1}$$

To obtain complete oxidation of methane several reaction steps are necessary whereby H_2 and CO and H_2CO occur as stable semi-finished reaction products.

At the end of the thermal degradation reactions carbon monoxide (CO) is present as a semi-finished reaction product.

The further oxidation of the CO to CO_2 during the incineration process can be described by the following equations:

$$H_2O + M \rightarrow OH + H + M \tag{2}$$

$$CO + OH \rightarrow CO_2 + H \tag{3}$$

where M = all components involved in the combustion reaction.

Vapour generated from the fuel is used for the oxidation of the CO to CO_2.

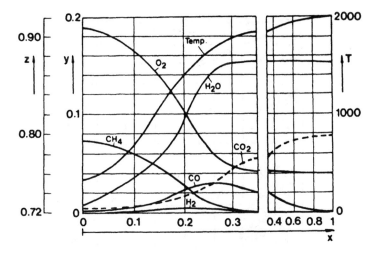

Figure 1. Temperature and gas composition along the flame after a pre-mixed gas phase during methane combustion: x = stream path of the flame; y = gas component (except O_2), molar fraction; z = oxygen, molar fraction; t = temperature (K).

Figure 1 illustrates the temperature and concentration course of the various stable molecule types during the combustion of CH_4 in a flame with a pre-mixed gas phase. The afterburning of the generated CO can be clearly observed. This process is called CO afterburning or second combustion phase. With gas-type fuels the CO conversion kinetics determine the rate of the complete process. If the flame suddenly cools down, which happens for example if the flame leads directly from the combustion chamber into the atmosphere, the CO to CO_2 reaction is interrupted and the CO emissions increase.

For a better understanding of the different steps of the process a combustion chamber can be ideally divided in two zones: the flame zone and the conversion zone. In the flame zone the fuel and air necessary for the oxidation are intimately mixed. In this zone highly reactive radicals, which are generated in chain and branch reactions, are present at a high concentration. Processes such as mixing turbulence cannot be easily described by mathematical models.

Thus the behaviour of the substances during thermal processes is estimated by referring mainly to the conversion zone. In this zone, however, besides the stable products from combustion, the highly reactive radicals are present at a much lower concentration than within the flame zone.

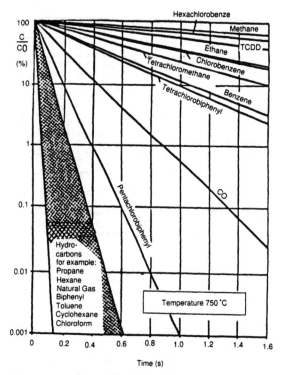

Figure 2. Thermal stability of different substances, in air, at 750 °C: C_O = initial concentration; C = concentration at time t.

As thermal conversion usually occurs under a surplus of oxygen (oxygen supersaturation), the general reaction regarding substance A can be described as follows:

$$aA + bO_2 \rightarrow \text{Products} \tag{4}$$

By this method, with the help of empirical data and considering thermodynamic aspects, the relationships between reaction time, temperature and residual concentrations of several substances have been obtained. The residual specific concentration C/C_0 (the actual concentration C at time t is related to the initial concentration C_0) of several substances is reported in Fig. 2 as a function of the reaction time t at a temperature of 750 °C, with a surplus of air.

From observation of the graphs in Fig. 2 the following comments may be made.

- All saturated hydrocarbons are unstabilized when heated and are reduced to carbon and hydrogen.
- As the temperature increases, the stability of the paraffins decreases faster than that of olefins.
- Methane is more stable than the other substances.
- CO, the classic product of an incomplete combustion process, is more stable than most hydrocarbons.

Although from the thermodynamic point of view a complete conversion of combustible compounds is possible with sufficient temperatures and reaction times, no conclusions can be drawn concerning the intermediate reaction products, reaction rate and the mechanisms of the reaction. These processes are, as described above, so complicated that it is not yet possible to describe them completely. Therefore, if unburnt fuel or other intermediate products are emitted from firing units, the reason for these emissions has to be looked at in terms of macroscopic oxygen deficit, bad fuel–air mixture or fast cooling of the substances involved in the reaction below the reaction temperature.

Air Requirement and Burnout Behaviour during the Incineration of LFG

Equation (1) can be used to determine the minimum air requirement for LFG combustion.

From this equation it can be seen stoichiometrically that 2 kmol of O_2 is necessary for the oxidation of 1 kmol of CH_4. The oxygen required is usually taken from the atmospheric air (assuming 21 vol.% oxygen and 79 vol.% nitrogen).

To obtain the combustion of landfill gas with a methane concentration of 55 vol.% the minimum air demand (A_{min}) can be calculated as follows:

$$A_{min} = \frac{O_{2min}}{0.21} = \frac{V_{LFG}\, X_{CH_4}\, 2}{0.21}$$

where A_{min} is the minimum air demand (Nm^3 air/Nm^3 gas), $O_{2\,min}$ is the minimum oxygen demand ($Nm_3 O_2/Nm^3$ gas), V_{LFG} is the volume of LFG (Nm^3), and X_{CH_4} is the fraction of CH_4 in landfill gas. Thus for the combustion of 1 Nm^3 of LFG, with 55% methane content, the minimum air requirement is 5.24 Nm^3.

To ensure a complete conversion of all combustible substances the processes should take place with a higher air supply than theoretically necessary. The ratio between the air that is actually needed to reach a complete incineration (A) and the theoretically required minimum amount of air (A_{min}) is called lambda (λ) and is characterized by the equation:

$$\lambda = \frac{A}{A_{min}} \tag{5}$$

The surplus of air should be optimized in such a way that the oxidation of the combustible components can be carried out up to the highest oxidation level (optimal combustion). If combustion is not complete, increased CO values appear in the exhausted gas. An increase in air volume is reflected in an increase of exhaust gas volume and a decrease of the flame and fuel temperatures. The highest flame temperature is obtained in conditions of complete combustion with $\lambda = 1$. To achieve the combustion of landfill gas with a minimum emission of noxious materials a homogeneous minimum temperature of 800 °C should be maintained in the flame. Furthermore, the exhaust gas should stay within the 800 °C zone for at least 0.3 s.

The higher the combustion temperature and the longer the detention time within this temperature (800 °C) under sufficient surplus of air, the more complete is the combustion and the grade of conversion of the organic substances.

Furthermore, it is desirable to maintain the required combustion conditions during all levels of operation even when variation of the amount and composition of the gas occur. Sufficient detention time and mixture intensity have to be properly considered in the flare design.

The combustion behaviour of the gas flare can be controlled and the incineration process can be optimized by controlling the combustion temperature and CO emission and, eventually, by adjusting the combustion air stream. Combustion in gas flares is usually carried out with $\lambda = 1.6$–1.8. This results in an oxygen content in the exhaust gas of about 8–9 vol.% O_2.

With lower air supply, the oxygen is insufficient for complete combustion. With higher air supply the flare may cool excessively and the combustion process may stop suddenly. In both cases the exhaust gases are characterized by high CO content.

EMISSIONS FROM LFG COMBUSTION

Limit Values for Emissions according to the German TA-Luft Regulations

According to the TA-Luft Regulations (27 February 1986) no specific limit values are provided for installations in which LFG is burned.

Regulations 3.1.6 and 3.1.7 deal with emissions in general. Only regulation 3.3.1.2.3 refers directly to the emissions from combustion processes. Moreover, in regulations 3.1.6 and 3.1.7 limit values for substance specific mass streams are given. If these values are exceeded the limit values for mass concentrations are applied. In most cases, the emissions from LFG combustion do not reach the limit mass-stream values given in regulations 3.1.6. and 3.1.7.

The TA-Luft aims to limit emissions as far as possible, according to the latest technological developments. It can be expected therefore that in the future flare installations will have to meet the standards given in Table 2.

Potential Emissions from LFG Combustion

The quality of LFG is important not only in the technical operation of the combustion facilities but also for the composition of the exhaust gas.

TABLE 2. Limit Values for Combustion Emissions According to the German TA-Luft regulations[a]

Reference substance	Concentration mg/Nm^3
Dust	5
Carbon monoxide (CO)	100
Nitrogen oxides ($NO + NO_2$ as NO_2)	200
Sulphur oxides ($SO_2 + SO_3$ as SO_2)	500
Hydrofluoric acid (HF)	5
Hydrochloric acid (HCl)	30
Organic substances (class I)	20
Organic substances (class II)	100
Organic substances (class III)	150

[a] The minimum oxygen content in the exhausted gas should exceed 3 vol.%. Organic substances are classified into classes I–III according to the hazardous and toxic potential of the various substances.

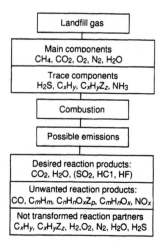

Figure 3. Classification of landfill gas components and potential emission from combustion process.

Problems with emission of noxious substances are caused not by the main components but by the trace components that are present in LFG. For this reason gas flares should also carry out a complete combustion of the trace elements, so that only CO_2, HCl, HF, SO_2 and H_2O should be found in the exhaust gas. Because of the very complex reactions that take place during the combustion, it is not possible to convert all the LFG components totally into the desired reaction products. Thus the exhaust gases will invariably contain traces of unwanted products or incompletely transformed reaction components. Figure 3 lists all possible emissions that might be caused by landfill gas combustion.

The possible emissions have been classified as follows:

- *Desired reaction products:* all products that are generated by the complete incineration of the fuel. The brackets around the substances SO_2, HCl and HF in Fig. 3 indicate that these products are desired from thermal point of view.
- *Undesired reaction products:* all products from incomplete incineration processes or from unwanted oxidation processes (for example NO_x).
- *Not converted substances:* residual fuel from incomplete combustion.

Emissions Connected to Fuel Consumption

The content of sulphur, chlorine and fluorine compounds in the exhaust gas, even in the case of complete and optimized combustion, depends entirely on the quality of landfill gas.

This relationship can be shown by the overall equation for the oxidation of halogen and sulphur compounds present in LFG:

$$C_xH_yZ_z + [x + \frac{1}{4}(y\text{-}z)]\,O_2 \rightarrow x\,CO_2 + z\,HZ + \frac{1}{2}(y\text{-}z)\,H_2O \quad (6)$$

(Z = sulphur, chlorine or fluorine)

$$2H_2S + 3O_2 \rightarrow 2SO_2 + H_2O \quad (7)$$

Thus the chlorine and fluorine components are transformed into HCl and HF. Hydrogen sulphide, present in landfill gas, is oxidized to SO_2.

Based on the limit values provided in the TA-Luft Regulations the minimum content of Cl, F and S in landfill gas can be calculated so that fixed emission values may be adhered to. For a raw landfill gas with 50% methane, 40% CO_2 and 10% N_2 the following maximum content of Cl, F and S can be accepted:

Cl	160 mg/m^3
F	25 mg/m^3
S	1400 mg/m^3

If a high concentration of these substances is present in the gas, purification treatment should be applied in order to fulfil the emission standard.

Emissions Dependent on Combustion Technologies

CO and NO concentrations observed in the exhaust gas are largely influenced by the combustion technology and particularly by the conditions under which the process takes place.

Parameters such as temperature, reaction time and sufficient oxygen supply, which influence CO and NO_x emissions, can be affected by the technical design of the flare. Increased CO concentrations in the exhausted gas reflect incomplete combustion of the landfill gas. As previously mentioned, where the technical dimension of the flare is sufficient this phenomenon could be related to an inadequate air supply (too high or too low) or to an insufficient mixture of the reaction components (landfill gas/air).

CO measurements in the flared gas are suitable for surveying the behaviour of the installation. Furthermore, the combustion process can be optimized by increasing or lowering the air supply. Investigations at different flare installations showed that when CO concentration was lower than 100 mg/Nm^3, all other critical substances dependent on the combustion process met the required level.

Thus it could be advisable to select CO as the control parameter for investigation of flare emissions and for optimization of the flare operation.

NO$_x$ emissions from LFG flaring units are mainly caused by so-called thermal NO$_x$ generation: this is a combination of the combustion temperature, the detention period and the oxygen concentration.

NO$_x$ emissions are generated at temperatures above 1000 °C and increase at temperatures above 1300 °C. At 2000 °C the NO$_x$ generation is at its maximum. High oxygen supply supports the generation of NO$_x$. The generation of NO$_x$ emissions is controlled by the air supply. At high air supply the flame temperature lowers and thus NO$_x$ generation decreases. The relationship between air supply, flare temperature and NO$_x$ emissions is shown in Fig. 4.

Dioxin-Related Problem

If compounds containing chlorine are present in the flare the possibility of the generation of polycyclic hydrocarbons in the exhaust gas has to be avoided. Polychlorinated dibenzodioxins (PCDD) and polychlorinated dibenzofurans (PCDF) are generated as follows.

- PCDD and PCDF are present in the fuel and are not converted during incineration.
- PCDD and PCDF are generated from the fuel and/or from chlorinated hydrocarbons (so called precursors, such as benzene, phenol and polychlorinated biphenyls) present in the combustion gases.
- PCDD and PCDF are generated by *ex novo* synthesis from other non-chlorinated organic substances.

Incineration processes are extremely complex. As yet, not all single steps of these processes can be completely described. Under

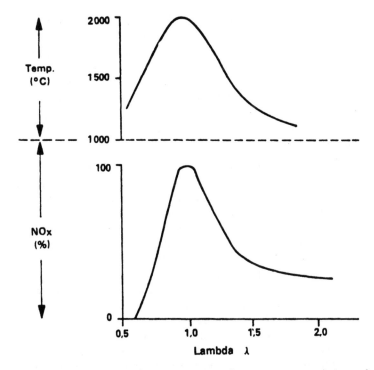

Figure 4. Relationship between NO$_x$ emissions, flare temperature and air supply (lambda values).

optimum conditions these processes lead to the thermodynamically more stable final products CO_2 and H_2O. Through pyrolytic processes further incineration products can be generated, especially aromatic hydrocarbons such as benzene, toluene and xylene, stable polycyclic aromatic hydrocarbons (PAH), etc. Laboratory experiments (Merz *et al.*, 1986) have shown that at low combustion temperatures (about 600 °C) substances chemically related to the initial components in the fuel, other than the main combustion products CO_2, H_2O and CO, could be detected.

On the other hand, at higher combustion temperatures (about 950 °C) more or less the same trace components appear even when chemicals or natural products such as wood or wool are used as fuel. From these results it might be concluded that the generation of PCDD and PCDF (such as PAH) takes place in so-called 'thermodynamic niches'. Up to now, however, it has not been clearly determined how the generation of PCDD and PCDF occurs, or by which steps chlorine or oxygen are introduced into the molecules.

On the basis of data available, it is currently assumed that (for the generation of PCDD and PCDF) several mechanisms exist. Up to a certain point, these generation mechanisms run parallel and, depending on the combustion process, lead to different amounts of PCDD and PCDF. Thus, it is assumed that at least four possible paths exist for the synthesis of PCDD and PCDF.

- *Direct preliminary stage*: It has been proven that direct preliminary stages, such as chlorinated phenols, polychlorinated biphenyls or chlorinated diphenylether used during combustion processes, could lead to the generation of PCDD and PCDF. Laboratory investigations carried out over the last few years, however, have shown that these preliminary stages are not necessary for the generation of PCDD and PCDF (Hutzinger and Fiedler, 1987).

- *Aromatization of chlorinated organic compounds*: By this process polychlorinated aromates are generated and PCDD and PCDF are produced by pyrolytic processes using chlorinated organic compounds as basic material.

- *Pyrolysis and chlorination of natural aromates*: Generation of preliminary stages of aromates from hydrocarbon compounds through pyrolytic processes during incineration, via substitution of the hydrogen atoms by chlorine atoms obtained from the combustion of chlorine containing substances.

- *Construction of PCDD/PCDF from C_2 units*: Generation of C_2 units ($C=C$) through pyrolytic processes of organic compounds and subsequent formation of aromatic rings using the C_2 units and chlorination by substitution reactions.

The above-described mechanism for generation and decomposition (reactions that occur in the gas phase during the incineration process), as well as the possibility of reactions with solid phases (dust particles), have been investigated in recent years. These studies were based on the presumption that dioxins and furans might be generated in the flue ash of filters or incineration facilities. First investigations (Vogg and Stieglitz, 1986) showed that the dioxin concentration in the flue ash for 2 h at 200 °C does not vary. At 300 °C the dioxin content increases by about 10% whereas at 600 °C a decomposition of the dioxins occurs.

Further investigations in this field finally led to the explanation of these processes with the help of the *ex novo* synthesis and to the knowledge that these processes are determined by the generation and destruction (dechlorination) of the PCDD and PCDF at a certain temperature range. The generation of PCDD and PCDF by the *ex novo* synthesis takes place at temperatures above 250 °C by using organic carbon in the presence of a chlorine source.

Through combustion of LFG in gas flares only gas phase reactions occur. Therefore, the generation of dioxines by *ex novo* synthesis is not likely to occur. Several investigations concerning off-gas from different LFG flaring units operated at optimum combustion have revealed no significant emissions of dioxines.

MEASUREMENTS AND OPTIMIZATION OF FLARE OPERATION

As described earlier, the performance of a flare unit can in principle be controlled by monitoring the CO emissions in the off-gas.

Table 3 summarizes several operation tests carried out by the authors on flares. The measurements were carried out on usual types of flares produced by different manufacturers.

For test A, a conventional low-temperature (LT) flare with internal combustion was investigated. The sides of the flare were not isolated: therefore there was a steep temperature gradient (up to 300 °C) from the centre to the side of the flare. As the combustion air could not be further reduced an optimization of incineration was not possible.

In test B, using raw LFG of the same quality as that used for test A, the performance of both a medium-temperature (MT) flare with internal combustion and isolated sides and that of a high-temperature (HT) flare was investigated. Both flares were equipped with installations for temperature measurement in order to adjust the air supply of the flare automatically. The first series of measurements carried out on the MT flare showed that the air surplus was too low, which also led to high temperatures in the off-gas. By changing the position of the temperature sensor and adjustment of the temperature set point, the performance of the flare has been optimized. Moreover, when operating on the HT flare at a temperature level 100 °C higher than the previous one a good thermal behaviour was observed.

TABLE 3. Performance of Various LFG Flaring Units during Different Operation tests

Operation test	Type of flare[a]	Raw LFG		Utilization level	Off gas			Distance to the burner nozzle
		CH_4 (vol%)	O_2 (vol%)	(%)	T (°C)	O_2 (vol%)	CO_2 (mg/m³)	(m)
A	LT	44	1.4	58.1	406	17.4	2001	4.3
		47	0.2	62.2	417	17.9	1614	4.3
B	MT	36	0.5	43.1	1046	6.7	106	3.8
	MT (optimized)	36	0.5	43.1	953	9.3	49	3.8
	HT	38	0.4	44.7	1058	7.8	74	3.8
C	LT with extended flare hat	37	1.3	64.5	980	6.8	>6000	3.0
		34	1.8	59.7	810	12.5	1850	3.0
		34	1.8	59.3	908	10.5	8570	3.0
D	HT	56	0	31.3	979	6.3	196	4.8
	HT (optimized)	56	0	31.3	958	10.3	100	4.8

[a]LT, MT, HT = low– medium– and high–temperature respectively.

Test C was carried out on an LT flare with an extended flare cap. In this test, unlike test A, the air supply could be regulated. Nevertheless, the thermal performance could not be optimized to an extent where the CO emission standard of 100 mg/m^3 could be guaranteed.

During test D the performance of the HT flare was optimized by changing the preadjusted temperature value, although with regard to combustion technology its utility was scarce.

All measurements were carried out by a portable instrument for the analysis of flue gas. The flue gas to be analysed was sucked up by a special probe and conducted to the analyser instrument. Thereafter the flue gas was purified by a condensation trap and filter. With the help of chemical measuring cells the parameters CO and O_2 were analysed in the flue gas. At the same time the off-gas temperature was determined by a NiCr–Ni thermocouple installed within the sampling probe.

CONCLUSIONS

In order to guarantee an environmentally safe disposal of landfill gas, in the future flare installations will still be necessary for the combustion of landfill gas.

Even if the landfill gas is used for energy production, gas flares are required in the case of breakdowns of the landfill gas utilization facility, to carry out thermic deodorization of the input of landfill gas. The composition of emissions generated by LFG combustion depends not only on the quality of the gas but also on the technical dimensioning of the flare installation. Owing to unsatisfactory combustion, flares with flames burning directly in the atmosphere should no longer be used for continuous operation. Investigations show that complete combustion cannot be achieved using these flare types, therefore resulting in high CO emissions. The conventional landfill gas flares, which have been further developed to a high technical standard, equipped with an extended (depending on the manufacture) partly or completely isolated flare cap and internal combustion within a defined combustion chamber, guarantee satisfactory performance.

Using a completely isolated flare cap, temperature distribution is more homogeneous and therefore avoids so-called 'cold air zones' etc., generating favourable combustion conditions. In order to adjust and optimize the thermal behaviour these flares should be equipped with an adjustable automatic air supply system. Air supply could be provided either by injection or forced air supply. If LFG to be treated contains high chlorine and fluorine concentrations, high-temperature flares should be installed because of the potential risk of dioxin generation. At regular intervals monitoring of the CO emissions in the flared gas should be carried out. According to measurement results, adjustment works should be carried out to optimize combustion. These measurements can be easily carried out using a portable instrument for flue gas analysis.

REFERENCES

Engel, H. and Rettenberger, G. (1988). New information regarding thermal treatment of LFG, in *Stuttgarter Berichte zur Abfallwirtschaft*, Bd. 29, *Zeitgemässe Deponietechnik II*, Erich Schmidt Verlag, Bielefeld, 267–319.

Hutzinger, O. and Fiedler, H. (1987). Development and existence of PCDD and PCDF in *VDI Berichte*, Bd, 634, 17–35.

Menz, W. *et al.*, (1986). Ein Verfahren zur Erzengang und analytischen Charakterisierung von Brandgasen, Fresenius Z. *Anal. Chem.*, 325, 449–460.

Vogg, H. and Stieglitz, L. (1986). Thermal behaviour of PCDD/PCDF in fly ash from municipal incinerators. *Chemosphere*, 15, 1373–1378.

6.2 Performance of Landfill Gas Flaring Systems

GRANT BALDWIN & PHIL E. SCOTT

Site Remediation and Waste Management,
National Environmental Technology Centre, Culham,
Abingdon, Oxfordshire OX14 3DB, UK

INTRODUCTION

Most of the work discussed in this chapter is related to experimental work undertaken on three different landfill gas (LFG) systems.

The first system investigated, Flare A, was a ground flare or 'flare pit', and consisted of 36 nozzles or jets arranged in three rows of 12. The second system investigated, Flare B, consisted of a ground base flare cluster in which the burner heads were covered by a series of ceramic tiles. The whole flare unit was enclosed in a large shield approximately 4 m high and 3 m in diameter and was supported on legs. This design results in a powerful updraft through the flare while the unit is in operation. The third unit studied, Flare C, was a commonly used, portable flare unit. The flare consisted of a single burner, mounted on a stack approximately 4 m high, supported on a tripod.

SAMPLING AND ANALYSIS

Sampling and analytical protocols capable of resolving the detailed composition of LFG have been developed and progressively refined throughout the period of the research programme. The analytical method involves the application of gas chromatography in

Landfilling of Waste: Biogas. Edited by T. H Christensen, R. Cossu and R. Stegmann. Published in 1996 by E & FN Spon, London. ISBN 0 419 19400 2.

combination with a variety of detectors, principally mass spectrometry. These have been specifically developed for this programme both at Harwell and in collaboration with Strathclyde Regional Council.

The approach adopted is considered to be 'state of the art' technology for quantifying the diverse range of trace organic compounds present in LFG.

A specialized probe suitable for use in high-temperature environments was employed to recover samples from the combustion zone of the three flare systems. This probe comprised variable lengths of silica glass tubing connected at one end to an inverted silica funnel (maximum diameter 12.5 cm), which served as the gas inlet to the probe. The downstream section of silica glass tubing was enclosed in a protective sheath of galvanized steel, which could be effectively supported in a purpose-built frame. The rear of the high temperature probe could be secured to the inlet of the standard sampling apparatus described elsewhere. Temperatures in the combustion zone above the flares were measured periodically throughout the course of gas sampling. This was carried out using a temperature lance comprising a NiCr/NiAl thermocouple fitted with a platinum resistance sensor. All vulnerable materials were enclosed in a galvanized steel sheath which also allowed the lance to be positioned manually. During sampling, all measurements were taken in close proximity to the inlet of the gas-sampling probe.

In addition to the measurement of trace components, measurements of the bulk components (N_2, O_2, CO_2, CH_4, H_2) were made throughout the programme.

TABLE 1. Concentration of Bulk Components in Raw and Flared Landfill Gas

	Average percentage concentrations (%)					
	Flare A		*Flare B*		*Flare C*	
Component	*Landfill gas before flaring*	*Flared gas as measured*	*Landfill gas before flaring*	*Flared gas as measured*	*Landfill gas before flaring*	*Flared gas as measured*
Carbon dioxide	32	3.5	20	2.9	21	15
Carbon monoxide	ND	ND	ND	ND	–	1.7
Hydrogen	ND	ND	ND	ND	–	1.1
Methane	51	ND	30	4.8	30	16
Oxygen	2.3	16	8.4	18	8.1	8.5
Nitrogen	15	80	41	75	41	59

The different designs of the three individual flare systems studied were such that it was not possible to sample flared gases from exactly corresponding points in each system. In each case, however, samples of input LFG were collected from a point as near to the flare burner as practical. Usually, this was a sampling point already installed in the system.

The sampling of input gases was coordinated as closely as possible with the sampling of the flared gases from each system. Thus the effect of variations in gas quality over time was minimized.

RESULTS

Bulk Component Analyses

Samples of LFG taken for the determination of bulk components (H_2, O_2, N_2, CH_4, CO_2, CO) were recovered from two locations on each flare system: the inlet gas line to each flare; and the combustion zone above the burner heads of each flare.

Flare A. Results from the analysis of samples for bulk components are presented in Table 1. The gas being utilized by this flare was surplus to the requirements of the on-site gas utilization scheme. This is reflected in the results of the bulk component analysis, which indicates an average methane concentration of 51%. There is evidence of air ingress (possibly through overpumping) into the gas abstraction system, as indicated by an average nitrogen concentration of 15.2% and an average oxygen concentration of 2.3%. There is a correspondingly lower than expected level of CO_2.

The results from bulk component analysis of the samples recovered from the combustion zone immediately above Flare A are also given in Table 1. The results suggest that all the methane present had been combusted to give rise to carbon dioxide, as no measurable methane was present in the off-gases. However, it is evident from the level of nitrogen and oxygen present in the sample of flared gas that a substantial dilution of off-gases with air was occurring during sampling. The dilution may have reduced low concentrations of (residual) methane and carbon monoxide to below the detection limit of the analytical method used. This view is supported by the results of the minor component analysis (where generally much lower detection limits are achievable), which indicate that some

TABLE 2. Generic Trace Component Groups Identified in Samples from Three Flares Studied

| | Total concentration (mg/m³) and as % of total trace components | | | | | |
| | Flare A | | Flare B | | Flare C | |
Generic group	Landfill gas before flaring	Flared gas adjusted for dilution	Landfill gas before flaring	Flared gas adjusted for dilution	Landfill gas before flaring	Flared gas adjusted for dilution
Alkanes (mg/m³)	920	0	370	39	510	1.2
%	21	0	34	21	41	2.5
Alkenes (mg/m³)	400	0	170	1.1	230	18
%	9.1	0	16	0.59	18	38
Alcohols (mg/m³)	180	2.1	1.3	55	18	6.1
%	4.1	6.4	0.12	29	0.59	13
Amines (mg/m³)	0	0	0	0	0	0
%	0	0	0	0	0	0
Aromatic hydrocarbons (mg/m³)	1600	6.1	380	58	350	2.6
%	36	19	35	31	29	5.3
Alkynes (mg/m³)	0	0	0	0	0	3.9
%	0	0	0	0	0	8
Cycloalkanes (mg/m³)	43	0	0	0	5.1	0
%	0.98	0	0	0	0.41	0
Carboxylic acids (mg/m³)	0.8	0	10	0	0	0
%	0.018	0	0.96	0	0	0
Cycloalkenes (mg/m³)	530	0	120	5.6	79	0.96
%	12	0	11	3	6.4	2
Dienes (mg/m³)	1.7	0	0	0.14	10	0
%	0.038	0	0	0.075	0.81	0

TABLE 2. *continued*

	Flare A		Flare B		Flare C	
	Total concentration (mg/m³) and as % of total trace components					
Generic group	*Landfill gas before flaring*	*Flared gas adjusted for dilution*	*Landfill gas before flaring*	*Flared gas adjusted for dilution*	*Landfill gas before flaring*	*Flared gas adjusted for dilution*
Esters (mg/m³)	290	0	0.2	0	0.5	0
%	6.6	0	0.018	0	0.041	0
Ethers (mg/m³)	1.8	0	0.4	0	0.2	0.32
%	0.041	0	0.037	0	0.016	0.66
Halogenated organic compounds (mg/m³)	320	1.9	32	18	39	3
%	7.3	5.8	3	9.6	3.2	6.1
Ketones (mg/m³)	120	1.6	2.9	2.2	1	0.43
%	3.7	5.0	0.27	1.2	0.081	0.89
Other nitrogenated compounds (mg/m³)	0	0	0	2.5	0	0.29
%	0	0	0	1.3	0	0.6
Other oxygenated compounds (mg/m³)	1.4	0	0.5	0	0.73	7
%	0.032	0	0.046	0	0.059	14
Organosulphur compounds (mg/m³)	19	21	2.6	6.7	3.9	4.6
%	0.43	64	0.24	3.6	0.32	9.6
Totals (mg/m³)	4427	32.8	1100	190	1200	48

compounds were not being combusted. The temperatures measured in the combusted zone during sampling, which were between 520 °C and 720 °C at the point of sampling, may also be conducive to the formation of thermally produced minor derivatives. The results for total trace components are given in Table 2 and discussed further.

Flare B. Results from the analysis of inlet and flared LFG for Flare B are given in Table 1. It is noticeable that the average methane and carbon dioxide contents of the gas being utilized by this flare were significantly lower than that seen for Flare A, and there is evidence of greater air ingress (i.e. average oxygen level of 8.4%, average nitrogen levels of 41%). At this site, the flare was not receiving gas surplus to the requirements of a utilization scheme, but was in use only as part of a gas control system. The relatively high levels of nitrogen and oxygen were therefore not unexpected and suggest that the migration control system was drawing in excessive air due to overpumping. The relative ratio of methane to carbon dioxide remains similar to that for undiluted LFG.

The analysis of the flared gas clearly indicates that the combustion conditions that prevailed at the time of sampling were giving rise to incomplete combustion of the input gas. This is evident from an average measured methane concentration in the flared gas of 4.8%. Temperatures were again measured in the combustion zone and were found to be in the range 620–720 °C (again, possibly conducive to synthesis of minor derivatives).

Flare C. Results of bulk component analysis of inlet and flare emissions are given in Table 1. The inlet gas quality observed for Flare C, which was also part of a gas migration control system, was similar to that seen for Flare B: i.e. methane and carbon dioxide levels were low owing to air ingress into the gas control system.

By contrast, the composition of the flared gas was significantly different from that observed for Flare B. Over 50% of the methane present in the inlet gas remained unburnt, and both carbon monoxide and hydrogen were present in detectable concentrations.

Carbon Mass Balance

A carbon mass balance per cubic metre of input and flared gas at a given temperature was calculated based on the compositional data given in Table 1. This indicates that the following dilution factors were in operation during the course of sampling each flare:

Flare A× 23.9
Flare B × 14.2
Flare C× 1.60.

The concentrations of components measured in samples taken from the respective flares have therefore to be increased by the appropriate dilution factor to produce estimates of *actual concentrations* at the 'point of combustion'.

Trace Composition of the Gas

Table 2 gives details of generic groups of organic compounds identified in the input LFG and the flared gas for each of the three flare systems sampled. Results are quoted as total concentrations and as a percentage of total trace components present. Flared gas composition has been adjusted for dilution.

The input LFG for Flare A contained the greatest diversity of individual compounds (105) and the highest concentration of total trace components (4427 mg/m³). Flares B and C were receiving LFG that, as stated earlier, appeared to be subject to considerable dilution as a result of air ingress into the system; the lower total trace component concentrations of 1100 and 1200 mg/m³ respectively for the raw gases for these flares appear to reflect this dilution effect. The diversity of individual compounds was also lower: 78 individual compounds in LFG to Flare B and 82 individual compounds in Flare C.

However, in the input LFG for *all three* systems, alkanes, alkenes, aromatic hydrocarbons, cycloalkenes and halogenated organic compounds were the predominant trace compounds, accounting for 85.4%, 99% and 97.6% of the total trace component concentrations from Flares A, B and C respectively.

Flaring of the gas did not lead to zero trace component levels but *did* reduce total trace levels (following adjustment for dilution) and diversity. It is particularly interesting to note that a number of compounds not previously identified in LFG were observed in the flared gases, suggesting that these were synthesized under the conditions of incomplete combustion that prevailed. These included low levels (i.e. < 10 mg/m³) of methyl cyanide, nitromethane, acrolein, ethylene oxide and various alkynes.

TRACE COMPONENTS AND ODOUR

Certain trace components in LFG are responsible for the malodours normally associated with landfill operations. The odour itself can represent a considerable nuisance and is probably responsible for the largest number of complaints made both to landfill operators and to regulators.

To assess the impact of individual trace components in terms of odour, the observed concentrations were compared with adopted odour threshold data, compiled after a thorough review of published data (Fazzalari, 1978; Mackinson *et al.*, 1981).

Previous research by ESC (Scott and Baldwin, 1990; Brookes and Young, 1983) indicated that a large proportion (between 39% and 64%) of the trace compounds typically observed in LFG are at levels exceeding adopted odour thresholds. However, only a very limited number of compounds were found in these studies to exceed their odour thresholds by three or more orders of magnitude. The majority of the compounds (which include organosulphur compounds, cyclic compounds, aromatic hydrocarbons, esters and carboxylic acids) are thought to derive from anaerobic microbial interactions and combine to provide landfill gas with its characteristic sickly sweet odour.

Table 3 presents ranked odour data for the ten most odorous components present in LFG, both before flaring and for the flared gases from each of the three flares investigated. The component that exceeded its odour threshold most significantly was ethyl butanoate, found in the input LFG to Flare A. This has previously been identified as a major contributor to the odour of LFG. From the table, it can also be seen that C_4 benzenes are particularly significant in terms of input LFG to all three flare systems. Other aromatic species, and sulphuretted compounds, were also present at levels significantly higher than odour thresholds in the input LFG to all three flares.

The results for the flared gas are also given in Table 3. The most obvious effect of flaring, in terms of odour, appears to be the general reduction in overall odour potential, as certain species are (at least partially) combusted. However, complete combustion of all odorous components was not achieved in any of the three flares sampled. Indeed, concentrations of some odorous components, e.g. dimethyl sulphide, actually increased in the flare emissions for Flares B and C.

TABLE 3. Major Odorous Compounds in Landfill Gas and Flared Emissions

	Landfill gas before flaring		*Flared gas adjusted for dilution*	
	Compound	*Odour factor*	*Compound*	*Odour factor*
Flare A				
1	Ethyl butanoate	3 200 000	Carbon disulphide	30
2	C4 benzenes	26 000	Toluene	6.6
3	Limonene	14 000	Xylenes	1.7
4	Butanoic acid	8 000	Ethyl benzene	1.2
5	C3 benzenes	5 000	Butan-z-one	0.6
6	Methyl butanoate	4 800	Methanol	0.3
7	Dimethyl sulphide	3 800	Acetone	0.15
8	Diethyl disulphide	900	Chloromethane	0.033
9	Toluene	710	Trichloroethylene	0.024
10	Propyl butanoate	550	Benzene	0.016
Flare B				
1	C4 benzenes	7 100	Dimethyl sulphide	1 800
2	C3 benzenes	330	C4 benzenes	740
3	Limonene	2 700	C3 benzenes	430
4	Hydrogen sulphide	2 600	Limonene	280
5	Dimethyl sulphide	410	Xylenes	26
6	Decanes	160	Decanes	21
7	Undecanes	160	Ethyl benzene	16
8	Xylenes	150	Toluene	14
9	Ethyl benzene	98	Undecanes	12
10	Other terpenes	67	Methanol	9.2
Flare C				
1	C4 benzenes	7 300	Hydrogen sulphide	3 300
2	Hydrogen sulphide	3 300	Dimethyl sulphide	650
3	C3 benzenes	3 300	Acetaldehyde	240
4	Limonene	800	Diethyl sulphide	53
5	Dimethyl sulphide	430	Dimethyl disulphide	22
6	Decanes	230	Pentenes	9
7	Undecanes	200	Butenes	7.6
8	Xylenes	140	C3 benzenes	4
9	Ethyl benzene	110	Propanal	4
10	Diethyl sulphide	80	Acrolein	2

TRACE COMPONENTS AND TOXICITY

Toxicological information relating to a wide range of trace compounds in landfill gas is currently limited. Even less is currently known about the influence of these compounds on living systems when they occur in mixed phases. The issue is compounded by the large potential for toxic synergism and potentiation in complex gaseous mixtures and the variation in response of different individuals to specific substances. An estimation of the toxicological hazards posed by the trace components of landfill gas is therefore considered to be a difficult task, and considerable further research is required in this area.

In the course of this and other studies undertaken by ESC into the composition of LFG, two largely empirical and simplistic methods have been employed to assess toxicity hazard. In the first, the observed concentrations of individual components have been compared to the published 8 h time-weighted average toxicity threshold limit values (TLVs) of the American National Institute for Occupational Safety and Health (NIOSH, 1981) the American Conference of Governmental Industrial Hygienists (ACGIH, 1988), and the Occupational Exposure Standards (OESs) of the United Kingdom Health and Safety Executive (HSE, 1989).

OSEs differ from TLVs in that they have specific legal implications under the Control of Substances Hazardous to Health (COSHH) Regulations.

Of a total of 125 compounds detected in raw and flared LFG during this study, only six were observed to exceed their relevant TLVs and OESs at any time. These were restricted to:

- benzene (39 mg/m^3) and toluene (500 mg/m^3) in the input LFG to Flare A;
- vinyl chloride (11 mg/m^3) in the input LFG to Flare B;
- carbon monoxide (19 330 mg/m^3), acrolein (1.6 mg/m^3) and ethylene oxide (4.32 mg/m^3) in the *flared* gas from Flare C.

The two compounds identified as exceeding their TLV/OES values in input LFG to Flare A, benzene and toluene, would require dilution in the order of × 1.3 to reduce their concentrations to below these values. Vinyl chloride, present in input LFG to Flare B at 11 mg/m^3, would require dilution of × 1.1 to reduce it below its TLV/OES.

The most potentially toxic compound observed (owing largely to its concentration) is carbon monoxide in the Flare C emission. This compound has an affinity for haemoglobin 210 times that of oxygen and can consequently lead to death from asphyxia. The TLV and OES of carbon monoxide are 55 mg/m^3; dilution of the order of × 351 would therefore be required to reduce the levels measured in the flared gas to below this value.

The levels of acrolein and ethylene oxide identified in the flared gas from Flare C would require dilution factors of × 6.4 and × 2.0 respectively to reduce them to below their respective TLV or OES.

Carbon dioxide, although not a trace component, is of interest because it is found in bulk (%) concentrations in the LFG and flare emissions and possesses a TLV of 5000 vpm (9000 mg/m^3).

The second approach referred to above attempts to assess the combined impact of components on specific organs and metabolic processes. Where there was no evidence relating to synergistic effects, the effect of individual trace components known to affect the same organ or body function was considered to be additive, i.e.:

$$F = \frac{C_1}{T_1} + \frac{C_2}{T_2} + \frac{C_3}{T_3} + \ldots + \frac{C_n}{T_n}$$

where F is the TLV factor for the body function, C_n is the concentration of compound n, and T_n is the individual TLV for compound n. The approach adopted is that recommended by ACGIH for application in environments where mixed gases are encountered, and is considered by HSE to be the 'most prudent course' of action.

An assessment of raw and flared gas data for Flare C is given in Figure 1.

Overall, the results suggest low toxicity factors for both raw and flared gases, with values remaining near or below unity. The most significant finding is the apparent 700-fold increase in potential blood toxicity factors for the flared gas ($F = 351$) compared with the input LFG ($F = 0.5$). This increase is almost entirely due to the presence of carbon monoxide in the former. The approximate six times increase in toxicity factor relating to the respiratory system is due primarily to acrolein in the flared gas, but not in the raw gas.

CONCLUSIONS

This study of three significant landfill gas flares indicates that:

- LFG flares, under certain conditions, may not achieve complete destruction of all the combustible (both bulk and trace) components of the LFG.
- The flares may also give rise to measurable quantities of other components *synthesized* within the flare.

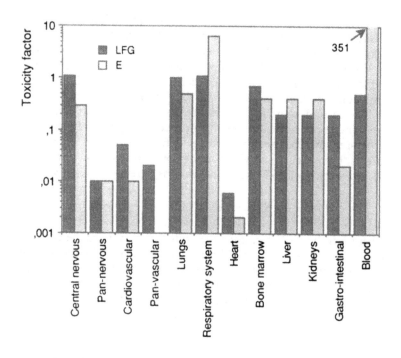

Figure 1. Toxicity factors for landfill gas (LFG) and for flare emissions (E), for different human body functions.

- The results for the flared gas do suggest that a reduction and/or modification of the odour is achieved following combustion. However, flaring does not appear to eliminate completely the potential for odour, and although a significant reduction in this potential appeared to be achieved for Flare A, there still appears to be the opportunity for odour nuisance from all three flare emissions studied.

- A full assessment of the potential toxicological hazard associated with both raw and flared gases was not possible based on the information available. While some potentially toxic species present in the LFG were destroyed by flaring, others were synthesized. A simplistic, empirical approach to assessment has been adopted, which indicates that flaring may actually increase certain toxicity factors if the flaring process is inefficient.

In general terms it is important that careful consideration is given to specifying the performance requirements of any particular flare system. If the nature and capacity of a system are closely matched with the gas control requirements of the landfill this is usually the main factor in obtaining the good combustion conditions required.

ACKNOWLEDGEMENT

The work discussed in this chapter was based on research funded by the Wastes Technical Policy Unit of the UK Department of the Environment. The opinions expressed in this chapter are those of the authors and do not reflect those of the Department of the Environment.

REFERENCES

ACGIH (1988). *Threshold Limit Values and Biological Exposure Indices for 1988–89*, American Conference of Governmental Industrial Hygienists, Cincinnati, OH.

Baker, J. M., Peters, C. J., Perry, R. and Knight, C. P. V. (1983). Odour control in solid waste management. *Effluent and Water Treatment Journal*, **23**(4), 135–8.

Bowscaren, R. (1984). Odours and deodorisation. 1 – Odorous products, their origin. *Tech. Sci. Munic*, **79**(6), 313–20.

Brookes, B. I. and Young, P. J. (1983.) The development of sampling and gas chromatography mass spectrometry analytical procedures to identify and determine the minor organic components of landfill gas. *Talanta*, **30**(9), 665–79.

Don, J. A. (1986). Odour measurement and control. *Filtration and Separation*, **23**(3).

Fazzalari, F. A.(ed.) (1978). *Committee E-L8 on the Sensory Evaluation of Materials and Products*, ASTM Data Series DS48A 05-048010-36, American Society for Testing and Materials, Philadelphia.

HSE (1989), *Occupational Exposure Limits*, Health and Safety Executive Guidance Notes EH/40/89, HMSO, London.

Mackinson, F. W., Stricoff, R. S. and Partridge, L. J. Dhew (eds) (1981). Pocket Guide to Chemical Hazards. US National Institute for Occupational Safety and Health/Occupational Safety and Health Administration, NIOSH Publication No 78–210.

Scott, P. E. and Baldwin, G. (1990). Methods used to characterize and assess the environmental impact of gaseous emissions from landfilled wastes.

Presented at the International Conference on Energy from Biomass and Wastes XIV, Florida.

Summer, W. (1971). *Odour Pollution of Air,* Leonard Hill, London.

Van Germert, L. G. and Netterbreijer, A. H. (1977). *Compilation of Odour Threshold Values in Air and Water,* TNO.

Verschueren, K. (1983). *Handbook of Environmental Data on Organic Chemicals,* 2nd edn., Van Nostrand Reinhold, New York.

6.3 Landfill Gas Treatment by Biofilters

RODRIGO A. FIGUEROA

Institute of Waste Management, Technical University of Hamburg-Harburg, Harburger Schlossstrasse 37, 21071 Hamburg, Germany

INTRODUCTION

Biological treatment is a solution for LFG purification that is both simple in technique and inexpensive. One advantage of this method is that the components of the exhaust air are not redistributed to other phases: when they are completely oxidized they are converted into water and carbon dioxide.

The following biological LFG treatment processes are known.

- **Biowashers** operate in a similar way to trickling filters or an activated sludge plant: the components are introduced into the water phase in a contact unit. The elimination of the substances takes place either in the biofilm of the percolating filter or in the activated sludge suspension. This technique is applied above all in the case of readily water-soluble components (Sabo, 1991).
- The **biomembrane** technique uses diffusion-permeable silicon hoses. Owing to the concentration gradients, the contaminants pass through the membrane and are oxidized on the inner side of the membrane by adapted micro-organisms in a biofilm. This method is especially suitable for removing components from the exhaust air that have low solubility or are very readily volatile (Fischer, 1991).
- In a **biofilter** the micro-organisms are settled on a solid filter material, over which the contaminated exhaust gas is pumped.

Landfilling of Waste: Biogas. Edited by T. H Christensen, R. Cossu and R. Stegmann. Published in 1996 by E & FN Spon, London. ISBN 0 419 19400 2.

The components are first sorptively bound to the filling material and are subsequently biologically degraded. The biological turnover of the components of the exhaust air regenerates the filter.

There are two basic processes that are of special importance for all methods of biological exhaust air purification. It must be possible to transport the contaminants from the gas phase to the environs of the micro-organisms: this can only happen if the contaminants are watersoluble. Moreover, it must be possible to biochemically convert the contaminants.

This chapter deals with the application of biofilters for LFG purification. Unlike the two other methods of biological LFG decontamination, the biofilter method has already been tested and applied in a working situation. As yet, the biowasher and the biomembrane reactor have had only limited application in the purification of LFG, and there is still a considerable lack of data.

DESIGN AND FUNCTION OF BIOFILTERS

The principle of design of biofilters is demonstrated in Fig. 1. The raw gas is normally pumped into the filter unit by means of fans. To

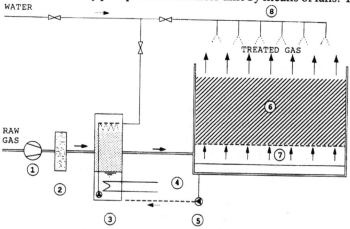

Figure 1. Basic design of a biofilter: 1, blower; 2, dust removal; 3, conditioning unit; 4, temperature control; 5, drain water; 6, filter material; 7, gas distribution system; 8, direct irrigation.

minimize losses in pressure, the air inlets should have smooth wall partitions. The material must be able to withstand the chemical – mechanical aggressiveness of the LFG components. To protect the filter material against clogging, dust and droplets must be removed from the raw gas. After moistening in the conditioning unit, the raw gas is pumped into the filter filling via a distribution system.

The biofilter consists mainly of filter material that influences the purification performance by its physical, chemical and biological properties. The filter material is a carrier for the bacteria cultures, and possesses sorption qualities; it is thus the most important part of the unit. Materials of predominantly biological origin (peat, compost from biowaste, heather, shredded bark and sawdust) are used. These filling materials show a high biological activity, good water-storage capacity and sufficient nutrient content. To improve the structure, inert materials can also be admixed (e.g. expanded clay, polystyrene, lava or active carbon).

To ensure optimum environmental conditions for the micro-organisms the following operational parameters must be controlled and adjusted:

- water content;
- temperature;
- pore volume or residence time;
- filter resistance.

Using filter material with a high water content has a positive effect on the productivity of the micro-organisms and on the sorption capacity of the filling material. A high capillary suction force and a large portion of organic material help to increase the water content.

The biofilter filling must be kept humid. Serious problems can arise in the biofilter if the filter material dries up: cracks occur in the filter material and the raw gas can flow through almost unpurified. The water balance can be adjusted by moistening the raw gas (Fig. 1). To yield favourable conditions for the micro-organisms in the biofilter material, the relative humidity of the raw gas should be > 95% and the temperature should be between 15 and 40 °C. Direct irrigation of the filter can lead to the occurrence of zones of very irregular humidity (Fischer and Bardtke, 1990), but used in conjunction with the moisturization of the raw gas this method is effective.

Seasonal variations in temperature also influence the productivity of the micro-organisms: higher temperatures usually have a

favourable influence on microbial degradation, although they reduce the solubility and the sorption of the gas components. It may
be useful to insulate the biofilter in order to achieve more constant
temperatures in the biofilter.

Microbial activity will increase the temperature in the biofilter.
An increase in the temperature of the gas in the biofilter will also
increase the water uptake of the gas stream, and the filter material
may become dehydrated. For this reason, a test re-run should be
done over a period of several months in order to investigate the
possible effects of temperature changes.

The pore volume of the filter filling is closely connected with the
water content of the material. It influences especially the gas flow
through the filter filling and the residence time of the gas in the
biofilter. By changing either the volume of the gas flow or the
residence time, the contact between the LFG and the filling material
can be influenced, which in turn affects the sorption of contaminants. Specific area loadings of 20–300 m^3 raw gas/m^2 filter per hour
and residence times of a few seconds in the biofilter are normal. The
residence times in soil filters are approximately 60 s with surface
load values from 1 to 20 m^3/m^2 h (Helmer, 1974; Bohn and Bohn,
1988).

Thorough control and care have a positive effect on the total
operational period of the biofilter. Generally, operation times of 1–3
years may be expected before it becomes necessary to change the
filling material; it is often sufficient to carefully remix the material
and put it back into the filter (Fischer and Bardtke, 1990).

In comparison with other techniques, the treatment of exhaust
air in a biofilter requires a relatively large filter volume because of
the limited degradation capacity. For this reason, in addition to
normal biofilters with filling heights of 0.5–1.5 m, more compact
and transportable types of biofilter (high filters, multiple screen
filters, and so-called tower filters) have been developed for industrial
application in recent years.

COMPOSITION OF LANDFILL GAS AND BIOLOGICAL DEGRADATION

A selection of substances quantified or detected in landfill gas is
shown in Table 1. The table also shows a qualitative presentation

of the degradability of some substances according to Standard 3477 *Biofilter* of the Verein Deutscher Ingenieure (VDI) (Association of German Engineers). In addition to giving general information on the design of biofilters, this standard contains a qualitative survey of all the individual components and groups of substances that had been tested for degradability up until 1989. The behaviour of these substances and groups of substances in the biofiltration process is discussed herein.

- For low methane concentrations, the biological oxidation of methane is a viable solution as far as both ecology and security are concerned. Methane-degrading micro-organisms have been known for more than 85 years (Söhngen, 1906). By 1970 more than 100 gram-negative, strict methane degraders had been isolated by Wittenbury *et al.* (1970).
- The aromatic hydrocarbons reach the landfills as solvents and as components of technical mixtures. Together with benzene they represent one of the most serious toxicological components in landfill gas. Compared with pollutants from the chemical industry, mineral oil production and from automobiles, however, these compounds cause only little ecological damage. In fact they are, in principle, biologically oxidizable (Greiner *et al.*, 1990).
- Compounds containing oxygen are produced during the primary phase of anaerobic degradation as an intermediate product of easily degradable organic components of the biomass. These compounds have only little toxic effect and do not cause specific ecological damage. The oxygen compounds reported in Table 1 can be biodegraded without difficulty.
- Besides specific solvents which are disposed of together with waste, hydrogen sulphide and organic sulphide contribute the most to odour emission (Laugwitz, 1988). A further group of intensively sweet-smelling compounds is formed by the low-molecular fatty acids and amines that occur in the acid fermentation phase. These compounds, however, are mainly present in the leachate. Landfill gas also contains alkyl benzenes and limonene, which cause the typical 'background smell'. Mercaptans and hydrosulphides are not only the most important but also, owing to their toxicity, the most dangerous odorous components of landfill gas (Young and Parker, 1983, 1984; Janson, 1988). Many of the odorous components

TABLE 1. Concentration of Different Organic Substances in LFG and Qualitative Degradation Behaviour of Individual Substances in the Biofilter According to VDI Standard 3477B

Group of substances or substance	Gas concentration MSW landfill		Qualitative degradation[a]
	Young (mg/m^3)	Old (mg/m^3)	
Aliphatic HC			
Methane	0–80% vol.		(+)
Pentane	detected		(+)
Hexane	detected		+
Cyclohexanes	detected		(+)
Aromatic HC			
Benzene	$0.6–15^b$	$0.3–10^b$	+
Toluene	$5–135^b$	$2–65^b$	+ +
Xylene	$20–226^b$	$0–51^b$	+ +
Styrene	$10–25^c$		+
Oxygen-containing compounds			
Alcohols	$16–1450^d$ (Ethanol)		+ +
Methanol	$2.2–125^d$		+ +
Butanol	$18–626^d$		+ +
Tetrahydrofuran	$<0.5–8.8^d$		+ +
Diethyl ether	$0.1–166^d$		(+)
Acetaldehyde	$0–150^e$		+ +
Acetone	$<0.8–4.1^d$		+ +
Butyric acid	$0.02–6.8^d$		+ +
Acetic ester	$<2.4–263^d$		+
Sulphur-containing compounds			
Dimethyl sulphide	$1.6–24^d$		+
Thiophene	$0.01–0.5^d$		+
Mercaptans	detected		+
Methylmercaptan	$0.1–430^d$		+
Carbon disulphide	$<0.5–22^d$		+
Nitrogen-containing compounds			
Trimethylamine	$<0.08–0.076^d$		+ +
Halogenated hydrocarbons			
Dichloromethane	$2–51^b$	$0–2^b$	(+)
Trichloroethylene	$0–14^b$	$0–1.1^b$	
Perchloroethylene	$0–10^b$	$0–0.8^b$	−
1,1,1,-Trichloroethene	$0–9.3^b$	$0–3.1^b$	−
Organic compounds			
Hydrogen sulphide	$0–633^e$		+
Ammonia	$0–100^f$		+

mentioned here, especially the sulphurous organic compounds, can be successfully degraded in a biofilter.
- Many of the trace components found in landfill gas belong to the group of chlorinated and fluorinated hydrocarbons (CFCs). The organic halogen compounds are used as detergents, coiling agents and propellants. Owing to their high chemical stability the CFCs are compounds that are difficult to degrade. The biological turnover of halogenated hydrocarbons depends not so much on microbial abilities but on external conditions, which can potentially have an inhibiting effect.

In summary, it can be said that alcohols, aldehydes, carbon acids, organic compounds containing sulphur, some nitrogen compounds and some aromatic substances can be successfully degraded. The degradation of ethers and halogenated hydrocarbons in biofilters is only partially successful.

An important criterion for the application of biofilters, besides the degradability of the substances to be treated, is the distribution of the individual components in the biofilter, i.e. to gas phase, liquid phase and solid phase. Four groups of substances can be roughly distinguished (Bardtke, 1987), as follows.

- Easily sorbable and degradable substances are well removed by biofiltration (e.g. methanol and butanol).
- Substances that are difficult to sorb but easy to degrade are also well eliminated (e.g. *n*-hexane and ethyl acetate).
- Substances that can easily be sorbed but are difficult to degrade are held back only to a minor degree (e.g. dioxan).
- Substances difficult to sorb and difficult to degrade were expected to show a low level of elimination in the biofilter. However, within the scope of Bardtke's investigations this could not be proved: whereas trichloroethylene, for example,

---

Source: After VDI 3477, 1989.
[a] + +, very good degradability; +, good degradability; (+), minimal degradability; −, no degradability.
[b] Laugwitz (1988).
[c] Rettenberger (1986).
[d] Young and Heasman (1985).
[e] Janson (1988).
[f] LAGA (1983).

was poorly eliminated, the elimination performance for dichloromethane was relatively high.

APPLICATION OF BIOFILTERS FOR LANDFILL GAS TREATMENT

Biofilters were used for biogas treatment at an early stage in their development. They were utilized for the first time in the USA in 1966 – then also known as soil filters – to deodorize sewage sludge digester gas (Carlson and Leiser, 1966). The field of application for biofilter techniques has been continuously expanding during recent years. At first, biofilters were used mainly in odour treatment, but recently their application for the treatment of landfill gas has been intensively investigated. Tabarasan *et al.* (1979) were the first to report the application of biofilters for landfill gas treatment. Besides deodorization, they investigated the degradation of hydrogen sulphide and methane on a laboratory scale. First investigations into the aerobic degradation of methane in biofilters were carried out by Mennerich (1986).

Depending upon the methane content in LFG and its utilization, biofilters are applied in the following cases.

- The LFG has a high methane content and is to be used as an energy source after treatment. In this case the components impeding further utilization must be oxidized in the biofilter by the addition of oxygen. Sulphur compounds make the further utilization of landfill gas difficult; these components must be removed, taking care that the methane content is not significantly reduced.
- The LFG is largely diluted with air and has only a low methane content and low concentrations of compounds; it can only be disposed of without the recovery of energy. In this case, the purpose of the biofilter is to remove both toxic and odorous substances or those harmful to the environment.

Treatment of Landfill Gas of High Energy Content

Owing to their corrosive and partly toxic properties, it is often necessary to remove compounds containing sulphur and chlorine from the LFG prior to utilization. At present, absorptive and adsorptive techniques are mainly applied in landfill gas treatment. As biofilter operation is simple and inexpensive, this method provides a good alternative to existing LFG treatment methods (see Chapter 7).

Extensive investigations into the treatment of high-energy LFG from a mixed landfill with *high* content of hydrogen sulphide (1.5–3 g H_2S/m^3) have been carried out (Sabo, 1991). Filling materials such as biowaste, or bark compost with the addition of expanded clay pellets, were investigated.

After a short adaptation period, the H_2S removal capacity lay at approximately 95%; the specific surface load amounted to 12–15 m^3/m^2h. After three weeks the pH value in the filter material stabilized at around 2.5, an ideal pH range for the genus *Thiobacillus*, a group of gram-negative bacteria that gain their energy by oxidation of reduced sulphur compounds.

With a decreasing pH value, the H_2S-removal capacity of the filter dropped to values below 90%. The increase in the sulphur content of the filter materials was seen to be the cause of the reduction of the purification performance. The high sulphur content of about 13% obviously impeded microbial activity. After 8–10 weeks the filter material was loaded with the end products of biochemical conversion (elementary sulphur and sulphate) to such a high degree that it was necessary to change it. However, LFG from traditional landfills normally shows lower H_2S concentrations, meaning that a biofilter used to decontaminate such gas can be expected to have a considerably longer operational life.

During winter it was necessary to insulate the pilot biofilter unit, as temperature has a considerable influence on the purification performance. For the conversion of H_2S the optimum temperature is between 28 and 30 °C.

A simultaneous conversion of organic compounds in the filter is hardly possible at such low pH values. To achieve this, a two-stage biofilter unit (connected in series) was necessary: in the first acid stage, the sulphur organic compounds were converted; in the second stage, the elimination of organic gas components in the neutral pH range was effected.

The purification effect on the individual gas components varied: the removal of alcohols (methanol and butanol) was very effective, whereas that of the aromatic hydrocarbons (toluol and styrol) was only relatively effective. For dichloromethane removal efficiency of

15–30% was achieved. Aliphatic substances and halogenated hydrocarbons were removed to a much lower degree.

Treatment of Landfill Gas of Low Energy Content

The gas disposal in aged, closed landfills is generally different from that in landfills either still in operation or only recently closed. In most cases, thermal treatment of aged, completed landfills is not possible because the methane concentrations are too low or the quantity of gas is not sufficient. Sufficient methane concentrations can be reached only by addition of natural gas (Roediger, 1991). The disposal of landfill gas by means of co-burning of fuel gas does not, however, appear to be reasonable from either an economic or an ecological point of view.

At the Technical University of Hamburg-Harburg, Germany, experiments were carried out using LFG of low methane content (< 4% CH_4) from old deposits. In a semi-technical test unit (height 1.0 m, diameter 0.32 m) the following filling materials were investigated: mature compost from biowaste with and without expanded clay pellets, as well as a mixture of peat and pine twigs. The most significant properties of the filling material, operational conditions and performance parameters observed in the experimental tests are reported in Table 2.

The gas was extracted in such a way that, by over-suction of the landfill body, the methane content of the landfill gas was reduced to below the lower explosion limit (Figueroa and Stegmann, 1992). The landfill gas, which was highly rarefied by air, had a relative humidity of only 45%; during the period of investigation its temperature fluctuated between 20 °C and 30 °C. The raw gas was saturated with water vapour by a preconnected humidification unit, and its temperature was slightly increased.

The filter material that was investigated showed different properties: the mixture of peat and pine twigs had, at lower bulk density, a higher waste absorptive capacity than the 6-month old biowaste compost; the water-absorptive capacity of this material could be increased by addition of expanded clay (3:1 vol./vol.). A disadvantage of using peat and pine twigs as filling materials is the low pH value, as only special micro-organisms prefer to live in an acid environment.

The specific surface load of the biofilters, which started at 30 m^3

TABLE 2. Properties of the Filling Material, Operational Conditions and Performance Parameters Observed in the Biofiltration Tests at the Technical University of Hamburg–Harburg

	Raw gas	Moistening	Biowaste compost	Peat-pine	Biowaste compost exp. clay
Bulk density (kg/m^3)			783	440	652
Water content (vol.%)			43–50	35–60	35–58
WAC-Max.[a] (gH$_2$O/100 g dry weight)			100	231	137
Pore volume (vol.%)			50–77	66–90	55–75
pH value			7.3	3.9–4.3	7.4
Gas humidity (%)	44.0 ± 8.1	99.2 ± 1.1	99.8 ± 0.2	99.9 ± 0.0	99.9 ± 0.0
Gas temp. (°C)	26.6 ± 6.5	28.4 ± 2.6	37.5 ± 4.8	31.5 ± 3.5	30.7 ± 3.1
Temp. mat. 15 cm height (°C)			30.4 ± 3.8	29.1 ± 2.8	34.8 ± 7.1
Temp. mat. 80 cm height (°C)			43.3 ± 5.2	35.2 ± 3.9	41.9 ± 7.4
Pressure loss (mBar)			6.4 ± 3.8	5.3 ± 3.0	4.9 ± 3.1
Odour (ou^b/m^3)	1.625	1.430	544	629	500
Methane (vol.%)	3.1 ± 0.3		2.5 ± 0.3	2.8 ± 0.3	2.6 ± 0.3
Oxygen (vol.%)	12.0 ± 2.8		11.4 ± 1.5	12.0 ± 2.3	11.6 ± 2.1

[a] Wac-Max: maximum water adsorption capacity.
[b] o.u.: odour unit.

landfill gas per m^2 filter per hour, was gradually reduced. After about 3 months a significant filter activity was observed at a surface load of 5 m^3/m^2 h: owing to the metabolic activity of the micro-organisms the temperature rose by approximately 15 °C over the whole filter height.

At this surface load, the system's capability to eliminate odorous substances ranged from 60% to 70%. The odour reduction was at first effected by leaching odour carriers in the humidification unit (approximately 10%); the odour concentration was then reached in the biofilter by a further 65% for the biowaste compost filling material and by 55% for the peat/pine twigs filling material.

The methane degradation in the biofilter is highly dependent on the specific flow, i.e. on the contact time between gas and filter material. The degradation of methane amounted to approximately 30% at a specific surface load of 5 m^3/m^2 h; this corresponds to a specific degradation rate of approx. 50 g CH_4/m^3 filter h. A further methane reduction was reached when a smaller surface load was chosen, as further investigations with the biowaste compost filling material showed: for a surface load of 0.5 m^3/m^2 h a complete methane degradation (CH_4 = 3 vol %.) was reached.

The existing level of gas dilution made it difficult to monitor the effectiveness of the biofilters with regard to the trace components: on the one hand, the initial concentrations of many trace components are close to the detection limit (0.5–5 mg/m^3) and, on the other hand, the possible adsorption of trace components onto water vapour of the air – gas mixture poses a further analytical uncertainty. It is necessary to use a sample system with which emissions with low concentrations and high humidity can be quantified (e.g. Teenax sample tubes).

The content of halogenated trace components in the raw gas decreased in the following order: vinyl chloride > dichloroethylene > trichloroethylene > tetrachloroethylene > dichlorodifluoromethane > trichlorofluoromethane. For the trace components dichloroethylene and dichloromethane a removal efficiency greater than 25% was detected in the biofilter.

In conclusion, there is very little difference between the purification effect of the various filter materials that were investigated. The filter filled with biowaste compost and expanded clay, however, showed not only good purification performance but also better operational properties: the flow resistance of the filling (pressure loss) was reduced by addition of expanded clay and the water adsorption capacity and long-term stability of the material was increased.

DISPOSAL OF THE FILLING MATERIAL

Besides mineralization of the organic material, an accumulation of non-usable intermediate and final products in the filter takes place during filter operation. Normally, the used filter materials can be classified as unproblematic to dispose of (Sabo, 1991). Used filter material of organic origin can normally be used in horticulture and landscape gardening to improve the soil quality. If the filter materials have a high sulphate content, care must be taken during disposal to avoid anaerobic conditions and to prevent new H_2S production and emission. Accumulations of landfill gas components that could not or only partly be degraded in the biofilter, i.e. the halogenated and some of the aromatic hydrocarbons, pose a critical problem.

The Preliminary Guideline of the Nordrhein-Westfalen Regional Council for Water and Waste forms the basis for the disposal of filter materials in Germany. If it is suspected that contaminants have accumulated in the filter material, examinations according to the VDI Standard 3477 must be carried out in order to ensure appropriate disposal. The scope of testing necessary is dependent upon the kind of substances or final products contained in the landfill gas, and should include the following parameters: pH value, sulphate ratio, chloride content, heavy metals, chlorinated hydrocarbons, PCB and AOX.

CONCLUSIONS

The use of biofilters for the treatment of landfill gas has become increasingly important in the last few years. The areas of use described here show what a wide field of application this relatively simple and environmentally friendly method of decontaminating gases has.

This method appears to be especially suited to the decontamination of landfill gases from aged, hazardous deposits: one of the major advantages that this method has to offer is the breakdown and neutralization of several of the contaminants present in landfill gases.

The use of biomembrane methods to decontaminate LFG-containing compounds with a low solubility in water (i.e. hydrocarbons) has shown promising results (Fischer, 1991). Owing to the combination of both biofilters and biomembranes a better elimination of CFCs could be achieved.

The same is true of the use of biowashers for the decontamination of landfill gas with a high H_2S content (Sabo, 1991). Treating highly contaminated landfill gases in a multi-step operation is a further possibility that shows promise.

REFERENCES

Anon (1989). *Materialien zur Ermittlung und Sanierung von Altlasten*, Landesamt für Wasser und Abfall Nordrhein-Westfalen (Hrg.) Düsseldorf, FRG.

Bardtke, D. and Fischer, K. (1987). *Biologische Verfahren zur Deponiegasreinigung*, Stuttgarter Berichte zur Abfallwirtschaft, Bd. 22, Erich Schmidt Verlag, FRG.

Bohn, H. and Bohn, R. (1988). Soil beds weed out air pollutants. *Chemical Engineering*, **25**, 73–76.

Carlson, D. A. and Leiser, C. P. (1966). Soil beds for the control of sewage odors. *Journal Water Pollution Control Federation*, **38**(5), 829–40.

Dernbach, H. (1991). Entgasung bei Altablagerungen auf kleinen Deponien, in *Proceedings Deponiegasnutzung '91*, 6–8 March 1991, Trier, FRG.

Figueroa, R. A. and Stegmann, R. (1992). Possibilities of controlled active gas extraction of old landfill sites. Poster presented at the 6th International Solid Waste Congress and Exhibition, 14–19 June, Madrid, Spain.

Fischer, K. (1991). Membranreaktor statt Biofilter? *Chemische Industrie*, FRG.

Fischer, K. and Bardtke, D. (1990). *Biologische Abluftreinigung*, Expert Verlag, Bd. 212, FRG.

Greiner, D., Kolb, M., Endler, J. and Faüst, R. (1990). Kinetik des biologischen Abbaus von Benzol im Biofilter. *Staub Reinhaltung der Luft*, **50**, 289–91.

Helmer, R. (1974). Desodorierung von geruchsbeladener Abluft mit Bodenfiltern. *Gesundheits-Ingenieur*, **95**(1), 21–26.

Janson, O. (1988). *Analytik, Bewertung und Bilanzierung gasförmiger Emissionen aus anaeroben Abbauprozessen unter besonderer Berücksichtigung der Schwefelverbindungen*, Stuttgarter Berichte zur Abfallwirtschaft, Bd. 32, Erich Schmidt Verlag, FRG.

LAGA (1983). *Merkblatt Länderarbeitgemeinschaft Abfall (LAGA)*.

Laugwitz, R. (1988). Entstehung und Verhalten von Spurenstoffen im Deponiegas sowie umweltrelevante Auswirkungen von Deponiegasemissionen, in *Hamburger Berichte zur Abfallwirtschaft*, H.1, Economica Verlag, Bonn, FRG.

Mennerich, A. (1986). Oxidation von Deponiegas auf biologischem Wege. *Muell und Abfall*, **18**(7), 271–7.

Ramanathan, V., Cicerone, R. J., Singh, H. B. and Thiel, J. K. (1985). Trace gas trends and their potential role in climate change. *Journal of Geophysical Research*, **90** (D3), 5547–56.

Rettenberger, G. (1986). Spurenstoffe im Deponiegas. Auswirkungen auf die Gasverwertung. GIT Supplement 1/86, 53–7.

Roediger, M. (1991). Entsorgung methan-armer Gase aus Altedeponien, in *Proceedings Deponiegasnutzung '91*, 6–8 March, Trier, FRG.

Rosenbusch, K. (1988). Zukünftige Entwicklungen auf dem Gebiet der Deponieentgasung und Deponiegasnutzung, in *Hamburger Berichte zur Abfallwirtschaft*, H.1, Economica Verlag, Bonn, FRG.

Sabo, F. (1991). *Behandlung von Deponiegas im Biofilter*, Stuttgarter Berichte zur Abfallwirtschaft, Bd. 47, Erich Schmidt Verlag, FRG.

Söhngen, N. L. (1906). Über Bakterien, welche Methan als Kohlenstoffnahrung und Energiequelle gebrauchen. *Zentralabl. Bakteriol. Parasitenkd. Infektionskr. Hyg. Abt.*, **2**(15), 513–17.

Tabasaran, O., Affoyon, L. and Rettenberger, G. (1979). Einsatz von Biofiltern zur Deponiegasdesodorierung. *Müll und Abfall*, **5**, 132–5.

Tabasaran, O. and Rettenberger, G. (1983). *Passive Entgasung von Abfalldeponien mit Hilfe von Kompostverfüllten Öffnungen in der Abdeckschicht.* Forschungsvorhaben im Auftrag des MELUF – Baden Württemberg, Final Report, unpublished.

VDI (1989). *VDI-Richtlinie 3477-Biofilter-*, Verein Deutscher Ingenieure, FRG.

Whittenbury, R., Phillips, K. C. and Wilkinson, J. F. (1970). Enrichment, Isolation and some properties of methane-utilizing bacteria. *Journal of General Microbiology*, **61**, 205–18.

Young, P. J. and Heasmann, L. A. (1985). An assessment of the odor and toxicity of the trace compounds of landfill gas, in *Proceedings from the GRCDA 8th Int. Landfill Gas Symposium*, April 1985, San Antonio, Texas, USA.

Young, P. J. and Parker, A. (1983). The identification and possible environmental impact of trace gases and vapours in landfill gas. *Waste Management & Research*, **1**, 213–26.

Young, P. J. and Parker, A. (1984). Origin and control of landfill odours. *Chemistry and Industry* 7, 329–33.

6.4 Landfill Gas Upgrading: Removal of Carbon Dioxide

MARTIN J. J. SCHEEPERS

GASTEC NV, Postbus 137, 7300 AC Apeldorn, The Netherlands

INTRODUCTION

Installations for landfill gas upgrading to natural gas quality are operational in the USA, Australia and the Netherlands. In Table 1 a comparison is made between landfill gas characteristics and specifications for the natural gas that is distributed in these countries.

The main process in an upgrading plant is the removal of carbon dioxide (CO_2). Because nitrogen (N_2) is hardly removed by any of the common CO_2 removal processes, the N_2 content in landfill gas is limited. The amount of CO_2 to be removed depends on the amount of N_2 in the landfill gas and the requirements of the product gas.

Table 1 shows that natural gas in the USA contains almost exclusively methane. The gross calorific value of the gas should exceed 38.6 MJ/m^3. As a consequence the amount of nitrogen in the landfill gas is limited to about 3%.

In the Netherlands natural gas contains 10–14% inert components (both N_2 and CO_2). In the upgrading process it is not the calorific value of natural gas that has to be met, but the Wobbe number. The Wobbe number is a measure for the interchangeability of gases with different compositions, and it is defined according to the following equation:

Landfilling of Waste: Biogas. Edited by T. H Christensen, R. Cossu and R. Stegmann. Published in 1996 by E & FN Spon, London.
ISBN 0 419 19400 2.

$$Wo = \frac{H}{\sqrt{d}}$$

where Wo is the Wobbe number of the gas in MJ/m^3, H is the calorific value of the gas in MJ/m^3, and d is the density of the gas relative to air. In practice the N_2 content in the Dutch landfill gas is limited to 7%.

In Australia the gas delivered to the grid should have a Wobbe number above 46.6 MJ/m^3. To achieve this value, first 0.5–1% LPG is added and next the gas is blended with natural gas. The landfill gas may contain up to about 4% N_2.

The specifications for oxygen, chlorine, sulphur, ammonia and water are related to corrosion of grid piping, appliances and related flue gas systems. For components alien to natural gas, such as halogenated hydrocarbons and ammonia, new specifications have been drawn up.

To produce a gas within the specifications listed in Table 1, the following purification steps are necessary:

- CO_2 removal;
- drying;
- purification of sulphur components;
- purification of halogenated hydrocarbons.

The last two purification steps can take place either before or after the CO_2 removal or in combination with it.

In the following section CO_2 removal processes will be presented and discussed. Removal of the other components will be discussed in Chapters 6.5 and 6.6.

CO₂ REMOVAL PROCESSES

For CO_2 removal five different processes are used.

A classic process is physical absorption in water (Kohl and Riesenfeld, 1985; Wezel *et al.*, 1988). The CO_2 is absorbed by the water at ambient temperatures and at a pressure of approximately 10 bar (Fig. 1a). In the stripper CO_2 is removed with air and vented to atmosphere. This process is applied at the LFG upgrading plant at the Tilburg landfill site, the Netherlands (Fig. 2).

TABLE 1. Composition of Landfill Gas compared with the Specifications for Natural Gas in Three Different Countries

		Landfill gas	Specifications for natural gas		
			Netherlands	USA	Australia
methane (CH_4)	(vol %)	55–60			
carbon dioxide (CO_2)	(vol %)	35–39		≤4	
nitrogen (N_2)	(vol %)	3–7		≤4	
oxygen (O_2)	(vol %)	0.03	≤0.5		≤0.1
sulphur – inorganic	(mg/m³)	0–100	≤5	≤6	
– mercaptans	(mg/m³)	0–10	≤10		
halogenated hydrocarbons (as Cl)	(mg/m³)	0–150	≤25		
ammonia (NH_3)	(mg/m³)	0–10	≤3		
water (H_2O)	(C*)	saturated	≤–10	≤–73	
calorific value	(MJ/m³)		43.46–44.41	>38.6	
Wobbe number	(MJ/m³)				>46.6

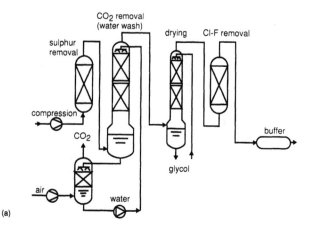

(a)

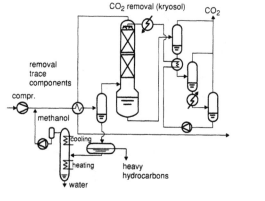

(b)

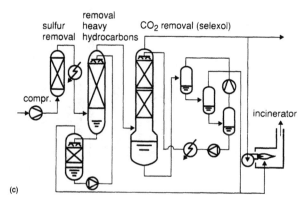

(c)

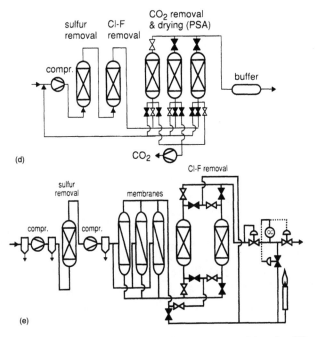

Figure 1. Scheme of different processes used for CO₂ removal from landfill gas: (a) water wash process; (b) Kyrosol process; (c) Selexol process; (d) pressure swing adsorption (PSA) process; (e) membrane process.

Another physical absorption process is the Kryosol process (Markbreiter and Weis, 1987). In this process the CO_2 is absorbed by methanol (Fig. 1b). The gas is compressed to a pressure of 20–35 bar. Before absorption the methanol is cooled down to a temperature of − 30 °C. The CO_2 is released from the methanol by a series of pressure let-down flashes and intermediate heat exchanges.

Heavy hydrocarbons (including halogenated hydrocarbons) and water are removed just before the CO_2 removal by methanol injected into the gas stream. Methanol with heavy hydrocarbons and water is separated from the gas stream. The heavy hydrocarbons and the water are separated from the methanol by a separator and a distillation column respectively. The removed CO_2 is vented to the air. The liquefied heavy hydrocarbons are disposed of as chemical waste.

The third physical absorption process used for CO_2 removal is

Figure 2. Plant for upgrading landfill gas to natural gas quality with CO_2 adsorption in water at Tilburg landfill site, the Netherlands.

the Selexol process (Anon, 1992) (Fig. 1c). The gas is compressed to 31 bar then cooled to − 12 °C. In a preabsorber the heavy hydrocarbons are absorbed by the Selexol solvent. The rich solvent from the preabsorber flows to a stripper. The stripper removes water, heavy hydrocarbons and some CO_2.

In the absorber the CO_2 is absorbed by the Selexol solvent. The rich solvent is regenerated by flashing off CO_2 in three flash drums in series: pressure flash, atmospheric flash and vacuum flash. The lean solvent is recycled to the absorber. The off-gas from the stripper and the off-gas of the flash drums are routed to a thermal oxidizer for destruction of malodorous hydrocarbons at a temperature of about 1000 °C. The product gas is used for supplemental fuel.

Another technique for CO_2 removal is pressure swing adsorption (PSA) (Scheepers *et al.*, 1991). The CO_2 is adsorbed on a molecular sieve. The molecular sieve is regenerated after depressuring with a vacuum pump. Three or four beds are operated automatically in a cyclic procedure (Fig. 1d). In the PSA installation water is removed

Figure 3. PSA plant for upgrading landfill gas to natural gas quality at Wijster landfill site, the Netherlands.

simultaneously with the CO_2. This technology is used at the landfill sites of Wijster (Fig. 3) and Nuenen, the Netherlands.

Carbon dioxide can also be removed by membranes (Connel, 1991; Rautenbach *et al.*, 1991). The gas is compressed to a pressure of 25–35 bar. CO_2 is separated from the landfill gas because in polymer membranes CO_2 and CH_4 molecules have different diffusion velocities. The gas flows through several membrane modules connected in parallel (Fig. 1e) or in series. The CO_2 stream, containing a small amount of CH_4, is flared off. As in a PSA installation, water in a membrane system is removed simultaneously with the CO_2.

When trace components cannot be removed by the CO_2 removal process additional processes are applied (see Chapters 6.5 and 6.6). In this case the treatment steps can be positioned before or after the CO_2 removal unit, depending on the CO_2 removal technique chosen.

The quality of the gas can be measured either by a Wobbe meter, a calorie meter, a gas chromatograph or a set of infrared analysers (CH_4 and CO_2). In the case of a gas chromatograph and infrared analysers the Wobbe number or calorific value is calculated from the measured composition of the gas. The measured value can be used for process control or verification.

If necessary, a buffer tank at the end of the upgrading plant can smooth out fluctuations in gas quality caused by the dynamic behaviour of the CO_2 removal process.

REFERENCES

Anon (1992). Information received from GSF Energy Inc., October 1992.

Connel, R. (1991). Commercial recovery of landfill gas for reticulated gas usage. Paper presented at the Research & Development Seminar of the Australian Gas Association, August 5–6, 1991.

Kohl, A. L. and Riesenfeld, F. C. (1985). *Gas Purification*, 4th edn, Gulf Publishing Company, Houston.

Markbrieter, S. J. and Weis, I. (1987). Conversion of landfill gas to pipeline quality gas. Paper presented at the American Gas Association Conference, Las Vegas, USA, 4–6 May, 1987.

Rautenbach, R., Welsch, K., Ten Berge, G. H. and Schoemaker, A. (1991). Deponiegasnutzung – Schadstoffbeseitigung und Methananreicherung durch Einsatz von Adsorptions- und Membranverfahren. *Gas, Wasser und Abwasser*, Abfall & Deponie.

Scheepers, M. J. J., Van Wingerden, A. J. M. and Hakkers, R. D. D. (1991). Biogas upgrading to pipeline quality by pressure swing adsorption technologies. Paper presented at the 18th World Gas Conference, Berlin. International Gas Union (IGU), Berlin.

Wezel, J. N. Van, Scheepers, M. J. J. and Heijkoop, G. (1988). Upgrading of landfill gas to natural gas quality. *Gas*, **108**, 298–303.

6.5 Landfill Gas Upgrading: Removal of Hydrogen Sulphide

GERHARDT RETTENBERGER

Fachhochschule Trier, Schneidershof, 54293 Trier, Germany

INTRODUCTION

The amount of H_2S that occurs in landfill gas (LFG) depends upon the composition of the deposited waste. Especially when household waste and demolition or construction wastes are deposited together, the content of H_2S in LFG (due to the gypsum content of the demolition and construction wastes) is relatively high.

The content of H_2S in LFG ranges between 0 and 20 g/m^3 (Mollweide, 1989).

H_2S is very toxic (see Chapter 2.4) and when in contact with water (LFG is usually saturated with water) gives rise to the highly corrosive sulphuric acid (H_2SO_4). Thus H_2S may affect the work environment quality and the sulphuric acid may cause corrosion problems in the gas utilization facilities. Because of these properties H_2S should be removed from landfill gas.

TREATMENT METHODS

The majority of currently available methods for the removal of H_2S from landfill gas have been developed for the treatment of natural gas or for gas treatment in industrial chemical processes. Because of these special applications, not all methods available are suitable

Landfilling of Waste: Biogas. Edited by T. H Christensen, R. Cossu and R. Stegmann. Published in 1996 by E & FN Spon, London. ISBN 0 419 19400 2.

for the treatment of LFG. Compared with natural gas or gas in chemical processes, LFG contains a high amount of aggressive and toxic components and is characterized by a very complex composition. A further problem is represented by the fact that LFG presents a varying and relatively small, compared with natural gas, total amount of gas.

An overview of all possible treatment methods for removal of H_2S is summarized in Fig. 1.

Depending on the different treatment processes, the removal of H_2S can give rise to the following end products:

- gas with a high H_2S content;
- pure sulphur;
- pure sulphur and ferrous sulphide;
- zinc sulphide;

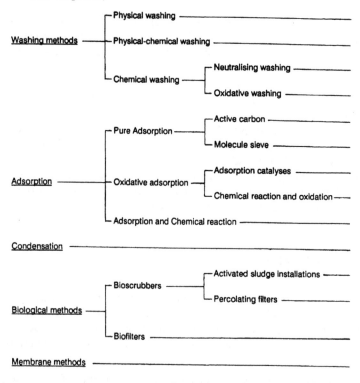

Figure 1. Overview of the methods which can be proposed for H_2S removal from landfill gas (Mollweide, 1989).

- a fluid (condensate) with a high H₂S content;
- sulphur- and sulphate-rich filter material.

Among these end products the first is by far the most problematic material. In fact gas with a high H₂S content, owing to its very high toxic potential, has to be specially treated. The other products (as long as no other trace components such as halogenated hydro-carbons are present) may be used as raw material (pure sulphur) for soil improvement, covering of landfills (sulphur and sulphate-rich filter material) or may be deposited in the landfill itself.

In Table 1 the main end products (with the exception of purified landfill gas) of the H₂S removal processes, as summarized in Fig. 1, are reported.

According to the specific requirements of H₂S removal from LFG, treatment methods should meet the following conditions (Mollweide, 1989):

- high reliability;
- low maintenance costs;
- resistance against other components in LFG such as halogenated hydrocarbons or other trace components;
- adaptability to varying amounts of gas;
- no generation of material or waste that requires further treatment;

TABLE 1. End Products Generated from Different Processes for the Removal of H₂S from Landfill Gas

Treatment processes	*End products*
Washing processes	
Physical washing	H₂S-rich gas phase
Physical–chemical washing	H₂S-rich gas phase
Chemical washing	
Neutralizing washing	H₂S-rich gas phase
Oxidative washing	Pure sulphur
Adsorption	
Pure adsorption	H₂S-rich gas phase
Oxidative adsorption	
Adsorption catalysis	Pure sulphur
Chemical reaction and oxidation	Pure sulphur/ferrous sulphide
Adsorption and chemical reaction	Zinc sulphide
Condensation	H₂S-rich condensate
Biological processes	Sulphur- and sulphate-rich fermentate
Membrane processes	H₂S-rich gas phase

- degradation of the H_2S to pure sulphur, liquid phase or chemical stabilization of the H_2S (for example as sulphate (SO_4);
- low energy consumption.

At the installation site of treatment facilities, the following should be present.

- The installation should be protected against environmental impacts such as rain, snow, wind and cold. Therefore installations should be situated in a room or under a roof.
- Necessary facilities, such as water supply and electricity, should be available.
- If necessary, special energy transport media such as hot and cold water, pressure air and steam should be installed.
- There should be installations for further treatment of end-products from treatment facilities.

Thus, of all the treatment methods reported in Fig. 1, the following processes are not suitable for treatment of LFG owing to the production of a gas phase with a high content of H_2S during the purification process (Mollweide, 1989):

- chemical washing;
- physical–chemical washing;
- neutralizing washing;
- adsorption (without any combination with other processes);
- membrane permeation.

The following processes are, in principle, suitable for the removal of H_2S from LFG.

Condensation. Condensation of the H_2S from LFG is possible, although the cost of cooling and compression is very high. This process is therefore not economical for LFG treatment.

Adsorption in combination with chemical reaction. This combination is usually used for the protection of catalysers. This process, owing to the high consumption of reaction material, can only be applied to the removal of H_2S of landfill gas if H_2S concentrations are low.

Oxidative washing. During oxidative washing the H_2S is dissolved in an alkaline washing liquid (H_2S is more soluble in an alkaline environment). Subsequently, with the help of other compounds, H_2S is transformed into elementary sulphur according to the following reaction:

$$H_2S + \frac{1}{2}O_2 \rightarrow S + H_2O$$

The process is highly selective, and purification of LFG up to concentration levels < 1 mg H_2S/m^3 is possible.

When compared with other processes the oxidative washing process, from the view of the process engineering, is relatively complicated.

Adsorption catalysis. With the help of oxidative adsorption, good purification grades of landfill gas can be achieved. The main advantage of this process is easy adaptation to various gas throughputs. Disadvantages are the high price of activated carbon and the high operative cost for regeneration of the adsorbers.

Chemical reaction and oxidation. The process using dry ferrous mass is the oldest applied to removal of H_2S from gases, and is actually the most common used in the purification of biological gases. The process is carried out in silos using preconditioned materials. During all chemical reactions and oxidation processes a high-content sulphur mass is created, which can be reactivated in special processes or deposited on landfills.

Biological treatment. Bioscrubbers and biofilters are new processes in the field of landfill gas purification. Both processes have been tested in pilot projects. Experiences gained to date provide hints that biofilters represent an economical alternative to the currently operating purification processes.

PRACTICAL EXPERIENCES

At the Gerolsheim landfill, where household, industrial and hazardous wastes are deposited, investigations and tests concerning the removal of H_2S of landfill gas were made.

The H_2S concentration in Gerolsheim landfill gas usually ranges between 1200 and 6150 mg/m³, but peak H_2S concentrations up to 15–20 g/m³ have been observed (Mollweide, 1989). The reason for the high content of H_2S in the gas phase of the Gerolsheim landfill is the high amount of gypsum from industrial neutralization processes that has been deposited in the past.

It has been estimated that the Gerolsheim landfill will produce high amounts of LFG over the next 20 years, which could, for example, be used in gas motors for the production of electricity. Therefore H_2S has to be removed. Two treatment processes for H_2S removal have been tested.

Chemical Reaction and Oxidation with Ferrous Material

The facility that has been installed (Mollweide, 1989) is based on reaction of H_2S with iron ore in the form of $Fe(OH)_3$, according to the following equations:

$$2Fe(OH)_3 + 3H_2S \rightarrow Fe_2S_3 + 6H_2O$$

$$Fe(OH)_2 + H_2S \rightarrow 2FeS_3 + 2H_2O$$

The regeneration of the iron is carried out by oxidation of the ferric and ferrous salts according to the following equations:

$$2Fe_2S_3 + 3O_2 + 6H_2O \rightarrow 4Fe(OH)_3 + 6S$$

$$2FeS + O_2 + 2H_2O \rightarrow 2Fe(OH)_2 + 2S$$

The main unit of the treatment facility is a reaction tower that has long been used in the purification of biogas from sewage plants. Figure 2 shows a scheme of the reaction tower.

In the reaction tower a pre-treated iron material provided by a chemical company was used. This reaction material is mainly composed of ferrous residues from the metal industry, a water-based binder and organic additives. This material, because of its grain size, showed a good permeability for gas.

At a given loading capacity of 35%, and an average content of H_2S in the landfill gas of 4 g/m^3, a mass consumption of 27.5 kg/m^3 of ferrous material was calculated (Mollweide, 1989).

As the average O_2 content of LFG in Gerolsheim is about 5–7 vol%, the installation could operate continuously. A special regeneration of the iron material by blowing into the reactor was not necessary (Mollweide, 1989).

The reaction tower was not insulated with mineral wool to keep the heat away from the exothermic chemical H_2S transformation

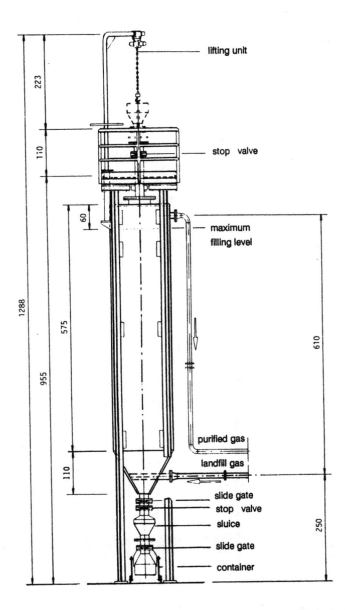

Figure 2. Scheme of the reaction tower installed at Gerolsheim Landfill site for H₂S removal by chemical reaction and oxidation with ferrous material (Mollweide, 1989).

processes in the reactor. Thus heating of the reactor in order to prevent ice generation in winter was not necessary. Only at the bottom of the reaction tower, where the exhausted iron material was taken out, was a heating system installed for wintertime (Mollweide, 1989).

Biofiltration

During the biofiltration process, the gas containing biological degradable substances flows through a filter on which microorganisms are supported. Thereby, the biologically degradable substances are adsorbed by the filter material, and thereafter converted by microbiological processes.

Figure 3 shows a cross-section of one of the biofilter containers installed at the Gerolsheim landfill for H_2S removal.

Purification of the gas phase is mainly dependent on the following conditions:

- pore size distribution;
- temperature;
- pH value;
- water content of the filter material;
- H_2S concentration of the gas phase.

To reach an appropriate pore size distribution in the filter material,

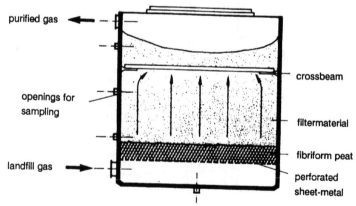

Figure 3. Cross-section of a biofilter used at Gerolsheim Landfill site for biological removal of H_2S from LFG (Sabo, 1991).

composted waste and a specially treated burned clay were mixed (Sabo, 1991).

As the optimal temperature for growth of the micro-organisms is about 28–30 °C, it was necessary to insulate the containers to keep the temperature as constant as possible and to hold the temperature at this level also during the winter (Sabo, 1991).

As a result of H_2S degradation, an acid environment is built up in the filter material. This is not a problem as long as only H_2S has to be eliminated from the gas. If other substances, such as chlorinated hydrocarbons, have to be degraded by microbiological processes, it is necessary to install a second container with neutral pH conditions (Sabo, 1991).

To reach optimal conditions for the growth of the micro-organisms, a minimum water content of at least 25–30% has to be maintained (Sabo, 1991).

With landfill gas containing high H_2S concentrations, such as that observed at the Gerolsheim landfill site, the surface of the filter material is soon covered by the products of the biological conversion of the H_2S (sulphur and gypsum, Ca_2SO_4) and has to be changed. Therefore in Gerolsheim, with a gas amount of about 150 m^3/h and an H_2S concentration of 2500 mg/m^3, the filter material of the container (about 18 m^3) has to be renewed every 8–10 weeks (Sabo, 1991).

CONCLUSION

Removal of H_2S from LFG can be conveniently performed by reaction with ferrous material.

Some problems can occur with the water content and conglomeration of the iron material; these can be solved by circulating the reaction material (Mollweide, 1989).

H_2S removal by biofiltration appears to be a possibility. Problems occur when high H_2S concentrations have to be eliminated from the gas phase. In this case, the biofiltration process, compared with other biological treatment methods such as bioscrubbers, can prove less economical. Biofiltration is suitable for H_2S removal from gas with a lower H_2S concentration, as in the case of LFG generated from most of the ordinary MSW landfill sites.

In this situation, the cost of the filter material is lower, and one filling of the biocontainer can last for up to 2 or 3 years (Sabo, 1991).

REFERENCES

Mollweide, S. (1989). Abschlussbericht zum Teilvorhaben AP IV.1; *Entwicklung von Schadstoffminderungstechniken für Deponiegas aus Abfalldeponien insbesondere zur Entfernung von Schwefelwasserstoff und Halogenwasserstoffen* (Development of techniques to minimize emissions from landfill gas especially the removal of H_2S and halogenated hydrocarbons) im Rahmen des Verbundvorhabens 'Neue Verfahren und Methoden zur Sicherung und Sanierung von Altlasten am Beispiel der Deponie Gerolsheim/Rheinland Pfalz', Stuttgart (unpublished).

Sabo, F. (1991). *Behandlung von Deponiegas im Biofilter* (Treatment of LFG by Biofilters), Stuttgarter Berichte zur Abfallwirtschaft, Bd. 47; Erich Schmidt Verlag, Berlin.

6.6 Landfill Gas Upgrading: Removal of Halogenated Hydrocarbons and Other Trace Organics

KLAUS-DIRK HENNING[a], MANFRED SCHÄFER[b] &
KLAUS GIESSLER[c]

[a]CarboTech-Aktivkohlen GmbH, Franz Fischer-Weg 61, 45307
Essen, Germany
[b]RWE Energie AG, Kruppstrasse 5, 45128 Essen, Germany
[c]CarboTech-Anlagenbau GmbH, Franz Fischer-Weg 61, 45307
Essen, Germany

INTRODUCTION

Before LFG gas can be utilized the removal of specific trace components is required either to meet health and legal regulations or for technical reasons. Of these components the halogenated hydrocarbons need to be carefully monitored owing to their environmental impact (ozone layer destruction, greenhouse effect) and possible negative effects in LFG utilization (corrosion, dioxin formation). Other trace organics, such as benzene and different aromatic substances, also need to be removed or destroyed, as they have proved to be carcinogenic.

If the landfill gas contains high concentrations of these substances the following problems may arise (Dernbach and Henning, 1987):

- possible emissions of polycyclic aromatic hydrocarbons;
- the risk of formation of highly toxic and persistent substances, such as polychlorinated dibenzodioxins (PCDD) and polychlorinated dibenzofurans (PCDF);
- corrosion by formation of hydrogen chlorides and hydrogen fluorides in the combustion chambers of gas engines;

Landfilling of Waste: Biogas. Edited by T. H Christensen, R. Cossu and R. Stegmann. Published in 1996 by E & FN Spon, London. ISBN 0 419 19400 2.

- destruction of the exhaust gas catalysts in the gas engines;
- chlorine and fluorine concentrations in the waste gas from flares and engines;
- concentrations of aromatics, fluorine, and chlorine beyond the statutory thresholds when methane enrichment for SNG (substitute natural gas) production is adopted.

This chapter focuses predominantly on the removal of halogenated hydrocarbons, as these substances are the most dangerous for landfill gas utilization. Other hazardous trace components, particularly aromatic compounds, are also removed in the same treatment processes.

Halogenated Hydrocarbons in Landfill Gas

The general term 'halogenated hydrocarbons' (HHC) covers a variety of compounds and a wide range of physical, chemical, toxicological and ecologically relevant properties and effects. The broad range of features of this group result from the chemical bond of carbon and halogen: that is, the reactivity is a function of the bond pattern (see Chapter 2.4).

The presence of the HHC in landfill gas is usually the result of dumped chemical residues, coolants, propellants agents and chlorinated solvents. Some of these compounds are produced by chemical reactions or by microbial degradation taking place in the waste mass.

An overview of the types and concentrations of these compounds detected in the gas from different landfill sites is shown in Table 1 (Lambert, 1985, 1987). From material balances it has been calculated that the chlorine content originating from identified and quantified substances can only account for approximately 60–90% of the total chlorine content.

Removal Process Concepts

For removal of HHC and other trace components from LFG it may be generally stated that the proposed processes for removing organic compounds from exhaust air streams are also suitable for removing HHC (Kohl and Risenfeld, 1985):

- absorption processes;
- adsorption processes;

TABLE 1. Halogenated Hydrocarbons Concentrations (mg/m^3) in Landfill Gas from Different Sites (Lambert 1985, 1987)

HHC	Landfill sites			
	Fresh kills (USA)	Ahrenshöft (D)	Braunschweig (D)	Karlsruhe (D)
Chlorodifluoromethane	3.1	–	1.0	0.01
Dichlorofluoromethane	0.9	–	11.0	0.01
Trichlorofluoromethane	–	33.1	47.8	12.0
Dichlorodifluoromethane	22.1	103.1	36.2	25.8
Dichloromethane	0.3	–	11.0	0.01
Trichloromethane	–	–	10.7	–
Tetrachloromethane	–	–	–	<0.01
Trifluorotrichloroethane	3.4	1.7	1.7	–
Tetrafluorodichloroethane	5.3	–	11.0	0.01
1,1-Dichloroethane	–	15.0	–	–
1,1,1-Trichloroethane	–	3.0	0.6	0.02
Chloroethene	0.6	–	9.5	–
1,1-Dichloroethene	–	–	7.8	–
1,2-Dichloroethene	3.9	3.9	294.4	–
Trichloroethene	1.8	12.3	31.7	<0.01
Tetrachloroethene	1.5	14.1	83.6	<0.01
Chlorine total	25.1	125.1	401.9	24.5
Fluorine total	11.9	37.5	18.5	9.8

- catalytic processes;
- high-temperature combustion;
- biological oxidation.

LFG purification, however, presents different problems from the normal industrial applications for the following reasons.

- Industrial exhaust air streams contain in most cases only one and, rarely, more than four different compounds.
- In landfill gas from 10 to 20 different chlorine and chlorofluorohydrocarbons are frequently found at the same time. In addition, a large variety of organic impurities, i.e. other material groups, are found, often with high concentrations.
- In industrial applications, the HHC concentrations can range between 1000 and 10 000 mg/m^3, while concentrations in LFG are in the range 5–500 mg/m^3 for each individual compound. It is only the sum of the large number of

individual HHCs that gives rise to the high HHC content in LFG.

ABSORPTION

General Aspects

In the absorption processes, an absorbent liquid removes the substance from the gaseous phase and forms the absorbate.

The rate and performance of the purification process are controlled by the following parameters:

- the concentration gradient between the absorbate in the gas and the absorbate in equilibrium in the liquid phase;
- the material transport surface;
- the material transport and diffusion coefficients.

For removal of small HHC concentrations and other trace components, only countercurrent absorption is suitable. Cylindrical, upright columns with packings are used as absorption reactors.

Water-soluble gas impurities are removed from the gas by using water as absorbent. After absorption the absorbate is treated in a purification plant as an aqueous solution.

Impurities that are not water-soluble can be removed only by organic absorbents with extremely low vapour pressures and moderate viscosity. In these cases the absorbate is separated from the absorbent (e.g. by distillation) either for recovery or destruction. The absorbent is then recycled to the absorption reactor.

An example of absorption technology is the Helasorp process (Schneider, 1989, 1991). The landfill gas is passed through one or two absorption columns, and trace components are washed out of the gas by the absorbent Helasorp, a blend of various hydrocarbons. The absorption liquid is continuously cycled between the absorber and the regeneration columns. In the regeneration column organic trace components are desorbed from the absorption liquid at a temperature of about 90 °C and a pressure of 5 hpa (abs.). The chlorine-containing desorbate is disposed of. Owing to the low regeneration temperatures (approximately 90 °C) no HCl or HF is produced from the halogenated hydrocarbons and thus the plant construction requires no corrosion-resistant materials. Chlorine removal efficiencies higher than 90% have been achieved.

Applications

Examples of full-scale application of the absorption process are given by the plants operating at the Braunschweig and Berlin-Wannsee landfill sites (Schneider, 1991). The Berlin-Wannsee one is the largest European plant: 3000 m^3/h of landfill gas are used in a combined heat and power (CHP) station to produce 4.5 MW of electrical energy and 6.5 MW of thermal energy. The raw gas contains about 1 g/m^3 of trace components and 35 mg/m^3 of halogenated hydrocarbons.

For gas purification, a countercurrent absorption reactor is used. Detailed information on its performance is given in Chapter 9.3.

Another example of HHC removal is given by the LFG-upgrading plant at the Tilburg landfill site, the Netherlands (see Chapters 6.4 and 7.5). This plant has been designed to convert LFG to SNG quality. The process includes CO_2 and sulphur removal steps. CO_2 is removed by a water absorption process, specifically designed for this purpose. Nevertheless, the halogenated hydrocarbons (C_{5+}) are largely removed in this process (Table 2).

For complete removal an activated carbon adsorption step was installed. The activated carbon unit consists of two adsorbers, which are regenerated in turn with hot CO_2 (flash vapour from the water absorption process). This automatic procedure has a cycle time of 24 h. The chlorine content in the product gas is constantly kept below 20 mg/m^3 (van Wezel *et al.*, 1988).

TABLE 2. Typical Removal Performance of Trace Components from LFG by Water Scrubbing and Activated Carbon Adsorption

Trace components	Landfill gas	After water scrubbing	After activated carbon adsorption	Specification SNG[a]
Halogenated hydrocarbons (mg Cl/m^3)	111	49	17	<25
Higher hydrocarbon (C^{5+} mol ppm)	50–250	25–125	–	No specification
Aromatics (mg mol ppm)	50–80	5–10	<5	<200

[a]SNG = Substitute natural gas.

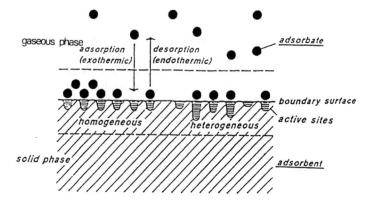

Figure 1. Graphic description of the adsorption and desorption process.

ADSORPTION

Fundamentals

In the adsorption process the enrichment of gaseous or dissolved substances (the adsorbate) on the boundary surface of a solid (the adsorbent) occurs. On the adsorbent surface there are 'active sites' where the binding forces between the individual atoms of the solid structure are not completely saturated. An adsorption of foreign molecules takes place onto these active sites (Fig. 1).

The adsorption process is generally exothermic. With increasing temperature and decreasing concentration the adsorption capacity decreases. For the design of the adsorption process it is important to know the adsorption capacity at constant temperature in relation to the adsorbate concentration.

Figure 2 shows typical adsorption isotherms, according to the Freundlich representation. The adsorption capacity of some halogenated hydrocarbons is plotted against their concentration in the liquid phase at equilibrium. The various substances are adsorbed differently, depending on the strength of the interacting forces between adsorbate and adsorbent.

Furthermore, in many cases, several adsorbates compete for available adsorption places. This occurs frequently in the application of the process in waste-gas and wastewater purification. In this case adsorption tests to determine the load of the activated carbon are necessary.

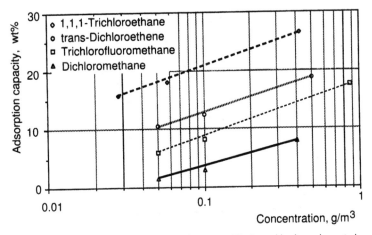

Figure 2. Adsorption isotherms at 20 °C for some chlorinated hydrocarbons (adsorbent: D47/4).

Adsorption Test for Single-Component System

A series of adsorption tests on landfill gas for HHC removal have been carried out under dynamic conditions by using a small-scale reactor (Schäfer, 1989; Schäfer *et al.*, 1991).

The experimental set-up for the test is summarized in Table 3. At the beginning of the test no HHCs were detected in the clean gas, downstream of the activated carbon reactor. After a certain period individual components broke through and were identified in the clean gas. When eventually the activated carbon was fully

TABLE 3. Experimental Set-up for Dynamic Fixed-Bed Adsorption Tests

Adsorber:	Height	1000 mm
	Diameter	50 mm
	Volume	2008 cm^3
	Gas flow	2.8 m^3/h
	Flow rate	0.4 m/s
Adsorbent:	Activated carbon (quality D 43/4)	
Concentration:	HHC (per adsorbate) water	100 mg/m^3 20% relative humidity
Temperature:	Adsorber	20 °C

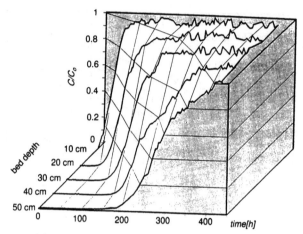

Figure 3. Breakthrough curves for trichloromethane.

saturated, the concentration in the cleaned gas (C), corresponded to the inlet concentration of the raw gas (C_0): i.e. the breakthrough ratio $C/C_0 = 1$.

This adsorption process is described, for trichloromethane, by the graph in Fig. 3. Figure 4 shows the breakthrough curves of various single-component systems for an activated carbon bed of 50 cm. The adsorption capacity values, observed for each component during the tests, are summarized in Table 4.

Adsorption Test for Multi-Component Mixtures

The tests for multi-component adsorption were carried out under similar conditions, adopting HHC mixture concentrations of 100 mg/m^3 per adsorbate (Schäfer, 1989; Schäfer *et al.*, 1991).

The breakthrough curves for multi-component mixtures show displacement effects. In the course of the competing adsorption, adsorbates with strong interaction forces displace, in the activated carbon, less firmly bound substances. These effects are shown for the binary trichloroethene/tetrachloroethene system in Fig. 5. In this case, trichloroethene and tetrachloroethene, contained in equal concentration in the mixture, were adsorbed completely onto the activated carbon surface at the beginning of the adsorption process. Later, the less firmly bound trichloroethene was found in increasing

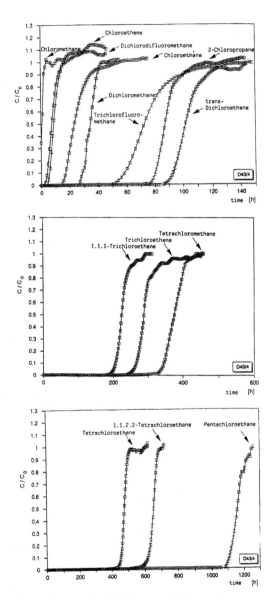

Figure 4. Breakthrough concentrations for adsorption of various halogenated hydrocarbons at different contact times, using activated carbon, (quality D43/4): (a) 0–150 h; (b) 0–600 h; (c) 0–1300 h.

TABLE 4. Adsorption Capacity Values under Dynamic Conditions Observed During Experimental Tests for HHC Removal by Activated Carbon Adsorption

	Adsorption capacity (g/100 g AC)				
Adsorbate/bed height (cm)	20	30	40	50	Average
Chloromethane	0.04	0.05	0.06	0.06	0.5 ± 0.01
Dichloromethane	2.32	1.93	1.98	2.25	2.12 ± 0.19
Trichloromethane	20.11	21.46	21.75	21.22	21.14 ± 0.72
Tetrachloromethane	23.79	24.78	25.39	25.24	24.80 ± 0.72
Chloroethane	1.55	1.62	1.75	1.74	1.67 ± 0.10
1,1,1-Trichloroethane	16.52	15.32	15.22	14.87	15.48 ± 0.72
1,1,2,2-Tetrachloroethane	43.01	42.96	42.51	42.41	42.72 ± 0.31
Pentachloroethane	82.94	82.69	80.62	79.01	81.32 ± 1.86
2-Chloropropane	5.31	5.67	5.83	5.87	5.67 ± 0.26
Chloroethylene	0.70	0.71	0.74	0.76	0.73 ± 0.03
trans-Dichloroethene	6.39	6.59	6.85	6.95	6.70 ± 0.25
Trichloroethene	21.26	19.31	19.97	19.63	20.04 ± 0.86
Tetrachloroethene	35.01	36.17	32.85	32.44	34.12 ± 1.78
Dichlorodifluoromethane	0.55	0.72	0.64	0.66	0.64 ± 0.07
Trichlorofluoromethane	6.11	5.13	5.20	5.23	5.42 ± 0.46

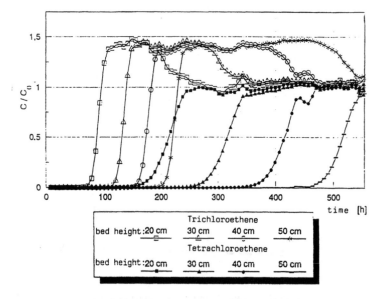

Figure 5. Displacement effect during simultaneous adsorption of tri- and tetrachloroethene, at 20 °C and 20% relative humidity (adsorbent D 43/4).

TABLE 5. Adsorption Capacity (g/100 g AC) at Equilibrium of Different HHC when Present in the Gas as Single Component (S) or in Binary (I + II) and Ternary (I + II + III) mixtures (M)[a]

			Bed height, (cm)							
			20		30		40		50	
Mixture	HHC Component	r.h. (%)	S	M	S	M	S	M	S	M
I	Dichlorofluoromethane	20	0.55	0.33	0.72	0.28	0.64	0.30	0.66	0.31
II	Trichlorofluoromethane	20	6.11	5.34	5.13	4.51	5.20	4.64	5.23	4.71
I	Trichloroethene	20	21.26	6.07	19.31	6.18	19.97	6.49	19.63	6.80
II	Tetrachloroethene	20	35.01	34.34	36.17	33.85	32.84	33.29	32.44	33.10
I	1,1,1-Trichloroethane	20	16.52	3.59	15.32	4.14	15.22	4.20	14.87	4.40
II	1,1,2,2-Tetrachloroethane	20	43.01	38.60	42.96	40.04	42.51	39.32	42.41	39.51
I	Chloromethane	20	<0.10	<0.10	<0.10	<0.10	<0.10	<0.10	<0.10	<0.10
		0	–	<0.10	–	<0.10	–	<0.10	–	<0.10
II	Freon 12	20	0.55	0.25	0.72	0.36	0.64	0.45	0.55	0.38
		0	–	0.59	–	0.62	–	0.53	–	0.55
III	Chloroethene	20	0.70	0.27	0.71	0.51	0.73	0.75	0.76	0.71
			–	1.14	–	1.12	–	0.98	–	1.03

[a] The values have been observed at different heights of the adsorber reactor. The tests have been carried out by using activated carbon (quality D43/4), at 20 °C and indicated relative humidity (r.h.)

concentrations in the clean gas. In the end the displaced component concentration was definitely higher than the inlet concentration. The more firmly adsorbed tetrachloroethene did not reach a concentration higher than C_0. At the end of the displacement adsorption the trichloroethene concentration stabilized at a value $C/C_0 = 1$.

Over long adsorption periods, these competing adsorption effects often lead to a significant increase in the actual adsorbate concentration of the partly displaced component. When planning adsorption plants on an industrial scale these phenomena are important and have to be carefully considered in the design.

Table 5 summarizes the adsorption capacity values observed at the equilibrium for single- and multi-component systems.

Influence of the Gas Humidity

The influence of the gas humidity on the adsorption behaviour can be noticed observing the adsorption capacity values in the ternary HHC mixtures reported in Table 5. The results show dry gas values higher by a factor of up to 2 for dichlorofluoromethane and for chloroethene: water vapour competes in adsorption and reduces the activated carbon adsorption efficiency. This efficiency reduction applies even more to high-volatility adsorbates. For them, no high adsorption by the activated carbon is to be expected. This effect will be even stronger if preliminary moisturized activated carbon – e.g. subsequent to water vapour desorption – is used for adsorption.

Process Development

Solvent recovery by adsorption is usually a batch operation involving multiple beds. At least one activated carbon bed remains on line while the others are being regenerated.

The adsorber feed is pretreated to remove solids (dust), liquids (droplets or aerosols) or high-boiling components, as they can affect performance. In most of the systems, the solvent-loaded airstream flows up through a fixed carbon bed.

Spent carbon is usually regenerated by backflowing low-pressure steam. This removes the adsorbed solvent, which is recovered by condensing the vapours and separating the solvent from water by either decantation or distillation. The countercurrent pattern of

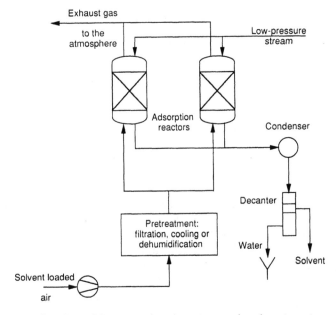

Figure 6. Flow sheet of the process for solvent recovery by adsorption.

adsorption and desorption ensures high removal efficiencies. A schematic flow sheet is shown in Fig. 6.

After steam regeneration the hot and wet carbon bed will not remove organics from air effectively, because high temperature and humidity strongly reduce adsorption. The activated carbon is therefore dried with hot air. Before starting the next adsorption cycle the activated carbon bed is air-cooled down to ambient temperature.

Pilot plants at a landfill in Braunschweig, Germany, have been set up to test modifications of this process of adsorptive removal of trace components from landfill gas (Dernbach, 1985; Olderdissen, 1988; Werner, 1988). The plants were equipped with an upstream filter for H_2S removal by catalytic oxidation with activated carbon. Subsequently, the desulphurized landfill gas was passed through an adsorber for removal of HHC and other organic components. The chlorine concentration of the raw gas varied between 400 and 600 mg/m^3. The HHC breakthrough into the clean gas was almost continuously monitored by an AOX unit or a specially developed analyser (Olderdissen, 1988). Desorption was run on steam. To select suitable adsorbents activated carbon types from different manufacturers as well as

TABLE 6. Results of the Field Trials at the Braunschweig Landfill Site

Adsorbents	Chlorine concentration in the clean gas (mg/m³)	Removal rate (%)	Specific throughput (after 10 cycles) (m³ gas/m³)	Specific chlorine load (after 10 cycles) (g Cl/kg)
Activated carbon	5–20	96–99	7700–13 000	10–13
Lignite coke	20	96	1700	1.8
Aluminium oxide	40	92	Not regenerable under test conditions	
Adsorber resins	10–30	93–98	2900–4700	1.9

adsorber resins, aluminium oxide, and lignite coke were tested over several adsorption/desorption cycles in each until constant results were achieved. Table 6 shows clearly that, as expected, activated carbon yielded the best purification and the highest specific chlorine loads. Furthermore, the load profiles of the activated carbon in the adsorber are interesting and typical (Table 7).

On the adsorber inlet side a large portion of the chlorine compounds was displaced by other gas components while on the adsorber outlet significantly higher loads were achieved. It is obvious that the activated carbon removes halogenated hydrocarbons but also other gas impurities with concentrations substantially higher.

Also, competing adsorption (as discussed before) is of industrial importance with respect to a well-timed switching of the raw gas stream to the second unloaded adsorber when HHC breakthrough starts in the first adsorber. If the load cycle is not stopped in time the clean gas may show higher HHC concentrations than the raw

TABLE 7. Activated Carbon Load Profile Observed at the Braunschweig Pilot Plant

Fraction (cm)	Activated carbon load	
	Chlorine load (g/kg ac)	Total load (g/kg ac)
0–20	2.2	296
20–50	2.1	287
50–80	2.3	209
80–110	3.1	173
110–140	7.5	180
140–170	38.1	183
170–200	33.2	169

TABLE 8. Description and Performance of Different Full-Scale Plants for LFG Upgrading with Removal of H_2S and HHC by Adsorption Processes

Landfill site	Gas quantity (m^3/h)	H_2S removal	HHC removal	Clean gas (H_2S mg/m^3)	Chlorine (mg/m^3)	Further treatment steps	Purposes
Wijster/NL	Max. 1825	Impregnated activated carbon	Activated carbon adsorber	<1	<5	PSA process for CO_2 removal	Immission into natural gas network
Nuenen/NL	Max. 1200	Impregnated activated carbon	Activated carbon adsorber	<5	<25	PSA process for CO_2 removal	Immission into natural gas network
Tilburg/NL	Max. 2000	Water absorption + iron oxide	Water absorption + activated carbon	<5	<25	CO_2 removal by H_2O	Immission into natural gas network
Vasse/NL	Max. 250	Impregnated activated carbon	Activated carbon adsorber Permeate gas desorption	<5	<25	Membrane process for CO_2 removal	Immission into natural gas network
Kaiserslautern /BRD		Impregnated activated carbon	Activated carbon adsorber Steam desorption	<5			Electricity production
Brandholz /BRD	1000	Impregnated activated carbon	Activated carbon adsorber Steam desorption	<5	<25		Electricity production

gas. Thorough drying of the desorbed wet activated carbon was found to be particularly important for purification performance and capacity. Possible cycle times are moderated mainly by the fraction of extremely high-volatile halogenated hydrocarbons (dichlorodifluoromethane, chloromethane, chloroethene) contained in the landfill gas, which are characteristically difficult to adsorb.

Application Examples

The process has to meet high standards with respect to removal of trace components if the landfill gas is to be processed to substitute natural gas (SNG) (see Chapter 8.1). The activated-carbon-based processes for LFG upgrading are well established in various industrial plants (Prinsloo *et al.*, 1974; Hinz and Schilling, 1987; Pilarczyk *et al.*, 1987), as summarized in Table 8.

Using carbon-containing adsorbents (activated carbon, impregnated activated carbon, carbon molecular sieves), the following process steps can be carried out (Table 9):

• H_2S removal;
• removal of halogenated hydrocarbons by adsorption onto activated carbon;

TABLE 9. Processes for Purification and Separation of Specific Contaminants from Landfill Gas

Processing by removal of	Adsorption process	Other processes
Hydrogen sulphide	Catalytic H_2S oxidation on activated carbon	H_2S-washing with bog iron ore
Ammonia	Conversion with acid-impregnated activated carbon	Water or acid washing
Mercury	Chemisorption of mercury on sulphur-impregnated activated carbon	Oxidizing washing
Halogenated hydrocarbons	Adsorption on activated carbon	Adsorption on other adsorbents, oil washing
Carbon dioxide	Pressure-swing adsorption using the carbon molecular sieve CMSC	CO_2 washing, zeolites, diaphragm processes

- ammonia removal by chemisorption onto activated carbon impregnated with sulphuric acid;
- mercury removal with activated carbon impregnated with sulphur;
- carbon dioxide removal with carbon molecular sieves.

Pressure-swing adsorption technology and diaphragm systems are well proven in upgrading landfill gas to pipeline quality. The removal of trace components is explained below using the example of the landfill site in Vasse, Netherlands.

Based on the results obtained with the pilot plant at the landfill site of Neuss-Holzheim, FRG, a commercial plant was constructed on the Vasse landfill site to upgrade landfill gas by a membrane process, feeding it into an SNG network (Rautenbach and Ehresmann, 1989a, b).

The process scheme includes the following units:

- H_2S adsorber
- diaphragm unit
- HHC adsorber
- flare for the residual-gas

In the first step H_2S removal is carried out: i.e. catalytic oxidation produces elemental sulphur, which remains within the pore system of specially impregnated activated carbon. Subsequently, the gas is compressed to 40 bar and methane-enriched to approximately 87% in a three-module membrane plant. HHC removal occurs by adsorption under pressure onto activated carbon in a twin-adsorber unit. With higher pressure, adsorbers may be smaller in size and thus a higher specific load is achieved. Desorption does not run on steam but on hot permeate gas, which is subsequently flared off. The Netherlands chlorine emission standard of 25 mg/m^3 has been constantly maintained.

CATALYTIC PROCESSES

Fundamentals

Work is in progress on the development of a catalytic process for the removal of halocarbons from landfill gas. Halocarbons are transformed by catalytic dehydrohalogenation/dehydrogenation into hydrocarbons (Prinsloo *et al.*, 1974; Dernbach and Henning, 1987).

TABLE 10. List of Catalysts Suitable for HHC Decomposition

Catalyst/catalyst carrier	Impregnated (10–20%) with:
Activated carbon	Calcium chloride
Aluminium oxide	Barium chloride
Silica gels	Copper salts
Molecular sieves	Cobalt salts

These reactions take place at temperatures ranging between 300 and 600 °C with suitable catalysts. The process of dehydrohalogenation can be explained taking chloroethane as an example.

Chloroethane is broken down by a suitable catalyst into ethene and hydrogen chloride:

$$CH_2Cl–CH_3 \rightarrow catalyst \rightarrow CH_2=CH_2+HCl$$

The resulting hydrogen chloride can be removed from the gas by calcium oxide:

$$CaO + HCl \rightarrow CaCl_2 + H_2O$$

Suitable catalysts for decomposition are listed in Table 10.

HCl removal may take place directly in the gaseous phase (alkaline impregnated activated carbons, alkaline products, such as calcium oxide), or by an adsorption process using alkaline aqueous solutions.

Table 11 shows test results for heterogenous catalytic

TABLE 11. Test Results for Heterogeneous Catalytic Dehalogenation of Specific Chlorinated Hydrocarbons

Chlorinated hydrocarbon	Cl concentration in the raw gas (mg/m^3)	Cl concentration in the treated gas (mg/m^3)	Conversion (%)
1,2 dichloroethane	590	2	99.7
trichloromethane	1150	1	99.9
trichloroethene	510	3	99.3
tetrachloroethane	490	1	99.8

Test conditions:
Temperature 450 °C
Catalyst: activated carbon A 35/3 + 5% BaCl$_2$
HCl adsorbent: Halexmasse, Grillo-Duisburg
Contact time: 4 s

dehalogenation of specific chlorinated hydrocarbons with conversions of >99%.

Process Development

If halocarbons with several halogen atoms in a molecule or mixtures of different halocarbons are present in the gas, a complicated reaction sequence with numerous parallel reactions can be expected. Tests with laboratory grade gasses containing defined amounts of chloro- and fluoro-hydrocarbons yielded encouraging results. An initial field trial was carried out using LFG from the Braunschweig landfill site.

Figure 7 illustrates the structure of the test plant and shows the results obtained. The landfill gas was pumped through a reactor, which was indirectly heated electrically and filled with catalyst and calcium oxide. It was possible to reduce the TOX content from about 750 mg/m^3 to values below 50 mg/m^3, thus achieving an elimination rate of more than 95% (Dernbach and Henning, 1987).

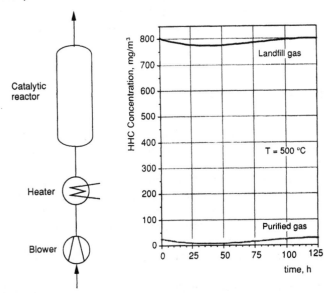

Figure 7. Flow sheet of the plant for catalytic HHC removal from gas at the Braunschweig landfill site, and process performance.

HIGH-TEMPERATURE COMBUSTION

In modern high-temperature flares, combustion temperatures of 1200 °C and residence times of 0.3 s can be reliably achieved. Thus formation of dioxins (PCDD, PCDF) from halogenated compounds can be avoided with certainty.

In contrast to thermal combustion, catalytic combustion involves oxidation of halogenated organic compounds on the surface of catalysts at temperatures ranging between 300 and 400 °C. The following catalysts may be used:

- precious metals (platinum, palladium) on organic support;
- precious metals (platinum, palladium) on metallic support;
- metal oxides (vanadium pentoxide, tungsten trioxide) on organic support.

For catalytic processes it must be ensured that catalyst poisons (such as silicone, phosphorus, arsenic, lead, or sulphur) are removed from the gas in order to prevent deactivation of the catalyst surface. This technology gave promising results in the concentration range of 1×10^{-9} g/m^3 for polychlorinated organic compounds, i.e. dioxins and furans in the exhausted gas from waste incineration plants.

BIOLOGICAL OXIDATION

Since the 1950s, biological oxidation has been used for purification of exhaust airstreams, particularly to remove compounds (such as alcohols and ketones) that can be easily biodegraded.

In landfill gas purification, a biological conversion takes place at the site, so that easily degradable compounds are converted to CO_2 and H_2O or to biologically stable intermediate products. Anaerobic biochemical degradation of tetrachloroethene via dichloroethene and the formation of vinyl chloride was observed. In the case of full degradation the production of hydrochloric acid is harmful to the micro-organisms.

For these reasons, even though investment and operation costs for biological gas treatment are low, the method is usually not used for the removal of trace components from LFG.

TABLE 12. Trace Components Concentrations in Landfill Gas and after Adsorption Observed at the Brandholz Plant

Components	*Landfill gas*	*After adsorption*
Hydrogen sulphide (mg/m^3)	max. 150	<5
Chlorinated hydrocarbons (mg Cl/m^3)	30–150	<25
Fluorocarbons (mg F/m^3)	10–50	<25
Aromatic compounds (mg/m^3)	max. 500	<25

DESIGN EXAMPLE FOR ADSORPTIVE GAS PURIFICATION

The Brandholz landfill site in Germany offers a good example of a two-step activated carbon process for H$_2$S and HHC removal (Krabiell and Giessler, 1991) treating 1000 m^3/h of gas for the production of electricity (Table 12).

The process comprises the following process steps:

1 catalytic H$_2$S removal onto impregnated activated carbon;
2 adsorptive removal of aromatics/halogenated hydrocarbons.

H$_2$S Removal

Hydrogen sulphide reacts on impregnated activated carbon in the presence of oxygen at low temperatures to give elemental sulphur and water (Henning and Schilling, 1987):

$$2H_2S + O_2 \xrightarrow{activated\ carbon} \tfrac{1}{4} S_8 + 2H_2O \qquad \Delta H = -444 \text{ KJ}$$

The sulphur is adsorbed onto the internal surface of the activated carbon. High sulphur loads have been achieved. In most cases the oxygen required for H$_2$S oxidation is already present in the landfill gas. For H$_2$S removal the gas is passed into a two-adsorber cascade filled with impregnated activated carbon at 70 °C. After a net operating time of 4.2 h the activated carbon of the first adsorber is loaded with sulphur. The gas is passed into the second adsorber first and then through the first adsorber filled with fresh activated carbon.

HHC Removal

The design data are outlined in Table 13. The H$_2$S-free gas from the first step is now passed into the HHC adsorbers after being

TABLE 13. Design Data of the HHC Adsorption Plant at the Brandholz Landfill Site

Two-adsorber plant	Design data
Adsorber	7 m^3
Activated carbon volume	7 m^3
Landfill gas feed	1000 Nm^3/h
Operating pressure	approx. 1.3 bar (a)
Landfill gas temperature upstream of the HHC step	approx. 25 °C
Rel. humidity of the landfill gas upstream of the HHC step	approx. 35%
Residence time of the landfill gas	approx. 30 s
Adsorption cycle time	approx. 50 h/cycle
Regeneration time	approx. 6 h
Steam consumption (regeneration)	approx. 280 kg/h (130 °C)
Drying period (activated carbon)	approx. 8 h
Drying agent/ambient air	2000 Nm^3/h
Steam consumption (for air pre-heating to approx. 100 °C)	150 kg/h (130 °C)
Cooling time (activated carbon)	approx. 4 h
Cooling agent/ambient air	2000 m^3/h
Electricity consumption	90 kW

cooled and the dew-point lowered. This purification step comprises two adsorbers made from alloyed steel, insulated, and filled with a special type of activated carbon. These adsorbers are arranged in parallel to allow continuous operation in which one adsorber (B03) is used and the other (B04) is regenerated alternately. This change-over from loading to regeneration and back takes place about every 50 h (Fig. 8).

The path of the partially purified landfill gas passes through the bottom of the K9 base into the B03 adsorber. While the gas flows through B03 the trace components are adsorbed onto the activated carbon. The purified landfill gas passes through the gates K11, K36, K37, K23, a dust filter F02, and through an anti-flashback system to the consumer. Once the activated carbon is HHC-loaded the partially purified LFG stream is switched over to adsorber B04. Adsorber B03 is then steam regenerated.

The steam flows, for regeneration, into the adsorber from the top via the pressure reduction unit D03 and gate K19. The steam warms up the activated carbon bed and leaves the adsorber B03 as steam/condensate mixture via gates K10 and K22. Heating causes desorption and the steam-condensate mixture arrives at condenser W03 for condensation. The condensate is passed by pump P04 to the condensate treatment plant.

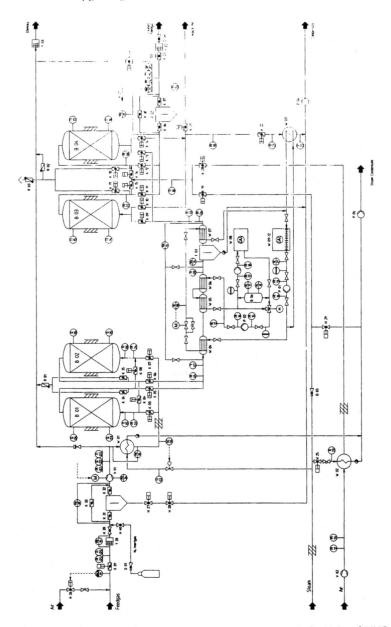

Figure 8. Flow sheet of the two steps activated carbon process for H$_2$S and HHC removal at the Brandholz plant.

The wet activated carbon is dried after regeneration by hot air (temperature > 100 °C) from the heat exchanger W02. Heated air flows through adsorber B03 via gates K20 and K15 countercurrent and leaves the adsorber by gates K10 and K21. This drying cycle runs until the outlet temperature (T1 14) corresponds approximately to the inlet temperature (T1 15). Before adsorber B03 is loaded again the residual air in the vessel needs to be flushed out by landfill gas. After 50 h of operation the other adsorber (B04) is loaded and switched with B03 to allow for the same regeneration process to be performed.

For monitoring the plant, several pressure gauges and thermometers are installed. The gas flow (by volume) and the HHC concentrations in the treated gas are measured continuously. The clean gas standards listed in Table 13 have been constantly maintained.

REFERENCES

Dernbach, H. (1985). Landfill gas utilization in Braunschweig – quality of gas and damages due to corrosion. *Waste Management & Research*, **3**, 149–59.

Dernbach H. and Henning K.-D. (1987). Purification steps for landfill gas utilization in cogeneration modules. *Resources and Conservation*, **14**, 273–82

Hinz, W. and Schilling, H. (1987). Die Reinigung und Aufbereitung von Deponie- und Abfallgasen Erdol und Kohle-Erdgas-Petrochemie vereinigt mit Brennstoff. *Chemie*, **40**(10), 442–6.

Kohl, A. L. and Riesenfeld, F. C. (1985). *Gas Purification*, 4th edn, Gulf Publishing Company.

Krabiell, K. and Giessler, K. (1991). Gereinigtes Deponiegas: Aus einem Ozonkiller wird ein verlaßlicher Engergieträger. *Wasser, Luft, Boden.*

Lambert, R. (1985). *Deponietechnik heute und morgen*, Erich Schmidt Verlag, Bielefeld.

Lambert, R. (1987). Reinigung von Deponiegas. *Stuttgarter Berichte zur Abfallwirtschaft*, **19**, 189–92.

Olderdissen, G. (1988). Monitor fur Chlorkohlenwasserstoffe in Deponiegas. *Technisches Messen*, **55**(5), 182–4.

Pilarczyk, E., Knoblauch, K. and Jüntgen, H. (1987). Erdgas aus Biogas mittels Druckwechseltechnik. *Gas-Erdgas gwf*, **128**, 340–5.

Prinsloo, J. J., van Berge, P. C. and Zlotnik, J. (1974). The catalytic decomposition of 1,2 dichloroethane with activated carbon catalysts. *Journal of Catalysts*, **32**, 466–9.

Rautenbach, R. and Ehresmann, H. E. (1989a). Gaslager der Zukunft – Deponiegas durch Membranen aufarbeiten. *Chemische Industrie*, **9**, 24–32.

Rautenbach, R. and Ehresmann, H. E. (1989b). Upgrading of landfill gas by membranes – process design and costs, comparison with alternatives, in *Proceedings 5th BOC Priestleg Conference*, Birmingham.

Schäfer, M. (1989) Thesis, Universitat-Gesamthochschule, Essen.

Schäfer, M., Schroter, H. J. and Peschel, G. (1991). Adsorption and desorption of halogenated hydrocarbons using activated carbon. *Chemical Engineering and Technology*, **14**, 59–64.

Schneider, J. (1989). Deponiegaserfassung und -aufbereitung. *Entsorgungspraxis*, **11**, 612–17.

Schneider, J. (1991). Reinigung von Deponiegas. *Entsorgungspraxis*, **4**, 61–166.

van Wezel, J. N., Scheepers, M. J. J. and Heijkoop, G. (1988). Upgrading of landfill gas to natural gas. *Gas*, **108**, 298–303.

Werner, M. (1988). Entfernung halogenierter Kohlenwasserstoffe aus Deponiegas. *Mull und Abfall*, **11**, 520–3.

FURTHER READING

Henning K.-D., Klein, J. and Knoblauch K. (1985). Schwefelwasserstoff-Entfernung aus Biogas mit einem Aktivkohle-Verfahren gwf. *Gas/Erdgas*, **126**, 19–24.

Henning K.-D., Schäfer M., Bongartz, W. and Knoblauch, K. (1987). Investigation of the adsorptive removal and recovery of halocarbons, in *Proceedings 2nd European Conference on Environment Technology*, Amsterdam, The Netherlands, 22–26, 1.

Laugnitz, G., Poller, T. and Stegmann, R. (1988). Entstehen und Verhalten von Spurenstoffen im Deponiegas sowie umweltrelevante Auswirkungen von Deponiegasemissionen, in *Deponiegasnutzung* (eds R. Stegmann and G. Rettenberger), Hamburger Berichte 1, Economica Verlag, Bonn, pp. 153–163.

7. GAS UTILIZATION

7.1 Options and Economics of Landfill Gas Utilization in the USA

SUSAN A. THORNELOE

US Environmental Protection Agency, Air and Energy Engineering Laboratory, Research Triangle Park, NC 27711, USA

INTRODUCTION

Because of the health and environmental concerns, the EPA has designated 'landfill air emissions' as a pollutant. The EPA has proposed Emission Guidelines for existing landfills and New Source Performance Standards for new landfills.

The regulatory alternative proposed by the Clean Air Act regulations would result in requiring 500–700 landfills to collect and control MSW landfill air emissions (Federal Register, 1991). It is hoped that those sites affected by these regulations will utilize the gas, rather than flaring it. The use of energy recovery for the control of MSW landfill air emissions has the potential to reduce secondary air impacts at coal-fired power plants and to decrease the use of fossil fuels. The focus of this chapter is to identify the different options for gas utilization that are considered in the USA, and the state of application of these options. Case studies are presented to illustrate the different options for landfill gas utilization.

UTILIZATION OPTIONS

Through a recent EPA survey of US landfill gas energy projects, landfill gas (LFG) energy recovery projects were identified and

Landfilling of Waste: Biogas. Edited by T. H Christensen, R. Cossu and R. Stegmann. Published in 1996 by E & FN Spon, London. ISBN 0 419 19400 2.

classified in the USA (Thorneloe, 1992; Thorneloe and Pacey, 1994). The detailed results of this EPA survey are being published in a specific EPA report. This survey was conducted in coordination with the Solid Waste Association of North America and used the results of other US LFG surveys (Anon, 1990; Berenyl and Gould, 1991).

Figure 1 provides a breakdown of the types of energy project in the USA. The majority of projects generate electricity, which is either used on site or sold to a local utility. Of the projects generating electricity, a net output of 350 MW of power is being produced with 60% of projects using internal combustion (IC) engines, 21 projects using gas-fed turbines, and five projects using steam-fed turbines. Pipeline quality gas is produced at four sites, and one site is processing landfill gas to produce diesel fuel. The most economical option for landfill gas utilization is direct use as boiler fuel. This is occurring at 20 sites.

Figure 2 provides a breakdown of the US landfill gas projects by state. The USEPA has estimated that landfills generate 8–16 Tg/yr of methane in the USA and 20–40 Tg/yr globally (Thorneloe *et al.*, 1994). Utilization of the methane results in conserving fossil fuel resources and offsetting by-product emissions (such as NO_x, SO_2, and mercury) from the use of fossil fuel. However, technical and non-technical barriers can limit the development of new LFG-to-energy projects (Pacey *et al.*, 1994). California has the largest number of landfill gas projects, partially because of State and local

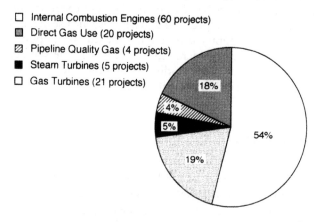

□ Internal Combustion Engines (60 projects)
▓ Direct Gas Use (20 projects)
▨ Pipeline Quality Gas (4 projects)
■ Steam Turbines (5 projects)
□ Gas Turbines (21 projects)

Figure 1. Breakdown of the types of LFG utilization project in the USA.

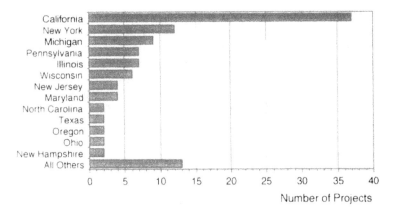

Figure 2. Number of LFG utilization projects per state in the USA.

requirements resulting in the collection and control of gas. However, many landfill gas energy projects have been initiated because of attractive economics, particularly in the early 1980s when the price of energy helped make this more economical.

Direct-Gas Use (Medium Heating Value)

The options for medium-HV landfill gas (i.e. 4.500 Kcal/Nm^3, 18.6 KJ/Nm^3) include use as boiler fuel, space heating and cooling, and industrial heating/co-firing applications. The majority of the 20 sites selling LFG for direct use are supplying fuel for boilers to produce steam. This is a particularly attractive option, as conventional equipment can be used with relatively little modification. In addition, boilers tend to be less sensitive to LFG trace constituents and consequently less gas clean-up is required compared with the other alternatives. A limitation in the selection of this option is that an LFG customer must be in close proximity; typically less than 2–3 km is considered desirable.

The other options for medium-HV gas include industrial applications such as lumber drying, kiln operations, and cement manufacturing. An advantage of many industrial applications is that fuel is required continuously, 24 hours per day. Landfill gas can also be used as supplemental fuel that meets a portion of the total demand. The use of landfill gas to produce space heating is in limited use, primarily owing to piping costs and difficulty in matching up the landfill gas energy output with nearby user needs. Depending on climate and other factors, heat energy supplied by 600 Nm^3/h landfill gas corresponds to heating needs of a 200 000–100 000 m^2 complex. Other difficulties include the fact that space-heating loads tend to be variable over time, both during the day and by season. One of the case studies demonstrates the successful use of landfill gas for producing space heating.

Electricity Generation

Of the 86 US LFG energy recovery projects that generate electricity, 70% generate it using internal combustion engines and 30% use turbines. Of the 310 MW of energy produced at these sites, 51% is generated using turbines and 42% is generated using internal combustion engines. The type of equipment is generally determined by the volume of gas available and the air pollution requirements of the area in which the project is located. Engines are typically used at sites where gas quantity is capable of producing 1–3 MW. Turbines are typically used at sites producing more than 3 MW.

Internal Combustion Engines

Reciprocating IC engines drive electrical generators to produce electrical power, which is typically sold to the local electric utility. Engines used in this application are sold by three manufacturing companies. Each of the three manufacturers has in place more than 20 engines at US landfill sites (GRCDA, 1989). These manufacturers design engines that are specific to LFG applications (i.e. adapted to medium-HV fuel and corrosive resistant). Typically, agreements are made that, to guarantee engine performance, the operator must agree to certain conditions regarding engine operation and maintenance.

Reciprocating engines used for LFG applications range from stoichiometric combustion to leaner combustion engines. The 'lean-burn' engines are a turbocharged design that burns fuel with excess air. The stoichiometric or 'naturally aspirated' engines are stoichiometrically carburetted with just sufficient air in the fuel–air mix to burn the fuel. The case studies provide four examples: two

LFG projects using 'lean-burn' engines and two LFG projects using naturally aspirated engines.

The lean-burn engines are typically used in areas where NO_x and CO emissions are of concern. The stoichiometric combustion can result in relatively high NO_x emissions, which can vary widely owing to carburettor setting and other variables. Waukesha suggested that 6–18 g/hp h is a typical range of NO_x emission from stoichiometric engines (Stachowicz, 1989). Lean-burn engines are available that minimize the production of NO_x and fuel consumption. At landfill sites with gas flows less than 250 Nm^3/h an operator could use one or two naturally aspirated engines. The NO_x emission would be less than 250 t/yr and the source would not be subject to new source review. At sites over 250 Nm^3/h, lean-burn engines are used to avoid new source review. The destruction performance of a lean-burn commercial engine at various NO_x emissions levels reported by the manufacturer is (Chadwick, 1989):

$NO_x(g/hp\ h)$	Destruction efficiency
2.0	98.3
5.0	98.7
10.0	99.1

A trade-off between low NO_x emissions and the reduction of non-methane organic compound can be observed.

Data from several internal combustion engines fuelled with landfill gas were collected by the EPA. The range at 15% oxygen of nitrogen oxides was 50–225 ppmvd (or 0.6–3.3 g/hp h). The range in ppmvd at 15% oxygen of carbon monoxide is 43–550 ppmvd (or 0.6–7.2 g/hp h). Emissions of sulphur dioxide were measured for only one internal combustion engine, and the concentration of sulphur dioxide at 15% O_2 was 1.5 ppmvd (Thorneloe and Evans, 1989; USEPA, 1991).

Gas Turbines

Gas-fired turbines are also in use at landfills to generate electricity. Gas turbines take large amounts of air from the atmosphere, compress it, burn fuel to heat it, then expand it in the power turbine to develop shaft power. This power can be used to drive pumps, compressors, or electrical generators (McGee and Esbeck, 1988). Gas turbines are used at 21 US landfills to produce 101 MW of power. It has been found that gas-fed turbines typically have an energy loss of 17%, compared with 7% for IC engines (Schlotthauer, 1991). A factor to consider is that turndown performance is poor in comparison with that of internal combustion engines. Turbines perform best when operated at full load and difficulties can occur when they are operated at less than full load. In addition, trace constituents have been reported to cause corrosion, combustion chamber melting, and deposits on blades. However, these difficulties can be overcome (Schlotthauer, 1991). A major advantage reported by those sites using gas turbines is that generally less day-to-day maintenance is required compared with the use of lean-burn engines.

Emissions of nitrogen oxides for seven gas turbines fuelled with LFG ranged from 11 to 174 ppmvd at 15% O_2. Emissions of carbon monoxide ranged from 15 to 1300 ppmvd at 15% oxygen, and emissions of sulphur dioxide ranged from 2 to 18 ppmvd at 15% O_2 (Thorneloe and Evans, 1989; USEPA 1991). Emissions of sulphur dioxide are not expected to be significant, as landfill gas typically contains relatively low amounts of sulphur compounds as compared with fossil fuels. The new source performance standard for gas turbines with a power output of 2.93–29.3 MW is 150 ppmvd at 15% O_2 (USEPA, 1977). Although the units tested were below this cut-off, six of the seven turbines did have less than 150 ppmvd of nitrogen oxides at 15% O_2. The data for NO_x emissions from a commercial landfill gas turbine ranged from 22 to 37 ppmv at 15% O_2, with a median of 30 ppmv (Maxwell, 1989).

Steam turbines are in use at five sites to produce about 70 MW of power. The largest landfill gas to energy plant is located at the Puente Hills Landfill in Whittior, California. This site began recovering landfill gas for energy utilization in November 1986. It is operated by the Los Angeles Country Sanitation districts. The facility consists of twin gas-fired steam generators. Each of the units fires 10.300 scfm of landfill gas, producing 210.000 lb of steam per hour at 1350 psig, heated to 1000 °F. This steam drives a turbine that generates approximately 50 MW net, which are sold to Southern California Edison. This is enough energy to supply electricity to 100 000 homes (Valenti, 1992).

High-Heating-Value Gas

Seven sites in the USA upgrade landfill gas to pipeline quality. This option was considered more attractive in the early 1980s when the price of oil and natural gas helped make it more economical. The sites that are producing pipeline-quality gas were initiated in the early 1980s, when gas prices on a heating value basis were comparable with those of oil. These sites have an average landfill gas flow rate of 6000 Nm^3/h, with the lowest gas flow rate being 1300 and the highest being 11.000 Nm^3/h. Stringent clean-up technology is applied to purify the gas to pipeline quality by removing the trace constituents and carbon dioxide. As for the medium-HV applications, a nearby natural gas pipeline is needed. Low natural gas prices in the late 1980s forced several previous projects to shut down, and continues to inhibit the development of new high-HV projects in the USA. However, sites in the Netherlands are finding more favourable economics (see Chapter 7.5).

A site in Pueblo, Colorado, is producing diesel fuel from landfill gas and other projects are planned to produce vehicular fuel.

Fuel Cells

In a fuel cell, hydrogen from the landfill gas is combined electrochemically with oxygen from the air to produce d.c. electricity and by-product water. The fuel cell is designed for automatic, unattended operation, and can be remotely monitored. Fuel cells are an attractive option for landfill gas because of higher energy efficiency, availability to smaller as well as larger landfills, and recognition for minimal by-product emissions. Other advantages include minimal labour and maintenance, and because there are no moving parts the noise impact is minimal.

The EPA initiated a project to demonstrate the use of fuel cells for landfill gas application. The type of fuel cell being demonstrated is a commercially available 200 kW phosphoric acid fuel cell power plant. The major issue associated with this demonstration is designing a landfill gas clean-up process that will remove the trace constituents from the landfill gas and at the same time not be cost prohibitive. As the composition of landfill gas varies over time, it is difficult to design a process that can allow for this variability. A clean-up process has been developed and evaluated, resulting in a patents application. The one-year demonstration at a California landfill has started operating recently. The fuel pretreatment system incorporates two stages of refrigeration combined with three regenerable adsorbent steps (Sandelli, 1992).

Given the higher energy efficiency and potential for minimal by-product emissions, fuel cells if proved to be successful may be the only alternative for some areas where there are stringent requirements for NO_x and CO emissions.

ECONOMICS

A major factor in helping to encourage landfill gas energy projects in the USA is the Public Utility Regulatory Policy Act (PURPRA). This requires that utilities purchase power that was generated from landfills at a price that was profitable to the generators: in effect subsidizing landfill power. In addition, tax credits have been available that also help to encourage renewable energy projects, such as landfill gas utilization.

The major capital costs of a landfill gas energy recovery project are presented (in terms of percentage of the total cost) in Table 1 for a 1 MW LFG plant, which is considered the minimum entry level.

TABLE 1. Estimate of Capital Costs (in US Dollars) for a 1 MW Landfill Gas-to-Energy Utilization Project

Item	$Cost^a$ $(\$10^3)$	Percentage	Range of value $(\$10^3/kWh)^b$
Extraction/collection system[c]	200	13	200–1000
Fees – planning/environment legal[d]	30	2	30–1000
Interconnect cost	76	5	20–500
Generating equipment	970	65	500–2000
Contingency	225	15	
Total	1500	100	850–4500

[a]These costs were provided by Laidlaw Technologies, Inc. (Jansen, 1992).
[b]Augenstein and Pacey (1992).
[c]The range in cost of gas clean-up systems is $10 000–$500 000/kWh.
[d]Legal fees are approximately 50% of the total (i.e. ~$15 000–$500 000/kWh).

The major cost and revenue components include administration and development costs, capital costs, operating and maintenance costs, royalty payments, tax credits, and energy-related revenues. A brief description is provided of each of these components.

- Administrative costs include legal fees, permit applications and contract negotiations, including gas lease agreements and power purchase agreements. These costs may vary widely depending upon the environmental issues, development considerations, and regulatory requirements.
- Capital costs include the cost of gas extraction and clean-up, energy conversion equipment, building, flares, and site modifications for construction.
- Operating and maintenance costs include the costs associated with the operation and maintenance of the energy project, labour, utilities, taxes and insurance.
- Royalty payments are proportional to energy output (or net or gross revenue) and are defined by the contract. Royalties are negotiated, and may be paid to the landfill owner, the owner of the gas extraction or delivery rights, or the initial project developer. Royalties can range from 5 to 20% of gross energy sales.
- Tax credits are benefits proportional to gas energy delivery, which have to be specially legislated by Congress. These credits are a direct offset to taxes and can only be used to offset a profit. The tax credits have to be extended through a significant period of time.
- Revenues for energy sales are usually based on prices of the 'competition' of equivalent energy sources (i.e. petroleum products). As the value of the energy base commodity can fluctuate, this can impact on profit. The early LFG projects were based on an established firm price for net energy, which provided a substantial degree of security for developers.

LANDFILL GAS UTILIZATION EXPERIENCES IN THE USA

Brown Station Landfill, Prince George's County, Maryland

Landfill gas is used to supply both the electrical and heating needs of a county building complex in addition to producing electricity for sale to the local utility. The energy equipment includes a gas clean-up and pumping station, a 3 km transmission system, three engine generators, and a boiler that supports the heating and hot water system of the 22 000 m^2 county correctional complex. The three engines are lean-burn turbocharged. Approximately 1200 Nm^3/h of gas is recovered from 4 Mt of landfilled waste. This sites produces 2.6 MW of power.

This site experienced typical difficulties during start-up, resulting in one of the three engines existing after less than 500 h of operation. Modifications were made to the engine to make it corrosion resistant including hardened valve guides, chrome valve stems, and a modified maintenance schedule including more frequent oil checks and changes. In addition, the site increased the clean-up of the gas. Since this occurrence, the site has had successful operation. Speculation about what contributed to the engine's seizing up suggests that higher levels of chlorinated organics (than is typical for landfills) may have caused the corrosion. Inspection of the engine showed evidence of extensive corrosion including deposit build-ups, which reduced piston clearances.

Central Landfill, Yolo County, California

Gas from this landfill is used to fuel three internal combustion engines to generate 1.5 MW. The generated power is delivered via an interconnect to the nearby high-voltage power lines. The owner of the landfill receives royalties based on net power and capacity sales. This energy recovery project began operation in November 1989. Approximately 1500 Nm^3/h of landfill gas is recovered from 3.1 Mt of refuse. This project has experienced equipment problems and other difficulties during the first years of operation. The type of engine being used was not specifically designed for landfill gas application. Modifications to the engine have been made to make it more corrosive-resistant for landfill gas applications. Because of the difficulties experienced at this site, the revenue from the royalties has dropped.

Otay Landfill, San Diego County, California

This site is recovering 2300 Nm^3/h of landfill gas to generate 3.4 MW using a lean-burn engine/generator set. The electricity is sold to the local utility. The Otay landfill contains approximately 6.1 Mt of refuse. The energy recovery project began operating in December 1986. The facility exports a net output of about 3400 kW, with a good gross power sale revenue.

Monterey Landfill, Marina, California

This site produces 1.2 MW from 1000 Nm³/h of landfill gas. This project was one of the first projects in the USA, having begun operation in December 1983. The installation is the result of persistence by participants, who were aware of the potential benefits of energy recovery from landfill gas. Approximately 3.7 Mt of refuse is buried at this site as of 1992. The Monterey Regional Waste Management District owns and operates the engines and catalysts for the outlet exhaust to reduce NO_x. Very little industrial waste is disposed at this site and the gas is considered relatively 'clean' when compared with other sites. Consequently gas clean-up is less stringent. It has been found that the type of oil used for the engines is critical to performance. This site can be considered one of the early pioneers in energy recovery from landfill gas.

Sycamore Canyon Landfill, San Diego, California

This site produces 1.7 MW from 1400 Nm³/h of landfill gas. The project began operating in December 1988 using two gas-fired turbines fuelled with landfill gas to generate electricity, which is sold to the San Diego Gas and Electric grid. It is reported that efficiency is reduced by 13% from the efficiency that would be obtained with the same turbine on more conventional pipeline gas or distillate fuels. This loss is almost entirely due to the greater parasitic compression load. This site has found that gas projections made prior to the start-up were not accurate. Consequently, the equipment is occasionally limited due to inadequate quantities of landfill gas.

Wilder's Grove Landfill, Raleigh, North Carolina

This site is an excellent example of how landfill gas can be used directly by an industrial client as medium-HV fuel. The gas from the Raleigh landfill is piped 1.5 km to a pharmaceutical facility for use as boiler fuel to produce 11 t/h of steam. This project began operating in December 1989 and is expanding. Approximately 1800 Nm^3/h of gas is recovered from approximately 3.5 Mt of refuse.

The management company pays royalties to the City of Raleigh based on a percentage of steam sales. This project also receives tax credits on the extracted gas.

CONCLUSIONS

Projects started in the early 1980s experienced difficulties due to the utilization of equipment primarily used for natural gas application. The US landfill gas industry has gained a great deal of experience over the ten years and new projects will benefit from this. An increase in the run of projects is expected due to the Clean Air Act regulation for MSW landfills.

The utilization of landfill gas is sensible in terms of economics, the environment and energy usage. The utilization of alternative energy sources such as landfill gas extends global fossils fuel resources. Not only are emissions directly reduced when landfill gas is collected and recovered for utilization, but emissions are also indirectly reduced when secondary air-emission impacts associated with fossil fuel use are considered. US landfills are currently recovering 1.2 Mt of methane and producing 344 MW of power (net output). The Clean Air Act regulations for MSW landfill air emissions are expected to result in additional emission reductions of 5–7 Tg/yr (Thorneloe and Pacey, 1994). Hopefully the owner/operators of those sites affected will consider energy utilization, which would result in increased benefits to economy, energy, resources and global environment.

REFERENCES

Anon (1990). Landfill gas survey update. *Waste Age*, March, 97–102.
Augenstein, D. and Pacey, J. (1992). Landfill gas energy utilization: Technology Options and Case Studies. EPA-600/R-92-116. Prepared for EPA's Air and Energy Engineering Research Laboratory by Emcon Associates, June 1992.

Berenyl, E. and Gould, R. (1991). *1991–92 Methane Recovery from Landfill Yearbook*, Governmental Advisory Associates.

Chadwick, C. (1989). Letter to Susan Wyatt, EPA, 21 February 1989, Response to request for information on internal combustion engines used to burn landfill gas, Caterpillar Inc.

Federal Register (1991). **56**(104), 24268–24528.

GRCDA/SWANA (1989). Engine and Turbine Panel Presentations, in *Proceedings of the GRCDA 9th International Landfill Gas Symposium*.

Jansen, G. R. (1992). The economics of landfill gas projects in the United States, presented at the Symposium on Landfill Gas/Applications and Opportunities, Melbourne, Australia, February.

Maxwell, G. (1989). Reduced NO_x emissions from waste management's landfill gas solar Centaur turbines, in *Proceedings of Air & Waste Management Association* 82nd Annual Meeting, Anaheim, California, June.

McGee, R. W. and Esbeck, D. W. (1988). Development, application, and experience of industrial gas turbine systems for landfill gas to energy projects, in *Proceedings of GRCDA's 11th Annual International Gas Symposium*, March.

Pacey, J. G., Doorn, M. and Thorneloe, S. (1994). Landfill gas utilization – technical and non-technical considerations, in *Proceedings of the 17th Annual International Landfill Gas Symposium*, Solid Waste Association of North America, Long Beach, California, 22–24 March.

Sandelli, G. J. (1992). Demonstration of fuel cells to recover energy from landfill gas. Prepared for EPA's Air and Energy Engineering Research Laboratory by International Fuel Cells Corporation, EPA-600-R-92-007, January.

Schlotthauer, M. (1991). Gas conditioning key to success in turbine combustion systems using landfill gas fuels, in *Proceedings of GRCDA/SWANA's 14th Annual Landfill Gas Symposium*, San Diego, California, March.

Stachowicz, R. W. (1989). Response to request for information on internal combustion engines used to burn landfill gas, Waukesha Engine Division, letter to S. Wyatt, EPA, 31 March.

Thorneloe, S. A. (1991). USEPA's global climate change program – landfill emissions and mitigation research, in *Proceedings Third International Landfill Symposium*, Sardinia, October.

Thorneloe, S. A. (1992). Landfill gas utilization – options and economics, in *Proceedings 16th Annual Conference Institute of Gas Technology on Energy from Biomass and Wastes*, Orlando, FL, 15 March.

Thorneloe, S. A. and Evans, L. (1989). The use of internal combustion engines of gas turbines as controls for air emissions from municipal solid waste landfills. EPA Memorandum to Susan R. Wyatt, Chief, Chemicals & Petroleum Branch, Emission Standards Division, Office of Air Quality Planning & Standards, 31 May, Docket A-88-09.

Thorneloe, S. A. and Pacey, J. (1994). Database of North American landfill gas to energy projects, in *Proceedings 17th Annual International Landfill Gas Symposium Solid Waste Association of North America*, Long Beach, CA, 22–24 March.

Thorneloe, S. A., Doorn, M., Barlaz, M. *et al.* (1994). Methane emissions from the management of solid wastes, in *EPA Report to Congress on International Anthropogenic Methane Emissions: Estimate for 1990*, EPA 230-R-93-010, January.

USEPA (1977). *Standards Support and Environmental Impact Statement, Volume 1: Proposed Standards of Performance for Stationary Gas Turbines,* EPA-450/2-77-017a (NTIS PB 292422), September.

USEPA (1979). *Stationary IC Engines. Standards Support and Environmental Impact Statement, Volume 1: Proposed Standards of Performance,* EPA-450/2-78-125a (NTIS PB83-113563), January.

USEPA (1991). Air emissions from municipal solid waste landfills – background information for proposed standards and guidelines. Prepared for the Office of Air Quality Planning and Standards, EPA-450/3-90-011 (NTIS PB 91-197061), March.

Valenti, M. (1992). Tapping landfills for energy. *Mechanical Engineering,* **114**(1).

7.2 Landfill Gas Use in Reciprocating Engines in England

HUGH D. T. MOSS

Shanks & McEwan (Energy Services) Ltd., The Cottage, Church Road, Woburn Sands, Milton Keynes,. Bucks MK17 8TA, UK

INTRODUCTION

Many landfill sites are located some distance from potential industrial energy users and may not have the commercial opportunity to sell gas. Under these circumstances, the generation of electricity is an attractive option, provided that a buyer can be found. In the UK until the Energy Act of 1983 there was no obligation on the part of the Area Electricity Boards to purchase the energy. Thereafter a fixed tariff was published each year, although because of the structure of the Bulk Supply Tariff submitted by the Central Electricity Generating Board (CEGB), only some 70% of the unit price paid to the CEGB could be paid to the private generator. Nevertheless, with schemes of sufficient size to gain the economies of scale, a commercial return could be achieved.

This chapter describes experiences with reciprocating engines carried out at Stewartby and Brogborough landfill sites in Bedfordshire, England.

INITIAL GENERATION

The compound used for the first commercial supply of gas was also used for the first phase of the electricity generation programme. The

Landfilling of Waste: Biogas. Edited by T. H Christensen, R. Cossu and R. Stegmann. Published in 1996 by E & FN Spon, London. ISBN 0 419 19400 2.

origins of the scheme are older than the 1983 Energy Act and, as such, restricted the scope of the scheme. The entire electrical output was employed within the boundaries of the gas abstraction plant. The gas plant itself operated through an off-load changeover switch, power being drawn either from the grid supply or alternatively from the generator. The system operated as a closed loop, power from the generator driving the main compressor, which provided gas under pressure for kiln firing. A fraction of this drove the gas engine itself.

The prime mover was an eight-cylinder spark-ignition Rolls Royce gas engine of 6.5 l capacity. On natural gas the engine is rated at 145 kW at 3750 rpm. The engine drove a Markon type B brushless generator of 75 kVA output through a 1.6:1 reduction gearbox; the prime mover ran at 2400 rpm and the generator at 1500 rpm. The engine was controlled by a Barber-Colman electronic speed governor, and gas was supplied to the engine by an Impco DG 200 'digestern gas' carburettor. The engine was started on propane, run up to full load and, when full landfill gas pressure was available, the fuels were changed over and the set continued on-load on landfill gas.

Experience

During on-load test runs (on a day hours only basis), oil consumption was found to be particularly high, and several modifications including the fitting of a supplementary oil tank were carried out. This enabled continuous unattended running to be achieved, and the test duration was increased to 100 h. Throughout these tests, rates of gas used and power generated were monitored, and from this data overall efficiencies were found to be in the range 17–19%.

Several minor breakdowns occurred during these test runs, one of the most persistent being 'exhaust valve recession' – a condition in which tappet clearances gradually reduce, causing severe loss in power output. This condition was monitored closely, and frequent maintenance tended to overcome it. Oil condition, in terms of acidity and metal contamination, was also checked and monitored closely.

Several other parameters were varied to maximize the efficiency of the unit. Ignition timing was found to be an important factor in the smooth running of the engine. Problems were encountered with instability in the carburettor at certain loads, and separate propane and landfill gas carburettors were tried to reduce this. Eventually a stage was reached where the excessive power demanded by the gas compressor, caused by wear, tended to overload the engine and shut down the system. After a series of these events, the generator was taken out of service.

However, this demonstrated that landfill gas could be used economically to generate electricity and thus allowed progression to the next stage of electricity generation.

STAGE II GENERATION

The original gas gathering, flaring and utilization compound was only large enough to control the gas in a small area of the Stewartby landfill. A second compound was planned, but because the only source of electricity would have been from an old and cramped substation, energy for the equipment was provided from a diesel-powered generator. A gas abstraction unit was commissioned, consisting of compressor, after-cooler and instrumentation. The cost of the power was relatively high and would have stayed high as the gas abstraction scheme was enlarged and the equipment extended. The decision was taken to design a new electricity-generating system that would remove the need to hire diesel generators, supply electricity for local use, and supply some power to the adjacent brickworks.

Installation

While the Dorman Diesel 12STCWG was not the cheapest unit, it had the advantages of some service life on digester gas and the manufacturers being relatively local to the site (Table 1). The scheme was laid out with three 275 kW units, leaving room for a fourth. The generator panels were supplied by the engine manufacturer, and the outdoor substation 6.6 kV ring main unit and 415 V distribution board were supplied by Yorkshire Switchgear. The building was designed in house and erected by a local firm. This allowed the original substation to be isolated and removed while maintaining supplies to existing points via up-to-date switchgear.

TABLE 1. Characteristics of the Generating Unit Moss at the Stewartby Landfill Site

---

Engine: Dorman diesel 12STCWG, 12 cylinder vee form turbocharged water-cooled engine

Bore	158.75 mm (6.25 in)
Stroke	190.50 mm (7.50 in)
Total swept volume	45.25 l (2761 in^3)
Sump oil capacity	102.3 l (22.4 gal)
Compression ratio	9.5:1
Speed	1000 rpm
Impco carburettor	
Altronic ignition system	
Radiator cooler with fan drive from engine	

Alternator: Newer Stanford

350 kVA 415 V brushless. self-exciting, drip-proof case
Operating between 0.8 and unity power factor
Two-thirds pole-pitch winding

Nominal out 275 kW

The capital expenditure was justified on the basis both of the savings made in diesel generator use and of the potential sales of electricity.

The three generating units are mounted with integral fan-cooled radiators on a single bed plate. Air is drawn through louvres at the back of the generator building over the machines, helping to cool them, and is pushed outside through the radiators. The fuel is supplied from the adjacent compound by moving vane compressors via two 15 μm filters. Each cylinder bank of the vee-form engine has its own gas regulators, Impco carburettor, air filter, turbo charger and intercooler. The load is controlled by a butterfly valve downstream of the carburettor. The exhaust has a single horizontally mounted silencer positioned vertically above each engine. Oil supplies are topped up continuously from a 'day tank', which itself is filled from a bulk storage tank.

Power is generated at 415 V and fed to a central board within the engine house. All the electrical protection is housed in the panels with the frequency-sensitive protection (ROCOF) connected to the outgoing circuit-breaker. The outgoing circuit feeds onto a 415 V distribution board, which is on the low-voltage side of the 6600/415 V transformer that connects the station to the local brickworks. An interlock on the ROCOF allows the power station and its local distribution board to run independently of the brickworks and hence the local 33 kV system, in cases of prolonged network or circuit outage.

Over a period of some six months, minor electrical faults were cured and fine mechanical adjustments were made to the engines. During this time, three separate faults occurred on the original compressor, the final fault shattering the moving blades because of the movement of a slug of water through the machine.

Lubrication

Probably the most important item in the care of spark-ignition engines is the monitoring of the lubrication oil. The process of blending makes it difficult to optimize total base number (TBN), ash and extreme pressure rating (EP). Experience of the different types of oil, particularly in Germany, had shown that different designs of engine were suited to different blends of oil. Initially, the oil manufacturer recommended a CRI 30, which was similar in characteristics to a number of oils used in continental Europe in similar applications. Simple alkalinity tests were undertaken at 100 h intervals to assess the remaining basic character of the oil. A full analysis was undertaken every 250 h which led to oil changes at the 500 h and 1000 h intervals. After this, sufficient experience had been gained to allow oil changes at 1000 h intervals with the same full analyses every 250 h.

After running for 5000 h, during routine maintenance lacquering was noticed in some of the cylinder bores and on the pistons. Also, valve rocker faces and valve stems showed high rates of wear. After consultation between engine and oil manufacturers, a change of oil specification led to the use of NG 30M oil with increased EP and TBN characteristics. At the same time the hardness of the valve seat inserts and valve faces was increased by using satellite surfacing and the engine water operating temperature was increased from 70 °C to 80 °C to allow the oil additives more scope for working. At the 10 000 h interval the effect of the new oil was scrutinized. The greater range in alkalinity allowed the confidence to extend the oil changes to 1500 h intervals, thus allowing the longer 5000 and 10 000 h planned maintenance periods to be performed at 6000 h and 12 000 h respectively.

Trace quantities of metals in the oil can be an indication of engine wear. In the Stewartby case, there were elevated levels of both

TABLE 2. Characteristics of Electricity-Generating Machines

Machine type	Typical efficiencies (%)	Typical life (h)	Typical sizes (MW)
Spark ignition	28–30	40 000	Up to 0.5
Dual fuel	35–40	150 000	1 to 10
Gas turbine	26–27	60 000	Above 2
Steam turbine	15–20	150 000	Above 2

copper and silicon. While copper in oil can be a sign that bearings are wearing, the main bearings in the engines are made of an aluminium and tin alloy, with the 'little ends' made of phosphor bronze. With no elevated levels of tin in the oil, bearing wear was discounted and the conclusion was that the copper came from the oil cooler tubes. With the silicon, a more extensive search was made. The air filters were modified on one engine, an oil centrifuge was put on the second engine, and a bypass filter put on the third. Despite these measures, the silicon level remained high. The problem has not really been resolved, although silicon grease and silicon rubber O rings are used in the engines.

Noise

This represents a fundamental problem with the legislation that controls substances hazardous to health (COSHH). It is particularly concerned with noise in the workplace, and although the engines are designed to run unattended, duty fitters visiting the generating plant could be liable to very high levels of exposure to noise. In cases of this nature, engines are recognized as noisy, and precautions are taken to reduce people's exposure to the noise. However, the building is constructed of hollow concrete blocks and on the upper half is a double-skinned and insulated wall of steel sheeting. The personnel door is standard but the 4 m wide roll shutter door is insulated. This leaves a warm environment inside in the winter without the problems of overheating in a hot summer. Whereas some installations have enclosed sound booths, a deliberate decision was taken not to enclose the engines. Maintenance is eased and there are fewer accumulations of oil and dirt.

Control

The installation is designed to run unattended for 24 h a day and 365 days each year despite planned outages for maintenance. Thus the normally comprehensive system of trips and alarms allows certain of these to be selected for use, either to shut down the plant, or to initiate a call to a central computer, which can be used as a 24 h call-out facility. The building is equipped with smoke and gas detection equipment. When activated, the sensors can shut down the complete plant and allow engineers to be called out.

The success of the installation lay in an average availability greater than 95%, even when 12 000 h maintenance intervals were counted. Thus by April 1989 the space left for a fourth engine was filled by an identical 12STCWG. At the fourth anniversary of the installation, well over 30 million units had been generated and the four engines had run for a total of 110 000 h.

STAGE III GENERATION

The Brogborough landfill site, only some 8 km from Stewartby, had an environmental gas control scheme. While the Stewartby generating unit was commercially successful under its own parameters, a new installation needed economies of scale to make a commercial scheme of selling electricity to an Area Distribution Board. To make this jump, the type of equipment needed to be changed, and hence plans were put in motion for a dual-fuel engine installation rather than spark-ignition engines. Table 2 shows the basic efficiencies of the two types of engine, which are joined by comparable data for gas turbines and steam turbines. In the Brogborough case, a fine commercial distinction lay between a gas turbine and a dual-fuel installation. At the time, the privatization of the electricity supply industry was mooted, with a promised better selling price than that obtainable under the old Energy Act of 1983. At this juncture a sensitivity analysis revealed that the higher-efficiency dual-fuel machine was the more prudent choice. Because of the uncertainty at the time of the price obtainable for electricity under the Non Fossil Fuel Obligation (NFFO), it was decided to install second-hand, refurbished engines. After an extensive search, two machines were found that had been used as standby plant in a gas works and as such had had little service and wear.

Installation

The two machines chosen are K7 Mirless Blackstone dual-fuel engines. They are capable of 2 MW each on landfill gas and a greater output on gas oil only. Whereas the Stewartby machines have integral coolers, all cooling on the Brogborough engines is remote. Oil coolers, jacket water coolers and oil centrifuges are remotely mounted, although the engines and alternators have been mounted on common baseplates. Because of the necessity to meter two fuels, the whole control philosophy is more developed, and a programmable logic controller (PLC) has been used for start-up and running.

The two seven-cylinder machines run at 500 rpm and generate electricity at 11 kV. Power is metered onto busbars from which the station auxiliary power is drawn, and the remainder is exported via an overhead line to a local combined 33 kV/11 kV substation, which forms part of the local distribution network. The line has been strung and insulated to be able to operate at 33 kV, as any expansion of the power station would require transmission to be at 33 kV instead of 11 kV.

Lubrication

Dual-fuel engines have a planned service life in excess of 20 years. For this sort of life to be achieved, great care must be taken to operate them and to maintain them in an adequate manner. Inevitably the lubrication regime plays a large part. As with spark-ignition engines, oil is sampled regularly and tested for acidity as well as the less frequent analysis for trace components. The quoted oil consumptions fall into a broad band, which ranges roughly from 0.2×10^{-3} l/kWh to 2×10^{-3} l/kWh. At the lower end of the range are clustered the smaller spark-ignition engines, while the dual-fuel engines tend to be at the upper end. The same compromise of oil characteristics must be made in selecting the correct grade. With the higher specific oil consumption shown by the dual-fuel engine, the make-up supply of oil is higher and so the contaminants within the oil tend to be diluted more quickly than would be the case with spark-ignition engines.

With similar amounts of contamination within the gas itself, the TBN of the oil does not become so important. To date no adverse signs have been seen, although a large amount of running is necessary before conclusions can be drawn.

TABLE 3. Characteristics of the Generating Unit at the Brogborough Landfill Site

Engine: Mirlees National KP7 major, 7 cylinder in-line turbocharged
water-cooled engine

Bore	25 in (635 mm)
Stoke	18 in (457 mm)
Swept volume	79 522 cu in (1303 l)
Sump oil capacity	320 gal (1455 l)
Speed	500 rpm
Fuel	75 l/hr fuel oil
	1325 m^3/hr LFG at 45% CH_4

Heinzmann KG90 electronic governor
Twin-circuit horizontal air-blast radiators
Alternator: brush

2250 kVA 11 000 V brushless, shunt wound
Operating between 0.8 and unity power factor
Nominal output: 2 MW (in dual-fuel mode)

FINAL REMARKS

The Brogborough engines were included within the first tranche of the NFFO. A third engine, this one of 6.7 MW rating, was also included, which with the first two will be able to use most of the landfill gas currently being flared. Because of the method of filling the site, it will be a few years before a fourth engine can be installed, at which point it would be possible to use the waste heat from all four engines to generate steam and run a steam turbine of some 2 MW.

Reciprocating engines have been shown to be efficient in the task of converting landfill gas into electricity. The type depends more on the quantity of gas available and the price obtainable for the electricity than on other parameters.

7.3 Landfill Gas Use in Reciprocating Engines and Turbines in Germany

PETER L. A. HENIGIN & ULRICH EYMANN

WAT Wasser- und Abfalltechnik Ingenieurgesellschaft mbH, Kleinoberfeld 5, 76135 Karlsruhe, Germany

INTRODUCTION

Active extraction of gases emanating from municipal solid waste disposal sites and subsequent thermal treatment is vital from an environmental point of view. Owing to the comparatively high content of energy in landfill gas, the combination of thermal treatment and a form of utilization would appear to provide a feasible solution. However, the heat produced during combustion can often not be fully exploited owing to the lack of purchasers of this form of energy. Landfill gas engines are thus often used to produce electrical current, as the use of this form of energy is largely independent of location.

The main idea behind the utilization of gas at the Wiesbaden and Karlsruhe disposal sites was to use the energy potential as far as possible for powering on-site equipment and units. In view of the increasing demands made upon landfill operation and equipment this concept makes excellent sense. Leachate treatment and waste pretreatment fall into the category of operations that require both thermal and electrical energy for processes involving drying, digging and driving.

The use of motors and gas turbines is still at an experimental stage in Wiesbaden; the landfill in Karlsruhe Ost has had nearly a year's operational experience.

Landfilling of Waste: Biogas. Edited by T. H Christensen, R. Cossu and R. Stegmann. Published in 1996 by E & FN Spon, London. ISBN 0 419 19400 2.

DESCRIPTION OF THE SYSTEM

Gas Quality and Quantities

Both landfills are used for the disposal of municipal solid waste (MSW) and similar industrial waste. In both cities measures to

TABLE 1. Gas Composition at the Landfill of Wiesbaden

Component	Landfill main street	Landfill section 1 (collection pipe)	Landfill section 2 (main drainage ditch)
	(vol. %)	*(vol. %)*	*(vol. %)*
Methane	30.2	61.2	60.9
Carbon dioxide	19.4	33.7	37.7
Oxygen	7.9	3.4	0.3
Nitrogen	41.7	4.7	1.5
	(mg/m^3)	(mg/m^3)	(mg/m^3)
Total chlorine	<1.0	5.5	102.0
Total fluorine	2.0	3.3	48.2
Total sulphur	10.5	19.7	14.9
Aliphatic hydrocarbons (without methane)	9.5	608.4	575.7
Terpene	<1.0	197.9	78.7
Aromatic hydrocarbons	<1.0	482.6	170.4
Propyl benzene	<1.0	452.8	14.5
Pentyl benzene			
Styrene	<1.0	<1.0	<1.0
Methyl styrene	<1.0	<1.0	<1.0
Chlorinated hydrocarbons	1.6	14.5	283.9
Alcohol	14.4	15.8	283.9
Ketone	<1.0	<1.0	<1.0
Ester	<1.0	<1.0	<1.0
Ether	<1.0	2.3	2.4
Aldehyde	4.3	4.5	1.2
Oxirane	<1.0	<1.0	<1.0
Tetrahydrofuran	6.4	4.2	5.0
Dioxan	<1.0	<1.0	<1.0
Dimethylfuran	<1.0	<1.0	<1.0
Hydrogen sulphide	<1.0	6.5	5.5
Amine	<1.0	<1.0	<1.0
Ammonia	25.6	17.8	<1.0
Arsine	<1.0	<1.0	<1.0
Phosphine	<1.0	<1.0	<1.0
Organic acids	<1.0	<1.0	<1.0

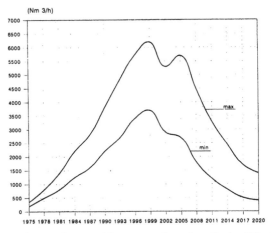

Figure 1. Prediction of extractable amount of gas from the landfill of Wiesbaden, Germany.

recycle reusable materials are taken (separate collection, composting of organic waste, recycling of building rubble etc.).

Extensive gas measurements were taken at the Wiesbaden landfill in 1987; the results are summarized in Table 1. It must be taken into consideration that a number of factors (i.e. technical measures such as surface capping, and further waste disposal measures such as separate collection of biowaste) make it difficult to predict the quantities of reusable gas with any accuracy. The gas extraction rates that are to be expected are shown in Fig. 1.

Throughout the observation period the gas quantity available from the Karlsruhe Ost landfill lay between 300 and 750 m^3/h. The gas analyses do not show any significant difference from gas from other landfills.

Treatment and Utilization Concepts

Karlsruhe Ost landfill. The landfill gas is used to provide the electrical and thermal energy necessary for driving the leachate treatment unit. The general concept is illustrated in Fig. 2. The two-stage reverse osmosis and subsequent evaporation/drying have a requirement of 1 t steam/h and a power consumption of approximately 270 kW. The end products of the leachate treatment are permeate (pure water) and approximately 0.8 t/d of dry matter.

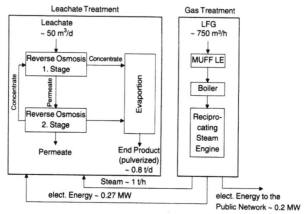

Figure 2. Scheme of LFG utilization plant at the Karlsruhe-Ost landfill site, Germany.

Wiesbaden landfill. At the landfill in Wiesbaden the landfill gas is used to supply energy for the treatment of leachate and sewage sludge. The concept is illustrated in Fig. 3. There are approximately 130 m^3/d of leachate and 50 000 t/a of sewage sludge to be treated. As described for the Karlsruhe Ost plant, the leachate is treated by two-stage reverse osmosis and subsequent evaporation/drying. The sewage sludge is treated by drying and storage in a designated area where it can be retrieved or utilized as combustion fuel. The gas treatment consists of a muffle with a connected waste-heat boiler and steam turbine. 1 MW of electrical current and 8 MW of steam are available for further use. As these quantities slightly exceed what is necessary for the treatment of leachate and sewage sludge, 0.3 MW of electrical energy is available either for further on-site consumption or for the mains.

SCHEMES OF THE UTILIZATION PLANTS

Karlsruhe Ost Landfill

The complete scheme of the landfill gas utilization plant is illustrated in Fig. 4. It consists of:

- a high-temperature muffle;
- a waste-heat boiler with feedwater purification;
- a reciprocating steam engine.

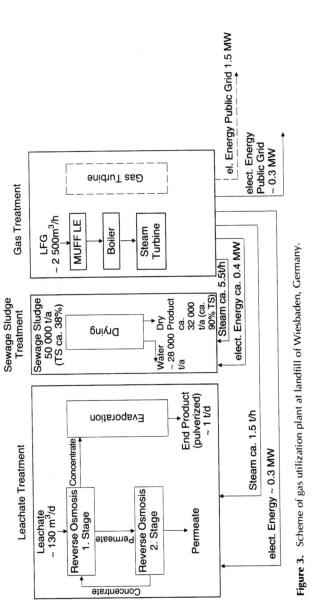

Figure 3. Scheme of gas utilization plant at landfill of Wiesbaden, Germany.

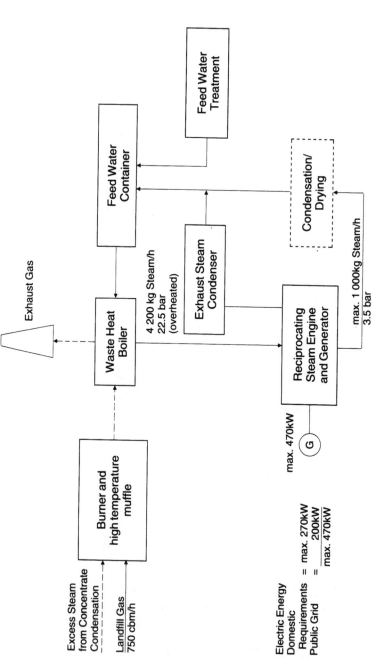

Figure 4. Scheme of the LFG utilization plant at the Karlsruhe Ost landfill site.

After precipitation of the condensate, the gas is compressed and combusted in the high-temperature muffle together with any non-condensable vapours left over after the evaporation of the condensate. The retention time is over 0.3 s and the temperature is above 1200 °C. In the waste-heat boiler the thermal energy is transformed into a maximum of 4200 kg steam/h at a pressure of 22.5 bar and 350 °C. The superheated steam is used to drive a reciprocating steam engine. After the pressure in the first two cylinders has been reduced to 3.5 bar, part of the steam is drawn off to a unit for evaporating the leachate condensate. In the next two cylinders, the pressure of the remaining steam is reduced to 0.1 bar. The maximum electrical power is approximately 470 kW. Exhaust gases are drawn out through an 18 m high chimney.

Wiesbaden Landfill

The process scheme involves:

- a high-temperature muffle;
- a waste-heat boiler with feedwater purification;
- a reciprocating steam engine.

The technique is similar to that used in the Karlsruhe Ost landfill: precompressed landfill gas in quantities up to approximately 2500 Nm^3/h are combusted in a high-temperature muffle. The steam produced in the conjunct waste-heat boiler (approximately 9 MW) is transformed into 1 MW of electrical current and approximately 8 MW steam, depending upon the requirement of the connected expander turbine. The electrical efficiency of the two-stage turbine is approximately 15%.

In addition to combustion in the muffle with a connected boiler, the use of a gas turbine was also considered in Wiesbaden.

The main features of the gas turbine (Fig. 5) are its compact construction, low-vibration operation, long life and a high level of availability (> 95%). In Germany, until now the use of such turbines has been restricted to the natural gas sector. This explains why there are as yet no data relating to the use of gas turbines with landfill gas, which can often have aggressive properties.

A simple gas turbine unit essentially consists of three systems:

1. The compressor compresses the surrounding air to a pressure between 12 and 18 bar.

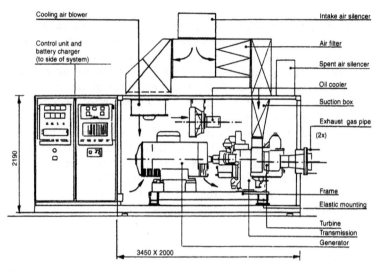

Figure 5. Gas turbine built in a container.

2. The compressed air flows into the combustion chamber, where the compressed combustible gas is injected through nozzles and combusted.
3. The hot flue gases are expanded to ambient pressure in the turbine.

In order to observe the operating conditions and emission behaviour of the gas turbine, it was tested over a three-month operational period in the Wiesbaden landfill. In parallel a lean gas engine was tested and the exhaust gas was also analysed. During this investigation, the main components of the gas supply, methane, carbon dioxide and oxygen, were measured continuously and nitrogen was measured discontinuously.

The exhaust gases from the gas turbine and those from the gas Otto engine (which ran simultaneously) were measured in the stack in three stages:

Stage 1: limiting conditions relating to emission technology (i.e. temperature, exhaust gas density, water content);
Stage 2: inorganic parameters (i.e. dust, nitric oxides, carbon monoxide, carbon dioxide, sulphur dioxide, oxygen volume, inorganic gaseous fluoride compounds, inorganic gaseous chloride compounds);

TABLE 2. Technical Data of the Gas Turbine and the Gas Engine with a Subsequent Oxidation Catalyst

	LFG – Otto Engine	*LFG – Turbine*
Type	MWM TBG 604 BV 8	Kawasaki/MWM S1T-02/ KT 215
Specialities	8 cylinder gas Otto engine with exhaust gas turbocharger and air cooling	LFG double turbine, reduction of the revs using planeitary gear
Power, mechanical	460 kW (625 PS)	427 kW (580 PS)
Revs, aggregate	1500 min^{-1}	53 000 min^{-1}
Power, electrical	430 kW	386 kW
Revs, generator	1500 min^{-1}	1500 min^{-1} (via gear)
LFG consumption	ca. 300 Nm^3/h	ca. 600 Nm^3/h
LFG quality	>45 vol.% CH$_4$	>50 vol. % CH$_4$
Exhaust gas treatment	Oxidation catalyst	Without
Waste heat utilization aggregate	Yes, for heating of the buildings in the entrance area of the landfill existing heating capacity: 450 kW	No
Waste heat utilization, exhaust gas	No, but possible to a degree (340 kW)	No. but possible (1550 kW)
Exhaust gas temperature	470 °C behind the turbocharger	500 °C
Necessary compression	>50 mbar (via LFG blower)	13 bar (via a separate LFG compressor)
Emissions	Via monitoring programme	

Stage 3: organic parameters (i.e. total amount of carbon, benzol/vinyl chloride, polychlorinated dibenzo-p dioxins and -furans, polyaromatic hydrocarbons).

The technical specifications of the gas turbine and the gas engine with connected oxidation catalyst are given in Table 2.

First Operational Results

Karlsruhe Ost landfill. Emission measurements have, in the meantime, been carried out for the high-temperature combustion unit at

TABLE 3. Emission Data related to 3% O_2

	Official maximum levels (mg/m^3)	Measured values (mg/m^3)
Carbon monoxide	50	<16
Nitrogen oxide	200	<100
Sulphur dioxide and sulphur trioxide	300	<30
Inorganic chlorocompounds	30	<15
Inorganic fluorocompounds (as HF)	5	<5
Dust	5	<4
Total carbon	5	<5
Dioxins and furans	0.1	<0.013

the Karlsruhe Ost landfill. Table 3 compares the measured values with the current official maximum levels. Very often the values fall significantly below the official maximum levels.

Operational experience has shown that it is necessary to take operation during minimal gas production into consideration.

In order to control the exceedingly complicated interaction between steam production and superheating of steam in the steam engine, with the possibility of extracting part of the flow – especially when starting up and shutting off the unit – it was necessary to modify the unit several times.

The extraction of steam in partial-load mode was possible only under certain conditions. Greater flexibility could be achieved by dividing the unit up into two independent steam engine units with separate generators.

First operational results from the Wiesbaden landfill. The three-month trial operation period of gas turbine and gas engine with connected oxidation catalyst took place in the summer of 1992. The exhaust gas analyses of the second reading at the end of the trial period are not yet available; for the present only the following statements can be made.

- After initial problems, various operational parameters (i.e. combustion chamber and air supply) were modified. This led to satisfactory test runs.
- The use of a gas turbine necessitates landfill gas with a methane content of at least 55%. The landfill gas should be dewatered and free of dust.

- At full capacity the gas turbine showed relatively good emission behaviour, meaning that existing maximum levels were not exceeded.
- The gas engine with oxidation catalyst also exhibited good operating characteristics. First readings confirmed a significant reduction of carbon monoxide. However, a conclusive report can only be given once the readings have been evaluated.

Ecological and Economical Aspects

The gas treatment plant at Karlsruhe Ost cost approximately DM 3.15 m (approximately US$ 2 m) including VAT. The resulting annual operational costs are approximately DM 700 000 (about US$ 440 000). The production of thermal and electrical energy resulted in savings of DM 730 000 (about US$ 460 000). This also includes the setting up of a standby generator, as the entire plant is run as an internal operation.

The treatment of gas at the Wiesbaden landfill required investments of approximately DM 11 m (about US$ 6 900 000). The annual operating costs are around DM 2.8 m (about US$ 1.8 m). If the profits made by the sale of electrical and thermal energy are taken into account then the total profit is DM 4.7 m (about US$ 3 m per year). This comparatively high profit is largely due to the optimum use made in Wiesbaden of thermal energy for leachate treatment and the dewatering of sewage sludge.

From an environmental point of view, the muffle technology used for the utilization of energy in Wiesbaden and Karlsruhe should be given preference over landfill gas engines. This refers to both the levels of total carbon, carbon monoxide and nitric oxide, and to the organic compounds in the exhaust gas.

A final evaluation of the gas turbine from an environmental point of view cannot be formulated as yet because the readings have still to be compared and interpreted.

CONCLUDING REMARKS

The experience gained during the planning and realization of the plants presented in this paper is summarized as follows.

- Owing to current laws and regulations, the initial authorization conditions (i.e. limit values, regulatory requirements) have not yet been specified. Thus it is advisable to plan and build in close cooperation with relevant authorities. This can lead to long delays, even without taking the legal time limits for application into consideration. It does, however, minimize the risk of unacceptable or unfulfillable requirements.
- In order to optimize the availability of the plant, especially if it is run as an internal operation, it is advisable to give preference to 'proven technology' with non-sophisticated equipment. It often pays to accept a lower level of effectiveness for a higher level of availability.
- The willingness of potential suppliers to try 'new methods' decreases with increasing economic situation. Only one supplier in Wiesbaden was prepared to supply a turbine geared to landfill gas, although initially quite a few companies had expressed great interest.
- In the future, the limiting conditions imposed by landfill technology (i.e. surface covering, separate collection and reuse of organic waste, pretreatment) must be taken into consideration for the layout and design of landfill gas utilization plants, as there is a great risk of oversizing the plant.

7.4 Landfill Gas Use in Reciprocating Engines: Operational Experiences

BERNHARD REINICKE

Energie-Versorgung Schwaben, Kriegsbergstrasse 32, 70174 Stuttgart, Germany

INTRODUCTION

The Energie-Versorgung Schwaben AG (EVS) supplies the south-western part of the Federal Republic of Germany with electricity, district heating and natural gas. Since 1983, the EVS has been working on the subject of converting landfill gas into energy.

The gas-collecting system is installed and operated by the local administrative district council. The EVS, on the other hand, is responsible for the exploitation of the landfill gas: i.e. converting it into electricity and – using a transformer – feeding this into the 20 kV medium-voltage network (Reinicke, 1989).

The first power plant was put into service in September 1984, at the landfill of Reinstetten, a small town near Biberach. Meanwhile, a total of 14 plants collectively generating about 4 MW have been installed (Table 1).

The electricity produced by the EVS from landfill gas does not exceed its total power generation.

These plants are installed in containers. Each container is an entirely independent unit. It contains the gas-driven reciprocating engine with all supplementary equipment such as cooling and ventilating system, exhaust pipes, oiling system and alternator, as well as electric control and surveillance (Fig. 1). The containers are

Landfilling of Waste: Biogas. Edited by T. H Christensen, R. Cossu and R. Stegmann. Published in 1996 by E & FN Spon, London. ISBN 0 419 19400 2.

TABLE 1. Data from Different Landfill Gas Power Plants in Germany (to December 1993)

Landfill gas power plant	Engine manufacturer	Electric power output (kW)	Beginning of operation	Total running hours (h)	Total generation of electricity (MWh)	Average availability (%)
Reinstetten I	MAN	210	Sept 1984	67 066	13 753	82.6
Reinstetten II	MAN	270	Nov 1988	36 841	8 495	82.3
Hechingen I	MAN	210	Aug 1986	52 482	11 090	81.3
Hechingen II	MAN	270	Nov 1988	35 136	9 109	78.4
Eberstadt I	KHD	430	Nov 1986	40 968	16 463	65.2
Eberstadt II	MWM	430	Apr 1989	25 995	10 685	62.5
Obermooweiler I	Jenbacher Werke	265	Nov 1986	41 195	9 925	65.6
Obermooweiler II	Jenbacher Werke	454	Nov 1990	10 175	2 858	36.8
Zaisersweiher	KHD	212	May 1987	34 806	5 448	59.9
Weiherberg I	MWM	258	Oct 1987	40 751	9 760	74.8
Weiherberg II	MWM	250	Oct 1991	15 140	3 673	79.4
Ringgenbach	Jenbacher Werke	454	Apr 1989	34 864	14 769	84.8
Litzholz	MWM	250	Oct 1991	17 782	4 354	91.8
Bengelbruck	MWM	250	Nov 1992	8 677	2 038	85.0
Total		4213		461 888	122 599	73.0[a]

[a] Weighted average.

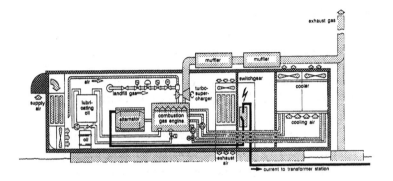

Figure 1. Cross-section of a container power plant.

already completely assembled when delivered to the landfill. On site, only a few tasks remain to be done, such as installing the exhaust flue, flanging the landfill gas pipe and connecting the control and power cables. Because of this concept, a gradual extension by deploying further units is possible – always depending on the gas supply.

PROBLEMS AND COUNTERMEASURES

Several series of experiments have been carried out to solve the problems that have occurred.

Corrosion

Even at the very first plant (Reinstetten), an unusual amount of corrosion was noticed on several engine parts, e.g. cylinder sleeves, piston rings, valves and connecting-rod bearings. This was caused by the quantities of chlorine and fluorine (as halogenated hydrocarbons) that the landfill gas contains. After extensive experiments with the participation of the engine manufacturer, several operating parameters were modified. Furthermore, re-cently developed materials were used for several parts of the engine.

Measures to Restrict Emissions

According to German federal regulation (4. BlmSchV, in force on 26 July 1988), landfill gas power plants have to be licensed by environmental standards. Therefore, they must adhere to the following limits of the amounts of carbon monoxide and nitric oxide emitted prescribed by the TA Luft (Technical Instruction, Air):

CO <650 mg/m^3;
NO$_x$ <500 mg/m^3 (referred to a residual oxygen content of 5%).

To fulfil these requirements, most manufacturers have developed lean-burn reciprocating engines, in which the combustion takes place with an air-excess factor (lambda) of up to 1.5. Thus the emitted exhaust gases can be kept within the above limits by technical measures on the part of the engine alone (Quirchmayer, 1987). This concept is especially well suited for operation with landfill gas as no catalyst is necessary. The limits on the mixture's leanness are imposed by eventual misfirings. The level of excess air at which these will occur depends mainly on the fuel, carburation, ignition system and the form of the combustion chamber.

The specifically lower energy content of the lean-burn gas–air mixture leads to a considerable loss of power in the engines. However, this can be compensated by using a turbocharger. The turbocharger has the following advantages:

- increased power;
- greater efficiency;
- homogenization of the gas–air mixture;
- more exact control of the gas-air dosage in the mixer.

A favourable plant price is obtained by using turbocharged spark-ignition gas engines. However, the turbochargers used have proved to be susceptible to faults.

Results of Exhaust Gas Emission Measurements

After converting all the engines to lean-burn operation, exhaust gas emission measurements were performed to verify whether the limits required by the TA Luft were maintained.

The following parameters were measured:

- carbon monoxide concentration;

- nitrogen oxide concentration (NO_x), given as NO_2;
- total organic carbon content;
- oxygen concentration.

TABLE 2. Measurements of Exhaust Gas Emissions at Different Landfill Gas Power Plants

Landfill gas power plant	CO (mg/Nm^3)	NO_x (mg/Nm^3)
Reinstetten	500	485
	640	450
Hechingen	470	300
	570	430
Eberstadt	410	260
	550	230
Obermooweiler	470	450
Zaisersweiher	180	475
Weiherberg	630	380
Limits prescribed by TA-Luft	650	500

Table 2 shows the results of the measurements carried out at the different plants.

QUALITY OF LANDFILL GAS AND ANALYSIS OF LUBRICATING OIL

Analyses of the landfill gas are carried out by chromatographic methods. Besides the methane extraction rates, carbon dioxide, oxygen and nitrogen, numerous chlorinated and fluorinated hydrocarbons as well as hydrogen sulphide are also specified. From this, the total amount of chlorine and fluorine contained in the respective compounds is calculated (Table 3).

For an internal combustion engine to operate properly, the landfill gas has to contain at least 40% by volume of methane. This necessary minimum percentage of methane can usually be matched as long as the landfill is not drained excessively of gas, and provided that no leakages exist in the collecting system.

To prevent corrosion on crucial parts of the engine (such as the bearings), the quantities of harmful compounds in the landfill gas should remain at a low level. The boundary value guaranteed by the engine manufacturers for the total amount of compound chlorine

TABLE 3. Example of an Analysis of the Landfill Gas Fed into One of the Power Plants Listed in Table 1

Oxygen	O_2	1.0	vol.%
Nitrogen	N_2	13.6	vol.%
Carbon dioxide	NO_2	35.3	vol.%
Methane	CH_4	50.1	vol.%
Chlorotrifluoromethane (F13)	$CClF_3$	<1.0	mg/m^3
Dichlorodifluoromethane (F12)	CCl_2F_2	21.9	mg/m^3
Trichlorofluoromethane (F11)	CCl_3F	7.6	mg/m^3
Tetrachloromethane	CCl_4	<0.1	mg/m^3
Trichloromethane	$CHCl_3$	<0.1	mg/m^3
Dichloromethane	CH_2Cl_2	1.4	mg/m^3
1,1,2-Trichlorotrifluoroethane (F113)	$C_2Cl_3F_3$	<0.1	mg/m^3
Tetrachloroethene	C_2Cl_4	1.0	mg/m^3
Trichloroethene	C_2HCl_3	2.9	mg/m^3
1,1-Dichloroethene (VD)	$C_2H_2Cl_2$	<0.1	mg/m^3
1,2-Dichloroethene	$C_2H_2Cl_2$	<0.1	mg/m^3
1,2-Dichloroethene	$C_2H_2Cl_2$	3.5	mg/m^3
Chloroethene (VC)	C_2H_3Cl		
1,1-Trichloroethane	$C_2H_3Cl_3$	1.4	mg/m^3
Hydrogen sulphide	H_2S	121	mg/m^3
Total chlorine		26.2	mg/m3
Total fluorine		7.9	mg/m3

and fluorine that the landfill gas may contain is about 75 mg/m^3, based on a content of 50% by volume of methane.

The analyses conducted so far prove that the refrigerants R11, R12 and R13 make up the major part of the halogenated hydrocarbons present. The range of chlorine and fluorine concentrations found in the landfill gas of each landfill is displayed in Fig. 2. The calculated total values remain – with the exception of the landfill in Weiherberg – well below the guaranteed boundary values.

Oil Analyses

The choice of a suitable gas-engine oil is especially important for engines driven by landfill gas.

Owing to the presence of chlorine and fluorine, an acid exhaust condensate is formed, which prematurely consumes the lubricating oil's buffering capabilities (Grahl, 1986).

For this reason, the large lubricant suppliers offer special motor-oil

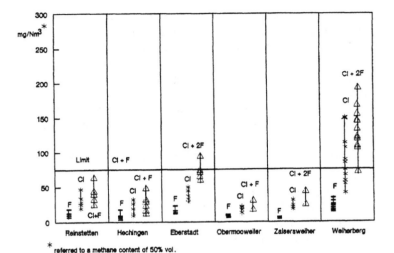

Figure 2. Chlorine and fluorine concentration in the landfill gas used in power plants at different landfill sites in Germany.

for landfill gas-driven engines. The principal difference from the lubricants used for engines driven by natural gas is the higher portion of sulphate ash, which delays the lubricant's acidification. On the other hand, the larger content of sulphate ash leads to additional deposits in the combustion chamber. Ignition by incandescence or increased wear of the valves can be the consequences.

RESULTS OF OPERATION

The landfill gas power plants are operating 24 h a day for baseload production of electricity. The annual average availability of the respective plants as well as the hours of operation on a yearly basis was about 83% (100% being continual operation) over an observation period of 4 years.

Problems of Operation

The landfill gas units are designed to operate without supervision. Daily inspection tours of the landfill personnel serve for the recording of specific data concerning engine and plant. The plant is shut down automatically when trouble occurs. A fault docket is dispatched to the nearest net-control station of the EVS. The number of shutdowns per plant averages about six per month. In most cases, our own operation personnel were able to put the plant back into operation. A distinction has to be made between different causes of trouble, and in particular between faults in the unit itself and problems with the gas collection, for which the EVS is not responsible.

In general, shutdowns are classified by the following causes:

- engine technics;
- control mechanisms;
- auxiliary units;
- gas collection;
- others (e.g. maintenance).

Figure 3 shows the causes of shutdowns, and the proportions that they constitute of the total periods of inactivity at each plant. At most plants, the main reasons for shutdowns are engine troubles and interruptions in the operation of the gas-collecting system. The latter often extend over a prolonged period of time (for example, the subsequent installation of a system for gas separation at the Weiherberg landfill). Recently installed plants, such as Reinstetten and Hechingen 2, owe much of their periods of inactivity to problems with control mechanisms. With continuing operation, the significance of this cause of shutdowns diminishes rapidly.

Some particular technical problems and their solutions are described below.

Reciprocating Engine

There was corrosion on the connecting rod bearings, cylinder sleeves, valves and piston rings of one engine. The extent of further corrosion was significantly decreased by modifications in the operating mode and improvements to the engines. In some of the plants, there were changes and displacements in valve clearance, requiring considerable maintenance. In some cases this resulted in the replacement of cylinder heads and the use of better quality valves. In one plant, damage to the turbocharger occurred several times. This was caused by both corrosion and material defects.

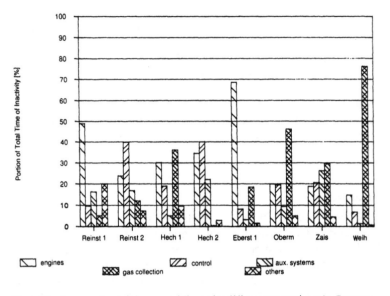

Figure 3. Reasons for shutdowns at different landfill gas power plants in Germany.

Problems with deficient exhaust gas collector lines were remedied by installing new developed water-cooled exhaust air collector lines.

As all the power generation sets are operated in the lean-burn mode, the ignition system problems are often caused by defective sparking plugs or defects in the electronic ignition system. The result is that the engines run irregularly and there is a drop in exhaust gas temperature depending on the cylinder. Deflagrations occurred several times in a number of plants, caused by ignition of unburnt mixture in the exhaust gas collector line. A relatively large deflagration occurred in the intake section of an engine belonging to a bank of stationary diesel engines as a result of carbon deposits. The pressure surges were compensated by the overpressure safety valves and the engine therefore suffered no major damage.

Control Mechanisms

Many faults result from the regulation and control systems. The power output regulation of several plants is unsatisfactory and this leads to power output fluctuations. The power limiter on one plant was defective.

In the O_2 control system, the λ-probe proved to be particularly sensitive to wear. Through the use of improved λ-probes, this fault source has been all but eliminated.

Faults frequently occur because of incorrectly adjusted or defective thermostats. Excessively high temperatures occurred in the switchgear cabinet in one plant owing to wrongly mounted or wrongly dimensioned protective motor switches. There were defects in a number of electronic components, resulting in plant outage. One plant went off stream for several weeks owing to a defective power switch and because of problems in obtaining spare parts.

Auxiliary Units

Problems occurred in the oil circulation system owing to blocked lines and defective or wrongly connected oil circulation pumps. Leakages in the cooling system of some plants had to be repaired several times. Damages to alternators in some plants were presumably due to winding short-circuits in the stator. There were also faults and defects, for example, to fans, cooling water pumps, pressure gauges, throttle flaps, and battery chargers.

Gas Collection

A number of plant outages can be accounted for by operational problems with the gas collection plant. Pressure fluctuations frequently occur when flare and engine are operated in parallel. If the minimum inlet pressure of approximately 50 mbar is undershot in the gas line to the engine, the plant shuts down owing to insufficient gas pressure. This can be remedied by installing a back-pressure regulator in the supply line to the flare.

Other causes to date for long outages in the entire plant have been leakages in the gas collection plant, blocked or slumped gas lines, defective blocked or slumped gas lines, defective blowers and faults in the CH_4 monitor.

SUMMARY

Using landfill gas for the production of electricity by means of internal combustion engines is very convenient on sites on which thermal exploitation is not possible. Since the end of 1984, the EVS has installed and put into operation 14 small landfill gas power plants.

Problems of corrosion and emission limits required by law were solved satisfactorily.

Despite the problems that appeared, the overall results of the operation of the plants are positive. This is also demonstrated by the average availability attained of about 83%.

Occasionally, improvements of some components of a plant as well as of the control mechanisms are necessary.

While conceiving a gas-collecting system, the feasibility of separation of 'usable' and 'unusable' gas should be taken into account. This guarantees the long-term exploitation of 'usable' landfill gas.

REFERENCES

Grahl, H. (1986). Gasmotorenöl-Verhalten bei verschiedenen Gasarten (Behaviour of gas engine lubricant with different kinds of gas), presented at *Technical Symposium of ESSO AG, Hohenheim*, 10 April 1986. ESSO AG, Hamburg.

Quirchmayer, G. (1987). Der geregelte Magermotor – eine schadstoffarme Variante des Gas-Otto-Motors (The controlled lean-burn engine – a low-pollutant version of the internal combustion gas engine), *VDI-Report No 630*, pp. 119–132.

Reinicke, B. (1989). Experiences with the conception and operation of landfill gas power plants, IRC-Congress, Berlin, November 1989, Deponie-Ablagerung von Abfällen B. Karl J. Thomé-Kozmiensky. EF-Verlag für Energie und Umwelltecnik GmbH, Berlin, 207–25.

7.5 Economics of Landfill Gas Utilization in the Netherlands

MARTIN J. J. SCHEEPERS

GASTEC NV, Postbus 137, 7300 AC Apeldorn, The Netherlands

INTRODUCTION

During the 1980s about 50% of the municipal solid waste (MSW) in the Netherlands was deposited on landfill sites. The remaining part was incinerated (35%) or recycled by composting (15%). During the present decade this breakdown has remained more or less constant.

Most landfill sites are situated in the north, east and south of the Netherlands. In the western part, which is more urbanized, the major part of MSW is incinerated. During the last decade the number of landfill sites for MSW decreased, while the size of the sites increased. Dutch landfill sites are normally situated above ground level. Heights vary from 5 to about 40 m and the surface areas from less than 5 to over 100 ha.

The first gas extraction experiments on a landfill site in the Netherlands were carried out in 1980 (Hoeks and Oosthoek, 1981). Beside encountering odour problems and vegetation damage it appeared that extraction yields considerable amounts of gas. In 1982 potential landfill gas production from the Dutch MSW landfill sites (Anon., 1982) was calculated. With respect to the market value of the gas, utilization of the gas seemed to be economically feasible for 12 sites. A total amount of 150 million m^3 per year was predicted as the mean production rate for these landfills for 25 years. Recent calculations predict an amount of 270–420 million m^3 of landfill gas per year, which technically can be extracted from all Dutch landfill sites until the year 2000 (Anon., 1992).

Landfilling of Waste: Biogas. Edited by T. H Christensen, R. Cossu and R. Stegmann. Published in 1996 by E & FN Spon, London. ISBN 0 419 19400 2.

TABLE 1. Landfill Gas Projects in The Netherlands

Landfill	Application	Start date	Capacity production 1991	
			m^3/h	million $m^3/year$
Wijster	Electricity	1983	600	6.4
	'Natural gas'	1989	950	7.5
Ambt-Delden	Industry	1984	1000	5.6
Bavel	Industry	1984	850	5.6
Joure	Electricity	1984	400	2.7
Maarsbergen	Electricity	1987	95	0.5
Winterswijk	Electricity	1987	100	0.0[a]
Tilburg	'Natural gas'	1987	2000	5.0
Veendam	Electricity[b]	1989	1540	5.4
Nuenen	'Natural gas'	1990	1200	3.5
Borsele	Industry	1990	865	2.3
Linne	Electricity	1990	900	5.0
Hengelo	Heat	1990	130	0.2
Emmen	Electricity[b]	1990	400	
Vasse	'Natural gas'	1991	250	0.9
Delfzijl	Electricity	1992		
Total				50.9

[a] This project stopped in 1990.
[b] Combined heat and power.

Because the internal energy demand of a landfill site is virtually zero, the gas should be utilized outside the landfill site. The gas can be used for various applications in industries situated in the vicinity of the waste site. The first Dutch landfill gas projects that were commissioned (Table 1) deliver the gas to local industries for use in a brick kiln (Bavel) and for steam generation (Ambt-Delden). In Borsele, landfill gas has been delivered since 1990 to an aluminium factory. Exceptions are the landfills in Wijster and Hengelo, where the gas is used on site for co-generation in a waste recycling plant and a leachate treatment plant respectively.

Only for a small number of landfill sites is it possible to find a suitable customer for the gas. In many cases the following problems occur.

- The distance between the site and potential industrial customers is too large.
- Industrial customers can utilize only part of the gas. This affects the feasibility of the project.

- The offtake of the gas by an industrial customer is not guaranteed over a sufficiently long period of time.

The public gas and electricity distribution grids do not have these drawbacks. Dutch waste sites are almost always situated near these grids. The base load of a distribution grid ensures a continuous offtake of the gas for prolonged periods of time. For electricity supply to the public power grid, electricity can be produced on the waste site with a gas engine/generator unit. In Joure, Maarsbergen, Winterswijk, Veendam, Linne, Emmen and Delfzijl this option was chosen. In Veendam and Emmen the heat produced by the gas engines is also used: for an industrial process and for heating a building respectively. For injection into the gas distribution grid the landfill gas should be of natural gas quality. In an upgrading plant landfill gas is treated by removal of the carbon dioxide and purification of some trace components. In Tilburg, Wijster, Nuenen and Vasse landfill gas is upgraded to natural gas quality.

In Chapters 6.4, 6.5 and 6.6 different processes for upgrading of landfill gas to natural gas quality (pipeline gas) are discussed.

LANDFILL GAS MARKET

Figure 1 shows the market values in the Netherlands for the three different utilization alternatives during the last 10 years. The market values are determined as follows.

- For industrial applications the market value matches the selling price of natural gas to industrial customers with an offtake of 1–10 million m^3 per year.
- For electricity delivered to the public power grid the market value is equivalent to the fuel costs for power stations. From 1988 onwards a capacity compensation fee of 2.8 ct per m^3 is added.
- For upgraded landfill gas the market value is equivalent to the price that gas distribution companies pay for natural gas.

At the end of 1986 energy prices collapsed. This had considerable impact on the planned landfill gas projects. The market values of the gas for industrial applications and electricity production were most affected (both minus 60%). The decrease of the market value for upgrading to natural gas quality was smaller but also substantial

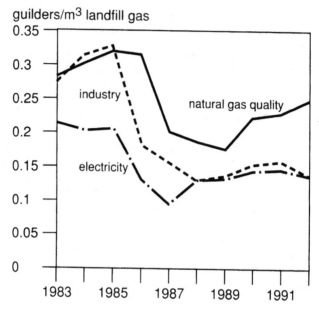

Figure 1. Market value of landfill gas.

(minus 45%). After 1986 the market value for landfill gas did not really recover. Some planned projects were delayed and others changed their utilization option. Many planned projects, however, have been stopped in spite of the economical feasibility even at lower energy prices (Voorter, 1987).

The economical advantage of landfill gas production disappeared, but the environmental problems remained. Even new environmental problems were distinguished: landfill gas emissions contribute to the greenhouse effect.

Because of the latter reason landfill gas production is again taken seriously. There are, however, some changes in the approach, as follows.

- Landfill gas extraction is carried out with the intention to reduce the emissions. By doing so, air could be sucked into the landfill. This influences the quality of the landfill gas (lower calorific value) and makes the gas less suitable for utilization.
- Stringent measures to prevent penetration of rainwater into the landfill are impeded by the gas emissions from the landfill. A landfill extraction system is necessary.

- An extraction system should be required for optimum landfill management (reduction of the emissions). A project for utilization of the landfill gas will then not be charged with the investment costs of the extraction system.

EVALUATION OF UTILIZATION OPTIONS

When the utilization of landfill gas is considered often the following question is raised. Which is the best utilization option to choose: upgrading to natural gas quality or electricity generation?

The choice between upgrading landfill gas or producing electricity not only depends on which option yields the highest revenues, but is also a matter of financial risks and flexibility. The financial risks are influenced most by the following aspects:

- the investment and operating costs (see below);
- the market value of the landfill gas (see Fig. 1);
- the amounts of gas that can be extracted from the site during the lifetime of the project. The gas amounts will decrease after closure of the waste site. The rate of this decrease is difficult to estimate.

In contrast to upgrading to natural gas quality, electricity generation is more flexible, for the following reasons.

- Gas engine/generator units can be placed in a container or on a skid. When the amounts of gas are too low or the project has failed economically, the unit can be moved easily to another landfill or be sold.
- When more gas is available, an additional unit can be installed without disproportionate extra investment costs.

REVENUES

With these considerations in mind it is possible now to have a look at the revenues that can be obtained by the two alternatives. The revenues can be calculated from the cost price and the market value of the gas. The cost price is related to the investment costs and operating costs.

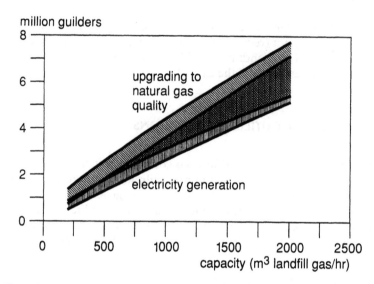

Figure 2. Investment costs.

The investment costs for upgrading to natural gas quality and electricity generation are shown in Fig. 2. This figure gives the investment costs for the installations on the site, exclusive of the transmission pipeline or cable to the distribution grid.

Figure 2 shows that at all capacities the investment costs for electricity generation are slightly lower than the upgrading alternative. However, the distances between the landfill site to either the gas or the electricity distribution grid could have an impact on the difference between the investment costs of the two utilization options.

The operation costs for the upgrading plant consist of costs for maintenance, energy, auxiliary materials and control. The operation costs for the electricity generation consist of costs for maintenance and control only.

In Fig. 3 the cost price per cubic metre is given as a function of capacity. Capital costs are included on the basis of a discounting rate of 7% per year and a depreciation period of 10 years. Figure 3 shows that the cost price for upgrading is much higher and more dependent on the capacity than the cost price for electricity generation. However, Fig. 1 shows that the market value for upgraded landfill gas also lies above the market value for electricity generation.

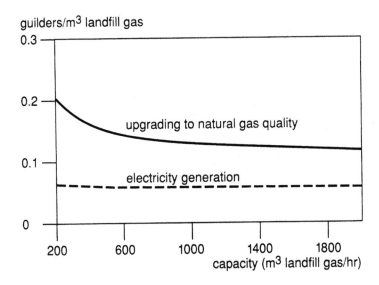

Figure 3. Cost price.

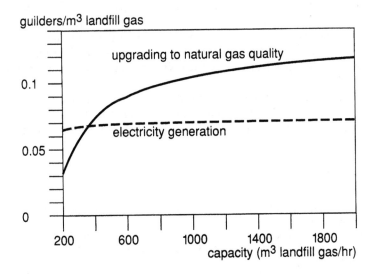

Figure 4. Profit margin.

m³ landfill gas/hr

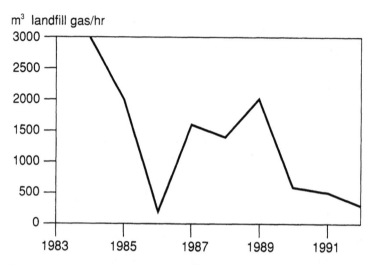

Figure 5. Breakeven point between upgrading to natural gas quality and electricity generation.

In Fig. 4 the profit margin (market value minus cost price) is shown as a function of the capacity. This figure is determined at a market value of NLG0.23 per m³ of landfill gas for upgrading and NLG0.14 per m³ of landfill gas for electricity generation. The figure shows that below a capacity of 250 m³ per hour upgrading is not economically feasible, in contrast to electricity generation, which is economically feasible at all capacities. At a capacity of 350 m³ per hour upgrading will give the same revenues as electricity generation (breakeven point).

The breakeven point will be influenced easily by changing cost price or market value. How the breakeven point varied during the last 10 years as a result of the changing market values is shown in Fig. 5.

REFERENCES

Anon., (1982). *Durable energy*, Report for the Minister of Economical Affairs. SDU, The Hague.

Anon., (1992). Prognosis by the Landfill Gas Advisory Centre, Apeldorn.

Hoeks, J. and Oosthoek, J. (1981). Gas production from landfills. *Gas* (101) 563–568.

Voorter, P. H. C. (1987). Production and utilization of landfill gas on 23 locations studied. *Energie en Afvalbeheer.*

7.6 Landfill Gas Utilization as Vehicle and Household Fuel in Brazil

JOSÉ H. PENIDO MONTEIRO

Rua dos Parecis 15, Cosme Velho, Rio de Janeiro, Brazil

INTRODUCTION

As a result of the petroleum crisis of the 1970s, several alternative energy sources, so far unfeasible, became more competitive and gained a new impulse towards development and improvement. Those more suitable for use in Brazil are aeolian energy, solar energy, hydroelectric power stations working on small water courses, and alcohol from sugar cane as vehicular fuel. The latter might be the most widespread and successful alternative energy source in this country, which currently has at its disposal a fleet of approximately 5 000 000 vehicles fed on this type of fuel.

The steadiness of petroleum prices, as observed in the 1980s, acts on the contrary as an inhibitor of the search for renewable energy sources, thus resulting in a progressive decrease of the planet's resources and consequently anticipating the energy crisis foreseen by international organizations, which will occur in the near future. Obviously this crisis will not affect all countries equally. The economically developing countries will suffer more than developed countries, which will continue to consume available energy clumsily and carelessly.

For the above reasons landfill biogas (LFG) may represent an important alternative energy resource for economically developing countries.

In this regard it is fundamental to develop and implement LFG recovery technologies that involve low investment, operational and

Landfilling of Waste: Biogas. Edited by T. H Christensen, R. Cossu and R. Stegmann. Published in 1996 by E & FN Spon, London. ISBN 0 419 19400 2.

maintenance costs, are labour intensive (basically for unskilled workers) and require low energy consumption, yielding a highly positive energy balance.

This objective could be reached using LFG, without any treatment, for simple household (cooking, lighting, refrigeration) or industrial (boilers and furnaces) uses, provided that users are located close to the landfill. These uses of LFG should be widely stimulated in all cities with more than 30 000 inhabitants that possess a landfill with a minimum thickness of 8 m, either well or badly operated and which is being or has been recently used. More sophisticated energy recovery options from LFG, such as use as vehicle fuel, can be adopted only after a careful economical feasibility study in areas where a good technological level can be assured for operation of the facilities.

In the following sections two examples of LFG utilization in Brazil (one sophisticated and one simple case) are reported and discussed in detail.

LFG USE AS VEHICLE FUEL

Comlurb – Compania Municipal de Limpeze Urbana (Municipal Public Cleansing Company) – is the public company responsible for solid waste management in the city of Rio de Janeiro, Brazil, including domestic waste collection (4000 t/d), street and beach cleansing, operation of four transfer stations, three recycling and composting plants and four controlled landfills.

Comlurb, with a workforce of about 12 000 employees, possesses a fleet of more than 1200 vehicles (collecting trucks, transfer trailers, dumping trucks, Dempster trucks, tractors, front loaders, small trucks etc.) with a monthly fuel consumption rate of 800 000 l of diesel oil, 30 000 l of gasoline and 50 000 l of alcohol.

Following the oil crisis in the 1970s, the Brazilian government has encouraged and subsidized research into alternative energy solutions, especially those originating from activated wastes, with the aim of reducing diesel and gasoline consumption, in both urban and road transportation. In the state of Rio de Janeiro, Comlurb together with CEG (Gas Company of the State of Rio de Janeiro) has pioneered the biogas recovery technology development of water landfills.

In 1977, Comlurb implemented the biogas collection project in the Caju landfill; the biogas was transported through a 4 km pipeline to the CEG gas plant – located in Sao Cristovao, a city section – and added to naphtha gas and subsequently to natural cracked gas, in order to be distributed to the residential areas of the city through the company's distribution network.

In 10 years of operation the CEG system has recovered approximately 20 million m^3 of raw biogas, with 5800 kcal/m^3 heat capacity, which has been combined with the gas produced by the plant (4200 kcal/m^3 heat capacity), without any kind of treatment whatsoever. The operational costs of the system were extremely low (one pump with a 2 hp blower, operated by three technical experts), thus confirming the excellent viability of the programme.

In 1980 Comlurb engineers started to work on studies concerning the utilization of biogas in their vehicles; a detailed project for the collection, purification, compression and supply unit was developed, and finally in 1985 work was started and since then the installations have been operational.

Characteristics of the Caju Landfill

The Caju landfill is located by the Guanabara bay shore, 8 km from downtown Rio de Janeiro. It was installed in 1935, at a time when waste was still collected using animal-traction vehicles. The landfill operated until 1977, when it was closed; it was considered one of the major landfills in the city. During its operational existence there was no control whatsoever over the quantity or quality of the wastes. In reality, it was an open dump, with irregular land cover, and had received about 30 million m^3 of waste. Today the Caju landfill has spread out over an area of approximately 1 000 000 m^2, totally covered by irregular layers of clay. Its highest spot stands approximately 20 m above sea level. The gas collection system occupies an area of approximately 250 000 m^2 and is located in the last filled areas.

It can be concluded that not only the high tier of organic materials existing in Rio de Janeiro waste, but also the high pluviometric and insolation rates, constitute ideal conditions for the obtaining of a high production of biogas, although the landfill has received no previous treatment with regard to future processing of wastes. More than 10 years after closure, the landfill gas production was still at acceptable levels although only a quarter of the total landfill area was utilized.

Project Description

The original project can be outlined through the subsystems of collection, transportation, purification, compression, supply and conversion of the vehicles, as follows.

Collection and transportation. In previously selected landfill areas, 12 test pits were drilled with 2 in (50 mm) diameters and at an average depth of 15 m.

The original project provided for the drilling of eight wells with 70 cm diameter and an average depth of 15 m, in sites considered adequate by the prospective studies. 'Benoto' drilling equipment was used, but it proved to be expensive and difficult to move. In 1987 seven other wells were drilled using simpler technology, as explained later on.

The collection pipe installed in the drilled well is a PVC tube of 4 in (100 mm) diameter, class 20, perforated with $\frac{1}{2}$ in (12.5 mm) bores along its length. The well ends were finished with special clay with a low permeability rate.

The wellhead is engineered as usual with a valve for gas sampling, a flow control bore plate, a flexible tube to connect the gas pipeline, and a flow control and locking valve. A syphon was installed in each well for the removal of condensates.

The pipeline for gas transportation was made of high-density polyethylene tubes of 2 in, 3 in and 4 in diameters (50, 75 and 100 mm). The tubes were placed in ditches with an average depth of 40 or 80 cm under the streets, and then filled with dirt.

Compression and purification. Two Sulzer five-stage compressors manufactured in Brazil, each with a 225 Nm^3/h capacity, working in parallel operation and driven by two 100 hp (75 kW) engines, were used.

Before reaching the compressor, the raw LFG runs through two steel wool filters for the removal of H_2S, with condensation drains and input pressure that does not exceed 2000 mm of water column. After the second stage, with a pressure of 14 kg/cm^2 and with a flow rate of approximately 225 Nm^3/h, the biogas is forwarded to the

Figure 1. Scrubbing and regenerating towers.

scrubbing tower for removal of CO_2, and returns to the third stage with only 5% of this gas (Fig. 1). The biogas is then compressed by the third, fourth and fifth stages, with a final pressure of 200 kg/cm^2 and a flow rate of approximately 160 m^3/h.

The scrubbing water is regenerated in the regenerating concrete tower with a suction fan at the top. The CO_2 is naturally released into the atmosphere through the air thrust in counterflow by water sprinklers.

High-pressure storage. The system comprises 20 chromium–molybdenum cylinders, with 80 l hydraulic capacity and 220 kg/cm^3 pressure. The cylinders, interconnected by manifolds, are assembled in three sets (two sets with seven cylinders and one with six cylinders). The sets are installed in an open and ventilated area, isolated from the compression house. Its discharge system is manual (cascade type) to allow a fast filling of vehicles (one at a time, by differential pressure).

Filling system.   The filling system is organized in central and decentralized units.

The central filling unit is located close to the biogas purification unit (Fig. 2). Gas transportation pipelines, as well as the valves, fittings and other accessories, are made of steel and are compatible with the operational pressure of 220 kg/cm^2.

The 20 filling stalls were designed to allow a slow filling, except for one stall for fast supply located next to the high-pressure central filling unit.

One of the greatest problems concerning utilization of a gaseous fuel in vehicles is the high price of transportation in high-pressure cylinders. In order to decentralize the supply, and because it was impossible to build a network pipeline throughout the city that could reach the Comlurb garages, where the vehicle fleet is concentrated, two forms of filling were designed and implemented, as follows.

Figure 2. Biogas filling unit.

- *Filling through differential pressure.* Five sets of 24 steel cylinders with 80 l hydraulic capacity and pressure of 220 kg/cm^3 were built, and placed in four horizontal superimposed runs, each formed of six cylinders. The cylinders nest in a rigid steel structure, with lifting devices and safety transportation by Dempster or Brooks trucks.
- *Decentralized filling unit.* This unit was constructed in one of the Comlurb garages that shelters approximately 20 light vehicles and 100 heavy vehicles; this station is capable of filling seven vehicles at a time. The project has foreseen the installation of a booster compressor with 30 Nm^3/h capacity with the purpose of transferring gas from the cylinder sets to the filling points without pressure loss.

Vehicle conversion.   The conversion of all vehicle engines to purified biogas was performed by Comlurb engineers in the purification unit workshop.

Basically, the conversion system consists of two parts. The first comprises high-pressure gas storage using special cylinders manufactured from seamless steel tube, closed by a hot spinning process (Fig. 3). The choice of the cylinder diameter, quantity and length depends on the type of vehicle to be converted and on the degree of autonomy desired. The gas is stored at a pressure of 220 kg/cm^2 in the high-pressure cylinder and released through the excess valve in steel tubes to the charger valve, which, when open, allows the gas to reach the pressure reducer. The pressure is reduced from 220 kg/cm^2 to atmospheric pressure in three stages and the gas is then released to the mixer as required by engine suction.

The second part of the system is a mixer specially designed for every type of carburettor in order to allow a correct and homogeneous mixture. The original supply of gasoline or alcohol to the engine is interrupted by means of a solenoid valve. The sector switcher selects the desired fuel to be used. A water circuit originating in the engine block, in the case of a watercooled engine, passes through the reducer heating it up, to avoid freezing or low temperatures, and then returns to the radiator. Owing to their peculiarities, diesel cycle engines cannot be run on gas alone, therefore diesel fuel can only be partially substituted; there is always the need for some diesel oil to allow a spontaneous flame and to cool the injectors. The optimum ratio to prevent engine damage is 70% methane and 30% diesel fuel.

Figure 3. High pressure gas storage cylinders.

Civil Works and Electric Substation

All civil works took into account area adjustment and were built on the Caju waste landfill in the context of an economical criterion aiming at a rational occupation of the area. The following buildings were erected.

- *Operational control unit.* Administrative offices, laboratory, toilets, cafeteria and warehouse are installed here.
- *Industrial area.* Located near the operational control unit, where the substation, compressors, scrubbing and regeneration towers are installed.
- *Central filling unit.* Next to the industrial area with a shedded parking lot for the vehicles, including circulation area.
- *Vehicle conversion workshop.* A very simple building equipped with all necessary tools for conversion of vehicles to purified gas utilization. Furthermore, in this workshop the maintenance of all converted vehicles is performed, such as pressure reduction valve repair, engine tuning, and high-pressure cylinder cleansing.

Alterations and Modifications to the Original Project: the Use of New Technology

Continued operation of the biogas purification unit and the need to expand the system prompted the development of new technologies aimed chiefly at the reduction of the costs of landfill gas prospection and collection and maintenance of conversion kits.

Drilling and installation of new wells. In the first phase of the project the wells were drilled with equipment employed in building civil works foundations at very high cost, from both an operational and installation point of view.

Because of these difficulties, when the second phase of the project was started a more economical and faster solution for the drilling of wells was considered. It was decided to use more conventional and simpler equipment, normally used to drill artesian wells. This equipment consists of a 12 in (300 mm) diameter steel drill, which works by alternate percussion and rotating movements. From time to time, the drill is removed from the well and the material resulting from the percussion is mixed with water and taken from the well through a steel tube equipped with a check valve in its lower end. When the drilling is finished the well walls are cleaned with a high-pressure spray to facilitate gas migration.

Evaluation of the Well Productivity

One of the major difficulties concerning implementation of the project is represented by evaluation of the productivity of a newly installed well, not only with regard to quantity but also to quality. This information is extremely important, because it is only after confirmation of the satisfactory characteristics of the well that it will be connected to the collection network. The absence of electric energy in the landfill makes operation more difficult; therefore the following simple and inexpensive system was developed.

A flexible hose, installed on the edge of the well, was connected to a domestic vacuum cleaner at the suction side. To the other end of the vacuum cleaner another flexible hose was attached, and then a flowmeter. In the outlet of this flowmeter fittings were installed. The flexible hose outlet of smaller diameter was connected to the carburettor of a small 5 hp (3.7 kW) engine originally fuelled by gasoline. Coupled to the flowmeter outlet a steel tube, flame resistant, of 30 cm diameter was installed, in case it became necessary to burn the gas. The combustion engine was installed to actuate a small 2 kVA generator, which would provide electric power for the vacuum cleaner. The process is initiated by starting the engine with gasoline; the generator belt is tightened and the vacuum cleaner is turned on. It starts sucking up the gas, whose flow is measured by the flowmeter. Part of the gas at the flowmeter outlet is diverted to the combustion engine, and after a short operating period it has its gasoline supply cut and starts to operate with biogas only.

Through engine acceleration, which affects the generator's rotation and thus the vacuum cleaner, one can evaluate the well production at different suction pressures.

Mobile LFG Compression and Storage Unit

Together with a company that manufactures domestic waste compactors, a portable skid-mounted system was built. This system can be hauled by tractors or front loaders over the landfills and is designed to purify, compress and store the gas at high-pressures, also allowing vehicular filling. The unit consists of the following equipment:

- VW engine, 1300 cc, adapted to raw biogas use;
- 20 kVA generator, coupled to the VW engine;
- precompressor of $10 \text{ Nm}^3/\text{h}$ at 10 kgf/cm^2, driven by an electric engine;
- high-pressure compressor of $7 \text{ Nm}^3/\text{h}$ at 200 kgf/cm^3 pressure;
- scrubbing tower to remove CO_2 from the gas;
- gas regeneration tower consisting of a fibre-cement water tank;
- water pump and water circulation closed circuit;
- four high-pressure cylinders, each with 40 l hydraulic capacity;
- hose for filling of vehicles.

The unit has operated satisfactorily, but still requires two gas wells: one to feed the VW engine and another for the purification system. The only disadvantage that renders operation unattractive is the low production of purified gas ($7 \text{ Nm}^3/\text{h}$). If the operational regime is 16 h a day, the system would be capable of filling only eight vehicles of the VW van type, and only very slowly. Power generated by the generator set largely exceeds the power consumed by the system's equipment. In this way one could use a larger compressor to supply the system without changing the basic characteristics: i.e. autonomy, compactness and easy transportability.

Operation and Costs

Currently, Comlurb possesses several vehicles that run with purified biogas. Operational data are reported in Table 1.

Investments necessary to implement the project have been kept to a minimum, for to the following reasons.

- The project developed by Comlurb engineers was designed for simplicity and ease of operation, utilizing material and equipment exclusively available on the Brazilian market.
- The most expensive item of the project is the high-pressure compressor. The units acquired by Comlurb were the first manufactured in Brazil, and perhaps this fact influenced the equipment price, which was rather low. Nowadays the cost for compressors can reach 50–60% of the investment costs.

TABLE 1. Operational Data Regarding the Utilization of the Caju Landfill Biogas as Fuel for Different Kinds of Vehicle

Type of vehicle	Total of converted vehicles	Monthly raw biogas consumption (Nm^3)	Monthly purified biogas consumption (Nm^3)	Monthly utilization (km)
Light vehicles (Otto cycle, alcohol or gasoline)	140	39 000	31 200	156 000
Collecting truck (Otto cycle, alcohol)	2	2 500	2 000	3 000
Collecting truck (diesel)	1	750	600	1 300
Dumper (diesel)	9	10 000	8 000	40 000
Dempster truck (diesel)	1	1 000	800	4 000
City taxi cars	100	44 000	35 400	–
Total	253	97 250	78 000	–

- No work was carried out in the Caju landfill regarding preparation for the project, nor was any investment necessary for utilities (electric power, water, communications), which were already available on the site.

The annual operational cost was twice the amortization annual rate for investment costs.

Considering the costs of other vehicular fuels available in Brazil (gasoline, diesel and alcohol), the use of fuel from LFG is a good economic alternative (Fig. 4).

The economical advantage with diesel oil and alcohol could have been greater, as these fuels are heavily subsidized by the government, so that their price is lower than the real cost.

LFG USE AS HOUSEHOLD FUEL

Natal is the capital of the state of Rio Grande do Norte, located in the northeast of Brazil, about 2800 km from Rio de Janeiro. It is a littoral city of around 600 000 inhabitants, largely frequented by tourists.

The city produces approximately 500 t of urban waste per day, with a high rate of organic matter. This waste is dumped in a controlled landfill located near a large sand dune, from where covering material is taken.

In 1983 the city administration decided to elaborate a project for the utilization of LFG including, in three stages:

- a community kitchen for the low-income residents of the community, installed next to the landfill;
- a gas distribution network connected directly to the 150

Type of fuel	Cost (US$/litre)
Gasoline	0.514
Diesel	0.354
Alcohol	0.417

Figure 4. Relative cost of vehicle fuel in Brazil.

habitations of the community, the majority of which used wood collected from the landfill as fuel for cooking;

- a gas-feeding link to a boiler in a cashew nut industrialization factory, capable of replacing that in use, which fed on timber that was becoming progressively scarce in the region.

This project was not financially supported by the government. However, the Natal administration asked for implementation of the community kitchen using its own resources, and this took place in 1986.

Project Description

Six production wells were perforated with manual drills of 6 in (150 mm) diameter and an average depth of 6 m (the waste layer was 7 m deep), one 20 m apart from the others. PVC perforated pipes, 4 in (100 mm) in diameter, were placed in the wells, and the space between these and the well walls was filled with gravel. Well-compacted clay of low permeability covered the top of the gravel heap up to ground level.

Flexible and transparent PVC pipes ($\frac{3}{4}$ in, 20 mm) for domestic use were used for gas transportation. The six well pipes were linked to a connector from where another 1 in (25 mm) diameter pipe led the gas to the compressor, 70 m away from the landfill.

For the suction of the gas from the wells and subsequent compression the burners, a 2 hp (1.5 kW) air compressor normally used for painting of vehicles, with a 150 l stock, was utilized. The pressure regulator, normally operated at 150 lbf/in^2 (1034 kPa), was readjusted to 3 lbf/in^2 (20 kPa), thus allowing gas storage in the compressor stock. In its outlet a valve was installed aiming to reduce pressure to 660 mm of water column, the recommended rate for feeding of the burners.

The link from the pressure reducer valve to the manifold was performed by means of a flexible PVC pipe of the same diameter. The manifold, made of galvanized steel, crosses the 19 boxes with a $\frac{1}{2}$ in (12.5 mm) outlet for each burner controlled by a small valve to open or close gas feeding.

The burners are similar to those used in industrial stoves, and were installed under a concrete top with a hole in the centre for the flames to go through, with crossed iron bars to hold pans and pots (Fig. 5).

The community kitchen is situated in a simple masonry building

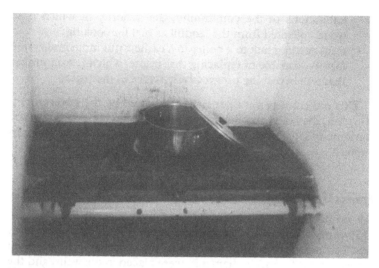

Figure 5. Burners used for landfill biogas.

covered with ceramic roofing tiles. There is a small room at the back where the compressor is installed and, at the front, the 1 m wide boxes with concrete tops. The building is entirely surrounded by a wall with a gate, through which entry is controlled.

The functioning of the community kitchen is controlled by the employees of the Municipal Cleansing Company. Its working period is from 1100 to 1400 h in the daytime and from 1600 to 1900 h at night. It has been utilized by a large section of the community, especially those with low incomes who do not have stoves fed on liquefied petroleum gas (LPG) at their disposal.

CONCLUSION

The studies carried out by Comlurb and by Natal municipality have proven the technical feasibility of utilization of the gas generated in the landfills. They have also shown that excellent levels of yield can be reached with the use of simple technologies without compromising the quantity and the productivity of the system, making viable an energy alternative that is available in the surroundings of all urban conglomerates, especially those of the economically developing countries.

8. GAS MIGRATION, EMISSIONS AND SAFETY ASPECTS

8.1 Landfill Gas Migration through Natural Liners

RODRIGO A. FIGUEROA & RAINER STEGMANN

*Institute of Waste Management, Technical University of
Hamburg-Harburg, Harburger Schlossstrasse 37, 21071 Hamburg,
Germany*

INTRODUCTION

In the Federal Republic of Germany several landfills are provided
with natural surface liners. In the case of closed landfill sites, where
remedial action must be taken, the caps are installed with the
primary aim of avoiding rainfall infiltration of the landfill body, thus
minimizing leachate production and reducing gas emission into the
atmosphere. Furthermore, according to a recommendation of the
LAGA (Lottner, 1991), municipal solid waste landfills in particular
should be provided with surface liners in order to optimize the
recovery of the landfill gas produced. This implies a further important purpose of the cap: it should prevent the intake of air through
the surface during active landfill gas extraction.

The surface liner has to meet several requirements: it must have
low long-term water and gas permeability; it must remain flexible
once installed so that no cracks occur with settling; moreover, in
order to avoid sliding, especially on high landfills, the liner materials
in particular for slope areas must have sufficient sheer strength
(Jessberger and Geil, 1989).

Until the present, the investigation of the behaviour, i.e. the
efficiency of natural liners, has for the most part been restricted to
the aspect of long-term watertightness. However, gas permeability
also plays an important role as regards gas-extraction facilities,

Landfilling of Waste: Biogas. Edited by T. H Christensen, R. Cossu and
R. Stegmann. Published in 1996 by E & FN Spon, London.
ISBN 0 419 19400 2.

water-vapour movement in the natural layer and the minimization of emission.

The authors have carried out long-term investigations concerning the gas components produced in the closed Georgswerder landfill in Hamburg. The behaviour of the surface liner installed in October 1989 has been investigated by means of test fields set up at a later date. Particular attention was paid to the extraction of landfill gas.

This chapter reports the first results obtained from field investigations at the test fields. It concentrates especially on the investigation of the gas permeability of the natural liner and its possible effects on the gas extraction operation and the long-term effectiveness of the cover system.

THE SURFACE CAP OF THE GEORGSWERDER LANDFILL

The closed Georgswerder landfill is situated about 5 km south of Hamburg city centre. Between 1948 and 1979 it was in operation for the deposition of construction waste, household waste and solid and liquid industrial waste. A total of 14 million tonnes of waste were disposed of in an area of approximately 44 ha, creating a hill of 44 m height.

In 1983 an extensive remedial programme was started to minimize emissions. An essential part of the remedial action was the encapsulation of the landfill body by means of surface lining. This was done in order to prevent the infiltration of water into the landfill body and to avoid gas migration out of the landfill. By October 1989 the upper part of the landfill (a third of the total area) was covered with a multilayered final cover (Fig. 1), consisting of the following components from bottom to top (Klenner *et al.*, 1989):

- a compensation sand layer (> 35 cm, 0.1–3.0 mm), which also performs the function of a gas-venting device (gas drain);
- 60 cm mineral liner compacted in three lifts of a 20 cm thickness of glacial till consisting of clay = 17%, silt = 23% and sand = 60% with max. $k_f \leq 10^{-9}$ m s^{-1} (Melchior *et al.*, 1990);
- a 1.5 mm HDPE membrane placed directly on top of the clay liner to protect it from roots and animal activities, provided with 6 mm spikes (The upper surface is structured with 2 mm

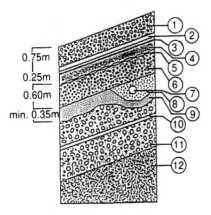

Figure 1. Construction of the upper lining system (Wolf, 1989): 1, cover soil; 2, new gas pipe; 3, geotextile; 4, sand drain; 5, HDPE membrane, $d = 1.5$ mm; 6, mineral liner; 7, gas pipe (slotted); 8, gravel; 9, sand gas drain; 10, former gas pipe; 11, existing cover; 12, waste.

straps to avoid sliding. In order to prevent the membrane liner from deterioration due to settling, it was only welded vertically and overlapped horizontally.);

- a 25 cm drainage layer to discharge surface water, $k_f \leq 10^{-4}$ m s^{-1};
- a protecting geotextile (450 g m^{-2});
- a vegetative layer as an erosion control component (>75 cm), with at least 20 cm topsoil.

The sealing of the lower part of the landfill is under construction and was due for completion in 1995.

GAS EXTRACTION TESTS (SUBSURFACE GAS DRAIN)

To reduce emissions, a gas extraction system consisting of 39 vertically rammed wells was completed in 1982. The gas collection system, which has been in operation since 1986, is currently extracting landfill gas at a rate of about 350 m^3 h^{-1}. This gas is utilized in high-temperature furnaces at a nearby industrial location (Stegmann, 1991).

First indications of a certain gas permeability of the till layer could be detected when the gas drain located directly below the layer was connected to the gas extraction unit for test purposes. Owing to the resulting vacuum in the gas drain, the oxygen content of the extracted landfill gas increased to a critical concentration. For this reason, this subsurface gas extraction system has not been used since. However, as the efficiency of the extraction system as a whole is not improved, this subsystem is not needed.

In November 1989 the gas drain area was intensively tested; the pressure conditions were continuously recorded (Fig. 2). The measurements clearly showed that the fluctuation of gas concentration in the subsurface gas drain is highly dependent on the existing pressure conditions. At slight sub-pressure the oxygen content of the gas increased considerably. These results show that changes in gas composition in the gas drain are mainly caused by a pressure-driven gas flow through the till layer. Furthermore, the pressure conditions are mainly influenced by the operation of the gas extraction unit. In addition to the active gas extraction, the barometric pressure is also of importance, as even small changes in pressure may affect the flow rate (Young, 1990).

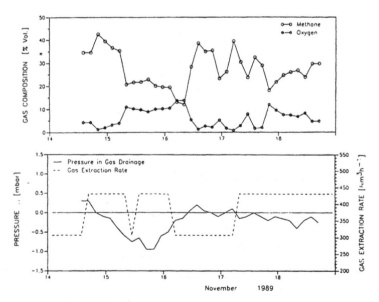

Figure 2. Pressure and composition of gas in the gas drain area for different extraction rates.

GAS TRANSPORT THROUGH NATURAL LINERS

The transport of gas through partly saturated, clayey soils is enabled by a coherent air pore system. Its structure can be characterized by a micromeric and macromeric pore system. When water is added to individual, air-dried grains of clayey materials they agglomerate, forming a cluster with a macromeric structure. The water bound to the grains forms a kind of membrane, sealing the micromeric pore system. The macromeric pore structure is of primary importance for the gas permeability of clayey soils (Barden and Pavlakis, 1971; Hartge, 1978).

Moreover, during emplacement of the clay liner, the compaction work increases the porewater pressure in the grain heap for a short period of time, thus enabling a deformation of the grains. As a result new contact areas between the individual grains are built up, forming a new macromeric soil structure. The intensity of these changes is mainly dependent upon the grain size distribution and the built-in water content.

The gas transport through porous bodies can be described by the processes of partially pressure-dependent diffusion and pressure-driven convective flow.

Including the compressibility of gases, the convective flow can be described by the Hagen–Poiseuille law. The laminar, convective gas flow may be described for *small* pressure gradients by Darcy's equation. Considering the possible changes of flow medium (gas) and temperature, the intrinsic permeability k_0 describes the *material-specific* property of porous media to cause gas flow dependent upon pressure differences (Hartge, 1967):

$$k_0 = \frac{\delta z \eta V}{A \, \delta p \, \delta t} \frac{p_h}{p_m} \tag{1}$$

where k_0 is the intrinsic permeability (m^2), δz is the flow way (m), A is the flow profile (m^2), η is the viscosity of the gas (N s m^{-2}), V is the gas volume (m^3), δt is the time interval (s), δp is the driving pressure difference (bar), p_h is the initial pressure (bar), and p_m is the mean pressure (bar). In the case presented here, the value of 5 mbar is not exceeded, making the factor p_h/p_m for the gas compressibility negligible.

With increasing water content in the macropore system, the possible gas flow through the partly saturated soil decreases and the diffusive gas flow gains importance. Close to the saturation limit only a very slow diffusive gas flow takes place (Barden and Sides, 1970).

For a porous body with known dimensions (δz, A), with a known intrinsic permeability k_0 and with a given driving pressure δp, the flow of a gas or gas mixture Q_{Gi} can be calculated using the following equation:

$$Q_{Gi} = \frac{V_{Gi}}{\delta t} = \frac{\delta z \eta_{Gi}}{A \, \delta p} \, k_0 \tag{2}$$

where η_{Gi} is the viscosity of the gas or gas mixture (N s m^{-2}).

EXPERIMENTAL INVESTIGATIONS

To characterize the behaviour of the gas permeability of the till layer, laboratory tests and field measurements were carried out.

Results from laboratory tests cannot be fully applied to site conditions. For this reason five test fields (TF) were installed half a year after the completion of the upper cover system (Fig. 3). Thus a general assessment of the completed surface liner and the characterization of the gas permeability behaviour became possible.

Laboratory Tests

The intrinsic permeability of disturbed samples was determined in modified triaxial cells. The material used for laboratory testing had been collected during installation of the test fields. It can be characterized as follows: consistency index, $l_c = 0.7$; plasticity index, $l_p = 6.5$; Proctor dry density = 2.00 g cm^{-3}; optimum moisture content 9.7% by weight. The samples were air-dried, and oversize grains >5 mm were removed. The material was compacted according to DIN 18127 by means of the standard Proctor compaction test at different moulding water contents (WC). (Samples: $H = 50$ mm, $\varnothing = 100$ mm).

As an example, Fig. 4 shows the permeability of disturbed till samples as a function of the water content and a degree of compaction = 0.6 MN m m^{-3}. The results show a significant dependence on the WC; the intrinsic permeability increases substantially at a

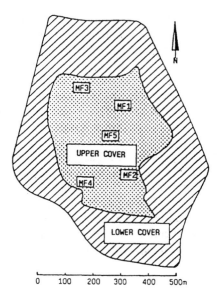

Figure 3. Ground plan of the lower and the upper liner system showing the different test fields.

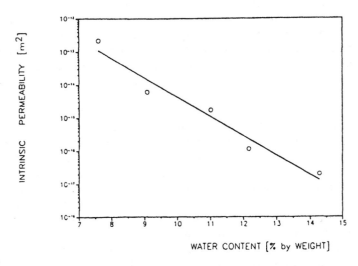

Figure 4. Permeability of disturbed till samples as a function of the built-in water content (WC). Degree of compaction = 0.6 MNm m^{-3}.

constant compaction rate by about one order of magnitude per 1.4% displaced WC. Most designs for compacted mineral liners recommend compaction equivalents of 95% of standard Proctor. An intrinsic permeability of 10^{-14} m^2 can be expected at a maximum sample density, with an optimal WC of 9.7% by weight (WC_{pr}).

Permeability tests were also carried out in the laboratory on undisturbed samples from the different test fields (H = 100 mm, $\varnothing$ = 100 mm). The results from the investigation of several samples from the same field and layer varied considerably ($k_0 = 1 \times 10^{-14}$ to 3×10^{-17} m^2). It has to be assumed that material sampling results in considerable changes in the macromeric structure. This problem, which is familiar from hydraulic conductivity quality assurance and control of natural liners (Farquhar, 1989), increases when the gas permeability is determined. The gas permeability of unsaturated compacted soils is several orders of magnitude higher than the hydraulic conductivity. It can only be reliably determined by using field tests.

Field Measurements

Gas boxes, positioned at various different depths in the glacial till layer, were placed in the test fields. The gas collection boxes were constructed of aluminium with a collection area of 1 m^2 and a height of 15 cm. They were filled with gravel (6/16 mm) and were open at the base. Each box was connected at two points with the exterior sampling pipe (diameter = 6 mm). The boxes were installed at 15, 40, and 65 cm below the surface of the till layer. One box was located in the gas drain area (Fig. 5).

By extracting gas from the collection boxes, the gas permeability of the *lowest lift of the till layer* was determined by means of the difference in volume flow, gas temperature and pressure between the two boxes and the gas drain. To ensure that the gas collection boxes were hermetically sealed, the main gas components methane, oxygen, and nitrogen were used as tracers. For this reason measurement could only be effected at times when the landfill gas extraction system was not working or when only small amounts of gas were being extracted from the total landfill.

In Table 1 the mean values for the intrinsic permeability measured on site are listed for four test fields.

The gas permeability readings varied considerably: for example,

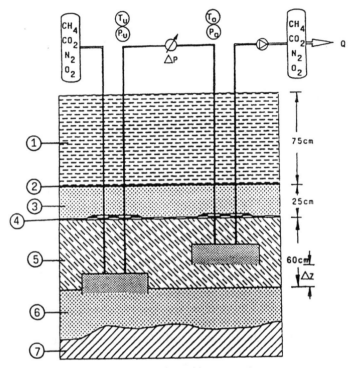

Figure 5. Scheme of the on-site permeability measurements: 1, cover soil; 2, geotextile; 3, sand drain; 4, HDPE membrane; 5, mineral liner; 6, sand gas drain; 7, existing cover.

the gas permeability measurements for field TF1 were 10^{-1} m^{-2} lower than for TF2, TF3 and TF4.

Column 5 of Table 1 shows the water content (WC) at the time of installation of the boxes for a dry weight basis of the clay in the various test fields. In each case the water content is higher than the optimum (Proctor) compaction value of $WC_{PR} = 9.7\%$ by weight.

When comparing the gas permeability measured in the laboratory with that of the till layer in the field (with the same built-in WC), the value of the intrinsic permeability of the laboratory sample is one to two orders of magnitude lower than the values shown in Table 1. The permeability measurements read *in situ* confirm the observations made when installing the test fields: the single till lifts showed

TABLE 1. Results of Field Measurements

Test field	δp (mbar)	Dynamic viscosity gas $(N\,s\,m^{-2})$	Intrinsic permeability (m^2)	Water content by weight (%)	Q_{GAS}^{e} $(m^3 m^{-2} s^{-1})$
TF1	2.0	1.7×10^{-5}	5.2×10^{-13}	12.7^a	5.2×10^{-6}
	4.4	1.4×10^{-5}	4.7×10^{-13}	13.3^b	5.4×10^{-6}
TF2	0.2	1.3×10^{-5}	3.3×10^{-12}	12.9^c	4.1×10^{-5}
	0.1	1.4×10^{-5}	7.9×10^{-12}	10.6^d	9.6×10^{-5}
TF3	4.2	1.8×10^{-5}	6.1×10^{-13}	10.8^c	5.7×10^{-6}
	2.6	1.8×10^{-5}	1.2×10^{-12}	11.13^d	9.5×10^{-6}
TF4	2.2	1.3×10^{-5}	1.3×10^{-12}	10.5^c	1.6×10^{-5}
	1.0	1.3×10^{-5}	5.2×10^{-13}		6.5×10^{-6}

Distances below upper edge of till layer:
[a] 46 cm; [b] 56 cm; [c] 50 cm; [d] 60 cm.
[e] According to equation (2) for $\delta z = 0.6$ m, $A = 1$ m^2 and $\delta P = 1$ mbar.

vertical microcracks. Moreover, it could be observed that the clay broke into coarse aggregate pieces when it was removed. It can be assumed that the natural structure of the glacial till had not been sufficiently changed by compaction with vibration plain-cover rollers. As a consequence, the interlaced macropores continued to exist after compaction, thus producing the higher gas flow rates through the till layer in the test fields.

Measurements on large test fields carried out by Melchior *et al.* (1990) on a clay liner without membrane cover, built in Georgswerder and using the methods and equipment as described above, show that the hydraulic conductivity ($k_f = 2.4$–8.9×10^{-9} m s^{-1}) was slightly below the performance requirements.

The results show that, even for natural liners with a reasonable hydraulic conductivity of about 1×10^{-9} m s^{-1}, a considerable gas flow can be expected (see Table 1).

CONCLUSIONS

Several gas extraction units are in operation at landfills that are covered with natural liners. Little is known about the influence of the extraction unit on the natural liner and vice versa. A convective gas movement through a natural liner can adversely affect the operation of the extraction system in several ways:

- When the gas extraction unit is in operation, a vacuum below the mineral liner enables the intake of air into the landfill body. This diminishes the quality of the utilizable landfill gas.

On the other hand, gas migration into the atmosphere may take place when the gas extraction rate is lower than the gas production rate or when the extraction system is not functioning properly.

In column 6 of Table 1 the specific gas flow (CH_4: CO_2 = 60:40 vol.%, T = 20 °C) through a 60 cm thick till layer at a pressure of 1 mbar has been calculated according to equation (2). Assuming that the till layer has a homogeneous structure, the total amount of landfill gas currently extracted at Georgswerder (ca. 350 $m^3 h^{-1}$) could migrate through the surface liner (18 ha). If sufficient oxygen is present, the methane can be biodegraded in the cover soil (Jones and Nedwell, 1989). This depends on the type of soil, the detention time of the methane in the soil, and the moisture content. Otherwise the gas may damage the vegetation planted in the cover soil.

- Depending on the long-term temperature gradients between atmosphere and clay, the mineral layer may become desiccated. This can result in the development of macropores. In order to be able to interpret humidity changes related to convective gas movement, long-term continuous recording and monitoring of pressure conditions, temperature and gas humidity in the individual collecting boxes is necessary. The instruments required for this will be installed later this year.

From the results presented above, further requirements for the operation of a gas extraction system and the construction of a landfill cover must be considered.

By maintaining a low vacuum below the mineral liner, the oxygen intake and/or the landfill gas migration out of the landfill body can be prevented. This can be achieved by installing a pressure and/or a CH_4 control unit into the extracted gas stream. In order to reduce the pressure influence on the mineral liner, the upper section of the wells must be gas-tight.

Furthermore, for a simple gas extraction operation, a welded membrane is placed on top of a mineral liner (composite liner). Using this system, we are currently investigating whether a vapour diffusion movement due to temperature gradients may contribute to the long-term desiccation of mineral layers underneath membrane linings (composite liner systems). For this reason the gas permeability of the clay should be kept to a minimum: i.e. either the interlaced pore system or the natural structure of the sealing materials should be destroyed. This could be effected by using either stamping or kneading machines, homogenized or refined material and the thorough combination of the lifts of mineral liner. The recommendation for compacted natural liners ('moisture contents of 0–5% of wet optimum') should be reconsidered, as the gas permeability of natural liners can also be reduced by means of a high built-in moisture content.

To reach the required low permeability values of the mineral liner, a quality control programme during implacement of the material is necessary. Laboratory-based measurements of hydraulic and gas permeability differ significantly from the test field results. This is due to the different compaction modes and the existing field heterogeneities. The installation of monitoring gas-boxes to characterize the gas phase in natural liners has proven to be a simple and effective means of control.

ACKNOWLEDGEMENTS

This project was funded by the German Federal Ministry of Research and Technology (BMFT) and the City of Hamburg.

REFERENCES

Barden, L. and Pavlakis, G. (1971). Air and water permeability of compacted unsaturated cohesive soils. *Journal of Soil Science*, **22**(3).

Barden, L. and Sides, G. R. (1970). Engineering behaviour and structure of compacted clay. *Proceedings of the American Society of Civil Engineers*, **96**, 1171–1200.

DIN 18127 (1987). *Proctor test; soil testing procedures and testing equipment*.

Farquhar, G. J. (1989). Overview of landfill liners using natural materials, in *Proceedings of the Second International Landfill Symposium*, 9–13 October, Porto Conte (Alghero), Italy.

Hartge, K. H. (1967). Der Zusammenhang zwischen Luft- und Wasserpermeabilität. *Zeitschrift für Pflanzenernachrung und Bodenkunde*, **111**, 97–107.

Hartge, K. H. (1978). *Einführung die Bodenphysik*, Enke Verlag, Stuttgart.

Jessberger, H. L. and Geil, M. (1989). Eigenschaften und Anforderungen an mineralische Abdichtungsmaterialien bei Oberflächenabdichtungen, in (eds V. Franzius, R. Stegmann and K. Wolf), *Handbuch der Altlastensanierung.* Ch. 5.1.1.0, R.v. Decker's Verlag, Heidelberg.

Jones, H. A. and Nedwell, D. B. (1989). Soil atmosphere concentration profiles and methane emission rates in the restoration covers above landfill sites: equipment and preliminary results. *Waste Management & Research*, 8, 21–31.

Klenner, P., Sokollek, V. and Zickermann, H. (1989). Oberflächenabdichtung der Deponie Georgswerder. *Wasser und Boden*, 41, 515–21.

Lottner, U. S. (1991). Stand der Fortschreibung des LAGA-Merkblattes: Die geordnete Ablagerung von Abfällen, in *Proceedings Symposium Deponieganutzung 91*, 6–8 March, Trier, Germany.

Melchior, S. *et al.* (1990). Testfeld- und Traceruntersuchungen zur Wirksamkeit verschiedener Oberflächenabdichtungssysteme für Deponien und Altlasten, in *Symposium Hydrogeologische Barrieren der Fachsektion Hydrogeologie der Deutschen Geologischen Gesellschaft*, Germany.

Müller-Kirchenbauer, H. *et al.* (1990). Wasser- und Gasdurchlässigkeit von Deponieoberflächenabdichtungen aus bindigen Erdstoffen. *Müll und Abfall* 1/90.

Stegmann, R. (1991). Remedial action at the Landfill Georgswerder, in *The Reclamation of Contaminated Soils*, Proceedings of Symposium, 16–18 January, Milan, Italy.

Wolf, K. (1989). Probleme und Möglichkeiten einer Gasnutzung am Beispiel der Deponie Gergswerder. *Müll und Abfall* 6/89.

Young, A. (1990). Volumetric changes in landfill gas flux in response to variations in atmospheric pressures. *Waste Management & Research*, 8, 379–85.

8.2 Retrofitting of Building Protection Measures to Developments on Landfill Sites

H. SIMON TILLOTSON, RICHARD TIPPING,
BERWICK J. W. MANLEY & SIMON BLACKLEY

Environmental Resources Management, Eaton House, Wallbrook Court, North Hinksey Lane, Oxford OX2 0QS, UK

INTRODUCTION

Landfill sites located in the urban environment are often viewed as a prime land resource with great potential for redevelopment. As a result, a number have been chosen for development of facilities such as motorway service areas, supermarkets and other light industrial purposes.

The location of buildings on such sites poses serious safety limitations, which must be well understood and adequately catered for if the safety of occupants and the public at large is to be assured.

This is especially important on sites where buildings have been constructed in which minimal or no gas protection measures are incorporated. In such cases a two-fold risk is present:

- the potential existence of accumulations of explosive quantities of flammable gases beneath the floor slab; and
- the potential for bases to enter the building interior and cause either an explosive or asphyxiant risk.

The retrofitting of control measures in such situations to reduce the risk to the buildings and their occupants poses a number of problems. Conventional gas control measure design is focused on the concept of migration control, where gases are prevented from migrating to offsite developments by the installation of engineered

Landfilling of Waste: Biogas. Edited by T. H Christensen, R. Cossu and R. Stegmann. Published in 1996 by E & FN Spon, London. ISBN 0 419 19400 2.

barriers, passive trenches and/or active abstraction systems via vertical/horizontal wells.

In the situations described, where buildings are sited upon the landfill itself, such systems will neither provide the means of excluding gas from building interiors nor prevent the accumulations of such gases beneath the buildings. The installation of gas wells around buildings to actively draw gases away from the building foundations cannot in itself provide protection. The gases withdrawn must, by mass balance considerations, be replaced by other gases. In such a scenario the gases are either replaced by other landfill gas (drawn to the area by the abstraction flow pattern) or, in the case of abstraction rates exceeding the landfill's capacity to generate gas, by air drawn preferentially from the surface. Therefore, either the explosive risk remains or it is replaced by a potential spontaneous combustion concern within the waste itself.

Such conditions were existing at the Duncrue Industrial Estate in Belfast, Northern Ireland. Here the development of a light industrial estate had occurred with only minimal protection measures incorporated. Subsequent attempts to provide protection measures had led to the installation of a system of gas wells to draw gases away from the buildings. After two years in operation, it had become increasingly evident that the risk to the buildings had not been alleviated by the measures installed. After auditing the existing protection measures, shortcomings in the existing system design, and in particular the philosophy behind the design, came to light.

BACKGROUND TO THE PROBLEM

The site, formally known as the North Foreshore Enterprise Zone, is situated to the north of Belfast on land reclaimed from the Belfast Lough by the landfilling of predominantly domestic wastes. Landfilling commenced in 1959, with the area on which the industrial estate is located being completed in late 1977 (Fig. 1). A total of approximately 2.4 million tonnes of waste was placed in this area with an average waste depth of 8 m.

Development at the site commenced during the late 1970s with the construction of a municipal waste incinerator.

The development of the remainder of the site began in mid-1982 in the south-west corner of the estate. During 1985 the Department

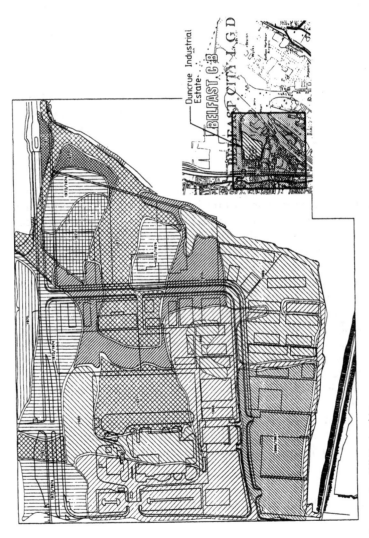

Figure 1. Duncrue landfill tipping history.

of the Environment (Northern Ireland) designated over 30 ha of the site as an Enterprise Zone. Given the financial benefits of such a zone, its close proximity to Belfast City Centre and the docks and good motorway links, the site has now over 190 light industrial units open for business.

In January 1987 a complaint concerning obnoxious odours at a newly completed warehouse on the estate was received by Belfast City Council. Initial investigations detected concentrations of methane gas within its flammable range in the general atmosphere of the office annexe to the warehouse. A notice was immediately served by Belfast City Council prohibiting the use of the office block, under Article 24 of the Health and Safety at Work (NI) Order 1978.

Checks were carried out on the remainder of the buildings on the site, and 13 premises were identified as having significant levels of methane at points of ingress. In none of the buildings had building protection measures against the ingress of gases been incorporated. It was realized that the significant settlement rates experienced under a number of the buildings exacerbated the problem by allowing the build-up of potentially explosive volumes of gas in the resultant void space beneath the floor slabs.

As a result of the problems identified, an active gas control system was installed during 1987/88, and a series of building protection measures were recommended for the existing buildings. This included the sealing of all identified gas ingress points, including service entry ways through the concrete slab floors. From 1988 all new buildings were constructed with some form of protection measures incorporated into their design.

Despite these provisions, continued settlement, in excess of 1 m in places, and inadequacies in the gas control system led to ongoing safety concerns.

These concerns culminated in the decision to audit the existing protection measures and find a consequent solution.

LIMITATIONS OF ORIGINAL CONTROL SYSTEM

The control system consisted of 3 km of collection pipework linked to 17 vertical gas wells and 10 horizontal wells. Gas was abstracted from the site and drawn to a central processing point, where the bases were vented to atmosphere. Originally the system was to have included 25 vertical wells to a depth of 8 m. However, the discovery of a high water-table, less than 3 m below ground level, on the extreme western area of the site led to the replacement of eight vertical wells by the horizontal wells described above.

Limited site investigations prior to the installation of the control system had indicated that a radius of influence of 50 m could be achieved using the vertical/horizontal arrangement. The number and position of the wells was therefore dictated by this finding.

On installation, several of the gas wells had to be redesigned on account of a rising water-table. Later it was found that substantial flooding of many of the wells was occurring, thus severely limiting their effectiveness. In many cases the high water-table left less than 1 m of the perforated well casing length unflooded. Subsequent investigations indicated that the actual radius of influence achieved was less than 20 m.

In order to attempt to achieve the design well capture radii, the excessive well vacuum used (greater than the initial 25 mbar) inevitably led to a large proportion of air being drawn into the collected gas mixture. Gas quality at the abstraction plant was measured at 6% CH_4, within the explosive range for methane.

Additional problems centred on:

- condensate problems resulting from collection pipework not having been laid to correct falls;
- the horizontal wells not being designed to ensure distribution of suction along the full length of the pipework;
- oversizing of the abstraction plant exceeding the landfill gas generation rates existent at the site; and
- no real allowance for a rising water-table.

It was clear from the systems audit that the excessive suctions applied to attempt to achieve design radii of influence could not be allowed to continue, given the large volumes of air entrained and the explosive mixtures passing through the abstraction plant. Additionally, given the achievable radii of influence, only a small number of the wells actually exerted influence on the buildings on the site. Consequently a greater part of the system was closed down (including the horizontal wells), and suctions applied at the abstraction plant were significantly reduced. At present six wells remain in operation exerting an influence on key buildings on-site enabling a gas quality of 16–18% CH_4 to be maintained.

OVERCOMING THE LIMITATIONS: AN INNOVATIVE APPROACH

Where buildings are sited upon the landfill itself, exhaust systems (such as gas-pumping schemes) will neither provide the means of excluding gas from building interiors nor prevent the accumulation of such gases beneath the buildings. The installation of gas wells around buildings to actively draw gases away from the building foundations cannot in itself provide protection. The gases withdrawn must, by mass balance considerations, be replaced by other gases. In such a scenario the gases are either replaced by more landfill gas (drawn to the area by the abstraction flow pattern) or in the case of abstraction rates exceeding the landfill's capacity to generate gas, by air drawn preferentially from the surface. The latter scenario was occurring at Duncrue.

This situation was further exacerbated by the high suctions employed and relatively high permeability of the material around the well. Therefore, using a conventional gas control system, either the explosive risk remains unchecked or it is replaced by a potential spontaneous combustion risk within the waste itself.

To overcome these limitations a new approach has been adopted. An 'inert buffer zone' was established beneath existing buildings to ensure that hazardous gases were prevented from accumulating or entering the buildings. It was envisaged that this could be achieved by simultaneously drawing landfill gas out along one side of the building foundation and reinjecting inert gas along the furthermost side (Fig. 2).

The protection system would involve horizontally placed collectors and injectors protecting individual buildings. The inert gas for such a system would be generated from the landfill gas abstracted using a water-cooled high-temperature combustion set. Collected and returned inert gases could be conveyed around the site by a system of separate ring mains and spurs, which would be coupled to the injector or collector pipes as appropriate.

Clearly, in such a system the collectors and injectors are a key component of the design. As such they should be located as close as possible to buildings and take advantage of nearby paved and road areas, which provided a generally impermeable cover reducing air ingress concerns.

The design of the collectors and injectors was not based on

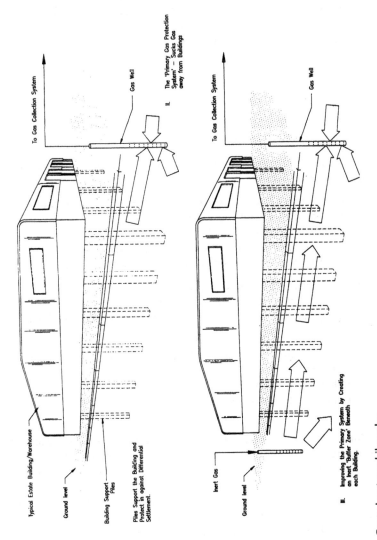

To Gas Collection System

Gas Well

Typical Estate Building/Warehouse

Ground level

Building Support Piles

Piles Support the Building and Protect in against Differential Settlement.

II. The 'Primary Gas Protection System' – Sucks Gas away from Buildings

To Gas Collection System

Gas Well

Inert Gas

Ground level

III. Improving the Primary System by Creating an Inert 'Buffer Zone' Beneath each Building.

Figure 2. Control system philosophy.

standard perforated drainage pipes, as calculations and proving trials had indicated that open areas were too large in standard drainage pipes to ensure suction effectiveness over long lengths. Therefore collectors and injectors had their open areas designed such that adequate pressure and flow control could be achieved over lengths of more than 50 m.

DESIGN DETAILS

Design of Collector and Injector Pipework

The key design parameters identified for the collector and injector pipes were considered to be as follows

- even distribution and collection of gases over the foundation area of the buildings requiring protection;
- a degree of gas flow directional control;
- ability to operate above the rising water-table;
- ability to be installed using accepted and available technology.

In considering the above requirements it became clear that a conventional type of vertical well, however shallow, or angled would be inappropriate. The deficiencies of a vertical well system included:

- the virtual impossibility of effective direction control;
- the tendency of such wells to failure through a rising water-table;
- failure through localized inert filling; and
- operational difficulties associated with the provision of an even coverage of flows over large areas.

Although the site had suffered some problems with horizontal wells, the performance characteristics of a correctly designed, horizontally laid, perforated pipe were considered to be well suited to the design objectives. The main deficiency in the existing system was the poor distribution of suctions along the pipe lengths, resulting in the need for large suctions to be applied to attempt to collect the gases. The large suctions resulted in the gas generation capacity of the waste in that area being exceeded and air being drawn preferentially from the surface.

The advantages of a correctly designed horizontal system were considered to be:

- control provided along the total length of the building;
- the potential for an evenly distributed 'inert buffer' zone under the buildings;
- that areas of inert or less permeable fill would not jeopardize total effectiveness of the system;
- the ability to lay such pipework close to the surface, minimizing the chances of potential flooding by a rising water-table; and
- its minimal interaction with other services and systems in the vicinity.

In achieving a suitable perforation configuration it was necessary to consider the likely gas flows into or out of the pipe. From on-site pumping trials, mass balance calculations and predictive computer modelling of the wastes in place, the landfill gas production rate was determined. The results showed that this rate was quite low and of the order of 2 m^3/t of waste in place/year or 540 m^3/h. Having determined this, the performance criteria for the whole of the system were set, and the collectors and injectors designed to accommodate a low flow over lengths up to approximately 100 m.

By way of background performance characterization, tests were performed on standard slotted casings, where the perforations were lagged with 'polyfelt' to simulate the resistance to gas flow for a horizontal well laid in a gravel-backfilled trench. The polyfelt employed had a specified water permeability from which a gas permeability of ~ 300 D was calculated. It was found that the applied vacuum along the well casing was limited by its total open area. The maximum total open area over which vacuum was effective was found to be in agreement with other works (Graziani and Crutcher, 1986; Leach and Moss, 1989). In conclusion, it was found that for relatively long lengths of horizontal wells to achieve an effective vacuum along their entire length (i.e. lengths greater than 30 m), then smaller open areas would be required than are available from standard slotted casings.

For the site in question, pipe flow calculations backed up by laboratory and field tests showed that injector and collector pipe diameters of 63 mm would be adequate. The holes in the pipe were chosen to be of sufficient size and spacing to ensure a gas flow over the required length, with a course polypropylene mat material preventing blockage by the gravel surround.

It was proposed that the pipes would be laid as close to the building as possible and at a depth to permit relatively unrestricted

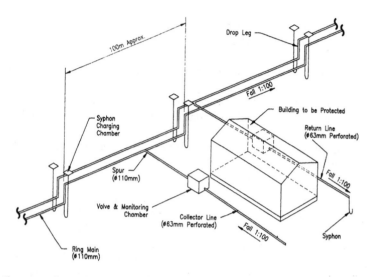

Figure 3. Sketch showing concept of pipe network (not to scale).

gas flow beneath the buildings. A schematic layout for a typical building is shown in Fig. 3. All the buildings on the estate are constructed using concrete piles and ground beams. For this reason all pipes will be laid below ground beam level at a minimum depth of 1.5 m. The depth of 1.5 m is considered a reasonable precaution against air ingress from the surface, given that low suctions and pressures were to be employed. The pipes were designed to be laid in a trench with a backfill of gravel measuring 300 mm square enclosed by a geofilter wrap.

Control Scheme Layout

Having established a concept for the new design, a detailed study of its execution was undertaken.

While the gas production rate had been established, it was considered important to determine where the gas was currently 'flowing'. By establishing this criterion, collector pipes could be laid in the general flow path, thus improving the collection efficiency and utilizing the flux effect to draw the injected gases beneath the buildings.

To perform this exercise the area of the site most at risk was divided into manageable segments, and further subdivided into paved and unpaved areas. As paved areas restrict surface emissions, lateral migration must occur towards unpaved areas. Thus the effective paved/unpaved 'centre of activity' for each segment was determined using moments of areas, to determine 'node points' for flow nets to be constructed. From the construction of such flow nets, the direction and relative magnitude of gas flow were determined, and from this a composite gas flow picture for the whole of the site was established. This indicated the potential dominant flows and the most suitable and effective positioning of the collector and injector pipes.

The proposed layout for the system is shown in Fig. 4. The area to the west of Duncrue Road was not included, for the following reasons.

- The average age of the waste was known to be greater than 25 years.
- The waste composition was considered to be mainly inert wastes.
- Calculations from data collected from the monitoring of the existing horizontal system indicated a low gas-generation potential.

Installation Detail

As can be seen from Fig. 4, the collection and reinjection of gases is achieved using a dual ring-main system, with spurs designed to connect horizontal collectors to the central ring mains. The use of a ring main was considered necessary to ensure adequate flows to all collector and reinjection pipework.

While the gas flows within the pipe network were expected to be quite low over the whole system (between 150 and 200 m^3/hr) precautions against condensate blocking were considered essential for trouble-free operation. These precautions included the following (Fig. 3):

- laying all pipes to falls;
- installation of syphons at low point;
- incorporation of a 'sawtooth' design for the ring main;
- installation of condensate collection tanks where the water-table prevents syphon operation.

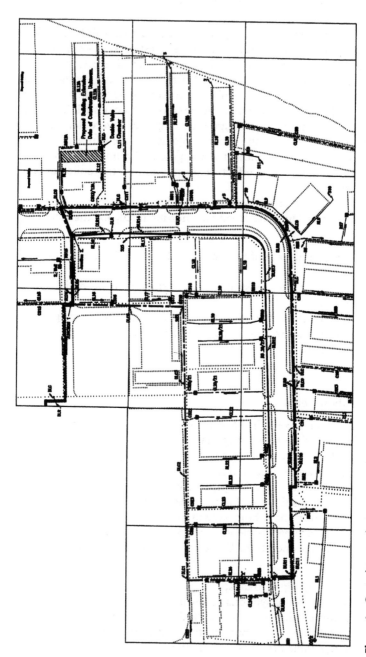

Figure 4. Control system layout.

Commercially available 6 bar rating, black PE pipe was specified for use throughout the design. Pipework of 110 mm OD was specified for the ring main and spurs and 63 mm OD for the collector and reinjector pipework.

The pipework will be laid in accordance with IGE/TD/3 (Institution of Gas Engineers, 1983) and CP 312 (British Standards). Control of collector and injector pipe gas and flows will be achieved by all-plastic construction valves, of the type used by British Gas.

The site covers a large area, so that in order to cover adequately all buildings at risk, almost 7 km of pipework will be required.

GAS CONTROL PROCESS DESIGN

Outline

Figure 5 shows a simplified diagrammatic representation of the processes involved in the control scheme. The concept is relatively

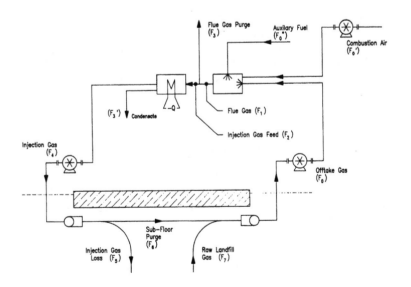

Figure 5. Process design stream definitions.

simple, and the plant is similar to that used for insert gas atmosphere generation in the steel industry. The process can be described as follows. Landfill gas will be extracted from the horizontal collector pipes and delivered via the ring main and an abstraction blower to the combustion chamber. The combustion chamber will be a refractory-lined water cooled unit, which burns the gas mixture with minimum excess air at a temperature of 1200 °C. A propane auxiliary fuel supply is available for system start-up and to augment the landfill gas supply as it reduces with time. The resultant exhaust gas will be predominantly nitrogen, with some carbon dioxide and water present. This forms the feed of inert injection gas to be redistributed around the site. As combustion increases the volume of the gas, approximately two thirds of the exhaust gases will be ejected to atmosphere in order to maintain an approximate mass balance of gases entering and leaving the volume beneath the buildings. The remaining third passes through a cooler to reduce the temperature to a more manageable level.

Dewatering of the gas is not considered necessary, as the injection gas lines have the same configuration of drop-legs and syphons as the collection pipework. The injection gas will then be delivered to the distribution system (reinjection ring main) by means of a booster unit, operating at approximately 75 mbar, and then fed to the individual injector pipes at each building. The process beneath the buildings is relatively complex, and therefore the rate of gas offtake and replacement will be the subject of frequent monitoring and adjustment at the commissioning phase.

Auxiliary and Emergency Systems

In the event of a combustion equipment malfunction or downtime due to maintenance, a number of operating modes are available, which should enable safety coverage of the site. These operating modes include:

- normal – landfill gas in, inert gas out;
- extraction only – both injector and collector mains under suction;
- external inert gas inlet (nitrogen generator or storage tank);
- standby and duty blowers operational.

BUILDING PROTECTION MEASURES

In order to maximize the safety of the buildings and their occupants, stringent gas control and safety measures must be incorporated into all building design. The objective of such measures is:

- to achieve the exclusion of flammable and noxious gases from the buildings; and
- to avoid the accumulation of flammable gases from beneath the buildings.

In order to achieve this a three tier protection scheme has been formulated:

1. the primary active gas control system described above;
2. building design features to provide secondary, passive protection to exclude gases from the building interiors; and
3. a monitoring system to provide operating data on the performance of the primary system and a means of monitoring the building interiors to ensure that flammable/asphyxiant gas trigger levels are not exceeded.

The provision of secondary building protection measures has required the development of a dual design strategy. Buildings constructed prior to ERM involvement had either no specific protection measures incorporated or had a number of features that were considered inadequate to provide the maximum secondary protection possible.

Detailed building audits defined areas of potential concern in all existing buildings: in particular, gas entry points to the building interiors and the absence of appropriate ventilation to confined space areas such as offices and store rooms. A programme of improvements to the majority of such buildings was defined by ERM and is being put into practice by Belfast City Council. The improvements include:

- sealing of cracks between the floor slabs and walls;
- sealing of service entry points; and
- provision of ventilation to confined spaces within the buildings.

For buildings developed since ERL's involvement a series of measures have been recommended to developers to be incorporated in the building design. These have included the following.

- An integral methane-resistant membrane should be incorporated within the floor slab, with a gas permeance of <0.2 ml/m^2/day.
- The cavity wall should be sealed with a methane-resistant damp-proof course.
- It should also be ensured that the cavity walls have open access to well-ventilated roof spaces.
- All entry points for services should be raised to above ground level. The service entry points should be not less than 200 mm above existing ground level.
- All openings that could provide a pathway for gas, e.g. service ducts, conduits, box-outs etc, must be sealed with a gas-resistant mastic or similar material.
- Operable windows should be incorporated into the building design to allow natural ventilation to all internal areas.
- All internal rooms and confined spaces should be adequately ventilated by means of natural permanent ventilation or by mechanical forced means.

Only by the development of an integrated site management strategy can the overall safety objectives be achieved. Prior to implementation of the strategy, regular gas-monitoring inspections of all buildings will be carried out to check against gas ingress to the building interiors.

CONCLUSIONS

The current situation at the Duncrue Site is one in which all buildings are potentially at risk from hazardous volumes of gas accumulating beneath floor structures. The aim of the control system is to reduce this risk by collecting a substantial proportion of this combustible gas mixture and replacing it with an inert gas mixture comprising nitrogen and carbon dioxide.

The risk associated with the flammable component is therefore, in our opinion, reduced to an acceptable value, but it may still persist in small isolated areas under the larger buildings. This small residual risk is effectively reduced still further by a secondary protection system consisting of floor slab sealing and room interior ventilation.

Tertiary protection may consist of gas-monitoring alarm systems in each building.

The reinjection of inert gas itself introduces a potentially asphyxiating environment beneath the buildings. However, without the flammability hazard, owing to the methane removal, the overall risk to building occupants is substantially reduced. As for the residual methane risk, the sealing of potential entry points into buildings and ongoing building inspection will reduce such risk to a reasonably low level.

Sound engineering principles and state-of-the-art gas technology have been applied to the design of this control system. While the risk to building occupants has been substantially reduced it has not been completely eliminated. Strict safety management procedures are essential to ensure that no potentially fatal incident occurs at the site relating to gas emissions.

REFERENCES

British Standards Institution. *Code of Practice for Plastics Pipework (Thermoplastics material)*, CP 312 Parts 1 and 2.

Graziani, W. and Crutcher, A. J. (1986). Landfill gas horizontal trench collection system, Keele Valley Landfill, in *Proceedings GRCDA 9th International Landfill Gas Symposium*, 17–21 March, Newport Beach, pp. 69–89.

Leach, A. and Moss, H. D. T. (1989). Landfill gas research and development studies: Calvert and Stewartby landfill sites. Contractor, Report No. ETSU B 1164, Department of Energy, UK.

The Institution of Gas Engineers (1983). *Recommendations on Transmission and Distribution Practice*, IGE/TD/3:ed.2:1983. *Distribution Mains*.

8.3 Safety Aspects of the Planning, Construction and Operation of Landfill Gas Plants

VOLKMAR WILHELM

Tiefbau-Berufsgenossenschaft Vollmoellerstr. 11, 70563 Stuttgart, Germany

INTRODUCTION

During the planning and design phase of landfill gas (LFG) plants, wide-ranging safety measures must be taken into account with respect to both the landfill buildings and the construction of gas extractors and utilization plants, particularly in the area of explosion prevention. In addition to this, LFG, because of its toxic quality and its capacity to displace oxygen, can be harmful to human beings, particularly in shafts and underground buildings. As the external conditions at LFG plants differ widely from those at public gas plants the technical safety measures required at LFG plants have been developed under the auspices of an R&D project (Müller and Rettenberger, 1986). The most important measures required at landfills are presented in the following.

FUNDAMENTAL ASPECTS OF EXPLOSION PREVENTION AT LANDFILLS

LFG consists largely of methane, carbon dioxide and water vapour. Mixed with air, methane can create a dangerous, explosive atmosphere. Methane has explosive qualities within the range of about 5–15 vol.%. Dangerous explosive mixtures are created by

Landfilling of Waste: Biogas. Edited by T. H Christensen, R. Cossu and R. Stegmann. Published in 1996 by E & FN Spon, London. ISBN 0 419 19400 2.

- the absorption of air through the landfill surface;
- faulty and broken pipes;
- drying out of water syphons;
- insufficient clearing of air out of pipes after installation or after alterations or repairs;
- penetration of LFG into buildings.

An explosion occurs in those cases when an explosive mixture is ignited by an effective ignition source. Effective ignition sources at landfills are mainly flares and blowers. In addition, ignition sources can be created in pipelines by the introduction of sampling points in pipes (measuring points for gas extraction, temperature, pressure and gas pressure). Static charges in plastics or the demolition of pipelines when sparks are caused can also lead to combustion. LFG penetrating buildings can be ignited by an electrotechnical installation, by mechanically created sparks (for example through personnel using tools), or via contact with open fires.

In order to avoid explosions, technical safety measures must be carried out that:

1. prevent or reduce the formation of dangerous, explosive atmospheres (primary explosion safeguards);
2. prevent the ignition of dangerous, explosive atmospheres (secondary explosion safeguards);
3. reduce the effects of an explosion to an acceptable limit (constructive explosion safeguards).

Priority is given to measures relating to primary safeguards. Such measures include the monitoring of gas compositions by means of gas warning devices and the monitoring of water syphons to ensure that they do not dry out, together with measuring devices to shut down gas extraction units when failures occur.

However, as will be discussed later, in many cases the gas extraction unit cannot be shut down in time. Nor can the ignition of dangerous explosive atmospheres by flares, blowers, static charges or plant failures by secondary safeguard measures be effectively prevented, so that constructive explosion safeguards are necessary. These consist, for example, of making the plant system blast-resistant, or explosion-proofing the plant units situated behind or in front of ignition sources (e.g. flares and blowers) in the LFG network. In practice a combination of all three explosion safety measures is employed.

In Germany, planned technical safety measures have to be

presented in the form of a technical safety plan, which must be approved by a safety inspector (expert).

CONSTRUCTION AND EQUIPMENT REQUIREMENTS

Plant Building

Buildings such as the plant building and the machine shed should be constructed at a sufficient distance away from the disposal site and, if possible, away from gas transportation areas in order to avoid hazards created by gas migration, the cracking of slopes, and settling. Basement rooms should not be constructed, to prevent LFG concentrations forming. However, if buildings are constructed in areas where LFG hazards are expected, penetration by LFG can be avoided by one or more of the following measures:

- the installation of de-gassing units in the vicinity of the plant building;
- extensive insulation of the area surrounding the plant building;
- construction of the building on firm, largely unfissured concrete platforms, which have been insulated by plastic sheeting and supplied with a gas drainage system;
- raising of the plant building above the ground to allow sufficient ventilation.

If penetration by LFG into landfill buildings cannot be prevented by these measures, the formation of noxious and dangerous explosive atmospheres must be prevented by the continuous monitoring of the interior atmosphere by means of permanently installed gas warning units and natural and technical ventilating equipment. The supply and waste pipes of the building must be laid and installed in such a way that LFG cannot penetrate into plant buildings through them.

Shafts and Underground Buildings

As LFG poses risks to people working in shafts and underground buildings these constructions should be built and equipped in such a way that plant personnel should not have to enter them regularly to carry out checks and repairs. This can be achieved, for example, by extracting leachate by means of pipelines; the amount of leachate can be calculated by means of inductive measuring devices, while flushing jets and television cameras can be introduced into leachate pipes from above, for example through curved pipes. Shafts with a height of more than 5 m should not be equipped with ladders. Instead, appropriate vehicles should be employed with which deeper levels can be reached without risk, and in case of hazard the building can be left quickly. Both the entry points and the underground buildings themselves should have such dimensions and be constructed in such a way that they can be entered and left with and without vehicles securely, work can be carried out in safety and personnel can be rescued in case of hazard. In order to avoid fires and explosions these buildings may be equipped only with explosion-proof installations and machinery. Working areas in shafts and sewers must be capable of being ventilated in such a way that no health noxious or dangerous explosive atmospheres can form. It is recommended that mobile ventilators be equipped with ventilation pipes in case of need.

Gas Extraction and Transportation

Risks to personnel from gas collectors are to be avoided. This can be achieved by sealing gas wells with cowls already during the operational phase, or raising the shaft rings or outer casing of gravel shafts to at least 2 m above the landfill surface and by making openings and seals between shaft rings airtight. The area around gas collectors must be made airtight for preventing escapes of concentrated LFG. Gas extraction units must be installed and operated at an early stage of the landfill development. Gas pipes must have flexible connections to gas collectors in order to compensate for expected settling. The tubing material must be non-corrosive. For extraction purposes HDPE has proved useful. Surface gas pipes connected to gas extraction, utilization and flare systems must be made of refined or zinc-coated steel. Damage by settling or from vehicles should be avoided. The gas piping, in view of the mechanical, chemical and thermal stress placed upon it, must be technically airtight. Therefore as little soluble material as possible should be present. Pipes must be explosion-proof. Risks of detonation by electrostatic charges should be avoided. Unit materials must be suitable for the gas composition present.

Dewatering units are usually constructions built on the syphon principle. It must be ensured that air is not unintentionally sucked in causing a dangerous, explosive atmosphere. The water syphon must be constructed to take the maximum expected suction pressure. The water syphons must be continuously in operation. It must also be possible to inspect and top them up without personnel having to enter shafts. The best way to ensure that the water syphons are full is to install a water level measuring device that automatically switches off the gas flow when the water level falls below the minimum. Condensate must be drained into the dewatering system.

Flame Traps

Flame traps are technical safety devices that aim to prevent blasts from penetrating adjacent sections of the plant after an explosive atmosphere has been ignited. They are situated before and behind potential ignition sources. The main sources of ignition are flares, gas engines and other gas utilization devices. Flame traps must have been issued with an official certificate of approval for their design and must be installed in accordance with the conditions laid down in the certificate of approval and the inspection report. Suitable flame traps are pipe explosion traps to arrest deflagrations occurring in pipelines, or detonation traps if the deflagration is in danger of turning into a detonation throughout the length of the pipeline. Usually pipe explosion traps of the type shown in Fig. 1 are employed.

If the L/D ratio (L = pipe length, D = interior diameter) given in the certificate of approval and inspection report cannot be adhered to, then detonation traps must be employed (Fig. 2). If plant conditions are such that, after a deflagration, gas burning is to be expected in the trap then the trap must be fitted with a temperature control, which in case it is destroyed by a fire in the trap must trigger a shutdown switch to interrupt the gas flow.

An additional measure employed on flares is dynamic flame traps. These prevent flame penetration by ensuring that explosive gas mixtures leaving an opening have a flow rate that is always significantly higher than the flame diffusion rate whatever the operational phase may be. Before the gas escape rate reaches a critical level, the gas flow must be switched off. It must be ensured that whatever the operational phase there is no overheating that could lead to combustions on the hot surface.

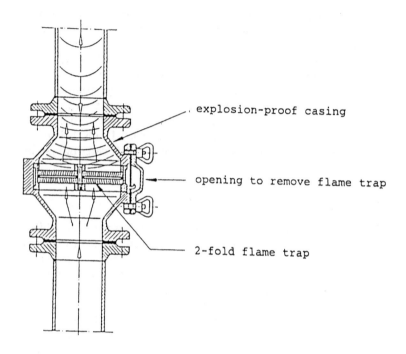

explosion-proof casing

opening to remove flame trap

2-fold flame trap

Figure 1. Deflagration trap.

Gas Extraction Systems

In order to prevent air entering the landfill plant on the extraction side and creating explosive atmospheres, the negative pressure differential should be as low as possible (50–100 hPa). Gas extraction devices are usually radial compressors with potential combustible sources (e.g. overheating arbors, sparks caused by rotors hitting the outer casing), which could ignite explosive atmospheres. For this reason gas pipes must be monitored on the extraction side by gas warning devices, which interrupt extraction so quickly that no explosive gas atmospheres can enter the extraction unit. When the extraction process cannot be interrupted quickly enough because of external conditions (e.g. gas collectors not comprehensively sealed) the following additional measures are necessary.

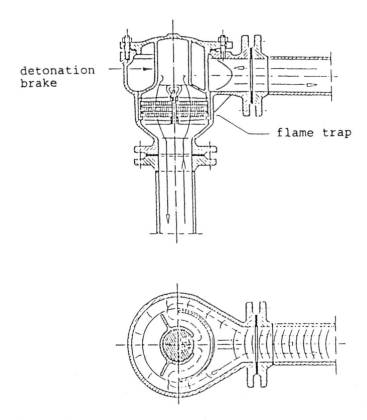

Figure 2. Detonation trap.

- The extraction unit must be suitable for the extraction of combustible atmospheres.
- Flame traps must be installed on the suction and compression sides.

Within the gas extraction unit, as a rule, temperatures over 160 °C should not occur. If necessary, gas extraction should be interrupted by a temperature control if the maximum level is exceeded.

Gas extraction units must, in view of the expected mechanical, chemical and thermal demands made on them, be technically airtight. To avoid vibrations, compensators must be installed, and the extraction unit itself should be erected in such a way as to ensure low vibration levels.

Gas Flares

Flares must be constructed and placed in such a way that personnel cannot be endangered by flames, hot components, unburned gas, flame blowbacks and explosions. This can be achieved if flare units:

- are constructed at least 5 m away from plant buildings and roads;
- are equipped with automatic ignition devices and flame controls;
- are equipped with flame traps using automatic temperature controls and automatic high-speed slide valves or dynamic flame traps.

In order to diminish emissions, medium- or high-temperature flares should be used. Investigations carried out so far on emissions from gas flares do not indicate that high-temperature flares should be made a fundamental requirement.

Gas Warning Devices

In order to switch off gas flow promptly, either methane or oxygen control devices are required. To measure methane, infrared gas analysers are required, for oxygen paramagnetic gas analysers. These gas warning devices must be inspected and tested for their functional suitability by an official testing laboratory. Gas extraction must be interrupted when the methane control shows that methane has fallen below 25 vol.% CH_4 or when the oxygen control indicates a level in excess of 6 vol.% O_2. It is recommended that a preliminary warning is given when the methane indicator shows lower than 30 vol.% and the oxygen indicator higher than 3 vol.% O_2.

If it is intended that the gas warning device is to prevent the formation of explosive atmospheres in the extraction unit (primary explosion safeguard), the unit must take into account the maximum flow speed of the gas, the maximum time-lag of the measuring device and the shutting time of the high-speed slide valve. This means that the distance between the gas sampling point and the slide valve must be great enough to enable the valve to block the pipe before the explosive mixture can reach the source of ignition. In order to calculate the distance the following formula should be used:

$$S_G = \frac{t_{90} V_{max}}{A} \tag{1}$$

where S_G is the distance covered by an explosive mixture (m), t_{90} is the time required by the gas measuring device (s), A is the cross-sectional area of the gas pipe (m^2), and V_{max} is the maximum gas flow (m^3/h).

For example:

$V_{max} = 500$ m^3/h
$A = $ DN 150 (150 mm diameter)
$t_{90} = 20$ s

$$S_G = \frac{20 \ s \times 500 \ m^3/h}{3600 \ s \times 0.01767 \ m^2} = 157.20 \ m$$

Thus the gas sampling point should be installed at least 157.20 m in front of the high-speed slide valve. In order to reduce this distance, piping with a greater diameter could be used, or a measuring device with a lower t_{90} time.

Although such large distances are sometimes achieved by installing S-bend piping in the vicinity of the extraction unit it is usual to rely on constructive explosion safeguards (blast-proof construction, installation of fire traps).

OPERATION OF LFG PLANTS

Inspection of LFG Plants

Before they go into operation for the first time, and after they have been modified or repaired, LFG plants must be approved for safety by an inspector. The inspection prior to initial operation evaluates the original technical safety design and tests whether the design has been fully implemented during construction. After initial operation, and at least annually, LFG plants must be inspected for safety by an official inspector. Additional inspections could become necessary, for example:

- after operational failure;
- after operation of the plant with explosive atmospheres approaching the critical level;

- after damage to the LFG plant as a result of landfill settling;
- after damage by external factors such as damage to gas collectors by vehicles.

The effectiveness of gas warning devices must also be examined by an inspector before initial operation and then after that in accordance with the requirements laid down in the inspection certificates. However, inspections must take place at least once a year. An inspection is also required after there has been contact with a high concentration of LFG.

The LFG plant can only go into operation for the first time, or after repairs or a major modification, after its airtightness has been established. The test for airtightness can be carried using inert gases (CO_2, N_2) or LFG. Written records must be kept on all inspections and results.

Operational Instructions

LFG plants are complex technical constructions, whose effectiveness and technical safety can only be assured by regular servicing, care and attention, gauging and functional monitoring of their component parts.

The necessary tasks must be regulated by detailed operating instructions; particular attention must be paid to the behaviour of staff during breakdowns and alarms. Here are some areas that require regulations in the operational instructions:

- activating the LFG plant;
- testing airtightness after repairs or modifications;
- reading, monitoring and gauging the gas warning unit;
- testing the effectiveness of condensate extractors, siphons and valves;
- determining intervals between jobs;
- entering and working in shafts and underground buildings.

Work on these plants may only be carried out by trained and qualified personnel. The monitoring, servicing and repair work must be documented. In this way changes in the operational conditions of the plant (e.g. reduction in LFG production, deterioration in the gas composition in the various collectors, leaks) can be determined in time and the necessary steps undertaken.

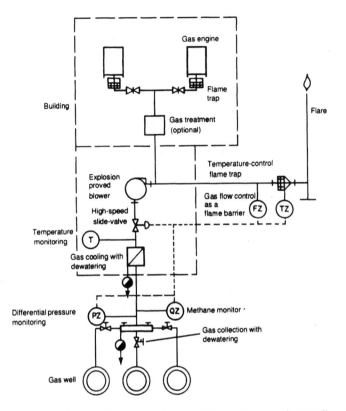

Figure 3. Example of a safety concept for a landfill gas utilization plant (Müller and Rettenberger, 1986).

SAFETY CONCEPT EXAMPLE

Gas is extracted from a sealed section of the landfill and transported to a gas collector. From here the LFG is transported to a flare and two gas engines. The gas pipes and the gas syphons are constructed to be explosion-proof. The gas wells have been made airtight towards the surface. The plant is monitored by a methane concentration analyser and a differential pressure gauge. Penetration by air is only occasionally to be expected, as there is regular inspection of the plant, the plant has been carefully constructed, and the gas wells have been made airtight. Possible ignition sources are the gas flare, the gas engines and the gas blower.

Primary combustion safeguards are the airtight construction of the pipeline system, as well as the methane sampler and the differential pressure gauge. The combustion sources were made safe by means of the following:

- *the gas flare*: temperature-controlled flame trap, gas flow control for the dynamic flame-trap (redundant);
- *the blower*: explosion-proof construction;
- *gas engines*: flame traps;
- *measuring units*: explosion-proof construction.

An example of a safety concept for a landfill gas utilization plant is given in Fig. 3.

REFERENCES

Müller, K. G. and Rettenberger, G. (1986). *Anleitung zur Entwicklung sicherheitstechnischer Konzepte für Gasabsaug- und Gasverwertungsanlagen aus Mülldeponien*, Forschungsbericht 1430293, BMFT und Umweltbundesamt, Berlin.

8.4 Explosion Protection of Gas Collection and Utilization Plants

GERHARD RETTENBERGER[a] & WOLFGANG SCHREIER[b]

[a]Fachhochschuler Trier, Schneidershof, 54293 Trier, Germany
[b]Ingenieurgruppe RUK, Schockenriedstrasse 4, 70565 Stuttgart, Germany

INTRODUCTION

In dealing with landfill gas, because of its composition and its chemical and physical properties, there is a risk of fire and explosion. Therefore measures have to be taken to avoid any risks in this regard. In this chapter, the fundamentals of explosion protection, the safety regulations and the steps to set up a concept for an explosion protection for landfill gas collection and utilization plants are presented. In addition, various technical solutions for an explosion protection system are described.

PROPERTIES OF LANDFILL GAS

Considering the technical questions of safety, it is necessary to know about the composition and the properties of the gas components, as well as the possible range of variations of the main components of landfill gas (LFG).

The gas that is produced in landfills consists in the first place of the combustible gas methane and the inert gas carbon dioxide, as well as water vapour. The content of methane can increase up to 66 vol.%. The water vapour may be removed in different kinds of

Landfilling of Waste: Biogas. Edited by T. H Christensen, R. Cossu and R. Stegmann. Published in 1996 by E & FN Spon, London. ISBN 0 419 19400 2.

dehydration units. This subject will here not be considered any further, as it is extensively handled in Chapter 5.6. Because of the lack of oxygen, landfill gas cannot explode by itself. Nevertheless, if it is mixed with air, an explosive gas mixture can develop. This mixture can be produced under certain circumstances by the process of extraction, transporting and flaring, as follows.

- The landfill itself does not produce any explosive gas mixtures, but under high vacuum, air may be sucked into the landfill via the surface. As a consequence there is an increase of the oxygen content in the gas, creating an explosive gas mixture.
- Air may penetrate into the gas system (which should be closed under normal circumstances), as a result of an operating breakdown, such as a fracture of a gas pipe, a broken gas transportation system, or a leaking valve.
- Landfill gas may escape from pipes, valves or blowers that are under pressure, entering into rooms and producing an explosive atmosphere.

Given specific proportions of methane and air, an explosive mixture may be produced. The inert carbon dioxide and nitrogen in the air counteract this explosive potential. There is an explosive mixture only if the components of the mixture have a certain proportion in their volumes.

The relevant proportions of the volumes of each gas component that are technically safe can be defined by the ternary diagram in Fig. 1. This diagram, for methane–air–carbon dioxide (CO_2) and methane–air–nitrogen (N_2) mixtures respectively, shows the explosive ranges as a function of the proportions of inert gas (N_2, CO_2). If the volume percentages of air, methane and inert gas as a function of the overall mixture are inserted on the diagram, the intersecting point will show whether the mixture is explosive. At the same time, the line representing the proportion of air shows by how much the percentage of each volume must be dispersed for an explosive mixture to be produced. Accordingly, there will be no risk of explosion with a volume percentage of air of less than 58%, which corresponds to a content of oxygen of 11.6%, as well as with any proportion of methane and inert gas.

As already mentioned, landfill gas is not an explosive mixture. If air should penetrate into LFG, so that the volume proportions are dislocated along the line representing the proportion of air, an explosive mixture may be produced if the air content reaches 81%

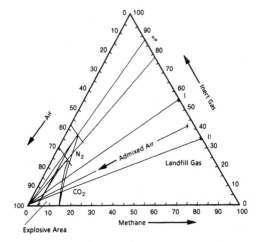

Figure 1. Ternary diagram for the explosive area of methane–air–carbon dioxide mixtures and of methane–air–nitrogen mixtures presented as a volumetric percentage of the concentration of the specific mixture related to the overall mixture.

of the gas mixture (corresponding to roughly 17% oxygen). If there is a sufficiently strong ignition agent, the mixture will explode.

As a result, certain safety thresholds of the proportions of the main components of landfill gas can be set up, beyond which there are no explosive mixtures:

- below 11.6 vol.% oxygen;
- above 15.0 vol.% methane;
- above 35.0 vol.% inert gas (carbon dioxide).

The threshold values can be deduced from the ternary diagram. Based on these data, values for setting alarm and shutdown of the plant have been fixed (Table 1).

As can be seen from Table 1, for explosion control the methane or the oxygen content should be monitored in the LFG. The risk of

TABLE 1. Threshold Values for Monitoring or Shutdown of a Landfill Gas Abstraction and Utilization Plant

Landfill gas component	Explosion threshold (%)	Shutdown (%)	Alarm (%)
CH_4 monitoring	12.5	25	30
O_2 monitoring	11.6	6	3

explosion must be limited by appropriate safety margins. One should consider not only the probability of an incident but also the possible effect of the damage resulting from an explosion.

TECHNICAL SAFETY REGULATIONS AND DIRECTIVES

At present there are no technical directives for landfill gas. It is therefore necessary to go back to technical regulations applied in similar fields. Basically, the *Richtlinien für die Vermeidung der Gefahren durch explosionsfähige Atmosphäre mit Beispielsammlung – Explosionsschutz-Richtlinien* (Ex-RL, GUV 19.8) (Directives for the avoidance of dangers from explosive atmosphere with an appendix of examples – explosion directives), edited by the Berufsgenossenschaft Chemie, should be respected (Anon., 1982a). Furthermore, for explosion control of electrical plants in closed rooms the *Verordnung über elektrische Anlagen in explosionsgefährdeten Räumen* (ElexV) (Decree concerning electrical plants in rooms exposed to the risk of explosion) should be followed (Anon., 1980a). Important advice for realizing explosion protection can be found in the *Verordnung für brennbare Flüssigkeiten* (TRbF, e.g. TRbF 100) (Decree for inflammable liquids), the *Gefahrstoffverordnung* (GetStoffV) (Technical rules for dangerous chemicals) and the decrees for the avoidance of accidents, especially the GUV 9.9 *Gase* (gases) as well as the GUB 17.4 *Sicherheitsregeln für Deponien* (Safety rules for landfills).

Following the technical rules that are commonly recognized is a prerequisite for the construction and operation of a landfill gas abstraction and utilization plant. Here, the DIN standard, the DVGW list with its directives and technical rules for gas installations, and the VDE regulations, especially the DIN 57100/VDE 0100 and DIN 57165/VDE 0165, are to be considered (Anon., 1980b)

Ex-RL and Measures for Explosion Control

The directive for explosion control (Ex-RL, GUV 19.8) deals with the avoidance of dangers from explosive atmospheres and the estimation of explosion risks when handling explosives, as well as the selection and realization of measures to avoid these dangers.

The same measures of protection may apply to landfill gas

extraction and utilization plants; the threshold of explosive atmosphere is only slightly exceeded. The term 'explosive atmosphere' includes explosive mixtures of gases, vapours, fogs or dusts with air under atmospheric conditions (pressures between 0.8 and 1.1 bar and temperatures from − 20 °C to + 60 °C).

Under the directive, measures can be distinguished into primary, secondary and constructive explosion protection. Primary explosion protection is the prevention or limitation of dangerous and explosive gas mixtures. In order to achieve this goal either the methane or the oxygen concentration in LFG is monitored continuously. If the given concentrations of these components are exceeded (Table 1), specific measures are initiated as alarm or shutdown of the plant.

As it cannot be guaranteed that the occurrence of explosive gas mixtures can be avoided in every case, additional measures of secondary and constructive explosion protection are to be taken.

The secondary explosion protection includes measures to prevent the ignition of an explosive gas mixture: for example by avoiding sparks being produced. The extent of the steps to be taken depends on the probability of the occurrence of an explosive gas mixture and the declaration of areas where this might happen (zones). The basis for explosion protection measures to be taken is the definition of those areas of the plant threatened by the risk.

In Ex-RL three zones are defined according to the risk of explosion:

Zone 0: areas in which there is a long-term or a permanent presence of dangerous explosive atmosphere;

Zone 1: areas in which there is an occurrence of a dangerous explosive atmosphere at times;

Zone 2: areas in which there is a low probability of occurrence of an explosive gas mixture; if it does occur only a short-term occurrence of dangerous explosive atmosphere is given.

In addition, the following distinctions should be made:

- inner explosion protection (the inside of pipes and the gas recovery plant);
- exterior explosion protection (the surroundings of parts and equipment of the gas extraction and utilization plant, such as blowers, valves, in rooms, manholes.

By using construction explosion protection the effects of a possible explosion can be neglected. This means that equipment (e.g.

blowers) is constructed in such a way that it resists the pressure build-up and shock caused by an explosion. If the methane–air mixture is ignited, pressures up to 7 bar can occur in a closed room or container. For a realization of constructive explosion protection, technical gas installations must at least respect the rated pressure value PN 6 (e.g. pipes). Additionally, the speed of explosion must be considered for an estimation of explosion processes. This is the speed at which the flame front expands within a pipe system. Closely related to explosion speed is explosion pressure, which goes up with increasing speed and can reach values that exceed the maximal acceptable explosion pressure in closed containers (7 bar).

Further characteristic values, which are important for the interpretation and the application of constructive measures for explosion protection, include ignition temperature, minimal ignition energy and the limit of gap width of the flammable or the explosive gas component. The limit of gap width is needed for estimating the penetration capacity of the ignition of combustion gas at narrow gaps. It is used to sort different gases into explosion groups and to select appropriate protection devices against flame penetration. The minimal ignition energy indicates the ignition capacity via electric sparks. The minimal ignition energy of combustion gases, including methane is in general so low that ignition must be expected even from electrostatic charges.

STEPS FOR SETTING UP A TECHNICAL SAFETY CONCEPT

As already mentioned, air can mix with landfill gas, which may result in the production of explosive mixtures. It is therefore necessary to work out a safety concept in order to avoid accidents. Such a technical safety concept includes the field of construction as well as operational requirements that define how the plant should be run and monitored. The steps for defining measures for the inner explosion protection of an LFG extraction and utilization plant are presented in Table 2.

As a first step, the whole system should be described; then possible sources of ignition must be pointed out. Examples of ignition sources include measuring devices, blowers and the utilization plant. Furthermore, measures, that can help to prevent the

TABLE 2. General Steps for a Formulation of Measures Concerning Inner Explosion Protection of an LFG Extraction and Utilization Plant

Step	*Remarks*
1. Basic technical safety investigations of the gas extraction system.	Definition of the gas extraction principle: active or passive. Concerning passive systems, it must be proven that no air can penetrate into the system. Furthermore, in the area of free outlets, measures for explosion protection can be taken.
2. Marking of possible sources of ignition.	Sources of ignition can be: • gas transportation facilities; • gas flares; • electric equipment; • stationary or non-stationary devices; • electrostatic discharge of components of the plant; • processing equipment.
3. Formulation of measures for primary explosion protection.	These are measures to prevent dangerous explosive atmosphere, e.g: • the technique and the constructive design of the plant; • the method of operation and monitoring; • the reduction of methane concentrations and inertization.
4. Estimation of the probability of occurrence of dangerous explosive atmosphere in remaining areas and the risk of explosion (current regulations are the basis).	The gas plant must be investigated regarding • the gas situation (e.g. close to the surface, migration resistance, etc.); • technical equipment for construction and installation; • the method of commissioning, maintenance and monitoring. The estimation leads to the definition of the zones according to Ex-RL
5. Definition of measures according to constructive measures as well as other measures (reduction of the potential of ignition of a dangerous explosive atmosphere).	The selection of measures is dependent upon the definition of zones 0, 1, 2; the appendix with examples of Ex-RL or the GUV 17.4 is to be consulted.
6. Definition of the final design and operation due to steps 3–5.	Possible modifications of the gas extraction and utilization plant design or the technical equipment respecting measures for explosion protection and operation.

occurrence of a dangerous explosive atmosphere (primary explosion protection) have to be examined. In case not all the problems can be solved, the probability that explosive atmospheres can still occur must be estimated. In the light of these considerations, additional measures for explosion protection must be taken.

EXAMPLES FOR EXPLOSION PROTECTION PLANS AT DEGASIFICATION PLANTS

In general, explosion control measures for landfill gas extraction and utilization plant will be realized by combining the measures for explosion protection presented above. The safety concept must apply to the whole plant and consider all the operation phases, including the starting phase, normal operation, operational incidents, and breakdown of the monitoring equipment. Because of its construction, the transport of gas in the pipes must always be considered as a potential explosive atmosphere, which cannot be made inoperative by secondary measures. Therefore in the safety rules for landfills (GUV 17.4, Anon., 1982b), two possible ways of explosion protection are defined. Under 5.16.2 the following requirements are to be met. Gas transport systems must be constructed in such a way that

- a dangerous explosive atmosphere cannot penetrate into the gas transport system (primary explosion protection according to Ex-RL);
- the harmful effects of an explosion can be minimized (constructive explosion protection according to Ex-RL).

The further possible sources of ignition mentioned in Table 2 under Step 3 can normally be made inoperative by constructive measures (secondary protection, according to Ex-RL).

In the following sections, plans are presented for primary and constructive explosion protection, as well as examples of technical solutions for explosion control. According to GUV 17.4 (Anon., 1982b), explosion control for a flare requires constructive measures. The gas flare must be equipped with an automatic ignition system. In addition, most flares are monitored by a UV sensor and a thermocouple. If the minimum temperature is not reached within a preset time, the plant will be shut down automatically.

Primary Explosion Protection Measures

As already mentioned, the aim of primary explosion control measures is to prevent a dangerous explosive atmosphere from penetrating air into the gas transporting system.

This can be realized by permanent monitoring of the gas composition; if the threshold values mentioned in Table 1 are exceeded the plant is shut down automatically. The accurate performance of the gas analysers that are in operation must have been tested by an accepted qualified institute. In the case of a breakdown or a failure of the performance of the analyser an alarm must be initiated automatically; as a result, the plant is shut down by itself. Furthermore, it is necessary to install a redundant gas monitor system. Redundancy in this context does not mean an automatic transfer from one gas monitor to the other, in the event of a failure but rather requires two separate analysers, each having its own feed system. The requirement for redundancy is fulfilled only if methane and oxygen are monitored in separate gas analyzers.

The gas flow must be interrupted before an explosive gas mixture reaches the potential source of ignition, which is the gas blower. The automatic closure of the pipe is provided by a fast operating valve. For this reason, this closure time must be compensated by an adequate flow time between the gas monitor and the blower. This flow time with respect to the necessary pipe length to compensate the time can be calculated by considering the maximum gas flow rate, the flow time of the test gas to the analyser, the t_{90} time of the analyser, and the operation time required for the valve to close. As the plant can be put into service again only if the explosive gas mixture has been removed, there should be an outlet with a valve or cock installed before the shut-off valve. If necessary, a special system for inertization can be built in. As the above measures excluded the possibility of there being an explosive atmosphere in the gas pipes that could be considered a potential source of ignition, explosion-proof construction of the gas pipes is not necessary.

The disadvantage of this concept is that leaks may result in air penetrating into the pipe located between the shut-off valve and the blower. For this reason this pipeline is constructed of steel without any joints.

Figure 2 shows a comparison of the flow patterns of the minimal

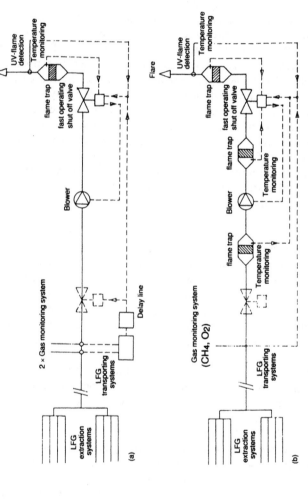

Figure 2. Flow patterns of primary and constructive measures for explosion protection of a landfill gas extraction and utilization plant (minimal requirements): (a) explosive mixture shall not enter a gas blower station; (b) the effect of an explosion in the blower shall be minimized.

realization of the plant according to GUV 17.4 (Anon., 1982b), following the primary and the constructive plans of explosion protection, respectively.

Constructive Explosion Protection Measures

In contrast to primary explosion control it is in this case assumed that an explosive gas mixture is transported through the extraction system. In this case, the possible effects of an explosion must be controlled and reduced to an unharmful extent.

This can be technically realized by an explosion-proof construction of the gas plant. To stop the flames from expanding towards the gas analysers, burner, blowers, flares, etc. there must be safety devices to avoid flame penetration into the various installations on the vacuum and on the pressure side. The construction of these safety devices against flame penetration must be approved by a qualified registered institute, and they must be installed according to the instructions of the manufacturer. If there is a fire in one of the safety devices for avoidance of flame penetration, this must be recognized at an early stage by means of a temperature meter; as a result the plant will be shut down automatically. In order to avoid piping an explosive gas mixture over a longer distance, the composition of the transported gas must be monitored permanently, according to GUV 17.4 paragraph 5.16.1. The same conditions applied for the primary explosion protection concept will also apply to the installed analysers. However, in this case, a redundant installation of the monitoring device will not be necessary.

Further Measures for Explosion Protection

In addition to primary and constructive explosion protection, further measures are necessary in order to achieve explosion control. These refer to internal as well as external explosion protection, and include measures to prevent the ignition of explosive gas mixtures or to avoid their production. The latter measures are, for example, installations for monitoring the liquid level of the condensate tank. According to GUV 17.4, paragraph 15.5 (Anon., 1982b), these condensate collection tanks must be constructed in such way that an unintentional suction of air can be excluded. As for condensate tanks with a hydraulic seal the constant effectiveness of the water seals must be guaranteed. Normally this requires an installation for the supervision of the liquid level connected to an automatic emergency shutdown system. The electrical components installed must be suitable for operation in Zone 1.

The measures mentioned above, which are to prevent the ignition of an explosive atmosphere, refer in the first place to the electrical components installed. During the operation of LFG plants electrical operating devices often have to be installed in Zones 1 and 2. In this context two different groups have to be distinguished: Group I (operating devices installed underground in danger of firedamp (e.g. coal mines) and Group II (operating devices for areas in danger of explosion).

Electrical operating devices that are operated in Zone 1 must be approved for their construction under ElexV (Anon., 1980a) by an officially licensed institute. In the case of a partial failure of the electrical operating device it must not produce any sparks or hot surfaces, so that there is no source of ignition.

Electrical operating devices in Zone 2 do not need a construction licence, but they must meet all the requirements of the VDE directive 0165 (Müller and Rettenberger, 1986). They must not produce any sparks during operation. Usually, plant components of landfill gas extraction and utilization plants that are in direct contact with landfill gas (e.g. gas analysers) will be constructed as self-protecting devices. The electrical current is measured in such a way that an explosive atmosphere cannot be ignited, either by sparks or by thermal effects. The minimal ignition energy of 0.28 MJ for methane must not occur in the construction device. Under VDE 0165 all electrical devices operating under conditions of Group II may also be used in Zone 1.

SUMMARY

When operating a landfill gas extraction and utilization plant, there is the danger of producing an explosive gas mixture. This may result from sucking air into the gas extraction system, through the landfill body (oversucking) and leaks or breakage of pipes, valves, etc. In addition, explosive mixtures may occur in areas where the LFG is under positive pressure and, enters because of leakages, buildings, manholes, etc. For these reasons, measures for avoiding the danger of explosion must be taken. In this chapter, the fundamentals of explosion control measures are described and illustrated by examples.

When developing a plan and taking measures for explosion protection, all the phases of operation have to be respected. This includes, for example, the transportation of the gas, the start-up and shutdown of the plant, failures of equipment and monitoring devices. Operating directives for the maintenance and technical safety aspects of the plant should ensure a safe and smooth operation of the LFG extraction and utilization plant.

Further information on measures for explosion protection of landfill gas plants can be found in GUV 17.4 (Anon., 1982b), the Ex-RL (Anon., 1982a), Müller and Rettenberger, (1986), VDMA-Arbeitsblatt, Rettenberger (1994).

REFERENCES

Anon (1980a). *Verordnung über elektrische Anlagen in explosionsgefährdeten Räumen* (ElexV) vom 27.2.1980, BGBl. Teil 1 vom 1.9.1980, S. 214, Erich Schmidt Verlag, Berlin.

Anon (1980b). *Verordnung über Anlagen zur Lagerung, Abfüllung und Beförderung brennbarer Flüssigkeiten (Verordnung über brennbare Flüssigkeiten – VbF)* vom 27.2.1980, BGBl. Teil 1 vom 1.3.1980, S. 229, mit dazugehörigen Technischen Regeln für brennbare Flüssigkeiten (TRbF) vom 1.7.1980, Carl Heymanns Verlag, Köln.

Anon (1982a). *Richtlinien für die Vermeidung der Gefahren durch explosionsfähige Atmosphäre mit Beispielsammlung – Explosionsschutz-Richtlinien – (Ex-RL)*, Ausg. 10.82, Berufsgenossenschaft der chemischen Industrie, Heidelberg, Druckerei Winter, Heidelberg.

Anon (1982b). *Bundesverband der Unfallversicherungsträger der öffentlichen Hand für Deponien*, GUV 17.4, Juli 1992.

Müller, K. and Rettenberger, G. (1986). *Anleitung zur Entwicklung sicherheitstechnischer Konzepte für Gasabsauge- und Gasverwertungsanlagen an Mülldeponien*, Forschungsbericht 1430293, BMFT und Umweltbundesamt.

Rettenberger, G. (1994). *Sicherheitstechnische Erfordernisse an Deponiegasanlagen aufgrund der GUV 17.4*, Trierer Berichte zur Abfallwirtschaft, Band 6, Economica Verlag.

VDMA-Einheitsblatt 24169, Teil 1, Beuth Verlag GmbH, Berlin und Köln.

8.5 Emissions from Landfill Gas Thermal Treatment Plants

KLAUS WIEMER[a] & GERHARD WIDDER[b]

[a] *Gesamthochschule Kassel, Fachgebiet Abfallwirtschaft und Recycling Nordbahnhofstr. 1a, 37213 Witzenhausen, Germany*
[b]*Amt für Abfallwirtschaft und Stadtreinigung, Mainzer Str. 97, 65186 Wiesbaden, Germany*

INTRODUCTION

In 1985 in Germany, the Federal State of Hessen (*Land* Hessen) decided to develop gas treatment facilities at all operating landfills and at all closed sites still producing significant gas volumes.

The choices for landfill gas (LFG) treatment and utilization were flares, boilers and gas engines. None of these three methods had been adequately tested with respect to their ability to meet federal air quality standards. In addition, the methodology for analysing gas and gas combustion emissions was not standardized. There was a lack of verifiable emission analyses.

The year 1985 was dominated by the discussion of the dioxin issue, and the newest scientific studies directly affected the evaluation and planning of facilities. Over subsequent years this discussion has grown in complexity, and has been supported by a large body of scientific studies. As a result, knowledge had to be gained into the formation (including the *de novo* synthesis) and destruction of dioxins, especially with respect to the most toxic form 2, 3, 7, 8 TCDD (Seveso dioxin). This isomer is used as the basis for comparing degrees of toxicity, which were expressed as toxic equivalents (TE) of 2, 3, 7, 8 TCDD.

Landfilling of Waste: Biogas. Edited by T. H Christensen, R. Cossu and R. Stegmann. Published in 1996 by E & FN Spon, London. ISBN 0 419 19400 2.

When the new LFG utilization technologies were examined regarding their potential to emit significant levels of pollutants, different theories came up. These were based on a small database and theoretical reflections:

- Landfill gas flaring results in incomplete combustion; as a result high levels of CO as well as other toxic pollutants such as vinyl chloride may be emitted. In addition there is a potential for dioxin formation.
- Boilers for heat recovery heated by LFG can have incomplete combustion, depending on their design. Potential problems regarding dioxin formation were expected along the relatively cool boiler walls.
- Gas engines are known to suffer from incomplete combustion. As a result high levels of NO_x, CO, as well as PAHs and dioxin concentrations can be expected in the exhaust gases.

The expected problems were sufficiently substantial for a testing and evaluation programme to be proposed in order to find out what the actual problems are. But instead of investigating the 'traditional technologies' alternative new technologies were chosen. Although these alternatives had not been used to treat landfill gas, sufficient experience using other fuels was available. The *Land* Hessen set up a programme to support financially these new technologies for LFG; it was mandatory to evaluate their operation and emissions on the basis of a monitoring programme.

Projects in all three of the following categories were eligible for funding.

- *Utilization of muffle furnaces instead of landfill gas flares.* In a muffle furnace, the entire combustion process takes place inside an enclosed combustion chamber at 1200 °C (later specifications set a range of 900 °C–1200 °C) with a minimum detention time of 0.3 s measured from the point of ignition.
- *Heat recovery through muffle furnaces with post-combustion boilers.* Combustion parameters remain as specified above.
- *Installation of an afterburner to burn the exhaust gases from a gas engine.* Owing to the chlorine levels in the LFG, the use of catalytic converters to reduce emissions such as CO and PAHs was not feasible. The lifetime of these converters would have been very short.

The announcement of this programme in the year 1985 had repercussions throughout West Germany. In the widespread debate about the programme, the State's intentions were misunderstood: apart from conventional flaring, it was not intended to prevent heat recovery and electricity generation from LFG but to evaluate new technologies for LFG that seemed best suited to minimizing emissions.

Regarding the necessity of improving the combustion processes during flaring, existing types did not seem to be adequate. For this reason – based on the discussion in Hessen – the technical specifications for muffle furnaces, or high-temperature flares or muffles, had been developed. As a result, new flares came onto the market in which the combustion takes place in a completely contained cylinder that is insulated in order to prevent cold zones along the inner walls.

The following results derive from the testing programme that was also supported by other states (*Länder*). The monitoring of landfill gas and gas burner emissions was carried out by Biogas Systems GmbH, Giessen. The programme investigated muffle furnaces with and without heat recovery, and gas engine emissions with and without afterburning in a muffle furnace. Sampling and analysis were carried out by the Workplace and Environmental Analysis Corporation, Muenster-Roxel, under the supervision of the authors and the State Environmental Agency in Hessen.

EMISSIONS FROM LANDFILL GAS COMBUSTION SYSTEMS

Emission levels were measured at five thermal treatment systems. To illustrate emission characteristics, various treatment systems and

TABLE 1. Mean Cl, F and S Concentrations (mg/Nm^3) Measured in LFG from Five Different Landfills in Hessen

Component	Landfill				
	A	*B*	*C*	*D*	*E*
Cl	115	80	25	13	16
F	>6	7	10	4	4
S	117	23	31	35	18

components were analysed in detail at one landfill and compared with data from other landfills.

The specific gas composition of this landfill is presented in Tables 1 and 2 (Landfill B). The pollutant levels in the gas from Landfill B represent common values, and thus the emissions are probably similar to those that can be expected from many other gas utilization plants.

The following tables present an overview of the monitoring results from the following five different gas combustion systems:

- muffle furnace emissions (combustion temperature approximately 900–1200 °C);
- muffle furnace emissions (combustion temperature approximately 1200 °C);
- muffle furnace emissions from the combustion of LFG together with the exhaust gas from an LFG-fired engine (combustion temperature approximately 1000 °C);
- muffle furnace emissions from the combustion of LFG together with the exhaust gas from an LFG-fired engine (combustion temperature approximately 1200 °C);
- exhaust gas from an LFG-fired engine.

Each of these five systems was always analysed for CO, CO_2, NO_x, SO_2, O_2, and total carbon, twice for PCDF/PCDD, PCB, PAHs and H_2O and six times for HCl, HF and particulates. All the results were corrected to 3% O_2.

Emissions from Muffle Furnaces

Table 3 shows emission levels from a muffle furnace on Landfill B under very different operating conditions at 900–1000 °C. These results are also produced under non-optimized conditions, including temporary shutdowns due to process control measures.

The CO concentrations reflect the degree of combustion: when it is complete, CO concentrations should be below detection limits. By measuring CO concentrations the emission of other components can be related to the level of combustion.

Table 4 shows an example of PCDF and PCDD detection in the exhaust gas of the same muffle furnace. The related other gas parameters during sampling P5.1.1. and P5.2.1 are presented in Table 3. The toxicity equivalent (TE) value is calculated by

TABLE 2. Hydrocarbon and Halogenated Hydrocarbon Concentrations (mg/Nm^3) in LFG from Five Different Landfills in Hessen (CH$_4$ between 54% and 60%)

Trace compoment	Landfill				
	A	B	C	D	E
n-Pentane	13.0	4.0	4.8	1.7	2.1
n-Hexane	16.7	5.5	7.5	1.7	0.9
Benzole	12.9	4.2	3.2	1.9	0.9
n-Heptane	46.0	18.1	12.9	2.3	0.1
Toluene	324.2	165.0	34.1	21.1	12.4
n-Octane	27.6	9.3	9.5	4.5	3.2
Ethylbenzol	127.4	74.0	40.5	30.4	23.3
m-/o-Xylene	296.5	158.0	64.6	45.7	59.8
o-Xylene	81.6	43.3	14.4	9.4	14.3
	0.0001	0.01	0.0001	0.0003	0.01
Cumol	199.3	102.0	29.0	20.1	26.2
Mesitylene	82.8	48.3	8.3	3.4	17.4
Butylbenzol	0.0001	12.1	0.0001	23.3	5.1
Dicyclopentadiene					
Sum of 13 hydrocarbons	1 227.9	643.8	228.8	165.6	165.7
Vinyl chloride	11.0	7.8	16.0	10.0	8.1
Trichlorofluoromethane	23.65	7.41	1.36	3.11	0.046
1,1,2-Trichloro-1,2,2-Trifluoroethane	2.30	1.33	0.11	1.32	0.006
Dichloromethane	3.52	0.001	14.09	0.001	0.005
Trichloromethane	0.28	0.04	0.03	0.02	0.0005
1,1,1-Trichloroethane	1.18	1.58	0.02	0.03	0.0004
Tetrachloromethane	0.0002	0.001	0.0002	0.001	0.0002
Trichloroethene	8.47	8.76	1.69	0.71	0.405
Tetrachloroethene	16.64	8.58	2.71	0.76	0.482
Sum of 9 halogenated hydrocarbons	67.04	35.50	36.01	15.95	9.129

TABLE 3. Exhaust Gas Quantity and Quality from LFG Combustion in a Muffle Furnace

Date	Time Begin	Time End	CO^a (mg/m³)	CO_2 (vol.%)	$NO_x^{a,b}$ (mg/m³)	SO_2^a (mg/m³)	O_2 (vol.%)	$Total^a$ (mg/m³)	T (°C)	$Volume^c$ (m³/h)	Norm-factor 3 vol.% O_2	Atm. pressure (hPa)
18 Jul 1989	14:23	14:53	5	13.1	35	175	5.1	(2	195	1485	1.132	1006.9
18 Jul 1989	14:53	15:23^d	208	9.7	105	209	8.7	54	193	1926	1.463	1007.2
18 Jul 1989	15:23	15:53	8	14.5	35	165	3.2	(2	197	1208	1.011	1007.4
18 Jul 1989	15:53	16:23	6	14.3	36	147	3.5	(2	192	1221	1.029	1007.4
18 Jul 1989	16:23	16:53^d	14	11.7	58	159	6.4	7	192	1725	1.233	1007.4
18 Jul 1989	16:53	17:23^d	98	13.2	47	159	4.5	20	193	1492	1.091	1007.4
18 Jul 1989	17:23	17:53	6	13.9	37	144	3.8	(2	193	1218	1.047	1007.3
18 Jul 1989	17:53	18:23	6	13.7	35	143	4.1	(2	195	1485	1.065	1007.0
18 Jul 1989	18:23	18:53	6	13.4	36	144	4.6	(2	192	1495	1.098	1007.0
18 Jul 1989	14:23	18:53	40	13.1	47	160	4.9	9	194	1473	1.130	1007.2
19 Jul 1989	09:53	10:23	12	14.5	34	140	3.4	(2	188	1201	1.023	1012.4
19 Jul 1989	10:23	10:53	14	14.0	37	136	4.0	(2	186	1206	1.059	1012.4
19 Jul 1989	10:53	11:23	16	13.7	38	136	4.4	(2	187	1474	1.084	1012.5
19 Jul 1989	11:23	11:53^d	127	14.5	47	149	3.1	2	189	1468	1.006	1012.5
19 Jul 1989	11:53	12:23^d	134	13.5	33	142	3.3	(2	187	1474	1.017	1012.5
19 Jul 1989	12:23	12:53	21	13.5	38	140	4.5	(2	188	1471	1.091	1012.2
19 Jul 1989	12:53	13:23	22	13.4	39	142	4.7	(2	196	1445	1.104	1012.0
19 Jul 1989	13:23	13:53	23	13.2	39	144	4.9	(2	212	1398	1.118	1012.0
19 Jul 1989	13:53	14:23	23	13.1	39	144	4.9	(2	210	1403	1.118	1012.0
19 Jul 1989	09:53	14:23	44	13.8	38	142	4.1	(2	194	1393	1.069	1012.3
19 Jul 1989	18:40	19:10	28	15.3	66	273	2.5	(2	257	3255	0.973	1010.9
19 Jul 1989	19:10	19:40	28	15.3	66	275	2.5	(2	257	3255	0.973	1011.0
19 Jul 1989	19:40	20:10	28	15.3	68	273	2.5	(2	258	3249	0.972	1010.9

TABLE 3. continued

Date	Time		CO^a (mg/m³)	CO₂ (vol.%)	NOₓ^{a,b} (mg/m³)	SO₂^a (mg/m³)	O₂ (vol.%)	Total^a (mg/m³)	T (°C)	Volume^c (m³/h)	Norm-factor 3 vol.% O₂	Atm. pressure (hPa)
	Begin	End										
19 Jul 1989	20:10	20:40	28	15.2	68	267	2.5	<2	261	3318	0.973	1010.6
19 Jul 1989	20:40	21:10	27	15.1	66	256	2.5	<2	261	3230	0.973	1010.6
19 Jul 1989	21:10	21:40	27	15.0	66	245	2.5	<2	260	3326	0.973	1011.0
19 Jul 1989	21:40	22:10	27	15.0	64	236	2.5	<2	253	3370	0.973	1011.1
19 Jul 1989	22:10	22:40	27	15.1	64	224	2.4	<2	247	3410	0.968	1011.5
19 Jul 1989	22:40	23:10	27	15.1	64	216	2.4	<2	245	3424	0.968	1001.5
19 Jul 1989	18:40	23:10	28	15.1	66	252	2.5	<2	255	3315	0.972	1011.0

Temperature in the muffle furnace 900–1000 °C.

Sampling time P5.1.1 on 18 July 1989 between 17:53 and 18:53 (Table 4).

Sampling time P5.2.1 on 19 July 1989 between 10:23 and 14:23 (Table 4).

[a] Based on: 0 °C, 1.013 hPa, dry, related to 3% O_2 by volume.

[b] Calculated as NO_2.

[c] Based on 0 °C, 1.013 hPa, dry.

[d] Plant shutdown from 3:00 to 3:10 pm and from 4:45 to 5:05 pm.

TABLE 4. PCDF/PCDD: Concentrations in the Exhaust Gas of a Muffle Furnace

	Sample	
PCDF/PCDD	*P 5.5.1* (ng/m^3)	*P 5.2.1* (ng/m^3)
Total tetraCDF	0.760	0.146
Total pentaCDF	2.587	0.358
Total hexaCDF	2.967	0.485
Total heptaCDF	2.376	0.366
OctaCDF	1.793	0.286
Total tetra-bis octaCFT	10.483	1.641
2,3,7,8-TetraCDF	0.017	0.005
1,2,3,7,8-/1,2,3,4,8-PentaCDF	0.174	0.027
2,3,4,7,8-PentaCDF	0.185	0.030
1,2,3,4,7,8-/1,2,3,4,7,9-HexaCDF	0.312	0.053
1,2,3,6,7,8-HexaCDF	0.329	0.057
1,2,3,7,8,9-HexaCDF	0.057	0.012
2,3,4,6,7,8-HexaCDF	0.531	0.092
1,2,3,4,6,7,8-HeptaCDF	1.571	0.249
1,2,3,4,7,8,9-HeptaCDF	0.163	0.026
Total tetraCDD	0.153	0.019
Total pentaCDD	0.635	0.099
Total hexaCDD	0.756	0.130
Total heptaCDD	1.125	0.219
OctaCDD	1.022	0.187
Total tetra-bis octaCDD	3.691	0.654
2,3,7,8-TetraCDD	<0.003	<0.001
1,2,3,7,8-PentaCDD	0.033	0.007
1,2,3,4,7,8-HexaCDD	0.029	0.005
1,2,3,6,7,8-HexaCDD	0.061	0.011
1,2,3,7,8,9-HexaCDD	0.038	0.007
1,2,3,4,6,7,8-HeptaCDD	0.606	0.122
Total tetra-bis octa-CDF/D	14.174	2.295
Toxicity equivalent according to BGA/UBA (1984)	0.265	0.045
Toxicity equivalent according to NATO/CCMS (1988)	0.281	0.048

Muffle temperature 900–1000 °C; concentrations are related to 1.013 hPa, dry gas, 3 vol.% O_2.

multiplying the measured concentration of the specific compound by the related factor representing the relative toxicity to 2378 TCDD; the different values are summarized to the TE value (see Table 5). Table 6 represents results from two exhaust gas samples where the muffle furnace temperature is 1200 °C.

Emissions from Lean-Gas Engines

In order to reduce NO_x emissions due to the set limits only lean-gas engines can be used (see Chapters 7.3 and 7.4). During the

TABLE 5. 2,3,7,8-TCDD Toxicity Equivalent Factors (TEF): Proposals of Umweltbundelsamt/Bundesgesundheitsamt (UBA/BGA) Germany and NATO/CCMS. TE = ΣTEF

Compound	Toxicity equivalent factors	
	UBA/BGA (1984)	*NATO/CCMS (1988)*
2,3,7,8-TetraCDD	1.0	1.0
1,2,3,7,8-PentaCDD	0.1	0.5
1,2,3,4,7,8-HexaCDD	0.1	0.1
1,2,3,6,7,8-HexaCDD	0.1	0.1
1,2,3,7,8,9-HexaCDD	0.1	0.1
1,2,3,4,6,7,8-HeptaCDD	0.01	0.01
OctaCDD	0.001	0.001
2,3,7,8-TetraCDF	0.1	0.1
1,2,2,3,7,8-PentaCDF	0.1	0.05
2,3,4,7,8-PentaCDF	0.1	0.5
1,2,3,4,7,8-HexaCDF	0.1	0.1
1,2,3,6,7,8-HexaCDF	0.1	0.1
1,2,3,7,8,9-HexaCDF	0.1	0.1
2,3,4,6,7,8-HexaCDF	0.1	0.1
1,2,3,4,6,7,8-HeptaCDF	0.01	0.01
1,2,3,4,7,8,9-HeptaCDF	0.01	0.01
OctaCDF	0.001	0.001
Total tetraCDD	0.01	0
Total pentaCDD	0.01	0
Total hexaCDD	0.01	0
Total heptaCDD	0.001	0
Total tetraCDF	0.01	0
Total pentaCDF	0.01	0
Total hexaCDF	0.01	0
Total heptaCDF	0.001	0

TABLE 6. PCDF/PCDD Concentrations in the Exhaust Gas of a Muffle Furnace

	Sample	
PCDF/PCDD	*P5.1.2* *(ng/m³)*	*P5.2.2* *(ng/m³)*
Total tetraCDF	0.150	0.060
Total pentaCDF	0.315	0.121
Total hexaCDF	0.487	0.167
Total heptaCDF	0.574	0.126
OctaCDF	0.395	0.099
Total tetra-bis octaCDF	1.921	0.573
2,3,7,8-TetraCDF	0.012	0.004
1,2,3,7,8-/1,2,3,4,8-Penta CDF	0.027	0.010
2,3,4,7,8-PentaCDF	0.028	0.009
1,2,3,4,7,8-/1,2,3,4,7,9-HexaCDF	0.052	0.018
1,2,3,6,7,8-HexaCDF	0.058	0.019
1,2,3,7,8,9-HexaCDF	0.023	0.005
2,3,4,6,7,8-HexaCDF	0.097	0.029
1,2,3,4,6,7,8-HeptaCDF	0.347	0.082
1,2,3,4,7,8,9-HeptaCDF	0.042	0.011
Total tetraCDD	0.029	0.014
Total pentaCDD	0.072	0.033
Total hexaCDD	0.100	0.061
Total heptaCDD	0.224	0.070
OctaCDD	0.275	0.079
Total tetra-bis octaCDD	0.700	0.257
2,3,7,8-TetraCDD	<0.008	0.001
1,2,3,7,8-PentaCDD	0.007	0.003
1,2,3,4,7,8-HexaCDD	<0.017	0.003
1,2,3,6,7,8-HexaCDD	<0.017	0.006
1,2,3,7,8,9-HexaCDD	<0.017	0.003
1,2,3,4,6,7,8-HeptaCDD	0.122	0.038
Total tetra-bis OctaCDF/D	2.621	0.830
Toxicity equivalent according to BGA/UBA (1984)	0.045	0.016
Toxicity equivalent according to NATO/CCMS (1988)	0.049	0.017

Temperature in the muffle: 1200 °C; concentrations are related to 1.013 hPa, dry gas, 3 vol.% O_2.

TABLE 7. Total Concentrations in the Exhaust Gas of a Lean-Gas Engine (sampling period P5.2.5); for Other Parameters See Tables 8 and 9

Date	Time Begin	Time End	CO^a (mg/m^3)	CO_2 $(vol.\%)$	$NO_x^{a,b}$ (mg/m^3)	SO_2^a (mg/m^3)	O_2 $(vol.\%)$	$Total^a$ (mg/m^3)	T $(°C)$	$Volume^c$ (m^3/h)	Norm-factor 3 vol.% O_2	Atm. pressure (hPa)
27 Jul 1989	07:34	08:04	1.209	11.2	428	141	7.0	1.298	416	1813	1.286	1008.0
27 Jul 1989	08:04	08:34	1.212	11.2	428	147	7.0	1.293	419	1828	1.286	1008.0
27 Jul 1989	08:34	09:04	1.216	11.2	436	143	7.0	1.322	419	1828	1.286	1008.0
27 Jul 1989	09:04	09:34	1.225	11.2	475	136	7.0	1.504	418	1831	1.286	1008.3
27 Jul 1989	09:34	10:04	1.237	11.1	487	133	7.1	1.608	419	1829	1.295	1008.5
27 Jul 1989	10:04	10:34	1.227	11.1	488	129	7.0	1.575	419	1829	1.286	1008.5
27 Jul 1989	10:34	11:04	1.218	11.1	457	121	7.0	1.393	413	1833	1.286	1008.2
27 Jul 1989	11:04	11:34	1.212	11.1	441	125	7.0	1.260	419	1817	1.286	1008.1
27 Jul 1989	11:34	12:04	1.219	11.1	441	118	7.0	1.234	418	1831	1.286	1008.0
27 Jul 1989	07:34	12:04	1.219	11.1	453	134	7.0	1 387	418	1827	1.287	1008.2

[a] 0 °C, 1.103 hPa dry gas, related to 3 vol.% O_2.
[b] As NO_2.
[c] 0 °C, 1.103 hPa, dry gas.

TABLE 8. PAH Concentrations in the Exhaust Gas of a Lean-Gas Engine

PAH	Sample	
	P5.1.5 (μm^3)	*P5.2.5* (μm^3)
Fluoranthen	1.373	1.116
Pyren	0.652	0.704
Benzo(b) naphtho (2,1-d) thiophen	0.172	0.120
Benzo(ghi)fluoranthen/Benzo(c) phenanthren	0.319	0.214
Cyclopenta (cd) pyren	<0.001	<0.001
Benz(a) anthracen	0.076	0.067
Chrysen/Triphenylen	0.425	0.302
Benzofluoranthene (b + j + k)	0.185	0.122
Benzo(e) pyren	0.035	0.030
Benzo(a) pyren	<0.003	0.001
Perylen	<0.001	0.002
3-Methylcholanthren	<0.001	<0.001
Indeno(1,2,3-cd) pyren	0.008	0.009
Dibenzanthracene (a, c + a, h)	<0.015	<0.010
Benzo(ghi) perylen	0.002	0.001
Anthanthren	<0.001	<0.001
Coronen	<0.016	0.006
Dibenzo (a,i) pryen	<0.001	<0.001
Sum 22 PAH	3.247	2.694

Sampling period for P5.2.5 see Table 7 (0 °C, 1.013 hPa, dry gas, 3 vol.% O_2).

measuring campaign, the exhaust gases from lean-gas engines were also investigated in order to find out whether afterburners are necessary or not. In Tables 7–9 the results from such a campaign are presented in which no afterburner has been installed.

Emissions from a Lean-Gas Engine with Afterburner

The combustion of gas engine exhaust gases in a muffle furnace aims to reduce pollutants such as CO, PAHs, dioxins and furans through thermal destruction at high temperatures.

Using this process, engines do not need to operate on the lean gas principle; they can instead be tuned for optimum combustion, as emission levels depend on the quality of thermal treatment in the muffle furnace.

The results given in Tables 10–12 are based on the post-combustion of the exhaust gas from the lean gas engine described

TABLE 9. PCDF-PCDD: Concentrations in the Exhaust Gas of a Lean-Gas Engine

	Sample	
PCDF/PCDD	*P5.1.5* (ng/m^3)	*P5.2.5* (ng/m^3)
Total tetraCDF	0.262	0.115
Total pentaCDF	0.179	0.055
Total hexaCDF	0.131	0.021
Total heptaCDF	0.099	0.021
OctaCDF	0.094	0.033
Total tetra-bis octaCFT	0.765	0.245
2,3,7,8-TetraCDF	0.014	0.013
1,2,3,7,8-/1,2,3,4,8-Penta CDF	0.016	0.004
2,3,4,7,8-PentaCDF	0.013	0.005
1,2,3,4,7,8-/1,2,3,4,7,9-HexaCDF	0.014	0.003
1,2,3,6,7,8-HexaCDF	0.015	0.003
1,2,3,7,8,9-HexaCDF	0.005	<0.001
2,3,4,6,7,8-HexaCDF	0.018	0.004
1,2,3,4,6,7,8-HeptaCDF	0.066	0.013
1,2,3,4,7,8,9-HeptaCDF	0.007	0.002
Total tetraCDD	0.012	b
Total pentaCDD	0.046	0.003
Total hexaCDD	0.052	0.008
Total heptaCDD	0.063	0.016
OctaCDD	0.072	0.030
Total tetra-bis octaCDD	0.245	0.057
2,3,7,8-TetraCDD	<0.003	<0.001
1,2,3,7,8-PentaCDD	0.004	<0.001
1,2,3,4,7,8-HexaCDD	<0.006	<0.002
1,2,3,6,7,8-HexaCDD	0.007	<0.002
1,2,3,7,8,9-HexaCDD	0.006	<0.002
1,2,3,4,6,7,8-HeptaCDD	0.035	0.008
Total tetra-bis octaCDF/D	1.010	0.302
Toxicity equivalent according to BGA/UBA (1984)	0.018	0.005
Toxicity equivalent according to NATO/CCMS (1988)	0.018	0.005

Sampling period for P5.2.5. see Table 7 & 8 (0 °C, 1.013 hPa, dry gas, 3 vol.% O_2).
[b] could not be detected

TABLE 10. Total Concentrations in the Exhaust Gas of a Lean-Gas Engine with an Afterburner

Date	Time		No	CO^a (mg/m³)	CO_2 (vol.%)	$NO_x^{a,b}$ (mg/m³)	SO_2^a (mg/m³)	O_2 (vol.%)	$Total^a$ (mg/m³)	T (°C)	$Volume^c$ (m³/h)	Norm-factor 3 vol.% O_2	Atm. pressure (hPa)
	Begin	End											
26 Jul 1989	15:38	16:08	P6.1.5	1.214	11.1	459	143	7.0	1.258	419	1872	1.286	1008.8
26 Jul 1989	16:13	16:43	P6.2.5	1.215	11.1	457	141	7.1	1.283	420	1869	1.295	1008.5
26 Jul 1989	16:45	17:15	P6.3.5	1.206	11.1	454	140	7.0	1.274	422	1863	1.286	1008.3
27 Jul 1989	09:35	10:05	P6.4.5	1.237	11.2	489	133	7.1	1.620	419	1829	1.295	1008.5
27 Jul 1989	10:07	10:37	P6.5.5	1.226	11.1	488	129	7.0	1.601	418	1832	1.286	1008.5
27 Jul 1989	11:07	11:37	P6.6.5	1.212	11.1	441	125	7.0	1.258	419	1828	1.286	1008.1

Temperature in the muffle furnace 1000 °C
a 0 °C, 1.103 hPa dry gas, related to 3 vol.% O_2.
b As NO_2.
c 0 °C, 1.103 hPa, dry gas.

TABLE 11. PCDF/PCDD: Concentrations in the Exhaust Gas of a Lean-Gas Engine Equipped with an Afterburner

	Sample	
PCDF/PCDD	*P5.1.3* (ng/m^3)	*P5.2.3* (ng/m^3)
Total tetraCDF	0.038	0.042
Total pentaCDF	0.040	0.041
Total hexaCDF	0.037	0.044
Total heptaCDF	0.025	0.055
OctaCDF	0.039	<0.014
Total tetra-bis octaCFT	0.179	0.182
2,3,7,8-TetraCDF	0.004	0.005
1,2,3,7,8-/1,2,3,4,8-Penta CDF	0.003	0.008
2,3,4,7,8-PentaCDF	0.003	0.007
1,2,3,4,7,8-/1,2,3,4,7,9-HexaCDF	0.004	0.007
1,2,3,6,7,8-HexaCDF	0.005	0.008
1,2,3,7,8,9-HexaCDF	0.002	<0.001
2,3,4,6,7,8-HexaCDF	0.005	0.007
1,2,3,4,6,7,8-HeptaCDF	0.015	0.032
1,2,3,4,7,8,9-HeptaCDF	0.003	0.005
Total tetraCDD	0.007	b
Total pentaCDD	0.013	b
Total hexaCDD	0.019	b
Total heptaCDD	0.022	0.028
OctaCDD	0.029	0.028
Total tetra-bis octaCDD	0.090	0.056
2,3,7,8-TetraCDD	<0.001	<0.002
1,2,3,7,8-PentaCDD	0.002	<0.004
1,2,3,4,7,8-HexaCDD	<0.005	<0.006
1,2,3,6,7,8-HexaCDD	<0.005	<0.006
1,2,3,7,8,9-HexaCDD	<0.005	<0.006
1,2,3,4,6,7,8-HeptaCDD	0.012	0.016
Total tetra-bis octaCDF/D	0.269	0.238
Toxicity equivalent according to BGA/UBA (1984)	0.004	0.006
Toxicity equivalent according to NATO/CCMS (1988)	0.0005	0.007

Temperature in the muffle furnace: 1000 °C, 0 °C, 1.013 hPa, dry gas, 3 vol.% O_2).
[b] could not be detected

TABLE 12. PAH: Concentrations in the Exhaust Gas of a Lean-Gas Engine Equipped with an Afterburner

PAH	Sample	
	P5.1.3 (μm^3)	P5.2.3 (μm^3)
Fluoranthen	0.101	0.073
Pyren	0.004	0.007
Benzo(b)naphtho (2,1-d) thiophen	0.001	0.002
Benzo(ghi)fluoranthen/Benzo(c) phenanthren	0.016	0.017
Cyclopenta (cd) pyren	<0.001	<0.001
Benz(a) anthracen	<0.001	0.003
Chrysen/Triphenylen	0.012	0.014
Benzofluoranthene (b + j + k)	0.002	0.006
Benzo(e) pyren	<0.001	<0.001
Benzo(a) pyren	<0.001	<0.001
Perylen	<0.001	<0.001
3-Methylcholanthren	<0.001	<0.001
Indeno(1,2,3-cd) pyren	<0.002	<0.001
Dibenzanthracene (a, c + a, h)	<0.004	<0.006
Benzo(ghi) perylen	<0.001	<0.001
Anthanthren	<0.001	<0.001
Coronen	<0.001	<0.008
Dibenzo (a,i) pyren	<0.001	<0.001
Sum 22 PAH	0.136	0.122

Temperature in the muffle furnace 900–1000 °C, 0 °C, 1.013 hPa, dry gas, 3 vol.% O_2).

earlier. The performance of this engine during the post-combustion test is given in Table 10.

DISCUSSION

Emission levels of selected pollutants are summarized and shown graphically in Figs 1–8. The sample numbers correspond to the following system variations:

- *P5.1.1 and P.5.2.1*: muffle furnace at 900–1000 °C;
- *P5.1.2 and P5.2.2*: muffle furnace at 1200 °C;
- *P5.1.3 and P5.2.3*: post-combustion of engine exhaust gas at 1000 °C;

- *P5.1.4 and P5.2.4*: post-combustion of engine exhaust gas at 1200 °C;
- *P5.1.5 and P5.2.5*: gas engine emissions.

PCDF and PCDD Emissions

The composition of landfill gas may give certain indicators for possible dioxin and furan formation. PCDD/F are destroyed with increasing temperatures (after about 300–350 °C) even after minimal dwell times. As gas viscosity decreases with increasing temperature, gas mixing may decline. This situation may result in channel formations, where zones of high (e.g. 1800 °C in engines, 1200 °C in muffle furnaces) and low temperatures may be close together. In the low-temperature zones PCDD/F and PAHs may form at low rates. CO concentrations in the exhaust gas reflect the degree of incomplete combustion.

Figure 1 shows the PCDD/F concentrations in the different exhaust gas samples. Note the wide range of values with peaks occurring during times of disturbed operation at low temperature (5.1.1), during the first sampling of the high-temperature combustion (5.1.2), and to a lower degree from gas engine operation.

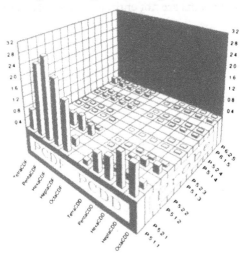

Figure 1. PCDF PCDD concentrations (ng/m^3) in the exhaust during different sampling periods P5.1.1–P5.2.5 (0 °C, 1013 hPa, dry gas, 3% O$_2$).

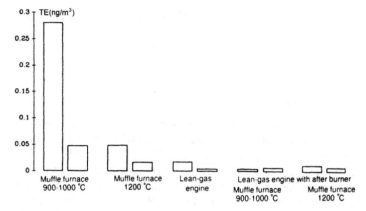

Figure 2. Toxicity equivalents (TE, NATO/CCMS, 1988; see Table 5) in the exhaust gas of different thermal LFG treatment plants.

A comparison of the different thermal LFG treatment plants, regarding their emissions as toxic equivalents, is presented in Fig. 2. This table shows also that, apart from periods of operating difficulties, the emissions of all plants were below 0.05 mg/m^3, and thus clearly below the expected standards for emissions from waste incinerators.

However, the results are not entirely satisfactory. Sample P5.1.1

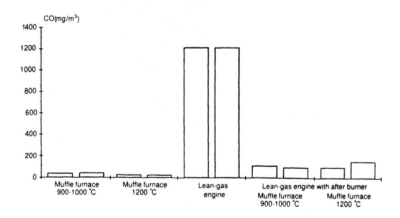

Figure 3. Comparison of average CO concentration in the exhaust gas of muffle furnaces and lean-gas engine.

shows that operating conditions can have a larger impact on PCDF/D emissions than the type of technology tested. The results indicate that muffle furnace operation was never optimized to the degree possible for this technology.

CO Emissions

As investigations at other plants have demonstrated, CO levels are below detectable limits when combustion is complete. CO concentrations may fluctuate widely during periods of operational difficulties, and the normal CO level of a non-optimized muffle furnace may be unnecessarily high.

The average CO emission levels of the various technologies are compared in Fig. 3.

PAH Emissions

As with PCDD/F emissions, PAHs in exhaust gases result from incomplete combustion. Figure 4 gives a graphical overview of the treatment systems where incomplete combustion may occur.

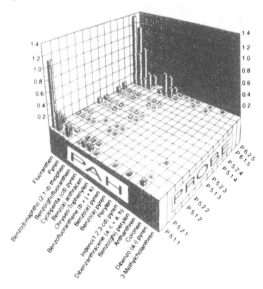

Figure 4. PAH concentrations ($\mu g/m^3$) in the exhaust gas during the sampling periods P5.1.1–P5.2.5 (0 °C, 1013 hPa, dry gas, 3% O_2).

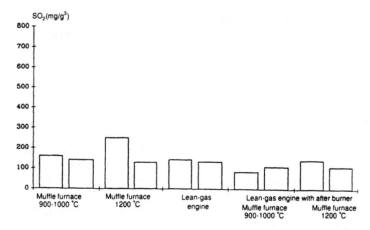

Figure 5. SO₂ concentrations in the exhaust gas of different thermal LFG treatment plants.

Such conditions occur in gas engines, as well as in optimized muffle furnaces during periods of operating difficulties. The post-combustion of gas engine exhaust gases reduced PAH concentrations, even when the muffle furnace was not optimized, to 4% of the initial levels (P5.1.3).

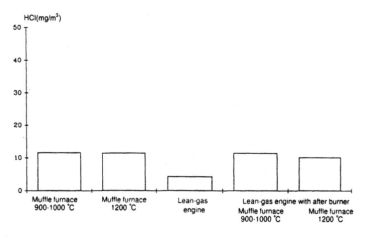

Figure 6. HCl concentrations in the exhaust gas of different thermal LFG treatment plants.

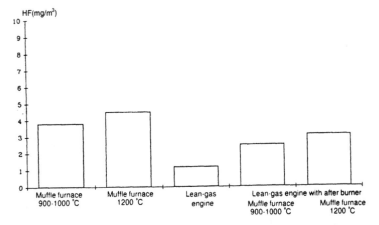

Figure 7. HF concentrations in the exhaust gas of different thermal LFG treatment.

SO$_2$-HCC-HF Emissions

The total quantity of these gases in the exhaust is determined primarily by the composition of the landfill gas, which varied from one sampling to the next. The emission levels shown in Figs 5–7 are thus specific to this landfill. The values in Figs 5–7 are below critical levels. Higher concentrations in the landfill gas can, however, lead to emission levels that require input gas purification or emission treatment.

NO$_x$ Emissions

The nitrogen oxides formed during combustion result either from atmospheric nitrogen transformed during the thermal process or from nitrogen present in the feedstock gas. For landfill gas combustion, the nitrogen in the gas is the less important source. Nitrogen oxides are formed at higher temperatures (especially peaks) such as occur regularly in gas engines.

Burner technology has been developed to the degree where peak temperatures higher than 1500 °C are largely avoided. In gas engines the temperatures in the combustion chambers can be reduced by increasing excess air levels (eg. a lambda of 1.6 for a lean-burn engine). As a result the engine is forced to operate from a combustion process perspective under marginal conditions. These

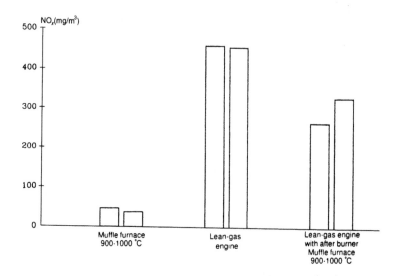

Figure 8. NO_x concentrations in the exhaust gas of different thermal LFG treatment plants.

conditions require particularly precise control and operation to maintain smooth operation.

The NO_x emissions of the muffle furnaces, <100 mg/m³, are far below the present standards. Lean-gas engines produce NO_x values just below the standard of 500 mg/m³ (see Fig. 8).

The post-combustion of the exhaust gas from lean-gas engines in muffle furnaces does not improve the emission values, as nitrogen oxides are not destroyed; these are rather diluted by the muffle furnace feedstock.

SUMMARY

The results presented in this chapter are based on an investigation programme in which the methodology for measuring emission levels of relevant pollutants from landfills and landfill gas incinerators was standardized. This programme monitored emissions at five different landfills. The study was contracted to Biogas Systems GmbH (Giessen), funded by the State of Hessen and carried out by the

Corporation for Workplace and Environmental Analysis GbR (GfA) Münster-Roxl. The programme was coordinated by the authors, as well as the Hessen State Environmental Agency.

Analysis of landfill gas from five landfills demonstrated that attention must be given to the carcinogens benzene and vinyl chloride. The landfill gas testing programme investigated muffle furnaces, gas engines, and the post-combustion of gas engine exhaust gases in muffle furnaces.

Muffle furnace technology is viable, regardless of whether or not heat recovery in the form of an added boiler or steam turbine is included. Under normal operating conditions, with temperatures ranging between 900 and 1200 °C, all relevant emissions are below the levels set by the federal air emission regulations (TA-Luft).

Landfill gas engines produce high levels of CO, which must be reduced. PAH concentrations in exhaust gases were higher than in the stack gases from waste incinerators. Dioxin and furan levels were below 0.05 mg/m^3, and thus far below the draft emission standards for waste incinerators. Nitrogen oxides could be reduced below federal standards through operational measures.

Afterburning the exhaust gases from lean-gas engines in muffle furnaces reduced CO emissions to 4% of the initial levels. Optimizing muffle furnace operation could probably eliminate CO. PAH as well as PCDD/F levels were reduced by the post-combustion. The dilution with muffle furnace feedstock decreased NO_x levels. Other emissions, such as HCl, HF, SO_2, PCB and particulate, were not significant.

The results from the test programme showed that landfill gas treatment plants require differentiated examination regarding the minimization of pollutant emissions in the exhaust gases. The muffle technology has been introduced on the basis of these investigations when biogas is flared or thermally utilized.

REFERENCES

Anon (1989). Messprogramm zur Ermittlung der Massenkonzentrationen relevanter Schadstoffe im Deponiegas und im Abgas von Deponiegasverbrennungsanlagen. *Schriftenreihe der Hessischen Landesansalt für Umwelt*, Hft Nr. 88.

Wiemer, K. and Widder, G. (1987). Emissionsminderung durch thermische Deponiegasbehandlungsanlagen. *Beiheft zu Müll und Abfall*, Heft 26.

8.6 Emissions from Landfill Gas Power Generation Plants

CHRIS P. YOUNG & NICK C. BLAKEY

WRc Environmental Management, Medmenham, Buckinghamshire SL7 2HD, UK

INTRODUCTION

The use of landfill gas as a renewable fuel source is not restricted to steam raising in boilers, firing kilns or space heating, but includes the generation of electrical power via reciprocating motors or gas turbines.

Monitoring has been undertaken at three installations, two of which (Sites A and B) employ reciprocating motors to drive generating sets, with a gas turbine being the power source at the third site (Site C). Brief details of the power plants and generating capacity at the three sites are given in Table 1.

Pretreatment of landfill gas at the sites using reciprocating engines (A and B) consisted of preliminary drying and filtering, followed by compression and then a final cooler/chiller phase for

TABLE 1. Summary of Power Plants and Generating Capacity

Site	Type of engine	Number of engines	Generating capacity (MW)
A	Dual-fuel diesel	3	3 (total)
B	Spark-ignition gas engine	3	0.74 (total)
C	Gas turbine	1	3.03

Landfilling of Waste: Biogas. Edited by T. H Christensen, R. Cossu and R. Stegmann. Published in 1996 by E & FN Spon, London. ISBN 0 419 19400 2.

completion of drying. In the case of the gas turbine (Site C), the raw gas was passed initially through a wet-scrubber to remove sulphide and acid gases, was then fed to a three-stage compressor with interstage cooling to remove moisture and condensable vapour, filtered and heated to 70 °C before combustion.

Each of the installations was located at major, established land-fills, receiving in excess of 100 000 t per year of controlled waste deposits.

MONITORING PROGRAMME

A phased monitoring programme was followed at each site. In Phase 1, samples were obtained for semi-quantitative analysis of minor organic components by gas chromatography-ion trap detection (GC-ITD), together with samples for quantitative analysis of bulk gas components from Sites A and B. During Phase 2, samples were taken of inlet and exhaust gases for quantification of haloforms, polysulphides and thiols at WRc. The quantification of sulphur dioxide, hydrogen chloride, oxides of nitrogen (NO_x) and selected dioxins and dibenzofurans was subcontracted to the Warren Spring Laboratory. Although time-averaged values for certain components of exhaust gases were measured during Phase 2 monitoring, the programme completed generally involved only two discrete sets of

TABLE 2. Extent of Monitoring Programme

Determinations	*Site*		
	A	*B*	*C*
Bulk gas	✔	✔	✔
GC/ITD 'fingerprint'	✔	✔	✔
Dioxins	✔	✔	✔
Dibenzofurans	✔	✔	✔
SO_2	✔	✔	✔
NO_x	✔	✔	✔
Total hydrocarbons	✔	✔	✔
CO	✔	✔	✔
Haloforms and polysulphides	–	✔	✔
Thiols	–	✔	✔
Particulates	✔	✔	✔

TABLE 3. Major Components of Bulk Gas Analysis (as Percentage by Volume, Dry Gas)

				Gas		
Site	*Hydrogen*	*Oxygen*	*Nitrogen*	*Methane*	*Carbon dioxide*	*Carbon monoxide*
A	<0.1	1.4–1.9	24.4–34.6	39.4–45.2	24.6–28.5	<0.1
B	<0.1	0.6	1.9	59.8	37.7	<0.1
C	<0.1	3.0	22.0	45.0	30.0	<0.1

samples from each site, separated in time by between 10 and 16 months. The programme followed at each of the three sites is outlined in Table 2.

The methods used to collect, prepare and analyse gas samples have been described by Young and Blakey (1990).

RESULTS

Input Gases

Analyses for major gas components, following dewatering and compression, showed significant differences between sites (Table 3), related to the age and degree of stabilization of the waste mass. In

TABLE 4. Trace Components in Inlet Gases (Semi-Quantitative Analysis, mg/m^3)

Group of compounds	*Site*			*Other landfills (Scott, 1990)*
	A	*B*	*C*	
Alkanes	0.25–5.0	<0.25–100	<0.25–10	0.3–1738
Alkenes	0.25–5.0	<0.25–100	0.25–10	0.1–305
Cycloalkanes	<0.25–5.0	0.25–5	0.35–50	0.4–56
Cycloalkenes	0.25–1	1–50	0.25–10	0.3–303
Aromatic hydrocarbons	1.0–50	0.25–150	0.25–50	0.6–528
Halogenated compounds	1.0–5.0	<0.25–10	5–50	0.2–1342
Alcohols	–	<0.25–10	<0.25–5	0.2–5053
Esters	0.25–10	<0.25–1	<0.25–1	0.03–320
Organosulphur compounds	<0.25	<0.25–1	0.25–1	0.2–519
Other oxygenated compounds	–	<0.25–10	0.25–50	0.18–171
Miscellaneous	–	<0.25–5	0.25–5	–

TABLE 5. Quantitative Trace Component Concentrations in Unburned Landfill Gas (mg/m³)

Compound	Site			
	B		C	
	Raw	Processed	Raw	Processed
Aliphatic hydrocarbons				
trans-1,2-dichloroethane	22.2	5.6	ND	0.6
dichloromethane	33.3	30.8	ND	<0.005
trichloromethane	41.6	12.4	ND	5.4
1,2-dichloroethane	<0.05	<0.05	ND	0.1
1,1,1-trichloroethane	7.0	3.4	ND	0.5
tetrachloromethane	<0.05	<0.05	ND	0.003
bromdichloromethane	<0.05	<0.05	ND	0.08
trichloroethene	92.0	4.7	ND	1.5
1,1,2-trichloroethane	<0.05	<0.05	ND	<0.005
dibromochloromethane	<0.05	<0.05	ND	<0.005
tetrachloroethene	16.8	0.4	ND	0.05
1,1,1,2-tetrachloroethane	<0.05	<0.05	ND	<0.005
tribromomethane	<0.05	<0.05	ND	<0.005
1,1,2,2-tetrachloroethane	<0.05	<0.05	ND	<0.005
Volatile sulphur compounds				
sulphur dioxide	<0.1	<0.1	ND	ND
dimethyl sulphide	<0.1	<0.1	ND	<0.1
diethyl sulphide	<0.1	<0.1	ND	<0.1
dimethyl disulphide	30	54	ND	12
ethylmethyl disulphide	25	49	ND	<0.1
methylpropyl disulphide	45	82	ND	<0.1
dirpropyl sulphide	<0.1	<0.1	ND	<0.1
diethyl disulphide	2.6	1.3	ND	<0.1
dimethyl trisulphide	75	76	ND	6.1
ethylisopropyl disulphide	7.5	4.0	ND	<0.1
ethylpropyl disulphide	1.4	<0.1	ND	<0.1
ethylmethyl trisulphide	2.0	1.4	ND	<0.1
ethylisobutyl disulphide	0.5	<0.1	ND	<0.1
butylpropyl disulphide	3.3	<0.1	ND	<0.1
methylpropyl trisulphide	16	7.7	ND	<0.1
ethylpropyl trisulphide	4.3	9.4	ND	<0.1
dipropyl trisulphide	5.2	17	ND	<0.1
Thiols				
1-ethanethiol	0.4	0.4	ND	0.25
2-propene-1-thiol	1.6	0.7	ND	2.3
2-propanethiol	0.3	0.5	ND	0.10
1-propanethiol	<0.2	<0.2	ND	<0.05
2-butanethiol	<0.2	<0.2	ND	<0.05
2-methyl-1-propanethiol	<0.2	<0.2	ND	0.05
1-butanethiol	<0.2	<0.2	ND	<0.05
1-pentanethiol	<0.2	<0.2	ND	ND

ND = not determined

particular, Site B was found to be at an advanced methanogenic state, with only minimal residual nitrogen in the gas.

Semi-quantitative identification of a wide range of compounds was made on inlet gases from each of the sites. The ranges of concentrations found were generally comparable between sites and also with the quantitative data given by Scott (1990) for three other landfills in the United Kingdom. The data are compared in Table 4.

Quantitative analyses of chlorinated aliphatic hydrocarbons, volatile sulphur compounds and thiols were completed on inlet gases at two of the sites (B and C), with comparisons being made also between concentrations of those components in raw, unprocessed landfill gas and after dewatering/compression. The results are presented in Table 5.

Exhaust Gases

Chlorinated aliphatics, sulphur compounds and thiols. As with inlet gas analysis, quantitative measurements of aliphatic hydrocarbons, volatile sulphur compounds and thiols on exhaust gases were completed at Sites B and C, while at Site A semi-quantitative estimates were made, indicating the presence of a variety of hydrocarbons including heptane, 2-methyl nonane, toluene, *m*- or *p*-xylene, dimethyl sulphide, texanol isobutyrate and a number of non-quantifiable components including benzene, ethyl benzene, styrene and carbon tetrachloride. With the exception of the

TABLE 6. Summary of Organic Compounds Detected in Exhaust Gases, Sites B and C (mg/m^3)

Compound	Site	
	B	C
Chlorinated aliphatic hydrocarbons		
trans-1,2-dichloroethane	3.8	0.007
1,1,1-trichloroethane	0.4	0.01
trichloroethene	0.6	0.2
tetrachloroethene	0.3	0.003
Volatile sulphur compounds	None above <0.1	
Thiols		
1-ethanethiol	<0.2	0.2

TABLE 7. Comparison of Gas Flow and Exhaust Particulate Flow Data

Site	Unit emission (g/h)	Particulate concentration (mg/Nm³)	Exhaust gas flow (m³/s)	Exhaust temperature (°C)
A	5.262	4.34	5.40	480
B	53	125	0.51	565
C	74	9	8.45	558

isobutyrate and 2-methyl nonane, the concentrations were found to be significantly lower than 1 mg/m³, with the isobutyrate being less than 10 mg/m³.

A high proportion of compounds measured in inlet gases (Table 5) were found to be below detection limits in the exhaust gases from Sites B and C, and Table 6 lists only those that were measured above detection.

Particulates and gas flow. Reporting of analyses of exhaust gas components is in accordance with the convention of reducing values to a dry cubic metre of gas at standard temperature and pressure (shown as mg/Nm³), and to a degree of dilution in the gas stream at 9% of carbon dioxide, by volume. Mean values of data from the three sites are compared in Table 7.

Major exhaust gas components. Measurements of principal components were completed at all sites. The comparative data in Table 8 are mean values of site measurements, converted to normalised gas volumes, where appropriate, in Table 9.

TABLE 8. Mean Exhaust Gas Concentrations

Site	CO (ppm)	CO_2 (%)	O_2 (%)	THC (ppm)	NO_x (ppm)	H_2O (%)
A	592	7.59	11.84	37.75	547	ND
B	1.24	23.8	0.09	1175	980	ND
C	7.3	5.42	14.65	18.5	31	15

THC = total unburnt hydrocarbons.

TABLE 9. Mean Exhaust Gas Concentrations, Normalized to mg/Nm^3, 9% CO_2

Site	CO	THC	NO$_x$
A	800	22.0	795
B		Not determined	
C	14	15.0	61

Acid gas. The considerable between-site variability in gas compositions was found to be present also for acid gases and exhaust streams (Table 10).

Dioxins and Dibenzofurans

Measurements of particulate and vapour phase emissions of dioxins and furans were completed for a large number of isomers. A substantial number of the isomers measured are of low toxicity, and those that are considered toxic are assigned toxicity equivalents (International Toxic Equivalent Concentrations – ITEQ) relative to the most toxic, 2, 3, 7, 8 tetrachlorodibenzo-p-dioxin (TCDD). The equivalents are summed and the overall ITEQ factor for an emission may then be compared with the appropriate guideline values. The ITEQ values for dioxins and furans at the three sites examined are given in Table 11.

TABLE 10. Mean Sulphur Dioxide and Hydrogen Chloride Concentrations in Exhaust Emissions

Site	Concentrations (mg/Nm^3, 9% CO_2)		Unit emissions (mg/s)	
	SO$_2$	HCl	SO$_2$	HCl
A	51	<12	275	<61
B	21.5	14.8	10.8	7.2
C	5.7	38	49	330

TABLE 11. Summary of Results of Dioxin and Furan Survey in Exhaust Gases

Site	2,3,7,8 TCCD ITEQ concentrations (ng/Nm³)	
	Dioxins	*Furans*
A	<0.23	<0.20
B	<0.41	<0.27
C	0.63	1.20

DISCUSSION

Significant differences noted between the proportions of major components in the raw landfill gases at the three sites are ascribed to the state of maturity of the waste mass from which the gas has been extracted. The bulk compositions of gases from Sites A and C are comparable, with significant residual nitrogen (20–30% by volume), suggesting an early stage of methanogenesis, which had not succeeded in flushing residual nitrogen from the wastes. In comparison, the gas at Site B contained methane at a volumetric concentration close to the theoretical maximum (c. 60%) and low residual nitrogen, indicating a fully developed, stable methanogenic phase. Examination of the analysis of trace components in the raw gases (Table 4) suggests that the levels of halogenated compounds, which may contribute to acid gas formation during combustion, were higher at Site C than at either A or B. Such variations in trace compound levels in the gas may be a function of the types of waste accepted at a particular site, and also of the age of the wastes from which gas is extracted.

It has been shown that significant decreases in total concentration of trace compounds in gaseous emissions occur within a few years of the initial deposition of waste which, in the case of halogenated compounds, was found to reduce to 50% of initial values within about 2 years (Scott, 1990). An effect of increased halogenated compound levels in the raw gas may be to accelerate the rate at which the buffering capacity of lubricating oils (total base number, TBN), is consumed, bringing about the need for more frequent oil changes in reciprocating motors. This effect was reported from site B (Hornsby, 1990), where the commissioning of additional gas wells in a recently filled part of the site led to a decrease in the TBN of the oil and a need for more frequent changes.

TABLE 12. Comparison of Mean Emission Levels of Principal Exhaust Gas Components between Sites and with Municipal Waste Incineration and EC Directive (mg/Nm^3 except dioxins and furans)

Gaseous component	Site			Typical UK incinerators	EC Directive
	A	B	C		
Particulates	4.3	**125**	9	400	30
CO	**800**	**c.10000**	14	330	100
THC	**22**	**>200**	15		20
NO$_x$	795	c.1170	61		
HCl	12	15	38	700	50
SO$_2$	51	22	6	400	300
T4 CDDs	0.4 ng	0.6 ng	0.6 ng	130 ng	
T4 CDFs	0.4 ng	2.7 ng	1.1 ng	160 ng	

Parameters exceeding Directive level are in bold.
THC = total unburnt hydrocarbons.

Reductions in the concentrations of trace components present in the raw gas were noted as a result of the dewatering/compression stage (Table 5), and significant reductions in the concentration of quantified organic compounds were found in the exhaust gases, compared with the inlet analogues. However, comparison between the exhaust gases, the emission levels typical of UK municipal solid waste incinerators (Woodfield, 1987) and the standards included in the European Union Directive on municipal solid waste incineration (Table 12) has indicated possible difficulties.

Overall, the exhaust emission from Site C (gas turbine) satisfied probable standards, although in the case of hydrogen chloride the recorded values ranged from 12 to 63 mg/Nm^3, indicating that the process may not always meet the EU incinerator standard. The increased mean hydrogen chloride levels at that site may be related to the higher concentrations of halogenated components in the inlet gas than at the other sites.

At both the sites employing reciprocating engines, high carbon monoxide, particulate and unburnt hydrocarbon emissions indicated possible carburation problems. This was particularly so at Site B, where low oxygen levels were taken as evidence of over-rich running (Hornsby, 1990).

Combustion in turbines is more uniform than in reciprocating engines because of the continuous gas flow. Turbine combustion temperatures are typically in the range 1000 to 1100 °C, while combustion chamber temperatures in reciprocating meters may reach 2500 °C.

At such temperatures it is possible that dissociation of carbon dioxide, a major component of the inlet gas, occurs. In the case of Site B, the mean carbon monoxide concentrations of 12 400 ppm represent a value significantly greater than the 8 h Occupational Exposure Limit Value (Health and Safety Executive, 1989). Improvement of carburation would be expected to alleviate the carbon monoxide problems encountered but could, at the same time, lead to increased NO_x emissions, which might subsequently require additional abatement measures.

CONCLUSIONS

Monitoring has been carried out of the composition of the feed and exhaust gases from three electrical power producing plants that utilize landfill gas. The raw gases show compositional differences related to the types of waste. Significant removal of trace components was found at the dewatering stage, with further removal during combustion. In general, the highest emission standard was achieved by a gas turbine unit.

The data presented have been gained from a survey of a small number of sites, with relatively infrequent sampling. The extent of short- or long-term variations from the mean values estimated must remain a matter of conjecture until such time as further data become available.

ACKNOWLEDGEMENTS

The work reported here has been carried out under funding from the Energy Technology Support Unit but the opinions expressed in this chapter are those of the authors. It is published with the permission of the Director, WRc Environmental Management.

REFERENCES

Health and Safety Executive (1989). *Occupational Exposure Limits*, UK Guidance Note 40/89 (January).

Hornsby, M. R. (1990). Lessons learned from the Stewartby and Packington Projects, in *Landfill Gas: Energy and Environment '90* (eds G. E. Richards and Y. R. Alston), Harwell Laboratory, Oxon, pp. 455–468.

Scott, P. (1990). Studies investigating the composition and evolution of trace gases from landfill, in *Proceedings Brunel Landfill Conference: Problems and Solution*, 15–18 May.

Woodfield, M. (1987). *The Environmental Impact of Refuse Incineration in the UK*, Warren Spring Laboratory, Stevenage, UK.

Young, C. P. and Blakey, N. C. (1990). Gases in, gases out. Monitored emissions from power generation projects, in *Landfill Gas: Energy and Environment '90.* (eds G. E. Richards and Y. R. Alston). Harwell Laboratory, Oxon, pp. 317–329.

9. CASE STUDIES

9.1 Landfill Gas Recovery and Utilization at Poplars, Staffordshire, UK

SAM ASHTON[a] & BOB COUTH[b]

[a]*Thomas Graveson Ltd, Keer Bridge House, Warton, Carnforth, Lancashire LA5 9HA, UK*
[b]*Aspinwall & Company, Walford Manor, Baschurch, Shrewsbury, Shropshire SY4 2HH, UK*

LANDFILL DESCRIPTION

Poplars landfill is one of the largest waste disposal facilities in England. It is operated by Staffordshire County Council's waste disposal company, Poplars Resource Management, and is located in a rural setting. Some statistics for the site are given in Table 1, while a general description is summarized below.

The site is a former opencast excavation; it offers a degree of natural containment, which has been supplemented by an engineered mineral liner. The quarry is large and shallow, above the groundwater-table, with the majority of waste disposal below ground.

TABLE 1. Landfill Parameters

Item	Parameter
1. Site area	98 ha
2. Landfill depth	10–20 m
3. Design principle	Containment
4. Start date for landfilling	1978
5. Anticipated completion date	2017
6. Annual input	250 000 t
7. Potential landfill void	12 000 000 m^3

Landfilling of Waste: Biogas. Edited by T. H Christensen, R. Cossu and R. Stegmann. Published in 1996 by E & FN Spon, London. ISBN 0 419 19400 2.

Poplars landfill currently accepts approximately 250 000 t of waste annually, of which approximately half is domestic, with the remainder being made up of varying amounts of industrial, commercial and inert wastes.

Landfill Gas Management System

The primary objective for the system is to extract gas as it is generated, preventing a build-up in pressure within the waste and uncontrolled migration. The secondary objective is utilization of the energy potential of the gas. In the particulars sent out to tenders for the gas management system it was stressed that any landfill gas collection scheme must have environmental protection as the priority, with particular emphasis on the prevention of gas migration. Although there would appear to be a potential conflict between the objectives of control and utilization, extraction can be optimized to complement utilization. Optimization of utilization is particularly relevant to engineered containment sites, such as the Poplars landfill, where volumes of air drawn in for perimeter migration control can be minimized.

INVESTIGATIONS AND LANDFILL GAS PREDICTIONS

The landfill gas management system is in the first four phases of the landfill, an area containing approximately 1.25 million m^3 of waste (Fig. 1). Initial impressions and a previous consultancy study suggested that the landfill was viable for utilization:

- The waste is contained, minimizing air ingress.
- Leachate levels are low, allowing gas recovery.
- There is a high percentage of biodegradable waste.
- The volume is over 0.5 million m^3.

The only negative aspects affecting the extraction of gas are:

- The phases have only a temporary 300 mm soil cap, potentially allowing air in through the surface.
- The waste is in places relatively shallow (10 m).

To appraise the utilization potential an assessment of gas production

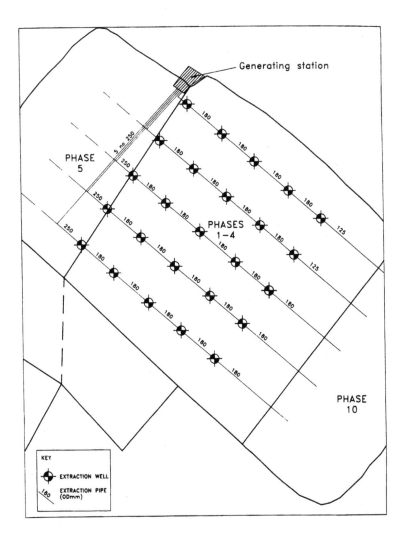

Figure 1. Poplars landfill: landfill gas management system.

was carried out using computer software and basic data about the landfill supplied by Staffordshire County Council. The information computed in the software includes: the total annual input for each year; the percentage of the various classifications of waste, e.g. domestic, civic amenity, industrial, inert, sewage sludge; and the temperature within the landfill (see also Chapter 4.1). The figures produced indicated that in the first four phases of the landfill there is sufficient gas to produce approximately 1 MW of electricity, but that this would increase in the future years as the site progressively fills.

No trial pumping was carried out to assess the yield of the waste prior to the construction of the system. It is often prudent to do so to avoid any embarrassment of the gas yield not meeting the fuel requirements of the engines. In this case, trial pumping was not considered necessary, for the following reasons.

- There was confidence in the yield predictions based upon the proven record of the computer software and experience on other projects.
- The modular design of the engines chosen allows flexibility: their number can be increased or reduced depending on yield.

DESCRIPTION OF GAS EXTRACTION SYSTEM

Figure 1 illustrates the layout of the extraction system, the main elements of which are described below.

Extraction Wells

The collection system comprises 25 wells in five rows spaced approximately 50 m apart. Experience has shown that extraction wells in this configuration of landfill are effective at 50 m centres, but that they require a closer spacing for migration control at the perimeter, while the spacing may be increased at deeper landfills.

For speed, the wells were constructed with two heavy-duty drills, one a piling auger on a telescopic 'kelly' and the other a barrel auger. The annulus of the well was not backfilled with gravel but allowed to collapse into the borehole, as it is our experience that gravel packs tend to prevent differential settlement. The wellheads are designed to be in-line and integral with the extraction pipework. The wellheads have an outer diameter of 315 mm and an aperture of 180 mm at their base, which slides over the borehole liner, allowing up to 2.5 m of differential settlement. They not only incorporate the

TABLE 2. Well Details

Item	Parameter
Number	25
Spacing	50 m approx
Depth	9.5–21.5 m
Volume of waste influenced	1.25 million m^3 approx
Diameter of borehole	325 mm
Well screen	160 mm perforated twin-wall pipe, to within 2.5 m of surface
Backfill	Collapse of annulus
Wellhead	Hofstetter GBA 'K', 6 bar

means to monitor and regulate the gas flow in the well and monitor flow in the lateral pipeline, but also contain an internal barometric leg for draining condensate away. The gas flow is regulated by a sleeve valve, which provides finer adjustment than a butterfly valve, particularly at low flows, and most importantly does not restrict condensate movement. The wellheads are sealed around with bentonite and protected by a lockable chamber (see also Table 2).

Extraction Pipework

Extraction pipework is arranged in five parallel lines (a spur system), with gas drawn through separate header pipes, which terminate at the generation station.

The pipe sizes were designed using the computer software in accordance with parameters for optimum gas speeds, relevant to the gradient of the pipeline, and based on the estimate of gas flows (see also Table 3).

Allowance has also been made for future expansion and extension of the system into other areas of the site as they become available.

TABLE 3. Pipework Details

Item	Parameter
Pipe material	Medium-density polyethylene (MDPE)
Pipe laying	In trenches, butt fusion welds
Pipe rating	6 bar
Diameter, extraction	125 and 180 mm OD
Diameter, headers	250 mm OD

The pipework is carefully laid to falls to ensure that gas and condensate are not in conflict.

Extraction Equipment

As a precaution against landfill gas, the pump, flare and generating station are situated in the open, on a sealed reinforced concrete base, on virgin ground adjacent to the landfill. The gas enters via a galvanized steel collection manifold, each inlet of which is fitted with a butterfly valve. From the single outlet the gas is drawn through a vertical cyclone centrifugal dewatering tank for the removal of particulates and the remaining condensate.

The two main elements of the extraction plant are the centrifugal pump and the flare. The centrifugal pump, which is constructed to handle the corrosive gas safely, has characteristics to enable the gas to be drawn from the furthest wells and pumped to the inlet manifold of the engines at the required delivery pressure. Radial and axial forces are taken care of by stainless steel compensators, fitted on both inlet and delivery pipelines, as are a pair of flashback arrestors. Pump shaft bearings are lubricated by grease nipples and the shaft seals are lubricated by an auto-lube system. A standby pump is not installed as it is not considered necessary, owing to very good reliability of the centrifugal pump, and also monitoring for preventative maintenance (see also Table 4).

From the pump the gas is delivered to the engines and the flare, via galvanized steel pipework. The flare performs a vital role, and for reasons of safety it has the capacity to burn all the extracted gas. It has automatic control and ignition, enabling it to modulate according to changes in the quantity and quality of the gas and the demands of the engines. It has a wide turn-down ratio so that when it starts up its minimum requirements do not cause a shutdown of the engines by starving them of gas. A small open-fronted building houses the

TABLE 4. Centrifugal Pump Details

Item	Parameter
Number	1 continuous duty
Capacity	1500 m^3/h
Differential pressure	180 mbar
Drive	24 kW electric motor to V-belt

control panel for the gas plant and also a second cabinet containing apparatus to continuously extract, clean, dry and cool a sample of gas for analysis for oxygen and methane quality, so that the percentage content of these gases is always known, and to enable the plant to shut down if safety limits are exceeded. The remaining equipment is ancillary: instrumentation and controls including flow measurement, suction and delivery pressure gauges, thermometer gauges etc. There is also monitoring of condensate level, suction pressure, pump temperature, flare temperature and flame condition. All these circuits are wired to warning lamps, shutdown controls and a telemetry link. The layout of the plant is designed so that its capacity may be simply duplicated when gas production has sufficiently increased.

DESCRIPTION OF UTILIZATION PLANT

For convenience, methods of utilization of landfill gas can be placed into three categories:

- direct end use as a source of heat;
- electricity generation;

Figure 2. Poplars power station. Brick building containing two compartments: producer's switchgear in one, area electricity board's in the other.

- indirect end use by processing to a higher-grade fuel.

Indirect end use is not technically or economically attractive in the UK. It is arguably most profitable to use landfill gas as a direct source of heat, but there are no single users who are situated near to the Poplars landfill of sufficient size and with a constant demand for energy. Consequently, electricity generation was considered the most suitable utilization option.

Three Jenbacher engines and generator sets were chosen for the Poplars project, based upon their successful use of the engines at other sites in the UK. The units are heavy-duty stationary spark-fired four-stroke gas engines with low air/fuel mixture turbocharging, self-regulatory three-phase synchronous generators with automatic power factor control, and automatic monitoring and engine management systems.

The sets are delivered ready packaged in a standard ISO 12 m container, complete with switchgear and control cabinets, enabling them to operate independently as shown in Fig. 2. Landfill gas is directly fed to each engine through a 3 µm in-line filter. The suitability of the gas and the requirement for processing was assessed

TABLE 5. Jenbacher Engine and Generator Details

Item	*Parameter*
Number of units	3
Gross output	
unit	463 kW
total	1389 kW
Total parasitic load	50 kW
Electricity export	1339 kW
Engine type	Spark ignition
Engine:	
bore	135 mm
stroke	145 mm
number of cylinders	16
nominal speed	1500
Required minimum gas pressure before air/fuel mixer	60–120 mbar
Generator, three-phase	415/240 V
Container	
length	12.2 m
width	2.4 m
height	1.6 m

with the manufacturers. There is no requirement to boost the pressure to the engines or for further processing of the gas. Details of the engine and generator sets are as follows (see also Table 5).

The electricity is generated at low voltage, i.e. 415 V. Cables from each generator take this to an adjacent transformer, where it is stepped up to 11 000 V. From the transformer a cable leaves the site to its connection to the local electricity board network via a small building divided into two compartments. The first contains the producer's meters, switchboard and contact breakers etc., and the second the electricity board's meters, switchgear, contact breakers etc.

The engines were selected for this scheme for the reasons summarized in Table 6.

Despite the many merits of the engine listed in Table 6, it is recognized that there are situations when other utilization

TABLE 6. Reasons for Engine Selection

- The supply of three engine modules offers flexibility for varying gas-generation rates. A number of units can be operated, and further units can be added with continued landfilling. Furthermore, if one engine has a problem, it does not cause a total loss of production.
- Spark-ignition engines do not require a brought-in fossil fuel supply (as do dual-fuel engines), nor a building or heavy foundations.
- The complete self-contained ISO units can be quickly and easily transported and erected.
- The engines employ lean-burn technology, which means that the exhaust emissions meet the stringent clean air standards that apply in Germany (TA-Luft) and the USA.
- The engines do not require compressors to boost the inlet pressure.
- The engines have a high rate of efficiency (35% electrical efficiency), which combined with a low parasitic load of 50 kW total (main engine pump, engine water pumps, cooling fares etc.) achieves a high rate of efficiency in terms of the energy produced from each cubic metre of gas. This maximizes the electricity exported, giving a high financial return.
- The engine design does not require high-wearing components such as stoichiometric sensors or catalytic convertors.
- The engines can automatically modulate within pre-set limits to cope with variations in methane quality and barometric pressure.
- The installation is designed for continuous unmanned operation; every parameter for the safe running of the engines is monitored. If limits are exceeded the engine is shut down, fail safe, and the local engineer is summoned via the telemetry link.
- Only one manufacturer has the responsibility to commission the package and thereafter supply the spare.
- The engines are quiet: a single engine in a standard container with standard exhaust system generates 65 dBA at 10 m.
- They have a proven track record with landfill gas.

equipment would be preferable. It should be noted that the chosen gas engines require a methane concentration of 38% or greater.

Installation and Commissioning

The entire project as described above was installed and commissioned in 53 days. This was achieved by two teams: one on the landfill, putting in the wells and collection pipework; and the second preparing the concrete base and assembling the equipment upon it. Electrical and mechanical engineers from the engine manufacturer were present for the commissioning. The engine containers are easily transported, with only the exhaust silencers and cooling system dismantled for transportation.

DESCRIPTION OF OPERATION

The date of commissioning was 13 September 1991, and the system has operated without major fault since. The gas collection system, i.e. the extraction wells, are monitored from time to time but little adjustment has been necessary despite the site having only a temporary cap. Moreover, the gas production has remained fairly constant, winter and summer. The gas generation and system performance, as detailed in Table 7, are in line with the gas prediction and thus serve to justify the planners' decision not to undertake a pumping trial.

The figures are in line with 'rules of thumb' gained from the second author's company's experience: i.e. gas wells in typical landfills in the UK might be expected to yield around 50 m³/h (20 m deep well) and sites typically yield 1000 m³/h per 1 000 000 m³ of waste.

The engines incorporate certain design features and operating procedures, based on past experience, aimed at getting the maximum efficiency from burning landfill gas and to prevent

TABLE 7. Operating Statistics

Item	Parameter
Gas extraction rate	1200–1250 m³/h
Methane concentration	47–48%
Mean yield per well	50 m³/h
Yield per 1000 m³ of waste	1.0 m³/h
Electricity generated first 595 days	16 750 000 kW
Efficiency of total generation capacity (1389 kW)	85%

corrosion. The engine manufacturers have learned that there is a need to provide adequate cooling in summer to compress the volume of the gas (increase the mass). The water cooling system is in two parts. The first is the jacket that cools the engine and exhaust manifold, and which operates in one particular temperature band; the second, operating at a lower temperature, is the intercooler, the object of which is to increase the mass of the incoming air and gas mixture. In addition, thermostatically controlled fans keep the interior of the container at a constant temperature throughout the year to avoid the formation of cold spots in the engines.

The designer and constructor of the plant worked closely with their lubricant supplier. Based on operating experience, the oil capacity of the engines has been increased from 165 to 500 l. Samples are taken for analysis at every 500 h up to 1000 h and then at 1300 h. Oil changes are now being carried out at 1500 and 1600 h, depending on the results of the analyses. The basic service period, which requires the engines to be stopped, has been extended to 1000 h. This has been made possible by such improvements as, for example, a gas-mixing system and special spark plugs.

It is the second author's experience, and also generally recognized, that engines function best when left to run continuously under constant load. Problems tend to occur in engines when they are frequently started and stopped for varying load conditions.

The corrosion potential of the landfill gas is monitored, but has not been significant in any of the installations to date.

ECONOMICS

The project is made financially viable by the UK Government's policy to encourage the generation of electricity from non-fossil fuel sources: i.e. wind, hydroelectricity, landfill gas, municipal and general waste, sewage gas and others. This policy is implemented through a scheme known as the Non-Fossil Fuel Obligation (NFFO) for Renewables and is operated by the Office of Electricity Regulation (OFFER). It is the Government's objective to underwrite a floor level of 1500 MW (declared net capacity) of new renewable generation in the UK by the year 2000. The NFFO scheme is designed to assist in the development of the non-fossil fuel utilization industry, to enable it to develop to compete on the open market.

The Poplars project was the subject of a successful application in the second tranche of this scheme, announced in 1991, which includes 27 other landfill gas products with a total output of 48 MW. The rate fixed by OFFER in the second tranche was 5.7 p ($0.09) per kW, but the current agreements under the tranche system only extend to 1998.

In the first 19 ½ months of operation, the gross income from the generation of electricity is almost £1 000 000. At this rate of production the engines are financially viable and might be expected to have a payback period of around 4 years. However, the viability would be questionable at the standard rate of 2–3 p per kW paid for electricity in the UK.

SUMMARY

Poplars is an active landfill site with the potential to take some 12 million m^3 of municipal waste. A landfill gas management scheme was constructed in 1991 for environmental control of gas from the first 1.25 million m^3 of waste. The scheme incorporated three 460 kW spark-ignition engines for the generation of electricity using landfill gas as a fuel. The engines have many merits: they are flexible, efficient, modular, do not require a secondary fuel nor the inlet gas pressure to be boosted by a compressor. Since installation, the engines have operated at 85% of their total generation capacity. Electricity generation using landfill gas has been made economically attractive by the United Kingdom's Government Non-Fossil Fuel Obligation (NFFO).

ACKNOWLEDGEMENTS

Thomas Graveson Ltd did the design and construction of the gas management system including utilization by electricity generation, and are the owners of the installation. Acknowledgement should be given in particular to Mr Peter Graveson, to Poplars Resource Management and to Staffordshire County Council. Aspinwall & Company supplied its experiences from other plants; it was not involved in the landfill gas project at the Poplars Site. The computer software for the gas prediction has been designed by Hofstetter of Switzerland and is owned by Thomas Graveson Ltd. The close cooperation of the lubricant manufacturer Mobil Oil was appreciated.

9.2 The Landfill Gas Plant in Viborg, Denmark

HANS C. WILLUMSEN

Hedeselskabet, Klostermarken 12, 8800 Viborg, Denmark

INTRODUCTION

The landfill gas plant in Viborg was established in 1985 by Viborg Municipality. The plant was supported by the Danish Energy Agency and by the European Commission as a demonstration project.

The landfill gas project is a typical example of landfill gas exploitation in small landfills, with a low filling height and a capacity below 0.5 million of waste. In Denmark, most of the landfills are small sites and are the property of municipalities.

The landfill gas is used as fuel in a boiler to produce heat, which is to be sold to the district heating company in Viborg, and also to produce electricity from a small combined heat and power production plant.

DESCRIPTION OF THE LANDFILL

The municipal landfill of Viborg was established in 1972. The site was previously occupied by a sand quarry. The type of soil in place is eluvial sand with an average depth of approximately 10 m.

The site receives waste from 83 000 inhabitants from five municipalities. The total mass of dumped waste is currently 60 000 t per year. The total amount of waste to date is about 600 000 t and the gas plant extracts gas from about 400 000 t of waste.

Landfilling of Waste: Biogas. Edited by T. H Christensen, R. Cossu and R. Stegmann. Published in 1996 by E & FN Spon, London. ISBN 0 419 19400 2.

The dumped waste consists of:

25% household waste;
35% industrial and office waste;
15% bulk and garden waste;
21% building waste;
4% other types of waste.

The waste has undergone a compaction pretreatment since approximately 1975. The estimated density of the compacted waste was 800 kg/m^3 and it is now 1000 kg/m^3. The dumped waste is daily covered with a 10 cm sand layer. The depth of the waste layer between two successive covers is about 2 m. The landfill has a synthetic liner in the bottom and a drainage system for the leachate.

PRE-INVESTIGATIONS

Since the end of 1981 the Danish Ministry of Energy has subsidized research activities related to the recovery and exploitation of landfill gas from Danish landfills.

The landfill can be considered as being representative of Danish conditions, in as much as the waste consists of household waste, industrial and gardening waste, and the quantity is around 600 000 t.

In 1983, a pilot plant for landfill gas exploitation was designed and constructed. The pilot plant included a landfill gas extraction system of eight wells, located in a 1 ha section of the municipal landfill of Viborg corresponding to 50 000 t of waste. The utilization of landfill gas consisted of a gas engine to generate electricity.

The pilot plant was equipped with measuring and analysing instruments so that the amount of gas from each well and the total amount could be registered. Furthermore, gas composition, temperature, pressure, meteorological parameters etc. were measured.

At that time no model for estimation of the gas quantity from the whole landfill was available, so the estimations were made from the results of the pilot plant.

Landfill Gas Quality

The composition of the gas is currently as follows:

- CH_4 (methane) 40–45%;
- CO_2 (carbon dioxide) 35–45%;
- N_2 (nitrogen) 5–15%;
- O_2 (oxygen) 0–2%;
- H_2 (hydrogen) 1–2%;
- Ar (argon) 0–0.4%.

Table 1 shows an analysis of the gas from the landfill, made by means of a PGS 100 mass spectrometer, which has been used in the analyses. The instrument can measure molecules with mass numbers 1–99, and in concentrations down to 10 ppm.

As can be seen, the values are under 10 ppm, and hence we have not been able to read them on the mass spectrometer. The Danish sanitary limit is given in the table, and as the limits for benzene, vinyl chloride and hydrogen sulphide are under 10 ppm, we have checked by means of Dräger tubes and found that we are below the limits that apply in Denmark.

THE LANDFILL GAS EXTRACTION PLANT

The structure of the landfill gas plant in Viborg is shown in Fig. 1. The landfill extraction system consists of 20 vertical wells drilled in

TABLE 1. Analysis of the Gas from the Landfill

Component	Content in ppm		
	Mass spectrometer	Dräger tubes	Danish sanitary limit
Benzene, C_6H_6	<10	<0.5	5
Dichloromethane, CH_2Cl_2	<10	–	50
Toluene, C_7H_8	<10	–	75
Vinyl chloride, C_2H_3Cl	<10	<1	1
Ethyl benzene, C_8H_{10}	<10	–	50
Xylene, C_8H_{10}	<10	–	50
Dichloroethylene, $C_2H_2Cl_2$	<10	–	200
Hydrogen sulphide, H_2S	<10	9	10

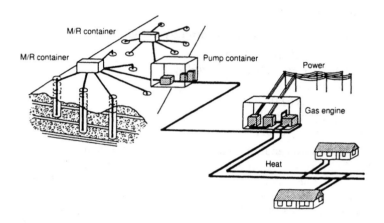

Figure 1. Structure of the landfill gas plant in Viborg, Denmark.

the Northern section and the oldest part of the landfill and another 12 vertical wells in part of the southern section. The wells have a diameter of 0.5 m. In each well, a vertical pit is constituted of a pipe of 110 mm diameter, made of high-density polyethylene (HDPE), which reaches the bottom of the landfill, and which is perforated from the bottom up to above 4 m below the landfill surface. The space between the pipe and the well is backfilled with gravel up to 3 m below the landfill surface. The last 3 m are backfilled with clay.

The head of each well is protected by a box. Each vertical well is connected to a horizontal pipe, which transports the landfill gas to one of two measurement and adjustment sheds. The horizontal pipes have a diameter of 40 mm and are made of medium-density polyethylene (MDPE). Several precautions are taken to avoid the blocking of the pipes by condensate, such as condensate receivers, slope gradients on horizontal pipes and by the possibility of reverse blowing. The horizontal pipes have their slopes directed towards the extraction wells so that any condensate water runs back to the wells.

The extraction system is controlled by an automatic analysis system, computer-operated, which adjusts the landfill gas flow rate of each well according to its quality through a system of valves on each extraction well. Each horizontal pipe arriving at the measurement and adjustment shed is connected to a stainless steel pipe of 40 mm diameter. The pipe is equipped with an adjusting butterfly valve and an automatic gas sampler.

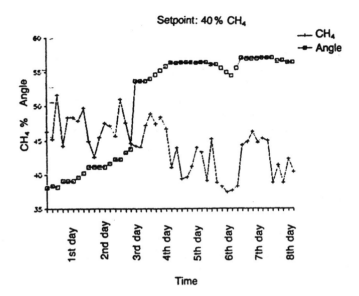

Figure 2. Adjustment of a valve and stabilization of the CH4 valve.

Figure 2 illustrates how one of the valves is automatically adjusted and the CH_4 valve is gradually stabilized.

The gas is sucked from the wells by means of pumps. The gas is routed from the measurement and adjustment sheds through filters first and then through the gas pumps. The removal of particles and water is required to avoid corrosion of pipes, valves, pumps and gas engines. The two pumps are rotary blowers, which run according to the Roots principle. This type of pump ensures a stable gas flow rate, independent of pressure. Its maximum capacity is 160 Nm^3 h at a maximum differential pressure of 80 kPa (0.8 bar).

The pump house also contains various pieces of metering equipment such as an oxygen meter, a temperature meter and a flow meter. The computer room is adjacent to the pump house. The computer compiles the various data.

The collected landfill gas is at a temperature of around 17 °C. The temperature rises to 50 °C after compression, and the gas requires cooling before transmission. The gas cooler sits on the roof of the pump house. It is a stainless steel convection cooler with an automatic condensate tapping of the separation of condensate from gas.

The gas pipe through which the landfill gas is sent from the pump house to the utilization plant has a diameter of 125 mm and is made of medium-density polyethylene (MDPE). The dimension of the transmission pipe allows a flow of 320 m^3 h. The pipe is buried in the soil with a slope of between 5 and 40% because of the presence of water vapour in the gas. The distance between the pump house and the utilization facilities is 2.5 km. Four automatic condensate separators are placed in the lowest parts of the transmission pipe as an additional precaution against condensate damage.

THE UTILIZATION PLANT

Most of the landfill gas collected from the Viborg landfill site is used as fuel to fire a boiler. A small part of the landfill gas has been used to test two types of engines: a Stirling and an Otto engine respectively.

The boiler unit, with a furnace, is a cast-iron boiler especially designed for burning landfill gas and completely isolated. It has a modulation burner, which ensures flexible control at variable gas flow rates, also following the oxygen concentration in the landfill gas. It burns gas at high temperatures. The boiler delivers hot water at 115 °C and 600 kPa (6 bar) of pressure into the district heating grid. The boiler has a capacity of 2.8 GJ h. The boiler output corresponds to a landfill gas demand of 241 Nm^3 h at 42% methane.

Two gas engines are placed parallel with the boiler unit. The project aims at testing the Stirling engine in comparison with a conventional Otto engine when running on landfill gas. Each engine is linked to an electricity generator. The cooling water from the engine and the exhaust gas from the engines are exploited in heat exchangers for preheating the incoming water to the boiler.

An advantage of the Stirling engine is that the engine can run at methane concentrations varying between 35 and 100%. The engine drives a 6 kW asynchronous ASEA electricity generator equipped with a heat exchange system, which recovers approximately 16 kW of heat. The efficiency of both engines was compared as a function of the methane content of the landfill gas supplied by means of an energy balance. Figure 3 compares the efficiencies of both engines regarding heat and electricity. The Stirling engine has not come up to expectations as far as an efficient exploitation of landfill gas with

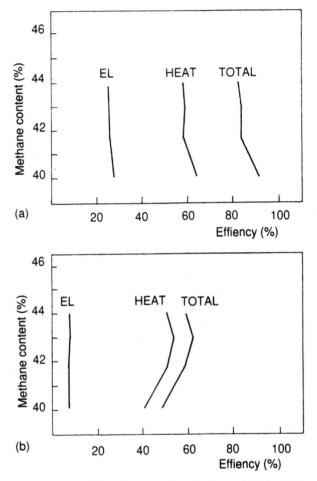

Figure 3. Comparison of the efficiencies of (a) the Otto and (b) the Stirling engines running on landfill gas, as a function of methane content.

45% methane content is concerned, and shows many other problems, so it has now been removed from the plant. The Otto engine showed a very stable energy production with efficiencies as high as 90%. The test also demonstrated the role of the automatic adjustment of the gas quality in the utilization of gas engines.

The Otto engine is a four-cylinder MWM gas engine, which is used as a reference to the Stirling engine. The engine drives a 27 kW asynchronous electricity generator and also allows the recovery of 54 kW of heat.

OPERATION AND EXPERIENCE

The landfill gas plant has been supervised by a consulting engineer. The supervision of the works involves mainly computer work, which includes monitoring of the works and checking a list of parameters relative to the various devices of the works.

At first, operation and maintenance of the works required more time than was anticipated, but after 3 years the figure is close to the anticipated estimate made before the start of the operation.

No corrosion nor collapse have been noticed with the vertical pipes of 110 mm diameter made of high-density polyethylene (HDPE). However, some distortion of the pipes occurred because of settlements in the landfill and often water impounded in the wells. These distortions made it difficult to remove the water from the wells with a pump and caused the gas extraction rate in these wells to be reduced.

To prevent the gas pipes from blocking by condensed water, at intervals landfill gas is routed with high pressure through the pipes in the opposite direction by a reverse blowing mechanism.

No corrosion has been noticed in the pipes or the valves of the pump house. After 3 years of operation, the particle filters have suffered no major problems. As the pump is the central component of the landfill gas extraction plant, it follows that only very durable materials should be used.

Since the beginning of the operation, only normal maintenance work has been executed on the pumps.

Many problems have been encountered with the flow meter, partly because it did not give the gas volume in normal cubic metres, and partly because of the presence of water.

During the first year of operation, water was found both in the filter and in the fuel cell of the oxygen meter, which caused false oxygen alarms.

The gas boiler is an essential component of the landfill gas utilization plant. After the first three years of operation, no corrosion of the boiler had occurred, but after five years corrosion required a change of parts of the boiler. The burner does not run normally when landfill gas flows at a variable rate to maintain the methane content between

35 and 55%, which is what occurs when the combustion process is controlled by the exhaust temperature. This device has now been replaced by a pressure control. A correct and optimum burning of landfill gas requires an exact oxygen percentage in the gas.

The Stirling engine has given rise to numerous technical problems, such as helium leakages, metal part breakages or gas regulator maladjustments. Following these numerous technical problems, it was decided in May 1988 to terminate the trial with the Stirling engine after a total of 5000 hours' operation.

Results from the pilot project anticipated the production of 3.5 m^3 of landfill gas per tonne of filled waste per year. This methane production rate should be maintained for 10–20 years after the landfilling of the waste. In the full-scale plant, it was hence anticipated to extract 2.5 m^3 of landfill gas per tonne of filled waste per year in the oldest part of the landfill and 3.9 m^3 of landfill gas per tonne of filled waste per year in the youngest part of the landfill. This corresponds to 160 m^3 h for the demonstration project, which involves 375 000 t of filled waste. In fact, the landfill gas extraction rate was stabilized after adjustment of the suction pressure to around 120 m^3 h.

The economy of the project is as follows.

* Investment exclusive of the measuring
  programme for the demonstration project   DKr 8 000 000
* Income per year for sale of energy   DKr ˙800 000
* Operating costs per year   DKr   300 000

CHANGES AND MODIFICATIONS

In 1989, the plant was extended on a new part of the landfill which contains approximately 80 000 tonnes of waste. In the meantime, the gas production from the oldest part of the landfill has decreased. But the total gas production has increased from 120 m^3/h to 150 m^3/h after the extension.

In the extended part of the extraction system the wells are changed because of the water problems mentioned under 'Operation and Experience'. The wells are now constructed either of two separate pipes or of a double pipe, one for gas extraction and one for pumping up the water from the wells.

MONITORING

The landfill gas plant is equipped with automatic measuring devices connected to the computer. This system monitors the landfill gas quality and also adjusts the landfill gas extraction flow rates as well as the utilization equipments. Continuous measurements are made and stored in the computer for each well in order to adjust the extraction flow rate of gas automatically according to the gas quality. The parameters measured are: the percentage methane in the gas, the gas flow rate, the gas and ambient temperature and the pressures.

In the case of gas leakage, alarm systems located in the measurement and adjustment shed as well as in the pump house automatically cut the extraction process. The methane content of the landfill gas is stabilized around 45%. The calorific value of the gas is thus 16 150 kJ/m^3.

During the measuring period, the average methane content was 42.4% and the gas production 120 m^3/h. The efficiency of the boiler was 78.9%. The actual working time has been over 90%.

The total efficiency of the Stirling engine varies between 49 and 63%, and that of the MWM (gas Otto) engine between 80 and 90%. Both power and heat are included here.

CONCLUSION

The results achieved until now have shown that the gas volume that can be expected in landfills with a relatively low filling height is in the neighbourhood of 2–5 m^3 landfill gas per year per tonne of waste over a period of 10–20 years from the time when the waste is discharged in the landfill.

The demonstration project has shown that it is technically feasible to exploit the gas from small-size landfills. A measurement and adjustment system has been made to function so that it fulfils its purpose and, among other things, controls the gas recovery in such a way that a constant methane percentage is attained at all times.

FURTHER READING

Commission of the European Communities, Directorate General for Energy. (1986). Flag brochure no. 58 for European Community Demonstration Projects for Energy Saving and Alternative Energy Sources, Project no. BM/741/83-DK.

Gendebein, A., Pauwels, M., Constant, M., Ledrut-Damanet, M. J., Willumsen, H. C., Butson, J., Fabry, R., Ferrero, G.-L. and Nyns, E.-J. (1991). *Landfill Gas. From Environment to Energy*, Office for Official Publications of the European Communities, Luxemburg.

Willumsen, H. C. (1988). *Recovery of Landfill Gas*, Final Report. Demonstration Project BM/741/83. For Commission of the European Communities (CEC), Directorate General (DG) XVII, CEC, Luxemburg.

9.3 Experiences with a Landfill Gas Purification and Co-generation Plant at Berlin-Wannsee, Germany

JOACHIM SCHNEIDER

DEPOGAS GmbH, Glienicker Str. 100, 14109 Berlin, Germany

INTRODUCTION

Landfilling of municipal and industrial waste started in 1954 at the Berlin-Wannsee site, which was a gravel pit of an average depth of about 10 m. Until 1980, 11 million m³ of landfill volume had been filled. When the site was closed, a hill of 1.5 km in length, 0.7 km in width, and 40 m height had been constructed within 26 years (Fig. 1).

During recultivation the site was covered with a layer of soil (mainly loam), where grass, trees and bushes were planted in 1985/86. Since then the former landfill has been used mainly as a public recreation area.

As a first step, investigations were done in order to receive data about the gas production rate of the site, the gas quality, and the

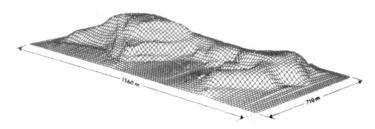

Figure 1. Graphical view of the Berlin-Wannsee landfill.

Landfilling of Waste: Biogas. Edited by T. H Christensen, R. Cossu and R. Stegmann. Published in 1996 by E & FN Spon, London. ISBN 0 419 19400 2.

efficiency of the gas collection system. At that time very few data existed in this field in the literature.

A gas collection pilot plant was constructed in 1982 on top of the site, serving an area about of 3 ha of the landfill. The extraction system consisted of 10 wells, a two-stage blower system, a flare, a computer-controlled gas-sampling and monitoring system, and a gas pipe to the Hahn-Meitner Institute, where the landfill gas was burned in one of the Institute's boilers. For several years, about 600 Nm^3/h of gas have been collected by the pilot plant. About 350 Nm^3/h were flared and about 250Nm^3/h had been burned in the central heating system (from May 1983 until May 1988) of the Hahn-Meitner Institute.

All the results and the experiences gained from this pilot plant, and from other scientific work, were incorporated into the design and construction of the full-scale co-generation plant built in 1987/88. This is especially true for the problems associated with the trace organics in the LFG as well as the exhaust gases from the burner.

Two major results of the scientific work that was mainly done during the period of 1982 to 1986 are: the detection of more than 500 organic hydrocarbon compounds including chlorinated components in the gas; and owing to the presence of chlorine in the raw gas, the identification of dibenzo-dioxin and dibenzo-furan isomers in the flue gas of the boiler.

DESCRIPTION OF THE PLANT

On the landfill, an extraction and collection system (130 gas collection wells, 15–25 m deep, about 15 km of subsurface pipes, and a compressor station with three screw compressors) have been constructed (Fig. 2).

A 1200 m gas line leads to a co-generation plant that consists of three combustion engines of 1.5 MWe each (Fig. 3). The first part of the plant is the gas purification unit, which is a two-stage liquid counterstream-absorption system, developed to reduce the organic impurities in the gas close to zero. Reducing the organic chlorinated hydrocarbon compounds in the LFG also minimizes the production of dibenzo-dioxin and dibenzo-furan isomers in the exhaust gas.

The electricity is sold to the Berlin utility company, while the waste heat from the engines' water cooling system and from the

Figure 2. Compressor station.

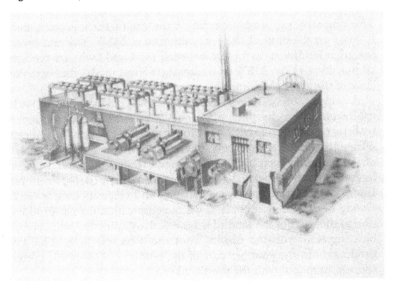

Figure 3. Gas purification and co-generation plant.

exhaust gas is used commercially to heat 135 apartments in the neighbourhood of the landfill and the Hahn-Meitner Institute.

The plant is automatically run by means of the fully computerized process control system (compressor station, gas purification and co-generation plant). Failures occurring during daytime (Monday to Friday from 7.30 to 16.00) are recognized by means of an alarm signal in the building; during other times of the day (or week) a radio signal is given to the staff. The plant is completely controlled, maintained and operated by a staff of three technicians. The period of operation to date covers 10 000–15 000 h for each of the major components of the plant.

OPERATION EXPERIENCE

The main results from 'routine' operation are described in the following sections.

Gas Production

The long-term gas production rate of the landfill site was calculated in 1983 on the basis of the yearly amount of MSW, soil and other materials landfilled on the site between 1954 and 1980, an average composition of the MSW and a mathematical model that was the state of the art at that time.

A comparison of the results from the gas prediction model with the results from the pilot plant shows that the model produces realistic results; the current gas production, however, is only about 40% of the predicted volume. The main reason is that in that part of the landfill, which had been operated from 1954 until 1974, a large amount of specific waste had been dumped (such as oil sludges, residues from ion exchangers, and electroplating waste). Owing to the different kinds of waste composition the gas production in that part of the landfill is far less than estimated and cannot be compared with the results from the 3 ha test field. The gas production in the younger part of the landfill (1974–1980) corresponds quite well with the predicted one.

Investigations regarding the influence of the landfilled oil on the gas production show a preservation effect on the organic matter of the garbage. Owing to the operation of the gas extraction system the concentration of preservative material that reduces the bioavailability may decrease in the future. As a result bacteria may have access to the MSW organics, which may result in a lower decrease of the gas production with time than estimated.

Gas Quality from the Collection System

The main goal of the operation of the gas extraction and utilization plant is to prevent as much gas as possible from leaving the landfill uncontrolled into the atmosphere. Respecting cost and benefit analysis of the gas collection system, the wells are constructed to obtain a zone of vacuum all over the site in a depth of 5–25 m.

Therefore, in the inner part of the landfill, a positive pressure exists according to the gas production and the free space to move. In the area affected by extraction (shell-shaped) a pressure of about 0.05 kPa below atmospheric pressure is present. But this means that the pressure difference between the inner part of the landfill and the shell-shaped area influenced by extraction changes with the change in the atmospheric pressure; as a result, the amount of air that is sucked into the landfill will change. As a result, oxygen from the air is reduced in the upper layer of soil or refuse to zero, and nitrogen mixes with landfill gas and lowers the concentration of methane and carbon dioxide. Depending upon the velocity of atmospheric pressure change the concentration of methane is reduced to values between 45 and 30%. As a consequence the whole system, including the power generation of the engines, has to be controlled automatically by monitoring the methane concentration in the gas in order to adjust the gas extraction rate.

In contrast to methane the concentrations of the trace components remain at the same level. This means that the nitrogen from the air is enriched with the organic hydrocarbons while it moves through the landfilled waste. By these means the amount of trace organics in the landfill is reduced.

Gas Purification System

Investigations regarding the trace organic components in landfill gas show a wide range of different hydrocarbon compounds, including organic metal compounds, organic silicon compounds, organic chlorinated and fluorinated compounds, aromatic and aliphatic compounds.

A problem that may occur when LFG is burned in energy recovery systems where heat is used (temperature below 250 °C) is the formation of dibenzo-dioxins and furan isomers. One of the necessary boundary conditions is the presence of chlorine in the flue gas and copper oxides in the dust layer on the surface of the heat exchanger. Both chlorine and copper can be found in trace concentrations in LFG. If the concentrations of chlorinated hydrocarbons in the LFG can be reduced close to zero the formation of dioxins and furans will be minimized.

As there is a wide range of experience using activated carbon for LFG purification, this process may be seen as the most favourable. But investigations that were done during the planning stage of an LFG purification plant in Berlin showed that, mainly because of the necessity of using high corrosion-resistant materials, the costs

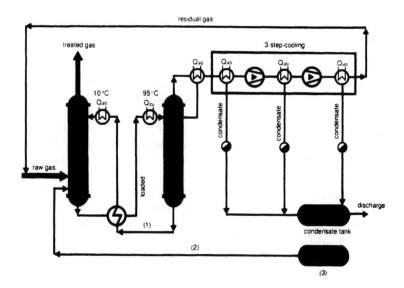

Figure 4. Flow sheet of the gas purification plant. Capacity 4000 m³/h; electrical energy supply 60 kW; efficiency > 98%; space needed 50 m²; height ≅ 8 m. 1, absorbent liquid; 2, absorbent liquid supply; 3, absorbent liquid-tank.

increase substantially. This is due to the steam regeneration of the activated carbon columns at a temperature of about 220 °C that has to be done regularly. As a result there is a cracking process of the chlorinated hydrocarbons, which leads to hydrochloric acid (HCl).

Based on these experiences, in 1987 the author operated a pilot plant that was constructed as a counterflow liquid absorption column. The experiments using this system showed good results of trace organic removal. As a next step a plant for a gas flow of 4000 Nm^3/h was developed and constructed in 1988/89 as a two-stage purification system (Fig. 4).

The efficiency of the gas purification plant is much higher than the estimated 90% in two stages; it actually reaches at least a 98% removal rate of chlorinated hydrocarbons in only one stage. During operation the chlorine concentrations in LFG of 25–30 mg/Nm^3 were reduced to values below 0.5 mg/Nm^3 in the treated gas.

In addition, the organic hydrocarbon compounds are reduced. With an average LFG concentration of about 1000 mg/Nm^3 organic hydrocarbon compounds, about 15 000 kg impurities per year were extracted from the LFG in the purification plant. The residues, with a chlorine concentration of about 2–3%, have to be treated as hazardous waste.

Combustion Engines

At the Berlin-Wannsee Site for the first time a combustion engine, type MWM TBG 441 BV 16 (16-valve turbocharged engine) with a maximum electric power generation of 1.5 MWe at the generator, has been fed with landfill gas. As a consequence some problems had to be solved.

One problem was the spark-ignition system, in which a specific spark plug had been developed to run the engine with a methane concentration as low as 30%. With some additional control of the engine with regard to the methane concentration, as there is control of the cooling water inlet temperature of between 70 and 80 °C, and also control of the turbocharger's compressed air temperature of between 30 and 40 °C, the engine can be operated with a methane concentration of 27% but with a reduced maximum power output of 1000 kW.

Varying methane concentrations in the extracted LFG create the problem of a constant fuel load of the engines. This has been solved by the controlled change of the gas pressure in relation to the air pressure from the turbocharger.

Other problems were cracks in the cylinder head and broken outlet and inlet valves. For a long time the gas quality (30–45% of methane) was regarded as the reason for an increase in the above-mentioned cracks. Investigations over a period of about 2 years finally showed three reasons for the damages; the gas quality was no problem. First of all, the construction of the cylinder head was not adequate. As a consequence, flue gas entered the cylinder head's cooling water system for a period of at least 12 months. This situation resulted in cracks in the pipes of the exhaust gas heat exchanger, and a loss of cooling water, which had to be replaced from the water treatment system (decarbonizer), could be detected.

In addition this system was not programmed in the right way. As a result about two-thirds of the cooling water consisted of untreated water with a high carbon concentration. The carbon and the carbon dioxide of the flue gas at the very hottest point in the engine built up thick layers of carbon and dust at the cylinder head (the hottest point) and isolated the main parts of the cylinder head from the cooling water. Thirdly, soil and sand in the cooling water pipes that originated from the initial period of construction increased the layer of isolating material in the cylinder heads.

SUMMARY AND CONCLUSIONS

In 1987/88 the biggest co-generation plant to treat landfill gas in Europe was constructed at the Berlin-Wannsee site. For the first time a landfill gas purification plant had been developed and constructed primarily to prevent the production of dibenzo-dioxin and dibenzo-furan isomers in the flue gas from a landfill gas utilization plant.

Experience with the Berlin-Wannsee gas collection system and treatment plant resulted in the following observations and conclusions.

- Despite the very intensive investigations on the landfill site the calculation of the long-term gas production rate was wrong. Because it was unknown, the wrong place for the location of the pilot gas extraction plant was selected, so that the low production rate from the older part of the landfill had not been respected. In addition, industrial waste had been landfilled in those parts, which may have had an influence on the gas production owing to inhibition and/or conservation processes.

- Operation of a highly optimized gas collection system with a high efficiency results in a wide range of methane concentration in the LFG. As a consequence an appropriately controlled gas treatment system had to be installed. The basic rule for the protection of the environment from LFG is such that the gas treatment system must encounter the conditions given by the landfill; the opposite, that extraction system of the landfill has to be operated in such a way that only high methane concentrations in the LFG are obtained, leads to high LFG emission rates into the atmosphere.

- The developed gas purification system – a counterstream liquid absorption system – operates at a very high efficiency even with low concentrations of hydrocarbon compounds in the raw gas. The final product is a 'clean' gas and a highly aggregated extraction liquid from the plant.

- Owing to the reduction of hydrochloric acid (HCl) in the gas purification plant, the intervals between the changes of the lubrication oil were prolonged by at least a factor of 10.

- In general, most of the technical problems of the LFG system did not originate from the landfill gas as might be expected, but mainly from mistakes made during construction.

- The investment for the whole system amounted to DM21.5 million, and the estimated payback time will be at least 17–18 years.

As a conclusion, if there is a problem all the possible reasons should be investigated, even the 'unthinkable' ones.

FURTHER READING

Schneider, J. (1989). Landfill gas from the Berlin-Wannsee site – environment protection due to energy production by a 13.5 MW co-generation power-plant, in *IRC 89*, International Recycling Congress Proceedings, Berlin.

Schneider, J. (1990). Protezione ambientale e recupero energetico con una centrale a energia totale di 13.5 MW. *Rifiuti Solidi*, **IV**(1).

9.4 Landfill Gas Recovery and Utilization at Modena Landfill Site, Italy

ADELIO PERONI[a], SANDRO PICCHIOLUTTO[b] &
ROBERTO PAPARELLA[a] (Italy)

[a]AMIU, Via Morandi 54, 41100 Modena, Italy
[b]Department of Environment, Municipality of Modena, Via Santi 40, 41100 Modena, Italy

INTRODUCTION

The landfill gas project at the Modena landfill site was developed in order to use LFG either as primary fuel (when needed) in an incinerator for industrial toxic wastes or to generate electricity by means of spark-ignited gas engines, thus supplying the industrial waste-treating plants located nearby.

The main advantage of this project is the use of a very low-cost fuel, which may produce less emissions than the usual auxiliary fuels. Problems are caused by the relatively poor reliability of the plant, particularly in relation to an effective danger of corrosion mainly located in the gas engine. Third-party investment is unlikely. On the other hand, lower energy expenses may promote lower refuse treatment charges and consequently better marketing chances.

Furthermore, the overall environmental benefit of the plant due to a clearing of noxious gases from the landfill areas combined with a lesser 'greenhouse' effect is important.

Landfilling of Waste: Biogas. Edited by T. H Christensen, R. Cossu and R. Stegmann. Published in 1996 by E & FN Spon, London. ISBN 0 419 19400 2.

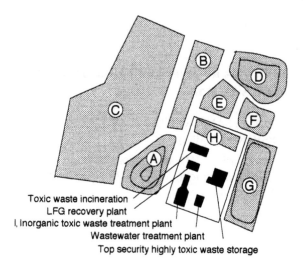

Toxic waste incineration
LFG recovery plant
I, Inorganic toxic waste treatment plant
Wastewater treatment plant
Top security highly toxic waste storage

Figure 1. Schematic plant of the waste management facilities at the Modena disposal site. Landfill zones: A, 1° MSW sector; B, 2° MSW sector; C, 3° MSW sector. D, demolition waste; E, incineration ash; F, immobilized toxic waste 1° sector; G, immobilized toxic waste 2° sector; H, toxic waste storage.

PLANT DESCRIPTION

The municipal solid waste landfill 1 (Fig. 1) of Modena is located on the outskirts of the city, nearly 2 km from the urban area limit.

The landfill was started in 1973 and closed in 1985. It has completely developed in height. It covers an area of about 70 000 m^2 with a volume of approximately 1 200 000 m^3, reaching a maximum height of about 37 m.

Between 1973 and 1984 about 600 000 t of MSW were disposed of. This exceeded the capacity of the municipal incinerators. The landfill also dealt with refuse of high organic content from a separate collection.

Initially, LFG was to have been directly burnt in a brick factory nearby. The decision to locate the Industrial Toxic Waste Treatment Plant (see Figure 1: F, G, l) in that area, plus the expectedly increasing amount of gas due to an expanding landfill site resulted in a different concept for the use of LFG.

A diagram of energy flow in the area is thus shown in Fig. 2.

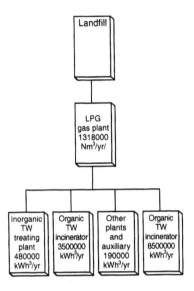

Figure 2. Energy flow scheme in the LFG utilization plant at the Modena disposal site. TW, toxic waste; kWh$_e$, kWh$_t$, electrical and thermal kWh.

During a test phase, 25 boreholes were drilled and connected to a collecting network, thus achieving a flow rate of about 210 m^3/h. The maximum extraction rate was calculated as 3.44 m^3/h per metre of the well.

LFG is used in two Deutz BGA 8 M turbocharged, water-cooled gas-fuelled Otto engines, each supplying 217 kW. One is equipped with a total-energy module in order to heat the control rooms in winter; the other has a simple generator configuration.

One engine usually runs continuously to guarantee basic power required by some priority area systems, which cannot afford a power blackout. Excess landfill gas is used either as primary fuel in the special organic waste incinerator (Fig. 1) or to produce additional electricity. Any fuel shortages due to extra-high burner consumption will automatically cut off the second (and/or the third) engine.

Particular attention has been paid to the computerized monitoring and control system of the extraction plant and the LFG engine.

The whole plant has been developed in order to be completely automatic: any alarm is radio-transmitted to the area main office.

The process computer operates the plant using data from natural gas and oxygen analysers. The computer compares the gas quality in each single well with the set-point (different in each well in order to get the most from both old refuse areas and recent ones). When the natural gas concentration is too low, the valve is automatically closed step-wise resulting in lower extraction rate. Conversely, the valve is opened step-wise when the CH_4 concentration is too high. This ensures highly constant gas composition. An automatic single valve shut-off will operate in critical situations.

A further exhaust gas oxygen meter operates through a dedicated computer in order to optimize the fuel–air ratio in connection with engine load fluctuations.

Gas flow-meters are installed for each user, while power meters measure the electricity produced by each gas engine. Other meters are placed before the plant distribution switchboard, which is connected in parallel with the local network, thus enabling energy production to be continuously monitored.

Figures 3–6 illustrate construction and installation details.

Figure 3. Gas engines building.

Figure 4. Collector pipe with flow regulation valves.

Figure 5. Blowers, main suction control valves and (right) gas filter.

Figure 6. Gas engines.

Technical Problems and Remedies

Only two major problems arose during the drilling phase.

- The auger sometimes jammed in old buried tyres, thus causing long delays in drilling, as the screw had to be removed, the nylon and/or steel cords cleaned and the equipment re-started. No solution was found to this problem: thus a decision was made to dispose of the tyres in restricted areas of the new landfill in future.
- Drilling through sludge pockets caused the well wall to collapse repeatedly. Drilling had to be repeated in adjacent areas. The landfill now in use consequently accepts sludges with an 80% water content. These are always mixed with refuse.

Considerable gas well flooding has also been noted after the plant has remained shut down over a long period of time and/or after rain storms. In the first case, the wells had to be pumped out.

No serious machinery problem has been recorded to date. On the other hand, there were some problems in the control section. Even if sampling gas is being cooled and dehumidified prior to the gas analysers, periodical outage of the O_2 infrared control

TABLE 1. Example of a Lubricating Oil
Analysis

Viscosity 100 °C		13.5
TBN (total base number)		9.08
H_2O		traces
Insolubles in penthane		
membrane	0.3 µm ppm	0.02
Fe	ppm	8
Cu	,,	65
Cr	,,	0.0
Si	,,	36
Pb	,,	1
Al	,,	1
Sn	,,	1

instrument due to the formation of water droplets has been noted. As this instrument has an on–off gas quality alarm function, this has resulted in faulty process control, causing the plant to stop. A further water-trap with humidity sensor and alarm had to be inserted.

To date, after 6000 h of operation, gas engine No. 1 has not yet presented any notable faults. The quality of the lubricating oil is analysed periodically. Table 1 shows an example of an oil analysis. The second engine has terminated its test run phase over a period of 1000 h and is closing down.

One major problem that occurred during this project, and which has not yet been completely solved, concerned authorization to run the power station from the National Electricity Board (ENEL), which is the only authority entitled by law to buy electric energy, and from the Manufacture Duty Office (UTIF) which in Italy collects a tax on electricity production.

In July 1989, an important bill that establishes part of the price for electricity when sold to any electricity board was approved in Italy. Unfortunately, the second part of the bill has still to be approved and provisional prices are being paid in the meantime.

OPERATION AND RESULTS

Operation History

The landfill gas extraction plant started operation in November 1987, while computerized data logging was installed in July 1988.

The cumulative gas production has been recorded from that date onwards.

Unfortunately, the gas then had to be almost completely flared because of the above-mentioned problems concerning delays in obtaining the licence to connect to the municipal network.

During 1988, we suffered a remarkable number of short-term plant breakdowns before we realized that they were caused by a high number of mains micro-interruptions and power fluctuation (due to in-line works), which negatively affected the functionality of the process.

Some other unexpected breakdowns were caused in 1989 and 1990 by the flooding of water separators (and consequent interruption of gas flow). This was due to automatic drain malfunctions.

The third year of management (1990) was mainly affected by a few major plant stoppages caused by pipe network enlargement works and installation of a second engine, plus a few outages due to oxygen meter flooding.

Even though it was installed in March 1990, the second engine has not yet been used for normal operation owing to bureaucratic delays.

Plant Performances

The extracted gas is currently used to produce electric energy. A system enabling the same gas to be used as auxiliary fuel for the incinerator for toxic waste of a mainly organic composition has also been set up. This system, built near the landfill, will be able to take advantage of this opportunity when the biogas pick-up plant is also activated for the third currently operating MSW landfill. The toxic waste incinerator will be supplied with about 450 Nm^3/h, thus covering about 30–35% of the plant fuel requirements.

In relation to the gas engine, electro/thermal functionality measurements have been taken with power set at 50%, 75% and 100% of the nominal rate. In the first case, we ascertained that with an electric power of 105 kW, the thermodynamic efficiency was 3129 kcal/kWh. At 75% of nominal electric power (165 kW) the efficiency increased to 2738 kcal/kWh, while at 100% of nominal power (electric power of 215 kW) the figure reached a maximum of

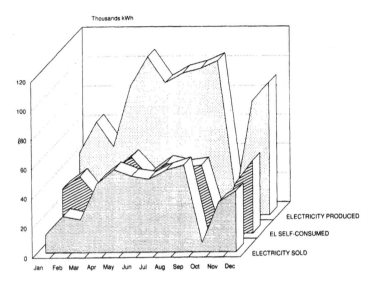

Figure 7. Electricity production and utilization in the year 1990.

2577 kcal/kWh. The average efficiency (yearly basis) is nevertheless fairly lower (about 3157 kcal/kWh) because of start-ups and speed fluctuation.

The electricity produced will be primarily self-used within the area around the clock. Nevertheless, a substantial surplus is supplied to the municipal network, owing to the delay in operating the incinerator.

As can be concluded from Fig. 7, the commercial production of electricity, which has been operating since the end of November 1989, gave rise in 1990 to a gross production of 928 000 kWh during 5700 h operation. Half of that was self-consumed by the complex.

The running time of the gas engine on a yearly basis (expressed as generator/extraction station running hours) was around 83%. This meant that the overall running time index (generator/theoretical running hours) was around 65%, which can be regarded as fairly satisfactory. One has to remember that about 20% of the outage was due to programmed plant stoppages to allow the second lot to be safely connected to the main station (Fig. 8).

Landfill gas extraction capacity was remarkably constant, with a highly reliable gas concentration.

Natural gas concentration in the landfill gas has been steadily

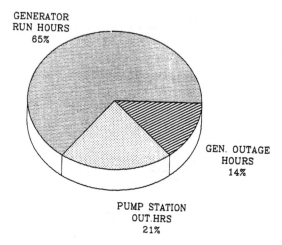

Figure 8. Availability of the different components of the LFG plant in 1990.

measured for more than two years by means of an infrared analyser. Further data have thus been collected, giving the averaged values listed in Table 2.

Further analyses on trace gas pollutants have been made of both the raw gas (Table 3) and of exhaust gases (Table 4), particularly in relation to halogenated hydrocarbons, which represent a probable source of corrosion initiation inside the gas engine.

Some tests have been conducted in order to determine the influence of precipitation rates on LFG production. Figure 9 shows an interesting correlation between energy productivity and rainfall.

TABLE 2. Average Gas Quality Data in LFG Extracted from the Landfill

CH_4	51%
CO_2	38%
CO	0.5%
N_2	5%
O_2	1%
H_2O	3%
H_2S (including mercaptans)	0.002%

TABLE 3. Analysis of LFG on Trace Components (All Values in mg/Nm^3)

Nitrogen oxides (as NO$_2$)	6
Sulphur oxides (as SO$_2$)	106
Aliphatic hydrocarbons	100
Aromatic compounds	75
Dichloromethane	ND
Perchloroethene	10
1-2 Dichloroethene	3
Trifluoromethane	ND
Trichloroethene	11
Dichlorotrifluoropropene	2
Dichlorofluoromethane	ND
Trichloromethane	ND
Trichloroethene	ND
Sulphurated compounds	30

ND = not detectable

CONCLUSIONS FOLLOWING OPERATION

Regarding individual components, the good performance of the gas engines and the reliability of the engines themselves are remarkable (Delhi *et al.*, 1989). On the other hand a higher dependability on the gas extraction system had been expected.

No major malfunction in relation to the monitor and control equipment has yet been recorded.

It is difficult to compare the theoretically expected economic results with the actual management data for the following reasons.

- Gas utilization had switched from a flexible usage for electrical co-generation and/or burning to 100% electricity production.

TABLE 4. Example of the Exhaust Gas Quality of the Gas Engine

Temperature (°C)	495
Carbon dioxide (vol.%)	9.7
Carbon monoxide (mg/Nm^3)	8655
Nitrogen oxides (as NO$_2$) (mg/Nm^3)	616
Sulphur oxides (as SO$_2$) (mg/Nm^3)	10.6
Chlorine (as Cl)(mg/Nm^3)	20.25
Fluorine (as F) (mg/Nm^3)	0.50

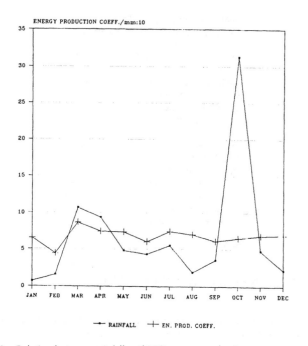

Figure 9. Relation between rainfall and LFG energy production.

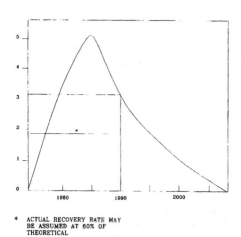

Figure 10. Prediction of the total gas production take with time and actual gas extraction rate in 1990.

- The plant started operating about 5 years after gas yield figures were drawn: this meant a considerable reduction in gas production, which can be calculated (and was experimentally checked) at 40% (Fig. 10).
- Even if the second phase of the project had been completed in spring 1990, owing to administrative delays the second engine was only allowed to run commercially in 1991, thus forcing about 42% of the gas extracted during 1990 to be flared.

Management Expenses

The maintenance and management costs are as follows (1990 year's-end figures):

		(US$)
Part-time labour:	1 analyst	
	1 electromechanic	
	1 mechanic	16 600
Gas engine overhaul (annual share)		3300
Oil and lubricants (ch. analysis incl)		3250
Plant maintenance		2670
Total current annual costs		25 820

Future of the Plant

The described plant represents only an initial phase in a more complex programme, the aim of which is to connect the operating power station to the gas produced in the second already completed MSW landfill.

The second-phase project has already been completed. As a result the gas extraction and utilization rate will be doubled.

The lifespan of the existing plant has been prudentially estimated on the basis of the lifespan of the gas engine.

Based on the manufacturer's experience with similar engines, a 13 year lifespan (100 000 h at 80% load) may be expected (Delhi *et al.*, 1989), giving an estimate of total operation time at full load of about 70 000 h.

CONCLUSIONS

This project represents an innovation, as it features an automatic plant control system between a computer and peripheral monitors.

Experience during two years management of the plant in Modena may be regarded as generally positive, even though the plant has not yet fully achieved its potential in gas utilization, meaning that an excessive amount of gas has to be flared.

Despite this, the economic results appear promising and back our choice to continue developing the plant in other landfill sites in the area.

Besides these considerations and other technical aspects indicated in the chapter, no specific instructions drawn from our experience could be given to anyone wishing to build a similar plant.

Supply of energy produced by LFG without a special governmental warranty should, however, be avoided.

REFERENCES

Delhi, M., Erb, P., Luik, K.-O. and Reinike, B. (1989). *Economic utilization of landfill gas from small and medium-sized waste dumps*, Commission of the European Communities demonstration project (Contract BM/001/84-DE) ECSC-EEC-EAEC, Brussels, Luxembourg.

INDEX